Electrical Machine Systems with High Efficiency, Reliability and Integration

Electrical Machine Systems with High Efficiency, Reliability and Integration

Guest Editors

Tao Wang
Chenwen Cheng
Yang Xiao

MDPI

Basel • Beijing • Wuhan • Barcelona • Belgrade • Novi Sad • Cluj • Manchester

Guest Editors
Tao Wang
College of Electrical Engineering
Zhejiang University
Hangzhou
China

Chenwen Cheng
School of Electrical Engineering
Southeast University
Nanjing
China

Yang Xiao
School of Engineering
University of Leicester
Leicester
UK

Editorial Office
MDPI AG
Grosspeteranlage 5
4052 Basel, Switzerland

This is a reprint of the Special Issue, published open access by the journal *Energies* (ISSN 1996-1073), freely accessible at: https://www.mdpi.com/journal/energies/special_issues/X16YU7WG9R.

For citation purposes, cite each article independently as indicated on the article page online and as indicated below:

Lastname, A.A.; Lastname, B.B. Article Title. *Journal Name* **Year**, *Volume Number*, Page Range.

ISBN 978-3-7258-6258-0 (Hbk)
ISBN 978-3-7258-6259-7 (PDF)
https://doi.org/10.3390/books978-3-7258-6259-7

Contents

energies

MDPI

Article

Influence of Winding Configurations and Stator/Rotor Pole Combinations on Field Back-EMF Ripple in Switched Flux Hybrid Excited Machines

Zhiyu Yang, Xiaoyong Sun *, Ruizhao Han, Ruyu Shang, Zhen Chen and Xiangdong Liu

School of Automation, Beijing Institute of Technology, Beijing 100081, China; 3220231079@bit.edu.cn (Z.Y.); 3120230863@bit.edu.cn (R.H.); 3120230862@bit.edu.cn (R.S.); chenzhen76@bit.edu.cn (Z.C.); xdliu@bit.edu.cn (X.L.)

* Correspondence: xiaoyongsun@bit.edu.cn

Abstract: Similar to armature back electromotive force (armature back-EMF), the back-EMF also exists in the field winding of hybrid excited machines. However, the existence of field back electromotive force (field back-EMF) is harmful to the safe and stable operation of machine systems, e.g., higher losses, lower efficiency, higher torque ripple, and reduced control performance. This paper systematically investigates the influence of armature/field winding configurations together with stator/rotor pole combinations on the field back-EMF ripple in hybrid excited machines with switched-flux stators. The two-dimensional (2D) time-stepping finite element modeling and prototyping experiments are used for the research. The investigated field and armature coil pitches equal to 1, i.e., non-overlapped windings. The influential factors that are investigated in this paper mainly include the number of layers of field/armature windings, the number of field/armature coils, and the stator/rotor pole combinations. The results show that the field back-EMF's harmonic order and peak-to-peak value are closely associated with field/armature winding configurations and stator/rotor pole combinations under various conditions. Finally, for validation of the results predicted by the finite element method, a prototype machine is built and tested. Overall, non-overlapped double-layer armature and field windings are recommended for the hybrid excited switched flux machines with various stator/rotor pole combinations to realize relatively lower field back-EMF under different conditions.

Keywords: field winding back-EMF; hybrid excitation; stator/rotor pole combination; winding configuration

1. Introduction

Due to high torque density, electrical machines with permanent magnets are extensively used in a variety of applications [1–6]. The variable-speed system is one of the most important fields for the applications of permanent magnet machines (PMMs) and it usually requires excellent flux-weakening capability. However, due to a fixed magnetic field, permanent magnet machines are always faced with relatively poor flux-weakening capability. Different from the electrical machines with pure permanent magnets (PMs), the electrical machines with both field coils and permanent magnets, i.e., hybrid excitation, have drawn considerable research interest due to the synergy of controllable magnetic fields and high torque density [7–12]. Hybrid excited machines (HEMs) can be broadly divided into stator-HEMs and rotor-HEMs according to the location of the excitation sources, i.e., the field coils and the permanent magnets.

Compared with the conventional rotor-HEMs with the excitation sources on the rotor side, the stator-HEMs with the excitation sources on the stator side are preferred for high-speed and high-reliability applications since the sliding contacts are not required and the rotor is purely iron-cored [13–16]. Among a variety of stator-HEMs, the hybrid excited machines with the switched-flux stator (HESFMs) are the most studied because of their wide flux-regulation range and high torque density [17–20]. However, as investigated in [21], HESFMs suffer from the field back-EMF ripple, which may introduce field current ripple, torque ripple, additional losses, etc. Moreover, the field back-EMF ripple may also bring challenges to the machine control and the field power supply, especially under high-speed operations.

The research activities for the field back-EMF ripple originated from the switched-flux wound-field machines with partitioned stators (PS-WFSFMs). In [22,23], based on the conditions of no-load and on-load, the field back-EMFs of the PS-WFSFMs are investigated and reduced by rotor pole pairing and rotor skewing, respectively. Moreover, the influences of five-phase and dual-three-phase windings on the field back-EMF in the switched-flux wound-field machines (WFSFMs) are presented in [24,25], respectively. Moreover, ref. [26] studies the field back-EMF in a doubly salient brushless DC (DSBLDC) generator and adopts the stator-damper winding to suppress the field back-EMF ripple. Apart from the electrically-excited machines, e.g., WFSFMs and DSBLDC generators, the hybrid excited machines inevitably suffer from the field back-EMF as well due to the adoption of field coils.

The field back-EMF ripple in hybrid excited machines is first presented in [21], in which the field back-EMF is reduced by the techniques of rotor skewing and unequal rotor teeth. Moreover, based on the HESFM, the influence of stator/rotor pole combination on field back-EMF is studied in [27]. Further, the field back-EMFs of hybrid excited machines with different switched-flux topologies are comprehensively compared in [28]. The multi-objective optimization of the HESFM by considering the field back-EMF ripple as one of the optimization objectives is investigated in [29]. Moreover, the design tradeoff between field back-EMF ripple and airgap flux regulation capability in the HESFM is also discussed in [30].

However, in the existing papers, all the HESFMs are equipped with double-layer (DL) concentrated armature and field windings (AC/DC windings). The influences of AC/DC winding configurations and stator/rotor pole combinations on field back-EMF at various conditions have not been systematically studied. From the perspective of machine design, a high winding factor, i.e., high output torque, is always the design criteria for the selection of winding configurations and stator/rotor pole combinations. However, when considering the effect of field back-EMF, there is a design tradeoff between a high winding factor and low field back-EMF ripple when choosing AC/DC windings and stator/rotor pole combinations. To date, there is no reference on how to select appropriate armature/field winding configurations to diminish the influences of the field back-EMF ripple while maintaining relatively higher output torque. Therefore, the main aim of this paper is to uncover the influences of the winding configurations together with the stator/rotor pole combinations on the field back-EMF ripple in HESFMs and provide guidance for selecting appropriate winding configurations to reduce the field back-EMF ripple and maintain reasonable torque performance. The influential factors to be investigated in this paper mainly include the number of layers of field/armature winding, the number of field/armature coils, and the stator/rotor pole combination.

The rest of this paper is organized as follows. The HESFMs to be studied in this paper are illustrated in Section 2. The HESFMs with various stator/rotor pole combinations and armature/field winding configurations are studied in Section 3 and four different conditions are considered. The experimental results are presented in Section 4, followed by the conclusion in Section 5.

2. Machine Topology

In this paper, the calculated/simulated results are obtained using the 2D time-stepping finite element analysis (FEA) method in the commercial FEA software (Ansys/Maxwell 18.2). The topology of the HESFMs to be investigated in this paper is shown in Figure 1 and their main parameters are summarized in Table 1. The investigated HESFMs originated from the switched flux permanent magnet (SFPM) machine by adding iron bridges for additional flux paths and reducing the volume of PMs to accommodate field coils. For simplicity, it should be noted that Figure 1 only presents four of the eight investigated stator/rotor pole combinations. The purpose of using iron bridges is to better regulate the flux of the airgap. As aforementioned, the winding configurations to be investigated in this paper mainly include the number of layers of field/armature winding, the number of field/armature coils, and the stator/rotor pole combination. The investigated stator/rotor pole combinations in this paper include 6/4, 6/5, 6/7, 6/8, 12/10, 12/11, 12/13, and 12/14, respectively. Both double-layer (DL) and single-layer (SL) armature/field windings will be studied in this paper but only the armature/field windings with one coil pitch are investigated. The number of field/armature coils is essentially determined by the stator slot number and the number of layers of field/armature winding. Figure 1 shows the HESFMs with various stator/rotor pole combinations and winding configurations. In Figure 1, the permanent magnets in red and blue color indicate the circumferentially magnetized PMs in opposite direction. The windings in blue and black color indicate field and armature wingdings, respectively.

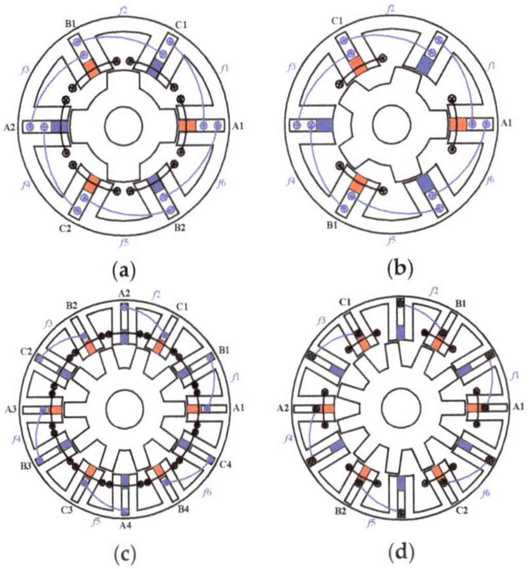

Figure 1. Machine topology of the investigated HESFMs with different winding configurations. (**a**) 6/4 with double-layer AC and DC windings; (**b**) 6/5 with double-layer DC and single-layer AC windings; (**c**) 12/10 with double-layer AC and single-layer DC windings; (**d**) 12/11 with single-layer AC and DC windings.

Table 1. Main parameters of the investigated HESFMs.

Items	Values
Stator outer radius	45 mm
Active axial length	25 mm
Rotor shaft radius	10.4 mm
Iron-bridge thickness	0.5 mm
Airgap length	0.5 mm
Lamination material	steel_1010
PM material	NdFe35
Rated speed	400 rpm
Armature copper loss	20 W
Field current density	5 A/mm^2
Slot packing factor	0.5

3. Field Back-EMF Ripple

To systematically illustrate the influences of the winding configurations together with the stator/rotor pole combinations on the field back-EMF ripple, the HESFMs with various stator/rotor pole combinations and armature/field winding configurations are studied in this part. Four different conditions are considered in the following analyses. It should be noted that the HESFMs with double-layer AC and DC winding configurations are taken as the baseline in Section 3 and their results are shown in each subsection for clear comparison. It is worth mentioning that the results presented in Section 3 are calculated using the time-steeping 2D finite element analysis method. The mesh type for the stator, the rotor, the airgap, and the PMs are element length-based refinement and the maximum length is 1 mm, 1 mm, 0.25 mm, and 0.5 mm, respectively. The choice of using 2D FEA and the corresponding mesh sizes is a tradeoff between computational efficiency and accuracy. All machines have the same stator outer diameter, i.e., 90 mm and active axial length, i.e., 25 mm. The mechanical speed of all machines is 400 rpm under different operating conditions. Moreover, the phase current is calculated based on the constraint of 20 W copper loss in armature winding and the field current is obtained based on the current density of 5 A/mm^2 in the field winding. Moreover, the packing factor in the armature and field slots is 0.5. The reluctance torque component in all machines is negligible since the d- and q-axis inductance is almost identical. Therefore, zero d-axis current control is applied in all the calculations.

3.1. AC and DC Windings Are Both Double-Layer

In this sub-section, the 6-slot and 12-slot HESFM counterparts with both double-layer armature and field windings are investigated. In [27], the relationships between stator/rotor pole combination and field back-EMF ripple are revealed, in which the AC and DC windings in the HESFMs are both of double-layer type. Therefore, to avoid the repetition of this part, the field back-EMFs and their FFT results are not shown here again instead, the harmonic orders (HOs) of the field back-EMFs with various stator/rotor pole combinations are illustrated in Table 2.

Table 2. Field back-EMF harmonic orders with double-layer AC and DC windings.

Slot/Pole	No-Load	Armature Current	On-Load
6/4	3N	3N	3N
6/5	6N	6N	6N
6/7	6N	6N	6N
6/8	3N	3N	3N
12/10	6N	6N	6N
12/11	12N	6N	6N
12/13	12N	6N	6N
12/14	6N	6N	6N

As can be found in Table 2, the stator/rotor pole combinations have a close relationship with field back-EMF harmonic orders under different conditions. This reveals that the field back-EMF ripple with higher harmonic orders always indicates a higher field back-EMF amplitude and thus its influence on machine performances will be more severe. Taking the 6-slot HESFMs as examples, the field back-EMF ripple's amplitude of 6/4 and 6/8 machines is large while it is small for the 6/5 and 6/7 machines at various conditions. This is due to the fact that for 6/4 and 6/8 machines, the field back-EMF harmonic orders are both 3N at various conditions and N should be a positive integer (N = 1, 2, 3...), while the field back-EMF harmonic orders of 6/5 and 6/7 machine are both 6N. As for the 12-slot HESFMs, the field back-EMF harmonic orders of no-load are 6N for 12/10 and 12/14 machines, but 12N for 12/11 and 12/13 machines. Therefore, 12/10 and 12/14 machines exhibit relatively higher no-load field back-EMF. However, under on-load, the field back-EMF harmonic orders are the same for all 12-slot machines, i.e., 6N, due to the armature reaction.

3.2. DC Winding Is Double-Layer While AC Winding Is Single-Layer

In this sub-section, the investigated HESFMs with various stator/rotor pole combinations are equipped with single-layer AC winding and double-layer DC winding. To clearly illustrate the relationship between the AC winding configuration and the field back-EMF ripple, Tables 3 and 4 comparatively summarize the induced EMF harmonic orders with different AC winding configurations for the 6-slot and 12-slot HESFM counterparts, respectively.

Table 3. Field back-EMF harmonic orders with double-layer DC winding.

Winding Configuration	Slot/Pole	No-Load	Armature Current	On-Load
Double-layer AC	6/4	3N	3N	3N
	6/5	6N	6N	6N
	6/7	6N	6N	6N
	6/8	3N	3N	3N
Single-layer AC	6/4	3N	3N	3N
	6/5	6N	3N	3N
	6/7	6N	3N	3N
	6/8	3N	3N	3N

Table 4. Field back-EMF harmonic orders with double-layer DC winding.

Winding Configuration	Slot/Pole	No-Load	Armature Current	On-Load
Double-layer AC	12/10	6N	6N	6N
	12/11	12N	6N	6N
	12/13	12N	6N	6N
	12/14	6N	6N	6N
Single-layer AC	12/10	6N	3N	3N
	12/11	12N	6N	6N
	12/13	12N	6N	6N
	12/14	6N	3N	3N

Table 3 comparatively shows the field back-EMF harmonic orders of the 6-slot HESFM counterparts when DC winding is double-layer while AC winding is double or single-layer. It can be found that the field back-EMF harmonic orders of different stator/rotor pole combinations remain unchanged under no-load since the DC winding configuration is unchanged and the AC winding is open-circuited. Different from the no-load condition, the induced EMF harmonic orders under armature current and on-load conditions are changed from 6N (with double-layer AC winding) to 3N (with single-layer AC winding) for 6/5 and 6/7 machines. For 6/4 and 6/8 machines, the number of AC winding layers does not influence the induced EMF harmonic orders under no-load, armature current, and on-load conditions.

Figures 2–5 show the field back-EMF ripples and their FFT results of 6/5 and 6/7 machines with double-layer DC winding under armature current and on-load, respectively. It can be found that by employing single-layer AC winding, the amplitude of the field back-EMF ripple is increased under on-load for 6/5 and 6/7 machines, although the induced EMF ripple under no-load is unchanged. This is due to the fact that the armature current field back-EMF harmonic orders of 6/5 and 6/7 machines are 3N when single-layer AC winding is employed. Therefore, the on-load field back-EMF harmonic orders of 6/5 and 6/7 machines are 3N as well since the on-load field back-EMF is the synthesis of no-load field back-EMF and armature current field back-EMF.

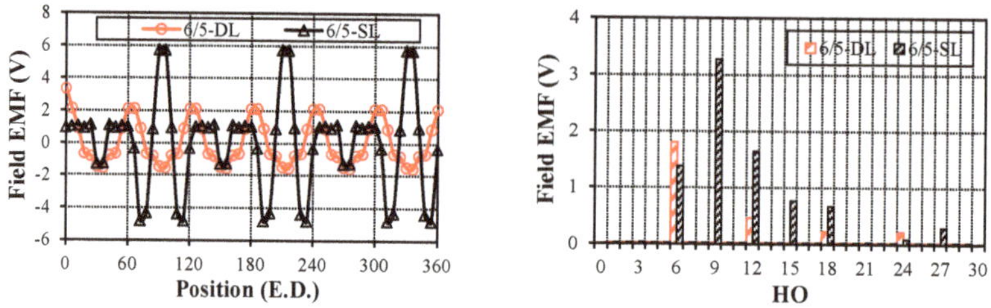

(a) (b)

Figure 2. Field back-EMF ripple of the 6/5 machine with double-layer DC winding under armature current (400 rpm). (**a**) Back-EMF; (**b**) FFT.

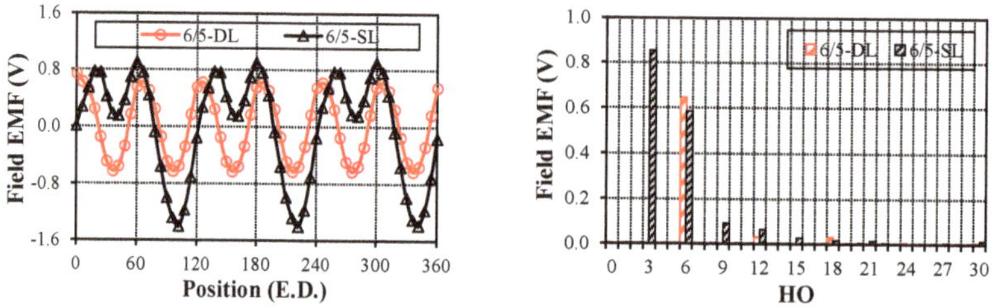

(a) (b)

Figure 3. Field back-EMF ripple of the 6/5 machine with double-layer DC winding under on-load (400 rpm). (**a**) Back-EMF; (**b**) FFT.

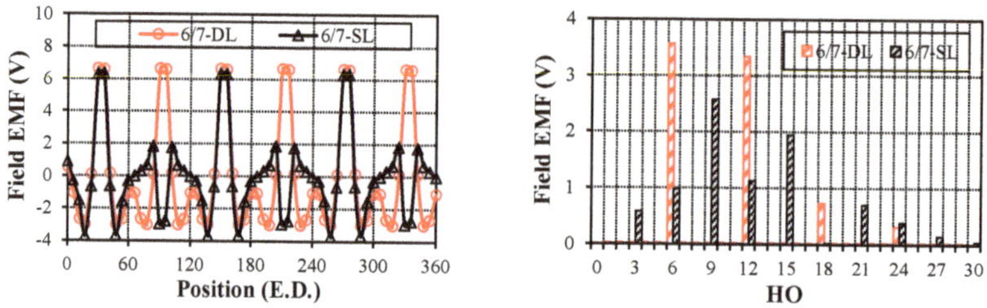

(a) (b)

Figure 4. Field back-EMF ripple of the 6/7 machine with double-layer DC winding under armature current (400 rpm). (**a**) Back-EMF; (**b**) FFT.

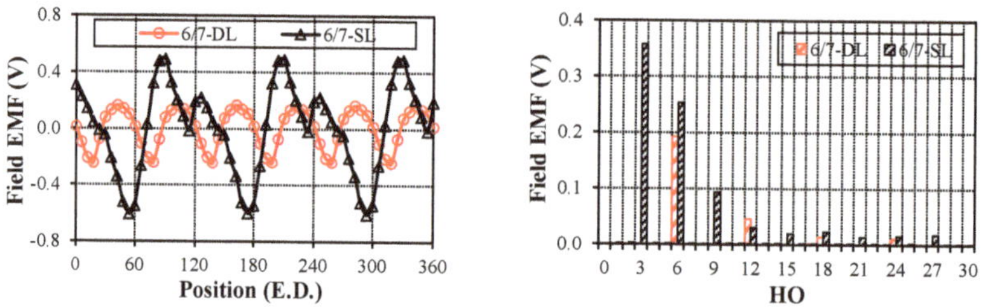

(a) (b)

Figure 5. Field back-EMF ripple of the 6/7 machine with double-layer DC winding under on-load (400 rpm). (**a**) Back-EMF; (**b**) FFT.

Table 4 summarizes the field back-EMF harmonic orders of the 12-slot HESFM counterparts when DC winding is double-layer while AC winding is double or single-layer. Similar to Table 3, the field back-EMF harmonic orders of different stator/rotor pole combinations remain unchanged under no-load since the DC winding configuration is unchanged and the AC winding is open-circuited. Different from the no-load condition, for 12/10 and 12/14 machines, the induced EMF harmonic orders under armature current and on-load conditions are changed from 6N (with double-layer AC winding) to 3N (with single-layer AC winding). For 12/11 and 12/13 machines, the number of AC winding layers does not influence the induced EMF harmonic orders under no-load, armature current, and on-load conditions.

Figures 6–9 show the field back-EMF ripples and their FFT results of 12/10 and 12/14 machines with double-layer DC winding under armature current and on-load, respectively. It is revealed that by employing single-layer AC winding, the amplitude of the field back-EMF ripple is increased under on-load for 12/10 and 12/14 machines although the field back-EMF ripple under no-load is unchanged. This is due to the fact that the armature current field back-EMF harmonic orders of 12/10 and 12/14 machines are 3N when single-layer AC winding is employed. Therefore, the on-load field back-EMF harmonic orders of 12/10 and 12/14 machines are 3N as well.

The field back-EMF harmonic orders of the 6-slot machines with an odd rotor pole number, e.g., 6/5 and 6/7, are influenced by the number of AC winding layers. Different from the 6-slot machines, the field back-EMF harmonic orders of the 12-slot HESFMs with an even rotor pole number, e.g., 12/10 and 12/14, are influenced by the number of AC winding layers. The employment of the single-layer AC winding leads to the lower induced EMF harmonic orders, i.e., 3N, for the 6-slot and 12-slot HESFM counterparts with specific stator/rotor pole combinations. Therefore, it can be concluded from the above findings that double-layer AC winding is recommended for the investigated machines with various stator/rotor pole combinations to achieve relatively lower field back-EMF ripple when the machine is equipped with double-layer DC winding.

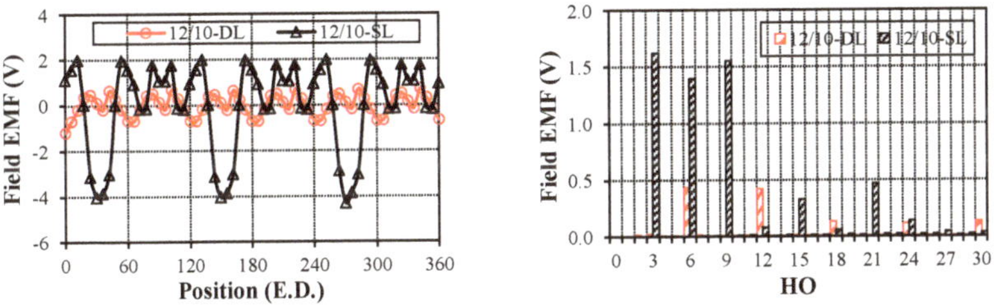

(a) (b)

Figure 6. Field back-EMF ripple of the 12/10 machine with double-layer DC winding under armature current (400 rpm). (**a**) Back-EMF; (**b**) FFT.

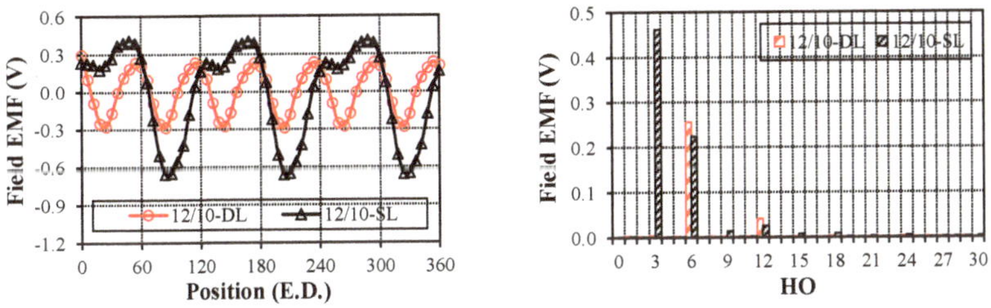

(a) (b)

Figure 7. Field back-EMF ripple of the 12/10 machine with double-layer DC winding under on-load (400 rpm). (**a**) Back-EMF; (**b**) FFT.

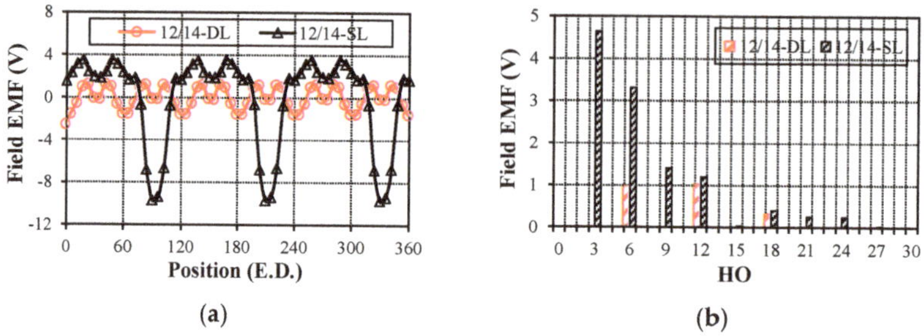

Figure 8. Field back-EMF ripple of the 12/14 machine with double-layer DC winding under armature current (400 rpm). (**a**) Back-EMF; (**b**) FFT.

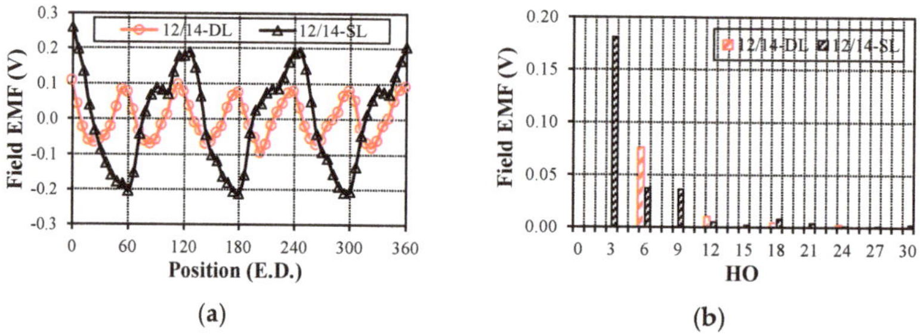

Figure 9. Field back-EMF ripple of the 12/14 machine with double-layer DC winding under on-load (400 rpm). (**a**) Back-EMF; (**b**) FFT.

3.3. DC Winding Is Single-Layer While AC Winding Is Double-Layer

In this sub-section, the investigated machines with various stator/rotor pole combinations are equipped with single-layer DC and double-layer AC windings. In order to clearly illustrate the relationship between DC winding configuration and field back-EMF ripple, Tables 5 and 6 comparatively summarize the induced EMF harmonic orders with different DC winding configurations for the 6-slot and 12-slot HESFM counterparts, respectively.

It can be found from Table 5 that the field back-EMF harmonic orders of the 6-slot machines with single/double-layer DC windings remain unchanged for each stator/rotor pole combination under no-load, armature current, and on-load conditions. Therefore, the number of DC winding layers does not influence the field back-EMF harmonic orders of the 6-slot machines under various conditions when double-layer AC winding is employed. Similar to the 6-slot machines, the employment of the single-layer DC winding does not change the field back-EMF harmonic orders of the 12-slot HESFMs with double-layer DC winding under various conditions, as shown in Table 6. Therefore, it reveals that by changing the number of layers of the field winding, the field back-EMF ripple in the 6-slot and 12-slot HESFMs with double-layer AC winding cannot be suppressed.

Table 5. Field back-EMF harmonic orders with double-layer AC winding.

Winding Configuration	Slot/Pole	No-Load	Armature Current	On-Load
Double-layer DC	6/4	3N	3N	3N
	6/5	6N	6N	6N
	6/7	6N	6N	6N
	6/8	3N	3N	3N
Single-layer DC	6/4	3N	3N	3N
	6/5	6N	6N	6N
	6/7	6N	6N	6N
	6/8	3N	3N	3N

Table 6. Field back-EMF harmonic orders with double-layer AC winding.

Winding Configuration	Slot/Pole	No-Load	Armature Current	On-Load
Double-layer DC	12/10	6N	6N	6N
	12/11	12N	6N	6N
	12/13	12N	6N	6N
	12/14	6N	6N	6N
Single-layer DC	12/10	6N	6N	6N
	12/11	12N	6N	6N
	12/13	12N	6N	6N
	12/14	6N	6N	6N

3.4. AC and DC Windings Are Both Single-Layer

In this sub-section, the investigated machines with various stator/rotor pole combinations are both equipped with single-layer DC/AC windings. To clearly illustrate the relationship between DC/AC winding configurations and field back-EMF ripple, Tables 7 and 8 comparatively summarize the induced EMF harmonic orders under different AC and DC winding configurations for the 6-slot and 12-slot machines, respectively.

Table 7. Field back-EMF harmonic orders with single-layer AC and DC windings.

Winding Configuration	Slot/Pole	No-Load	Armature Current	On-Load
Double-layer AC and DC	6/4	3N	3N	3N
	6/5	6N	6N	6N
	6/7	6N	6N	6N
	6/8	3N	3N	3N
Single-layer AC and DC	6/4	3N	3N	3N
	6/5	6N	3N	3N
	6/7	6N	3N	3N
	6/8	3N	3N	3N

Table 8. Field back-EMF harmonic orders with single-layer AC and DC wligodendroglioindings.

Winding Configuration	Slot/Pole	No-Load	Armature Current	On-Load
Double-layer AC and DC	12/10	6N	6N	6N
	12/11	12N	6N	6N
	12/13	12N	6N	6N
	12/14	6N	6N	6N
Single-layer AC and DC	12/10	6N	3N	3N
	12/11	12N	6N	6N
	12/13	12N	6N	6N
	12/14	6N	3N	3N

Table 7 comparatively shows the field back-EMF harmonic orders of the 6-slot machines when AC and DC windings are both single-layers. Similar to Table 5, the field back-EMF harmonic orders of the 6-slot machines remain unchanged under no-load since the AC winding is open-circuited. Different from the no-load condition, for 6/5 and 6/7 machines, the field back-EMF harmonic orders under armature current and on-load conditions are changed from 6N (double-layer DC/AC windings) to 3N (single-layer DC/AC windings). For 6/4 and 6/8 machines, the number of AC and DC winding layers does not influence the field back-EMF harmonic orders under no-load, armature current, and on-load conditions.

Figures 10–13 show the field back-EMF ripples and their FFT results of 6/5 and 6/7 machines with single-layer AC and DC windings under armature current and on-load, respectively. It is revealed that by employing single-layer AC and DC windings, the amplitude of the field back-EMF ripple is increased for 6/7 machines but almost unchanged for 6/5 machines under on-load conditions.

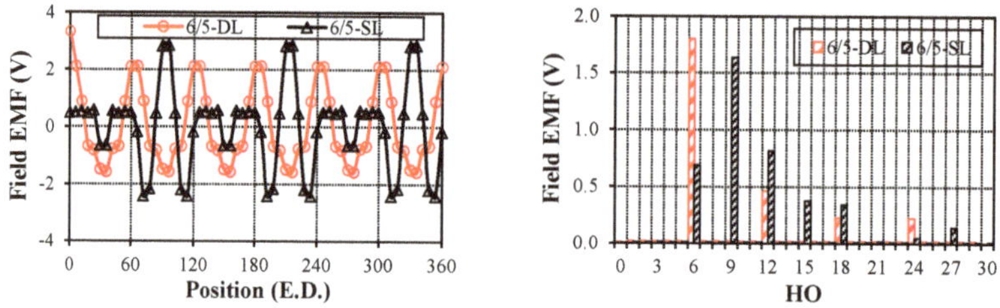

(a) (b)

Figure 10. Field back-EMF ripple of the 6/5 machine with single-layer AC and DC windings under armature current (400 rpm). (**a**) Back-EMF; (**b**) FFT.

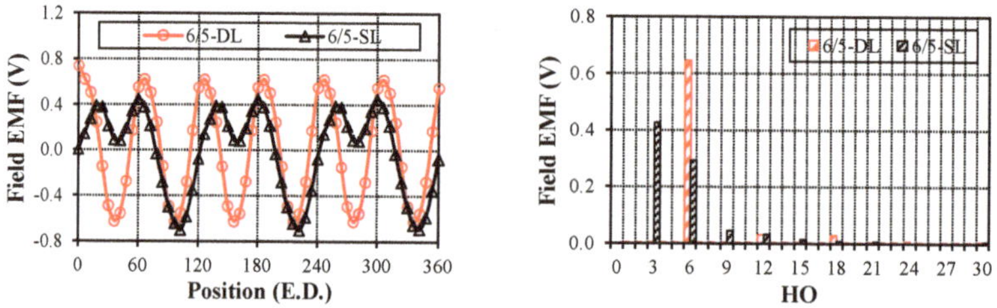

(a) (b)

Figure 11. Field back-EMF ripple of the 6/5 machine with single-layer AC and DC windings under on-load (400 rpm). (**a**) Back-EMF; (**b**) FFT.

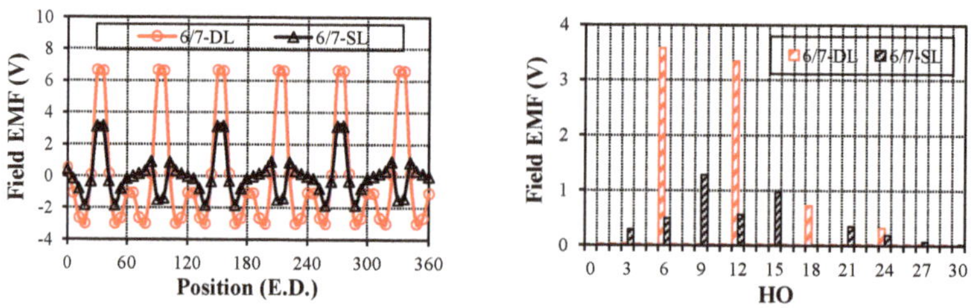

(a) (b)

Figure 12. Field back-EMF ripple of the 6/7 machine with single-layer AC and DC windings under armature current (400 rpm). (**a**) Back-EMF; (**b**) FFT.

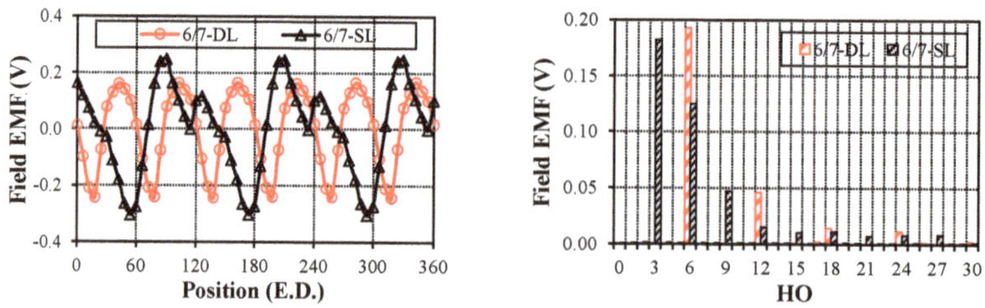

(a) (b)

Figure 13. Field back-EMF ripple of the 6/7 machine with double-layer DC winding under on-load (400 rpm). (**a**) Back-EMF; (**b**) FFT.

Table 8 summarizes the field back-EMF harmonic orders of the 12-slot machines when AC and DC windings are both single layers. Similar to Table 6, the field back-EMF harmonic orders of the 12-slot machines remain unchanged under no-load since the AC winding is open-circuited. Different from the no-load condition, for 12/10 and 12/14 machines, the field back-EMF harmonic orders under armature current and on-load conditions are changed from 6N (double-layer DC/AC windings) to 3N (single-layer DC/AC windings). For 12/11 and 12/13 machines, the number of AC and DC winding layers does not influence the induced EMF harmonic orders under no-load, armature current, and on-load conditions.

Figures 14–17 show the field back-EMF ripples and their FFT results of 12/10 and 12/14 machines with single-layer AC and DC windings under armature current and on-load, respectively. It can be found that by employing single-layer AC and DC windings, the amplitude of the field back-EMF ripple is almost unchanged for 12/10 and 12/14 machines under on-load although the field back-EMF harmonic orders under on-load are different.

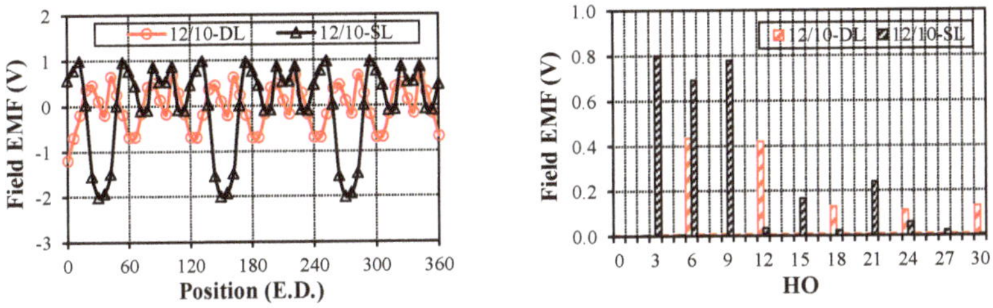

(a) (b)

Figure 14. Field back-EMF in 12/10 HESFM with single-layer AC and DC windings under armature current (400 rpm). (**a**) Back-EMF; (**b**) FFT.

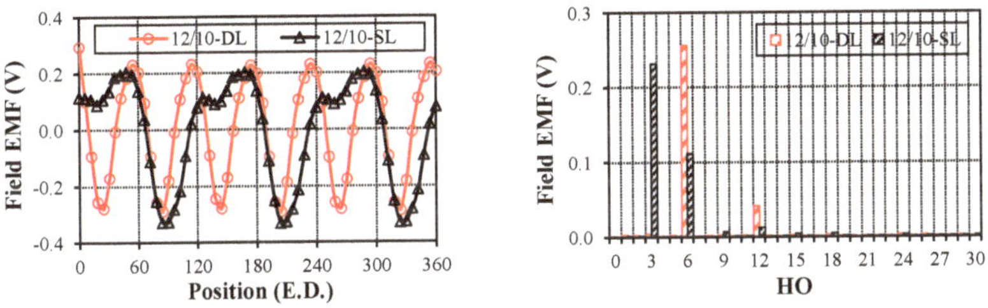

(a) (b)

Figure 15. Field back-EMF ripple of the 12/10 machine with single-layer AC and DC windings under on-load (400 rpm). (**a**) Back-EMF; (**b**) FFT.

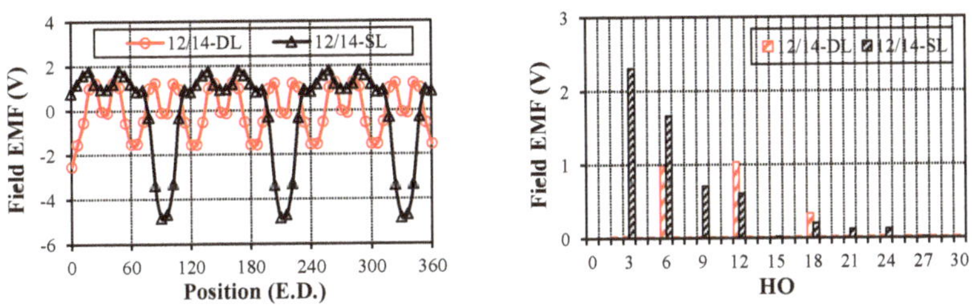

(a) (b)

Figure 16. Field back-EMF ripple of the 12/14 machine with single-layer AC and DC windings under armature current (400 rpm). (**a**) Back-EMF; (**b**) FFT.

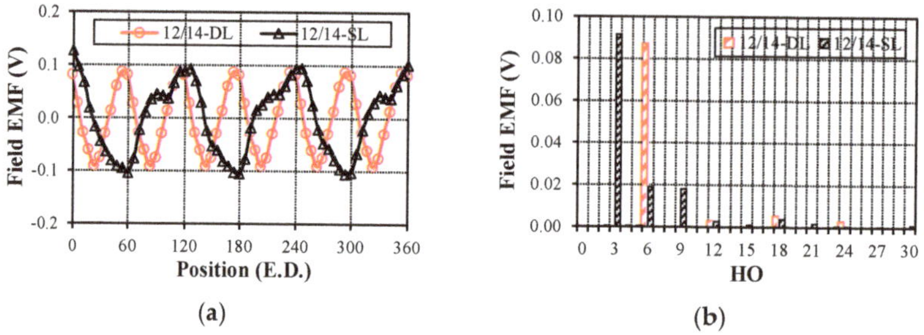

Figure 17. Field back-EMF ripple of the 12/14 machine with single-layer AC and DC windings under on-load (400 rpm). (**a**) Back-EMF; (**b**) FFT.

The field back-EMF harmonic orders of 6-slot HESFMs with an odd rotor pole number, e.g., 6/5 and 6/7, are influenced by the number of AC and DC winding layers. Different from 6-slot machines, the field back-EMF harmonic orders of 12-slot HESFMs with an even rotor pole number, e.g., 12/10 and 12/14, are influenced by the number of AC and DC winding layers. The employment of single-layer AC and DC windings leads to lower induced EMF harmonic orders, i.e., 3N, for the 6-slot and 12-slot HESFM counterparts with specific stator/rotor pole combinations. Therefore, it can be concluded from the above findings that double-layer AC and DC windings are recommended for the investigated machines with various stator/rotor pole combinations to achieve relatively lower field back-EMF ripple when the machine is equipped with double-layer DC winding.

Based on the simulated results presented in this section, some observations can be summarized as follows. Firstly, the peak value of field back-EMF with lower harmonics is higher than those with higher harmonics. Under no-load, the armature windings are open-circuited and thus the field back-EMF is not influenced by AC winding configuration but only related to DC winding configuration. The number of field coils, determined by stator pole number and field winding layers, is a key factor that influences the no-load field back-EMF. The field coil back-EMF may contain many harmonics with various orders due to the magnetic gearing effect. However, when the field coils are connected together in series, a lot of field coil back-EMF harmonics disappear due to phase difference between each field coil back-EMF waveform and only certain harmonics, e.g., 3N, 6N, 12N, may exist in the field winding. Under armature current, the mutual inductance harmonics and the corresponding initial phase between phase winding and field winding may change under different combinations of AC/DC winding configurations together with stator/rotor pole numbers. As such, the resultant on-load field back-EMF is determined by the no-load and armature reaction.

4. Experimental Verification

As can be found from the theoretical analyses shown in Section 3, non-overlapped double-layer armature and field windings are recommended for the hybrid excited switched flux machines with various stator/rotor pole combinations to realize relatively lower field back-EMF under different conditions. Moreover, the HESFMs with higher stator slot numbers, e.g., 12-slot, exhibit lower field back-EMF amplitude due to higher field back-EMF harmonics. Therefore, to verify the theoretical results calculated using the finite element method, a 12/10 HESFM with double-layer DC/AC windings is prototyped to balance the torque capability and the field back-EMF ripple, as shown in Figure 18. For the prototype machine, the stator contains the PMs, DC and AC windings. Similar to switched reluctance machines, the rotor is very simple and only composed of laminations. The AC and DC windings are both of concentrated and double-layer type and therefore the number of AC and DC coils is 12 in the prototype machine. The test rig for performing the experiments of the prototype machine is illustrated in Figure 19. For the experimental verification, zero d-axis current control is applied in all the tests due to negligible reluctance torque and the mechanical rotor speed is 400 rpm.

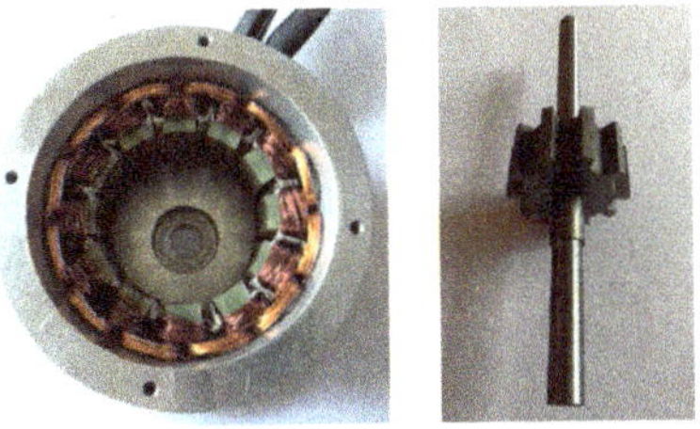

(**a**) (**b**)

Figure 18. The 12/10 HESFM prototype with double-layer DC and AC windings. (**a**) Stator; (**b**) Rotor.

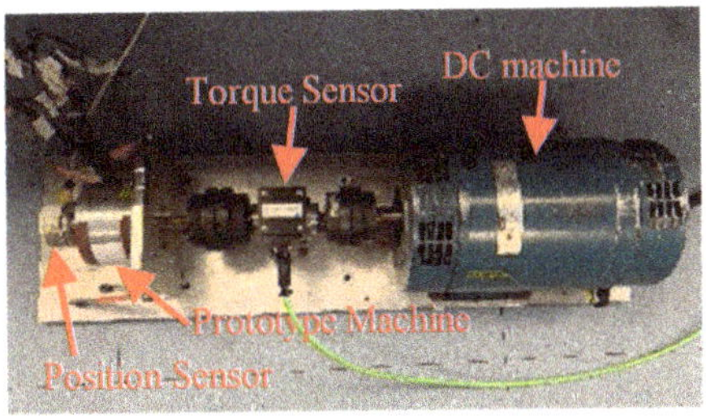

Figure 19. Test bench of the HESFM prototype.

The armature back-EMFs and their FFT results of the prototype machine are comparatively shown in Figure 20, where the rotor rotates at 400 rpm and the field current is removed (I_{dc} = 0 A). In Figure 20, the permanent magnet is the only excitation source and armature/field currents are both zero. It is shown that the test results agree well with the results calculated using the 2D finite element method and the error is mainly attributed to the 3D end-effect.

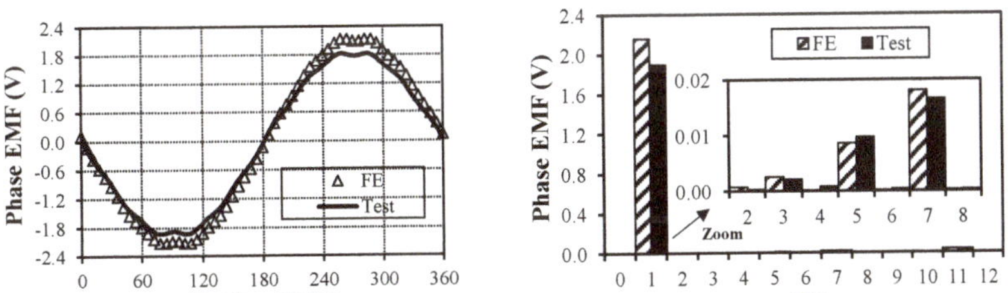

(**a**) (**b**)

Figure 20. Phase back-EMF of the prototype machine without field current (400 rpm, I_{ac_rms} = 0 A, I_{dc} = 0 A). (**a**) Back-EMF; (**b**) FFT.

Under various conditions (no-load and on-load), the test and FE results of the field back-EMFs are shown in Figures 21 and 22. In Figure 21, the phase current is zero and a constant DC current of 2 A is injected into the field winding. In Figure 22, the root–mean–square (RMS) value of the sinusoidal phase current is 2 A and the DC field current is also 2 A. It reveals that the measured field back-EMF harmonic orders are 6N under various conditions, which validates the theoretical results in Section 3.1. Figure 23 shows the measured and FE results of the on-load torque, in which i_d = 0 control is employed for the on-load calculations and tests since the reluctance torque component is very small. In Figure 23, the testing conditions are exactly the same as that in Figure 22. The observed errors are mainly caused by manufacturing imperfections and assembly tolerances, measuring errors, and the 3D end-effect, etc. As can be found in Figures 20–23, differences can be observed between simulated results and testing results, which are attributed to several factors. Firstly, the simulated results are obtained using the 2D FEA method and the effect of end-winding is not considered in the 2D FEA model. Therefore, overall, the testing results are relatively lower than the simulated results. The reason for using the 2D FEA method is to balance the computational efficiency and accuracy. Moreover, manufacturing tolerance and assembly error are common factors that lead to the observed differences. For instance, for the manufactured 12/10 HESFM prototype, the magnet position tolerance is a very typical assembly tolerance but it is very difficult to be controlled [31]. Moreover, the measuring error due to the precision of the measuring equipment, i.e., the bandwidth of torque and speed sensors, is also a common factor that leads to the differences. Therefore, it should be noted that the accuracy and reliability of the testing results presented in this paper are subject to the aforementioned factors.

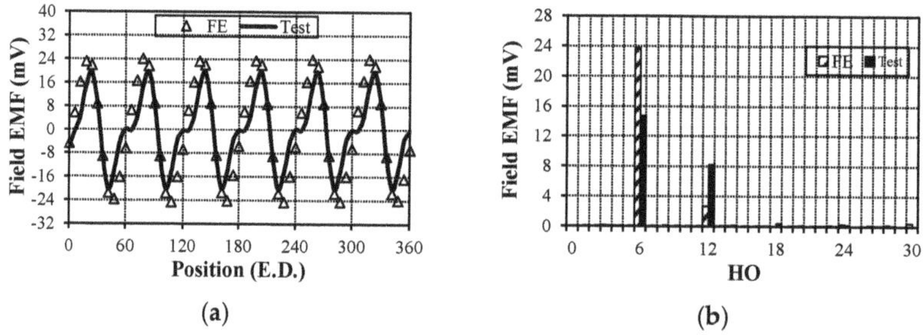

Figure 21. No-load field back-EMF ripple (400 rpm, I_{ac_rms} = 0 A, I_{dc} = 2 A). (**a**) Back-EMF; (**b**) FFT.

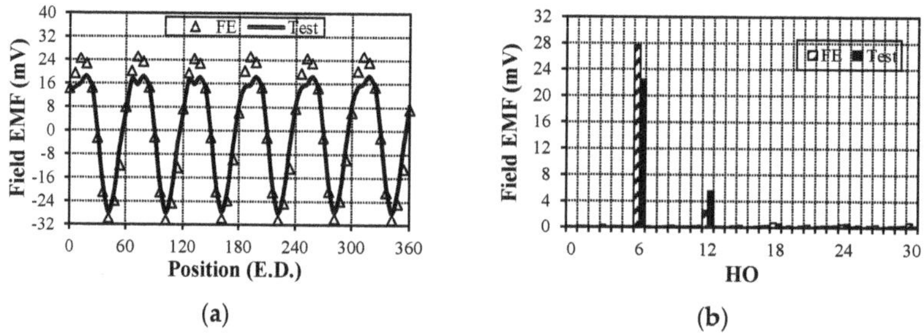

Figure 22. On-load field back-EMF ripple (400 rpm, I_{ac_rms} = 2 A, I_{dc} = 2 A). (**a**) Back-EMF; (**b**) FFT.

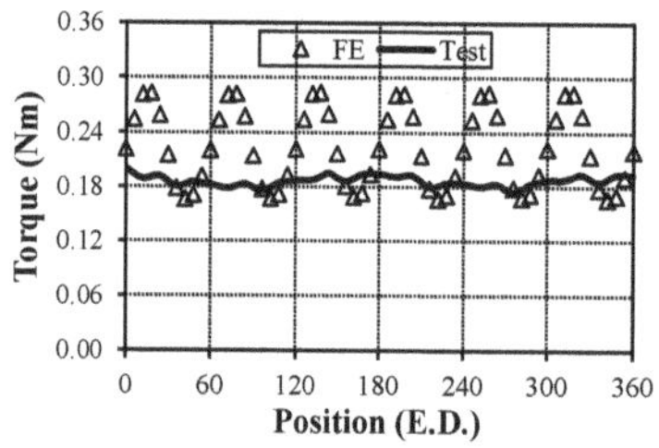

Figure 23. Torque waveform of the HESFM prototype (400 rpm, I_{ac_rms} = 2 A, I_{dc} = 2 A).

5. Conclusions

The paper aims to reveal the intrinsic relationship between armature/field winding configurations together with the stator/rotor pole combinations and the field back-EMF ripple in HESFMs. The number of layers of field/armature windings, the number of field/armature coils, and the stator/rotor pole combination are considered for the investigations. However, the coil pitch of field and armature coils equals 1, i.e., the non-overlapped windings. In this paper, four different combinations of the number of AC and DC winding layers were comparatively illustrated for the investigated machines with various stator/rotor pole combinations.

The peak value of field back-EMF with lower harmonics is higher than those with higher harmonics. Under no-load, the field back-EMF is not influenced by AC winding configuration but only related to DC winding configuration. The number of field coils, determined by stator pole number and field winding layers, is a key factor that influences the no-load field back-EMF. Under armature current, the mutual inductance harmonics and the corresponding initial phase between phase winding and field winding may change under different combinations of AC/DC winding configurations together with stator/rotor pole numbers. As such, the resultant on-load field back-EMF is determined by the no-load and armature reaction.

The harmonic order and amplitude of the HESFMs with double-layer armature and field windings are taken as the baseline for comparisons. It is revealed that the utilization of single-layer AC or DC windings cannot reduce the amplitude of the field back-EMF ripple and, in contrast, its amplitude may increase for specific stator/rotor pole combinations due to lower field back-EMF harmonic orders. Consequently, double-layer armature and field windings are recommended for various stator/rotor pole combinations to achieve relatively lower field back-EMF under different conditions. The variable-speed system is one of the most important fields for the applications of hybrid excited machines, e.g., electrical vehicles and hybrid electrical vehicles (EVs/HEVs). However, the field back-EMF ripple significantly affects the overall performance of hybrid excited machine systems in these applications, e.g., torque capability, efficiency, flux-regulation capability, and torque-speed characteristics. Therefore, the findings and conclusions presented in this paper are important references for designing hybrid excited machine systems for EVs/HEVs.

It should be noted that the investigations of the field back-EMF ripple in this paper are all restricted to the windings with one coil pitch field/armature coils. For future work, the investigations can be extended to explore stator/rotor pole combinations with multiple winding pitches. Moreover, skewing, notching, shaping, etc., can also be combined to further reduce the field back-EMF ripples. Moreover, the findings in this paper can serve as a reference for other HEMs.

Author Contributions: Z.Y.: Conceptualization, Methodology, Validation, Formal analysis, Investigation, Writing—original draft preparation. X.S.: Conceptualization, Methodology, Validation, Formal analysis, Investigation, Writing—review and editing, Visualization, Funding acquisition. R.H.: Validation, Formal analysis, Investigation, Writing—review and editing, Visualization. R.S.: Resources, Data Curation, Writing—review and editing, Visualization. Z.C.: Resources, Data Curation, Supervision. X.L.: Resources, Data Curation, Supervision. All authors have read and agreed to the published version of the manuscript.

Funding: This work was funded in part by the National Key Research and Development Program of China under Project 2023YFB2406400, in part by the National Natural Science Foundation of China under Grant 52407035, in part by the Hebei Natural Science Foundation under Project E2023105080, and in part by the Beijing Institute of Technology Research Fund Program for Young Scholars under Project XSQD-202207004.

Institutional Review Board Statement: Not Applicable.

Informed Consent Statement: Not Applicable.

Data Availability Statement: The data that support the findings of this study are available from the corresponding author upon reasonable request.

Conflicts of Interest: The authors declare no conflicts of interest.

References

1. Zhu, Z.Q.; Howe, D. Electrical machines and drives for electric, hybrid, and fuel cell vehicles. *Proc. IEEE* **2007**, *95*, 746–765. [CrossRef]
2. Chan, C.C. The state of the art of electric, hybrid, and fuel cell vehicles. *Proc. IEEE* **2007**, *95*, 704–718. [CrossRef]
3. Cheng, M.; Hua, W.; Zhang, J.; Zhao, W. Overview of stator-permanent magnet brushless machines. *IEEE Trans. Ind. Electron.* **2011**, *58*, 5087–5101. [CrossRef]
4. Chau, K.T.; Chan, C.C.; Liu, C. Overview of permanent-magnet brushless drives for electric and hybrid electric vehicles. *IEEE Trans. Ind. Electron.* **2008**, *55*, 2246–2257. [CrossRef]
5. Lee, C.H.; Chau, K.T.; Liu, C.H. Overview of magnetless brushless machines. *IET Elect. Power Appl.* **2018**, *12*, 1117–1125. [CrossRef]
6. Li, Y.X.; Lu, Q.F.; Zhu, Z.Q. Unbalanced magnetic force prediction in permanent magnet machines with rotor eccentricity by improved superposition method. *IET Electr. Power Appl.* **2017**, *11*, 1095–1104. [CrossRef]
7. Amara, Y.; Vido, L.; Gabsi, M.; Hoang, E.; Ahmed, A.; Lecrivain, M. Hybrid excitation synchronous machines: Energy-efficient solution for vehicles propulsion. *IEEE Trans. Veh. Technol.* **2009**, *58*, 2137–2149. [CrossRef]
8. Li, Y.; Yu, Z.; Meng, H.; Wang, J.; Jing, Y. Design and optimization of hybrid-excited claw-pole machine for vehicle. *IEEE Trans. Appl. Supercond.* **2021**, *31*, 1–4. [CrossRef]
9. Xu, L.; Xu, D.; Zhu, X.; Zhang, C.; Zhang, X. Multimode excitation analysis and design of a new hybrid excited modular stator permanent magnet switched reluctance machine. *IEEE Trans. Magn.* **2023**, *59*, 1–6. [CrossRef]
10. Xia, C.; Feng, Y.; Jia, M.; Gao, Y.; Huang, S. A novel counter-rotating axial-flux hybrid-excitation permanent magnet machine with dual-rotor. *IEEE Trans. Magn.* **2023**, *59*, 1–5.
11. Palka, R.; Cierzniewski, K.; Wardach, M.; Prajzendanc, P. Research on Innovative Hybrid Excited Synchronous Machine. *Energies* **2023**, *16*, 6600. [CrossRef]
12. Mörée, G.; Leijon, M. Overview of Hybrid Excitation in Electrical Machines. *Energies* **2022**, *15*, 7254. [CrossRef]
13. Cai, S.; Zhu, Z.Q.; Mipo, J.C.; Personnaz, S. Investigation of novel doubly salient hybrid excited machine with non-overlapped field winding. *IEEE Trans. Energy Convers.* **2021**, *36*, 2261–2275. [CrossRef]
14. Hlioui, S.; Gabsi, M.; Ahmed, H.B.; Barakat, G.; Amara, Y.; Chabour, F.; Paulides, J.J.H. Hybrid Excited Synchronous Machines. *IEEE Trans. Magn.* **2022**, *58*, 1–10. [CrossRef]
15. Dupas, A.; Hlioui, S.; Hoang, E.; Gabsi, M.; Lecrivain, M. Investigation of a new topology of hybrid-excited flux-switching machine with static global winding: Experiments and modeling. *IEEE Trans. Ind. Appl.* **2016**, *52*, 1413–1421. [CrossRef]
16. Jiang, J.; Niu, S. A novel high-order-harmonic winding design in hybrid-excited reluctance machine for electric vehicle application. *IEEE Trans. Magn.* **2023**, *59*, 1–7. [CrossRef]
17. Hoang, E.; Lecrivain, M.; Gabsi, M. A new structure of a switching flux synchronous polyphased machine with hybrid excitation. In Proceedings of the 2007 European Conference on Power Electronics and Applications, Aalborg, Denmark, 2–5 September 2007; pp. 1–8.
18. Chen, J.T.; Zhu, Z.Q.; Iwasaki, S.; Deodhar, R.P. Influence of slot opening on optimal stator and rotor pole combination and electromagnetic performance of switched-flux PM brushless AC machines. *IEEE Trans. Ind. Appl.* **2011**, *47*, 1681–1691. [CrossRef]
19. Hua, W.; Zhang, G.; Cheng, M. Flux-regulation theories and principles of hybrid-excited flux-switching machines. *IEEE Trans. Ind. Electron.* **2015**, *62*, 5359–5369. [CrossRef]
20. Farahzadi, M.; Abbaszadeh, K.; Mirnikjoo, S. Electromagnetic-thermal analysis of a hybrid-excited flux switching permanent magnet generator for wind turbine application. *IEEE Trans. Energy Convers.* **2023**, *38*, 1962–1973. [CrossRef]
21. Sun, X.Y.; Zhu, Z.Q. Investigation of DC winding induced voltage in hybrid-excited switched-flux permanent magnet machine. *IEEE Trans. Ind. Appl.* **2020**, *56*, 3594–3603. [CrossRef]
22. Wu, Z.Z.; Zhu, Z.Q.; Wang, C.; Mipo, J.C.; Personnaz, S.; Farah, P. Reduction of open-circuit DC winding induced voltage in wound field switched flux machines by skewing. *IEEE Trans. Ind. Electron.* **2019**, *66*, 1715–1726. [CrossRef]
23. Wu, Z.Z.; Zhu, Z.Q.; Wang, C.; Mipo, J.C.; Personnaz, S.; Farah, P. Analysis and reduction of on-load DC winding induced voltage in wound field switched flux machines. *IEEE Trans. Ind. Electron.* **2020**, *67*, 2655–2666. [CrossRef]
24. Wu, Z.Z.; Zhu, Z.Q.; Hua, W.; Akehurst, C.; Zhu, X.F.; Zhang, W.T.; Hu, J.; Li, H.Y.; Zhu, J.M. Analysis and suppression of induced voltage pulsation in DC winding of five-phase wound-field switched flux machines. *IEEE Trans. Energy Convers.* **2019**, *34*, 1890–1905. [CrossRef]
25. Wu, Z.Z.; Zhu, Z.Q.; Wang, C. Reduction of on-load dc winding-induced voltage in partitioned stator wound field switched flux machines by dual three-phase armature winding. *IEEE Trans. Ind. Electron.* **2022**, *69*, 5409–5420. [CrossRef]
26. Yu, L.; Zhang, M.; Zhang, Z.; Jiang, B. Reduction of field-winding-induced voltage in a doubly salient brushless dc generator with stator-damper winding. *IEEE Trans. Ind. Electron.* **2022**, *69*, 7767–7775. [CrossRef]
27. Sun, X.Y.; Zhu, Z.Q.; Cai, S.; Wang, L.; Wei, F.R.; Shao, B. Influence of stator slot and rotor pole number combination on field winding induced voltage ripple in hybrid excitation switched flux machine. *IEEE Trans. Energy Convers.* **2021**, *36*, 1245–1261. [CrossRef]
28. Sun, X.Y.; Zhu, Z.Q.; Wei, F.R. Voltage pulsation induced in DC field winding of different hybrid excitation switched flux machines. *IEEE Trans. Ind. Appl.* **2021**, *57*, 4815–4830. [CrossRef]

29. Zhang, W.; Wu, Z.; Jin, L.; Fan, Y.; Hua, W.; Cheng, M. Analysis and multi-objective optimization of the hybrid excitation switched flux machine. In Proceedings of the 2023 IEEE 6th International Electrical and Energy Conference (CIEEC), Hefei, China, 12–14 May 2023; pp. 3371–3376.
30. Zhang, W.; Wu, Z.; Hua, W.; Fan, Y.; Xu, Z.; Cheng, M. Design tradeoff between flux regulation capability and DC winding induced voltage in hybrid excitation switched flux machine. In Proceedings of the 2023 IEEE Energy Conversion Congress and Exposition (ECCE), Nashville, TN, USA, 29 October–2 November 2023; pp. 5060–5066.
31. Zhu, Z.Q.; Thomas, A.S.; Chen, J.T.; Jewell, G.W. Cogging torque in flux-switching permanent magnet machines. *IEEE Trans. Magn.* **2009**, *45*, 4708–4711. [CrossRef]

energies

MDPI

Article

Optimal Rotor Design for Reducing Electromagnetic Vibration in Traction Motors Based on Numerical Analysis

Seung-Heon Lee [1], Si-Woo Song [2], In-Jun Yang [2], Ju Lee [2] and Won-Ho Kim [3,*]

1 Department of Next Generation Energy System Convergence, Gachon University, Seongnam 461-701, Republic of Korea; dltmdgjs153@naver.com
2 Department of Electrical Engineering, Hanyang University, Seoul 133-791, Republic of Korea; thdtldn93@naver.com (S.-W.S.); dlswns78@naver.com (I.-J.Y.); julee@hanyang.ac.kr (J.L.)
3 Department of Electrical Engineering, Gachon University, Seongnam 461-701, Republic of Korea
* Correspondence: wh15@gachon.ac.kr; Tel.: +82-10-9856-1535

Abstract: Interior permanent magnet synchronous motor (IPMSM) for traction applications have attracted significant attention due to their advantages of high torque and power density as well as a wide operating range. However, these motors suffer from high electromagnetic vibration noise due to their complex structure and structural rigidity. The main sources of this electromagnetic vibration noise are cogging torque, torque ripple, and radial force. To predict electromagnetic vibration noise, finite element analysis (FEA) with flux density analysis of the air gap is essential. This approach allows for the calculation of radial force that is the source of the vibration and enables the prediction of vibration in advance. The data obtained from these analyses provide important guidance for reducing vibration and noise in the design of electric motors. In this paper, the cogging torque and vibration at rated and maximum operating speed are analyzed, and an optimal cogging torque and vibration reduction model, with rotor taper and two-step skew structure, is proposed using the response surface method (RSM) to minimize them. The validity of the proposed model is demonstrated through formulations and FEA based entirely on numerical analysis and results. This study is expected to contribute to the design of more efficient and quieter electric motors by providing a solution to the electromagnetic vibration noise problem generated by IPMSM for traction applications with complex structures.

Keywords: IPMSM; traction motor; noise; vibration; RSM; cogging torque; taper; two-step skew

1. Introduction

Increasing energy consumption and stricter environmental regulations around the world are driving the need for high-performance and efficient electric motors. Traction motors, in particular, require high torque and power density, and interior permanent magnet synchronous motor (IPMSM) models are the right choice to meet these needs. The pursuit of higher power density, lighter weight, and a wider speed range for traction compared to other applications presents the challenges of a more complex structure and lower structural rigidity, resulting in higher vibration and noise. Vibration and noise generated during operation in motors can hinder driver comfort and negatively affect the performance and lifetime of the motor [1–3]. Research to reduce vibration and noise involves a variety of factors, with three main electromagnetic causes being addressed [4,5]. These sources are as follows. First, there is cogging torque due to changes in the reluctance of the stator slots as the rotor rotates [6–8]. Second, there is torque ripple as a result of the fifth and seventh harmonic components arising from the temporal and spatial harmonic components when current is applied [9–11]. Finally, there are radial and tangential forces that occur during rotation [12–14]. An international standard established in 1974 is the only standard that evaluates the vibration generated by electric motors and is shown in Table 1 [15,16]. This standard is divided into Class 1 through Class 4 depending on the capacity of the motor and is evaluated as four grades: A (good), B (slightly bad), C (bad), and D (extremely bad) based on the RMS value of the vibration speed generated in the motor housing. This study presents a theoretical study on the modification of rotor geometry to minimize vibration and noise at rated and maximum operating speeds for a 16-pole 24-slot traction motor [17,18]. The main objectives of the study are as follows. The first is to minimize vibration noise by applying tapering to the rotor core to minimize the cogging torque due to the change in reluctance of the stator slot with the rotation of the rotor. Second, an asymmetric two-step skew is applied to the tapered rotor by dividing the rotor into upper and lower parts to adjust the cogging torque and back electromotive force (EMF) to minimize vibration and noise. The RSM is used to derive the optimal model for cogging torque and vibration noise reduction [19–21]. To calculate the electromagnetic vibration in the proposed model, finite element analysis (FEA) with flux density analysis of the air gap is essential. In this process, the radial force, which is the source of the vibration, is obtained and the vibration velocity is predicted [22–25]. The results of these studies are verified using formulas and FEA, which will play an important role in improving the performance of electric motors as well as the convenience and satisfaction of users.

Table 1. Standard for the vibration grade ISO 2372 (ISO 10816-1).

<table>
<tr><th>RMS Vibration Velocity [mm/s]</th><th>Up to 15 [kW] Class I</th><th>15 to 75 [kW] Class II</th><th>>75 [kW] (Rigid) Class III</th><th>>75 [kW] (Soft) Class IV</th></tr>
<tr><td>0.28</td><td rowspan="3">A</td><td rowspan="4">A</td><td rowspan="5">A</td><td rowspan="6">A</td></tr>
<tr><td>0.45</td></tr>
<tr><td>0.71</td></tr>
<tr><td>1.12</td><td rowspan="2">B</td></tr>
<tr><td>1.8</td><td rowspan="2">B</td></tr>
<tr><td>2.8</td><td rowspan="2">C</td><td rowspan="2">B</td></tr>
<tr><td>4.5</td><td rowspan="2">C</td><td rowspan="2">B</td></tr>
<tr><td>7.1</td><td rowspan="5">D</td><td rowspan="2">C</td></tr>
<tr><td>11.2</td><td rowspan="4">D</td><td rowspan="2">C</td></tr>
<tr><td>18</td><td rowspan="3">D</td></tr>
<tr><td>28</td><td rowspan="2">D</td></tr>
<tr><td>45</td></tr>
</table>

Zone A: The vibration of a newly installed machine usually falls within this zone. Zone B: Machines that vibrate within this zone are generally considered to be acceptable for long-term operation without restrictions. Zone C: Machines that vibrate within this zone are generally considered unsatisfactory for long-term continuous operation. Zone D: Vibration values in this zone are generally considered to be severe enough to cause damage to the machine.

2. Theoretical Analysis

2.1. Analyze the Cause of Cogging Torque

There have been countless studies and significant research on the causes of cogging torque. Cogging torque is an irregular torque that occurs when the rotor of an electric motor interacts with the stator. It interferes with the smooth rotation of the motor and causes vibration and noise. Cogging torque can be expressed as:

$$T_{cog} = -\frac{1}{2}\varnothing_m^2\frac{dR}{d\theta} \quad (1)$$

where $\varnothing_m$ represents the magnetic flux from a permanent magnet and the remaining term is the change in reluctance. Figure 1 shows an example of the geometry of how cogging torque occurs. As the rotor rotates, the magnetic flux from the permanent magnet tends to flow toward the side. The smaller reluctance is due to the difference in reluctance between the stator slot opening and the stator tooth, resulting in a change in reluctance and cogging torque. In this study, rotor tapering and a two-step skew were applied to reduce this change in reluctance.

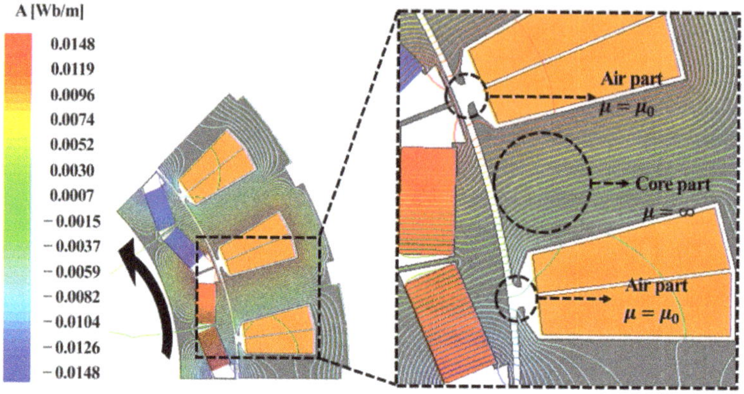

Figure 1. Example explanation of cogging torque occurrence.

2.2. *Analysis of Electromagnetic Vibration of Motor*

2.2.1. Electromagnetic Vibration Source (Radial Force)

Electromagnetic vibration is a radial force that propagates into the air. This is transmitted from the stator back to the housing, and the radial force that contributes to external noise can be calculated as the square of the air gap flux density. The torque density of an electric motor is proportional to the air gap flux density, and IPMSM types with higher torque densities exhibit greater electromagnetic vibration. In addition to electromagnetic performance, it is necessary to design an electric motor that considers vibration noise for high quality, and to do so, the basic design should consider not only torque, power, and efficiency, but also vibration caused by electromagnetic forces. Vibration velocity is the radial vibration speed generated by an electromagnetic radial force acting on the stator shoes or teeth and transmitted to the stator outer diameter. It can be expressed as vibration acceleration through differential calculation, but vibration velocity is used in this paper. In general, the radial force received in a unit area is equal to the square of the total flux density minus the square of the tangential component of the flux density [22–24]. The radial force is expressed as follows:

$$p(\theta^m,t) = \frac{1}{2\mu_0}\left[B^2(\theta^m,t) - B_t^2(\theta^m,t)\right] \cong \frac{1}{2\mu_0}B^2(\theta^m,t) \tag{2}$$

where θ, t are the components in space and time, μ_0 is the permeability of the air gap, and $B(\theta^m,t)$, and $B_t^2(\theta^m,t)$ are the total and tangential air gap flux density. Since the permeability of the iron core is much higher than the permeability of the air gap, the flux path is almost perpendicular to the rotor and stator core. Therefore, the air gap flux density in the tangential direction can be neglected because it is a very small value compared to the radial direction. The flux density $B(\theta^m,t)$ in Equation (2) is the sum of the flux densities occurring in the stator and rotor windings and is equal to:

$$B(\theta^m,t) = B^s(\theta^m,t) + B^r(\theta^m,t) \tag{3}$$

By substituting Equation (3) into Equation (2), we derived the following expression, which can be expressed as:

$$\begin{aligned} p(\theta^m,t) &= \tfrac{1}{2\mu_0}\left[B^s(\theta^m,t) + B^r(\theta^m,t)\right]^2 \\ &= \tfrac{1}{2\mu_0}\left[\{B^s(\theta^m,t)\}^2 + 2\{B^s(\theta^m,t)B^r(\theta^m,t)\} + \{B^r(\theta^m,t)\}^2\right] \end{aligned} \tag{4}$$

Equation (4) is expressed in three parts: the radial force due to the flux density of the stator winding, the radial force due to the stator and rotor flux density, and the radial force due to the rotor flux density. If the current applied to the motor is assumed to be sinusoidal, which is an ideal condition, the time harmonic component of the stator flux density is only the fundamental wave component, and the rotor flux density generates harmonic components in addition to the fundamental wave.

2.2.2. Stator System Mechanical Properties for Vibration Calculations

The electromagnetic vibration generated in the stator is influenced by the circumferential mode of the radial force, its frequency, and magnitude. Therefore, the radial force as a source of electromagnetic vibration must be separated into its circumferential modes and analyzed for quantitative and qualitative characteristics in terms of magnitude and frequency. In addition, the mechanical system must be taken into account to predict electromagnetic vibrations in motors. However, it is very difficult to make a perfect prediction of the vibration considering the mechanical structure. Therefore, assuming that the radial force affects the inner part of the stator, the shoe part, a simple model of the stator is used to analyze the vibration displacement and vibration velocity mathematically with some assumptions. First, the stator is assumed to be a ring-shaped cylinder, as shown in Figure 2. Since vibration velocity and vibration displacement are generated in the circumferential direction by the radial force, which is an electromagnetic force, it is necessary to find the vibration mode m and the natural frequency of the stator system in the circumferential direction to analyze it.

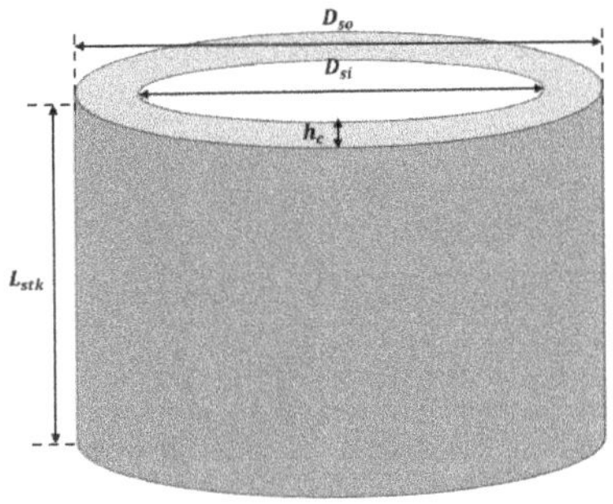

Figure 2. Simplified stator design.

The vibration displacement A_m caused by the mth order vibration mode in the circumferential direction is expressed by the circumferential mode of the radial force with $r = m$, as shown in Equation (5):

$$A_m = \frac{F_m / M_c}{\sqrt{(\omega_m^2 - \omega_r^2)^2 + 4\zeta_m^2 \omega_r^2 \omega_m^2}} \quad (5)$$

where M_c is the mass of the stator core [kg], ω_r is the angular frequency of the radial force, ω_m, ζ_m are the natural frequency in mode m and the damping ratio. The magnitude of the force experienced by the stator is then:

$$F_m = \pi D_{si} L_{stk} P_{mr} \quad (6)$$

where D_{si} is the stator inner diameter, L_{stk} is the stack length, and P_{mr} is the magnitude of the radial force in mode m.

According to Equation (5), as the angular momentum ω_r of the radial force approaches the natural angular momentum ω_m of the mode m, the vibration becomes larger, and the vibration of the mode m is maximized when $\omega_r = \omega_m$. Reformulating Equation (5) with the magnification factor h_m, yields:

$$h_m = \frac{A_m}{F_m/(M_c\omega_m^2)} = \frac{1}{\sqrt{\left[1-(f_r/f_m)^2\right]^2 + [2\zeta_m(f_r/f_m)]^2}} \tag{7}$$

where f_r is the vibration frequency of the rth order vibration mode and f_m is the natural frequency of the mode m. As $\Delta f = |f_r - f_m|$ and the damping ratio ζ_m decrease, the magnification factor h_m increases. Especially when $f_r = f_m$, the magnitude of the vibration depends on the mechanical damping of the device structure.

However, the mechanical damping ratio is often difficult to predict with a formula, so it is usually analyzed empirically through experiments. The empirical damping ratio is a practical approach to approximate complex mechanical damping that is difficult to describe theoretically. In small- and medium-sized electrical devices, it is used to express damping ratio as a function of frequency or other design variables based on experimental data to predict the dynamic behavior of the device and perform design and optimization. The empirical damping ratio for small- and medium-sized electrical devices is as follows [26]:

$$\zeta_m = \frac{1}{2\pi}\left(2.76 \times 10^{-5} f_m + 0.062\right) \tag{8}$$

From Equation (8), it can be concluded that as the natural frequency f_m increases, the damping ratio ζ_m also increases. Equation (5) is the vibration displacement due to one mode, and to obtain the total vibration displacement, the vibration displacement due to all modes must be calculated. Assuming that the natural frequency of each mode is distributed in such a way that they do not affect other modes, the vibration displacement caused by each mode can be calculated by substituting the natural frequency of each mode into Equation (5). Reformulating Equation (5) with the previously obtained expressions for the magnitude of the radial force P_{mr} and the magnification factor h_m, yields:

$$A_m = \frac{F_m}{M_c w_m^2} h_m = \frac{\pi D_{si} L_{stk}}{M_c \omega_m^2} P_{mr} h_m \tag{9}$$

Since radial force is a function of time and space, the velocity of vibration V_m in mode m is the derivative of A_m with respect to time t. Therefore:

$$V_m = \omega_r A_m = 2\pi f_r \frac{\pi D_{si} L_{stk}}{M_c \omega_m^2} P_{mr} h_m \tag{10}$$

As shown in the above equation, the vibration displacement and vibration velocity are determined by the natural frequency ω_m and the magnification factor h_m. In addition, the natural frequency increases as the mode increases, so even if the magnitude of the radial force is the same, the effect of vibration is relatively small for higher-order modes, and conversely, the effect of vibration is large for lower-order modes.

2.2.3. Calculate the Natural Frequency of the Stator

It is very important to calculate the natural frequency of a stator in order to predict the vibration generated by the stator. In general, the natural frequency of the stator is calculated by considering the stator configuration, considering the core, windings, housing, etc. In this section, the natural frequency of the stator is calculated by assuming the mechanical geometry of the stator is a ring-shaped cylinder, as shown in the previous assumption. In general, the mode-specific natural frequency of the stator is as follows:

$$f_m = \frac{1}{2\pi}\sqrt{\frac{K_m}{M_m}} \tag{11}$$

where K_m is the stator stiffness in mode m and M_m is the stator mode mass. Therefore, the natural frequency f_m in mode m can be calculated by finding K_m and M_m. In Equation

(10), r = 0, 1, 2, 3... represents the order of the circumferential mode and corresponds to the spatial harmonic component of the radial force, which has different effects on the vibration depending on the order of the circumferential mode. Figure 3 shows the results of 3D FEA modal analysis of the stator geometry, without the windings of the conventional model, conducted in this paper with Ansys Workbench, and the results for the modal geometries (m = 2, m = 3, m = 4) were obtained. The material of the stator is Steel_1010, Young's modulus is 205 [GPa], Poisson's ratio is 0.28, and mass density is 7872 [kg/m^3]. The frequency of each circumferential mode obtained by modal analysis is relatively far away from the electrical frequencies of 2 fe (472 [Hz]), 4 fe (944 [Hz]), 6 fe (1416 [Hz]), and 8 fe (1888 [Hz]) of the conventional model, so there is less concern about resonance.

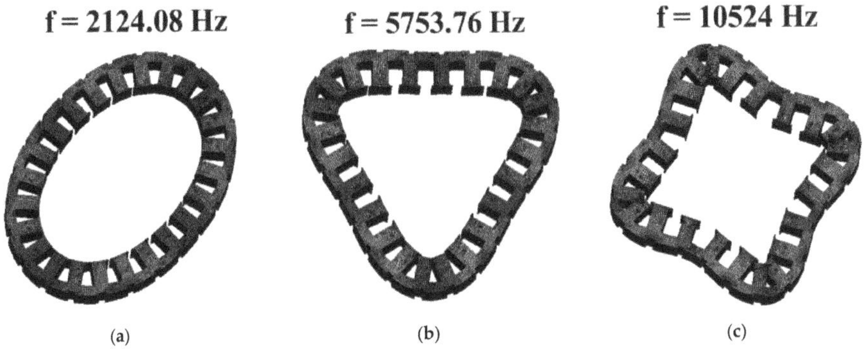

Figure 3. 3D FEA analysis results of natural frequency and corresponding modal shapes (**a**) m = 2, (**b**) m = 3, (**c**) m = 4.

2.2.4. Extraction Main Vibration Factor Harmonics Order

The radial force and vibration velocity have different time/space harmonic components depending on the position as the rotor rotates. For the conventional model in this paper, there are 16 poles and 8 pole pairs. Therefore, the radial force and vibration velocity are dominated by the temporal harmonic order components of the second order and the spatial harmonic order components of the eighth order.

In the case of temporal harmonic order, radial force and vibration velocity are considered up to the 10th order of the 2nd multiple because of their significant influence. For the spatial harmonic order, the comparison was conducted in eight orders of magnitude from 0 to 56. Figure 4 shows a comparison of the magnitude of radial force and vibration velocity at the rated speed of the conventional model (1770 rpm) with each temporal and spatial harmonic order component. In Figure 4a, the x-axis shows the spatial harmonic order from the 0th to the 56th in multiples of 8, the corresponding temporal harmonic order from the 2nd to the 10th in multiples of 8, and the y-axis shows the corresponding radial force density. The radial force density values show the greatest influence at the 16th spatial harmonic order because the motor is 16 poles, and the influence decreases as the order increases.

In this paper, we needed to extract the order with time/space influence, so we analyzed the influence of vibration velocity based on the same time/space harmonics, as shown in Figure 4b. In Figure 4b, we can see that the zero order spatial harmonic component of the vibration velocity is 0 [Hz], so it does not vibrate in the presence of radial force, so it is excluded from the harmonic order to be extracted. Figure 4a,b shows the extraction of the influential temporal and spatial harmonic orders. Table 2 analyzes the radial force and vibration velocity in time/space to extract the influential harmonic components. In this study, the radial force was analyzed through these extracted dominant harmonics [19].

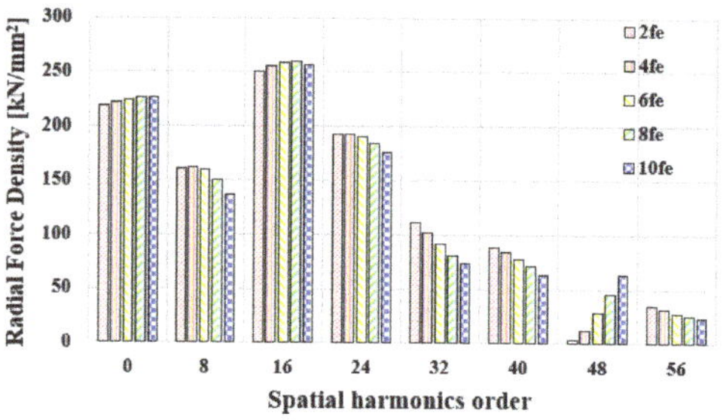

(a)

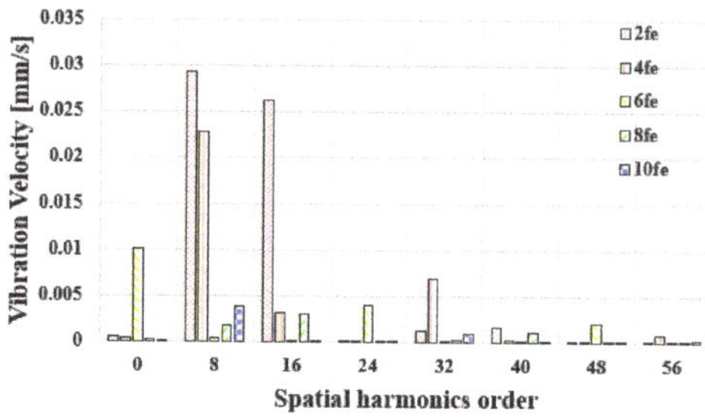

(b)

Figure 4. Time/spatial harmonic order separation of (**a**) the radial force density (**b**) the vibration velocity (@1770 rpm).

Table 2. Selected time/spatial harmonics order.

	Time Harmonics Order				
	2 fe	**4 fe**	**6 fe**	**8 fe**	**10 fe**
Spatial harmonics order	8, 16, 32, 40	8, 16, 32, 56	0, 24, 48	8, 16	8, 32

3. Conventional Model Specifications

3.1. Analyze Conventional Model Electromagnetic Performance

The conventional model selected in this paper is a 24[kW] traction motor, which is an IPMSM with 16 poles and 24 slots that uses a V-shaped rotor structure. The geometry of this motor was realized through Ansys MAXWELL 2024 R2 2D software, and the results are shown in Figure 5a. Since the model was too complex and difficult to analyze to include mechanical factors, the results were based on electromagnetic FEA. The meshing in the FEA is important to check the cogging torque and air gap flux density, so a more refined mesh was applied to the air gap between the stator and rotor. This resulted in a total mesh count of 17,320 elements in 2D. For the motor analysis, the nonlinearity problem, which takes into account the saturation component of the core and the demagnetization of the permanent magnet, was solved using the Newton-Raphson method with an energy convergence criterion to improve accuracy. Ansys MAXWELL 2024 R2 program was used as the analysis tool. To ensure the reproducibility and reliability of this FEA, the conventional model was referenced to the actual fabricated model [1], and the simulation results were verified by comparing them with the existing literature. This ensured the accuracy and reliability of the results.

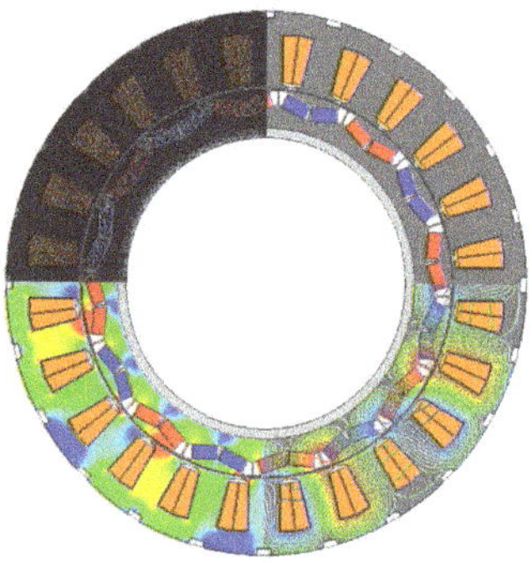

(a)

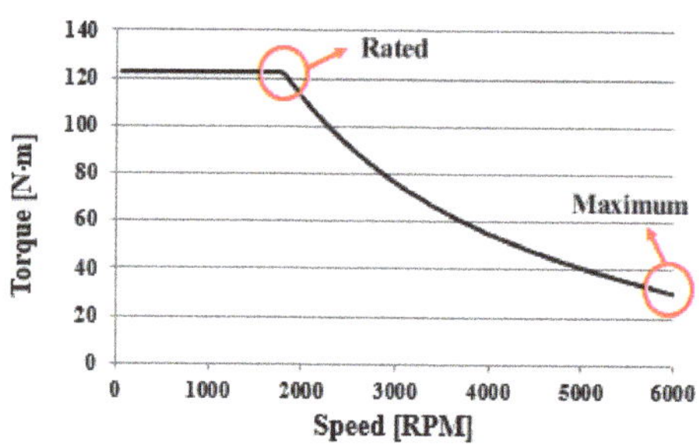

(b)

Figure 5. Conventional model (**a**) geometry (**b**) TN curve.

In addition, the torque-speed (TN) curve of this model is shown in Figure 5b. In this curve, the red circle on the left represents the rated operating condition at 1770 rpm and the red circle on the right represents the maximum operating condition at 6000 rpm. The major specifications for these two operating conditions are summarized in Table 3. For the rated and maximum operating conditions of the conventional model, referring to Equation (1), it can be seen that the amount of flux generated by the permanent magnets is constant, so the change in flux and reluctance is almost the same. This also maintains the cogging torque constant at 10.4 [N·m], which means that since there is no change in the rotor geometry, the reluctance remains constant and there is no significant difference in the cogging torque.

Table 3. Conventional model basic specifications.

Parameter	1770 rpm	6000 rpm	Unit
Stator & Rotor	Steel_10101		-
Coil	Copper		-
Magnet	N42UH		-
Pole/Slot	16/24		-
Stator outer diameter	280		mm
Rotor outer diameter	200		mm
No-load phase back EMF	76.8	260.4	V_{rms}
Cogging torque	10.4	10.4	N·m
Current	102.5		A_{rms}
Torque	128	26.3	N·m
Line Back EMF	202.9	205.5	V
Output	23.6	16	kW
Efficiency	96.8	85.9	%

3.2. Analyze Conventional Model Vibration

The vibration noise characteristics were analyzed under the rated (1770 rpm) and maximum (6000 rpm) operating conditions of the conventional model. Figure 6 shows the electromagnetic characteristics at the initial position ($\omega t = 0$) extracted by 2D FEA for vibration noise prediction. Figure 6a shows the flux density as a function of mechanical position at rated and maximum operation from the conventional model. Figure 6b is a radial graph of the radial force [kN/mm^2] at a specific position ($\omega t = 0$). The spatial harmonics are difficult to match perfectly because the rotational speed contains multiple temporal harmonics.

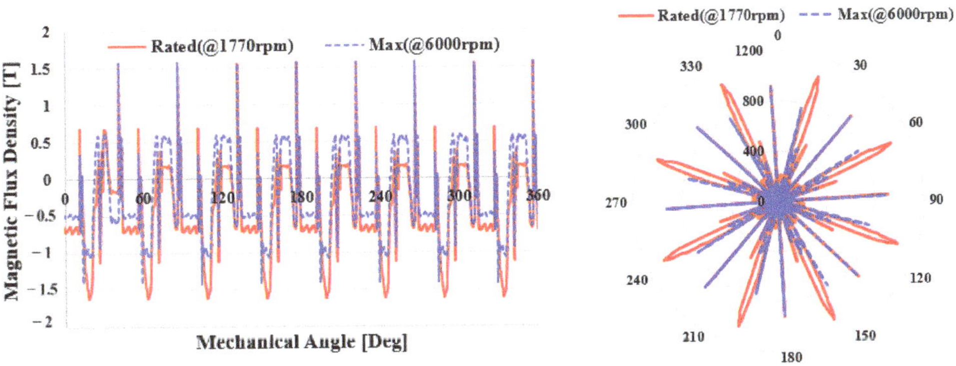

(a) (b)

Figure 6. Electromagnetic characterization of conventional models with 2D FEA ($\omega t = 0$) (**a**) magnetic flux density (**b**) radial force density [kN/mm^2].

Figure 7 shows the results of deriving the vibration noise characteristics from the temporal and spatial harmonic components extracted through 2D FEA. In Figure 7a, the

x-axis lists the spatial harmonic order extracted from each temporal harmonic component, which is used to compare the radial force density. Since the model has 16 poles, we can see that the largest values are found mainly at the 8th and 16th orders. The radial force density at the rated condition (1770 rpm) is larger than the maximum condition (6000 rpm), which is due to the fact that the beta angle control at maximum operation tends to generate a reverse magnetic field, resulting in a lower radial force density. Figure 7b shows a comparison of the vibration velocity for each time harmonic order. Since the zero-order time harmonic component is 0 [Hz], it can be seen that the presence of radial forces does not cause vibration. The vibration velocity RMS value at rated is 0.52 [mm/s], and at maximum condition, it is 2.94 [mm/s], which is higher than the rated value. This was analyzed as a result of the increase in vibration velocity due to the increase in speed and frequency during maximum operation.

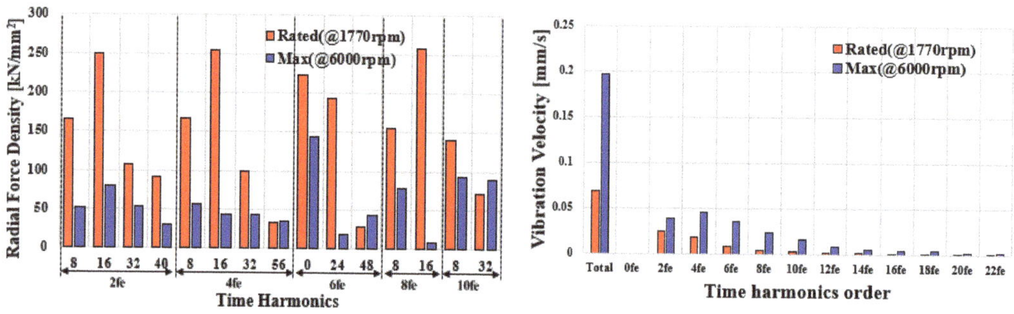

(a) (b)

Figure 7. Comparison of conventional model electromagnetic characterization at rated speed (1770 rpm) and maximum speed (6000 rpm) (**a**) and the radial force density distribution over the selected time/space harmonic order. Overall, higher radial force at rated operation than at maximum operation, (**b**) comparison of vibration rates for different harmonic orders. For the overall harmonic order, the vibration velocity is lower at rated operation than at maximum operation, with a significant increase in total vibration velocity at maximum operation.

4. Proposed Model

4.1. Main Variable Parameters

Figure 8 shows the rotor tapering parameters tapering thickness (T_t) and tapering angle (T_{deg}). T_t suggests tapering at the rib so as not to obstruct the flow of magnetic flux from the permanent magnet.

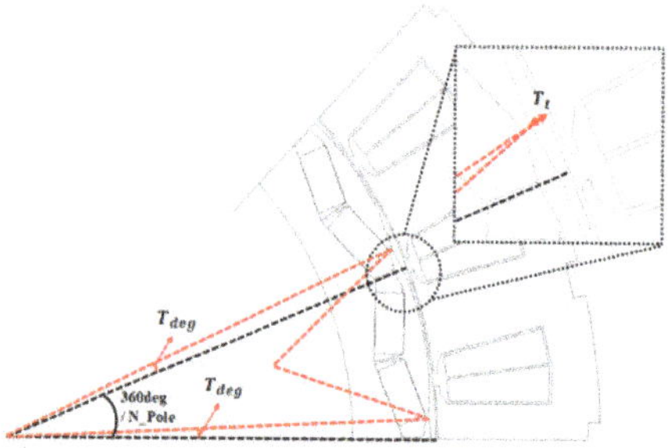

Figure 8. Definition of T_t, T_{deg}.

For manufacturability considerations, the length of T_t is formed such that the rib thickness is at least 0.6 [mm]. To reduce the effect on the back EMF, T_{deg} is given to the rib area, except the pole piece area where the main magnetic flux flows, to provide an ideal cogging torque reduction model. When applying the variable of T_{deg}, a two-step skew is applied by dividing the stack length in half and angling it in opposite directions while keeping T_t applied. This has the effect of canceling out cogging torque in any localized area subjected to a difference in reluctance.

4.2. Deriving the Optimal Model

In this paper, to perform optimal design using the variables in Figure 8, we used the response surface method (RSM), which can statistically analyze the influence of a response variable on multiple variables to derive an optimal model. In this paper, we considered the optimal design of the objective function using two variables. Based on a comprehensive consideration of the suitability, complexity, and compatibility with the experimental design of optimal design methods, we conclude that RSM provides an interpretable and simplified optimization approach compared to other methods. In particular, the second-order polynomial regression model is used as a mathematical tool to clearly represent the interaction between variables and effectively solve the optimization problem [27,28]. RSM is a method that statistically analyzes the change in response when multiple variables X act in combination to affect a specific response variable Y [19,20]. The experimental range of the proposed variables for the conventional model at rated operation (1770 rpm) is shown in Table 4. Equation (12) shows the fundamental equation of the second-order polynomial regression function, and Equations (13)–(15) show the regression functions for cogging torque, no-load back EMF, and vibration velocity, respectively.

$$y_i = \beta_0 + \sum_{i=1}^{p} \beta_i x_i + \sum_{i \le j}^{p} \beta_{ij} x_i x_j + \varepsilon_i \tag{12}$$

$$y_{Cogging\ torque} = 4.49 - 4.12(X_1) - 0.65(X_2) \cdots - 0.73(X_1)(X_2) \tag{13}$$

$$y_{Noload\ back\ EMF} = 73.9 - 2.16(X_1) - 2.24(X_2) \cdots - 1.96(X_1)(X_2) \tag{14}$$

$$y_{Vibration\ velocity} = 0.39 - 0.11(X_1) - 0.02(X_2) \cdots - 0.02(X_1)(X_2) \tag{15}$$

where $\beta_1, \beta_2, \cdots, \beta_k$ are the regression coefficients of each variable, $x_1, x_2, \cdots, x_k$ are the variables in the code, ε is the error term in the response, and $X_1 - X_2$ are the variables T_t, T_{deg}. Table 4 shows the ranges of design variables set to use RSM. Variable T_t was set to 0–1.8 [mm] to account for a minimum fabrication thickness of 0.6 [mm] and a range of no more than 1.8 [mm] without physically destroying the rotor geometry. Variable T_{deg} was set to 0–3.0[°], considering the range from the pole piece to the rib piece, excluding the area that does not affect the pole piece, which is the passage through which the main magnetic flux flows, while maintaining the same performance to the extent possible. Subsequently, the variables were selected based on the design of experiments (DoE) method, which systematically selects experimental variables and value ranges for optimal design. Based on the variables and ranges presented in Table 4, a design matrix including center points (0.9 [mm], 1.5[°]) and axis points (1.8 [mm], 3.0[°], etc.) was generated using a quadratic response surface method design and central composite design (CCD). The optimized increment size was then set, and the optimization was performed by substituting the optimal points derived from the RSM into the regression function, and the optimal points were derived as T_t = 1.8 [mm] and T_{deg} = 3.0[°]. Comparing RSM and FEA, we found that the error was less than 1[%]. The error rate was similar to the FEA results, which is enough to prove its effectiveness and reliability. The reason for the slight error is that the CCD was used to derive the optimal model based on the design matrix including the center and axis points. If the sampling points were set more precisely, the error rate in certain areas would be significantly reduced.

Table 4. Design variables and experiment range for RSM.

Contents	T_t [mm]	T_{deg} [°]
Range	0–1.8	0–3.0
Optimal point	1.8	3.0

In conclusion, we can conclude that RSM is valid. The results of the RSM results are shown in Figures 9 and 10, where the estimated values are plotted against the variables T_t and T_{deg}. Figure 9a shows the cogging torque at the rated speed of 1770 rpm, and it can be seen that the larger the values of T_t and T_{deg}, the smaller the reluctance change, which reduces the cogging torque. Figure 9b shows the no-load back EMF, and it can be seen that as T_t and T_{deg} become larger, the polypiece portion, which is the magnetic flux path, shrinks, resulting in a proportional decrease in the back EMF. Figure 9c shows the RMS vibration velocity, and, similar to the cogging torque, it can be seen that the vibration velocity decreases as T_t and T_{deg} increase. In this paper, it was determined that the conventional model can maintain the same performance even if the no-load back EMF is slightly reduced because the current density has a margin of about 10%. Therefore, the final model is proposed by selecting the optimum point of T_t = 1.8 [mm], T_{deg} = 3.0[°]. Figure 10 shows the results of each variable at the maximum operating speed of 6000 rpm, and the optimum model is derived in a similar way to Figure 9.

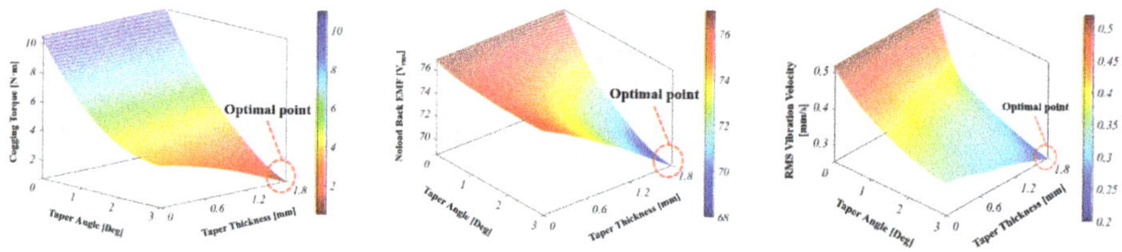

(a) (b) (c)

Figure 9. Results of deriving response variables based on variables a and b with RSM (rated @1770 rpm) (**a**) cogging torque (**b**) no-load back EMF (**c**) RMS vibration velocity.

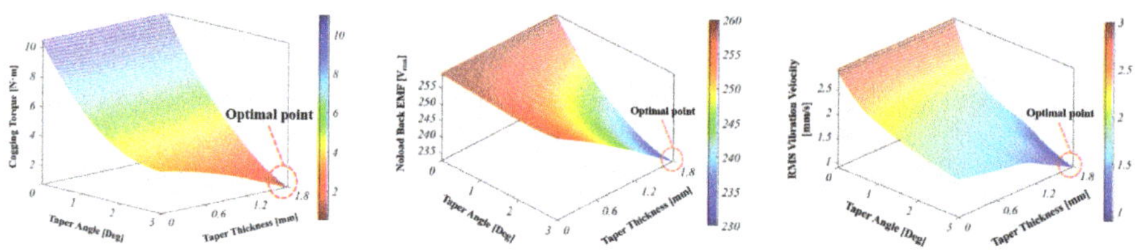

(a) (b) (c)

Figure 10. Results of deriving response variables based on variables a and b with RSM (max @6000 rpm) (**a**) cogging torque (**b**) no-load back EMF (**c**) RMS vibration velocity.

4.3. Comparative Analysis of Proposed Model Performance Characteristics

4.3.1. Electromagnetic Performance of Proposed Model

As explained earlier, the optimal proposed model with T_t = 1.8 [mm], T_{deg} = 3.0[°] was derived through RSM. As a result, the proposed rotor two-step skew geometry reduces the cogging torque by about 91.6[%] at nominal and about 91.8[%] at maximum compared to the conventional model while maintaining the same performance even though the no-load back EMF is slightly reduced. Figure 11a shows the 3D realization of the 1/8th cycle model of the conventional model rotor. Applying an asymmetric two-step skew to the rotor in the conventional model results in the geometry shown in Figure 11b. The waveforms of cogging torque and no-load back EMF are shown in Figure 12. Figure 12a shows that the total cogging torque can be canceled out in both sections through the asymmetric angle at the top and bottom, which reduces the cogging torque. Figure 12b shows that the asymmetric structure of the rotor at the top and bottom prevents the asymmetric structure of the back EMF, which enables the maximum back EMF and maximum torque.

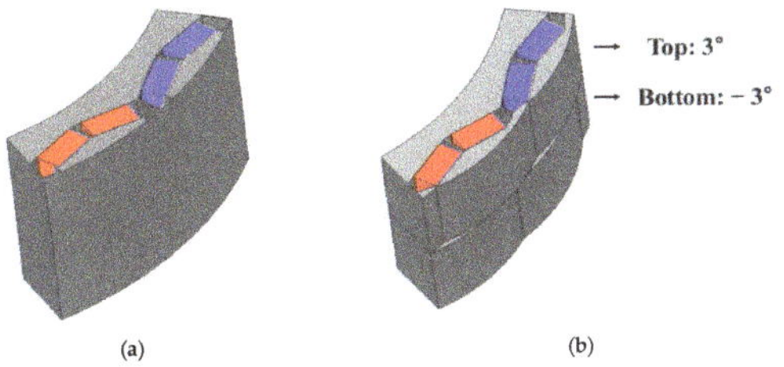

Figure 11. Model topology (**a**) conventional model (**b**) proposed model.

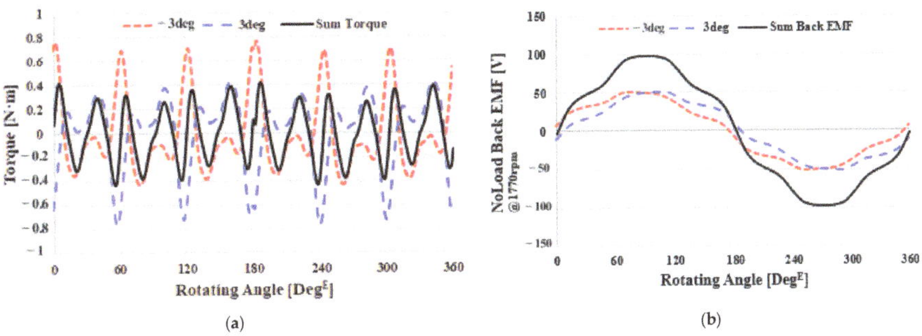

Figure 12. Sum waveform according to $T_{deg} = 3.0[°], -3.0[°]$ (**a**) cogging torque (**b**) no-load back EMF(@1770 rpm).

To determine the mechanical safety of the proposed model, Workbench 3D FEA was performed at maximum operation (6000 rpm) and the results are shown in Figure 13. Figure 13a shows the equivalent stress of the proposed model. Here, the maximum stress is generated locally between the permanent magnet and the rib due to centrifugal force as the rotor rotates. However, the analysis shows that the yield tensile strength of Steel_1010 is 78.45 [MPa], compared to 18.05 [MPa], which is low enough to be mechanically robust. Figure 13b also shows that the minimum safety factor is about 4.4. As an empirical value, a safety factor of 2 or more is considered safe. Therefore, the proposed model can be said to be mechanically robust at rated and maximum operation.

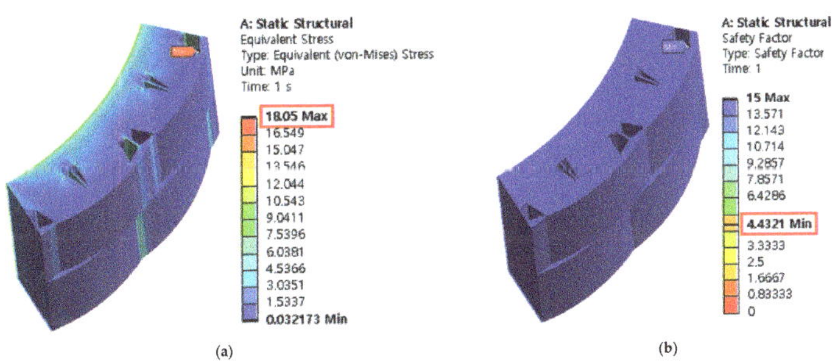

Figure 13. Result of mechanical stiffness analysis of the proposed model (**a**) equivalent stress, (**b**) minimum safety factor.

4.3.2. Electromagnetic Vibration Analysis

To validate whether the radial force and vibration noise are substantially reduced by reducing the cogging torque, one of the electromagnetic vibration noise sources, a comparative analysis of the extracted main temporal and spatial harmonic order components was performed. Figure 14 shows the analysis of radial force and vibration velocity at a rated speed of 1770 rpm. Figure 14a shows the radial force values from the extracted time/space harmonic components. Compared to the conventional model, it was observed that the radial force is reduced overall in each harmonic component by the proposed model. In Figure 14b, the vibration velocity tends to decrease overall from the conventional model to the proposed model, which verifies that the proposed model reduces the vibration noise by about 51.9[%] compared to the conventional model at the rated level. The analysis of radial force and vibration velocity at the maximum operating speed of 6000 rpm using the same method is shown in Figure 15. In Figure 15a, the radial force tends to decrease overall when the proposed model is applied from the conventional model to the proposed model, but there are some sections where the value increases, which is partly due to the phenomenon that the vacancy flux density is temporarily skewed due to the saturation of both ribs, resulting in a partial increase at high spatial harmonic orders. However, in Figure 15b it can be seen that the vibration velocity is significantly reduced from the conventional model to the proposed model, and even if the radial force is increased at some spatial harmonic orders, it is judged that the vibration is also reduced because it is reduced overall. In other words, at the maximum, the vibration of the proposed model is reduced by about 68.7[%] compared to the conventional model. By reducing the cogging torque, it was possible to reduce the radial force, which ultimately reduced the vibration noise. As a result, Table 5 shows that applying the proposed model to the conventional model reduces the cogging torque and vibration velocity at rated and maximum operation.

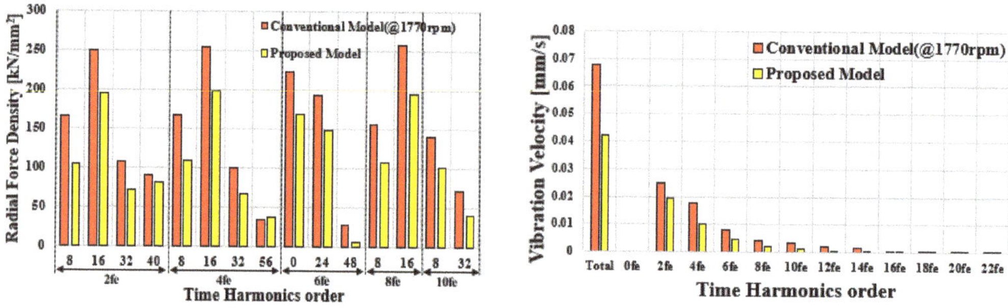

(a) (b)

Figure 14. Comparison analysis between the conventional and proposed models at the rated speed of 1770 rpm (**a**) the radial force density distribution over the selected time/space harmonic order. Overall lower radial force in the proposed model compared to the conventional model, which contributes to reduced vibration, (**b**) comparison of vibration rates for different harmonic orders. The proposed model reduces the total vibration velocity by reducing the overall harmonic order compared to the conventional model.

Table 5. Analyzed major results of the proposed model compared to the conventional model.

Contents	Conventional Model		Proposed Model		Unit
	Rated 1770 rpm	Max 6000 rpm	Rated 1770 rpm	Max 6000 rpm	
Cogging torque	10.4		0.7		N·m
No-load back EMF	76.8	260.4	68.7	232.8	V_{rms}
RMS vibration velocity	0.52	2.94	0.25	0.92	mm/s

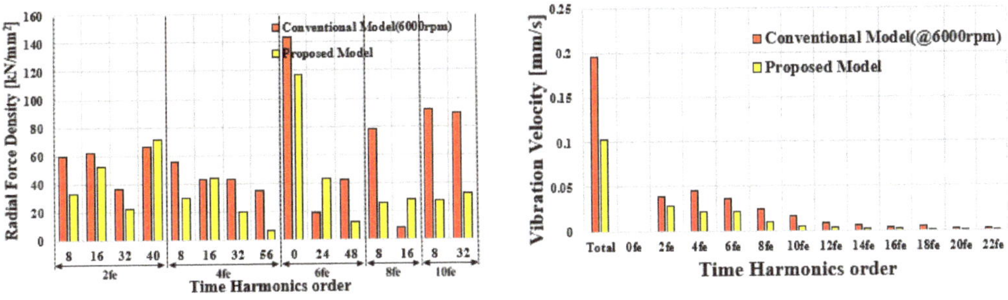

(a) (b)

Figure 15. Comparison analysis between the conventional and proposed models at the maxed speed of 6000 rpm (**a**) the radial force density distribution over the selected time/space harmonic order. Overall lower radial force in the proposed model compared to the conventional model, which contributes to reduced vibration, (**b**) comparison of vibration rates for different harmonic orders. The proposed model reduces the total vibration velocity by reducing the overall harmonic order compared to the conventional model.

5. Conclusions

The vibration and noise reduction proposal for traction motors proposed in this paper applies an asymmetric two-step skew model to the rotor to reduce the variation of magnetic resistance, resulting in a significant reduction of cogging torque. This approach significantly reduces cogging torque as well as vibration noise, contributing to improved operator comfort and minimizing negative impacts on the performance and lifetime of the motor. The reduction in cogging torque was significantly reduced to approximately 91.6[%] and 91.8[%] at rated and full operation, respectively. In addition, the vibration velocity was also reduced by approximately 51.9[%] at rated operation and 68.7[%] at full operation due to the overall reduction in radial force, as determined by extracting the dominant time/space harmonic components and identifying the corresponding radial force. The validity of the proposed model is demonstrated through formulations and FEA based entirely on numerical analysis and results.

Author Contributions: Conceptualization, S.-H.L. and S.-W.S.; Methodology, S.-H.L., S.-W.S. and W.-H.K.; Validation, S.-H.L., J.L. and W.-H.K.; Formal analysis, S.-H.L. and S.-W.S.; Investigation, S.-H.L., S.-W.S. and I.-J.Y.; Resources, I.-J.Y.; Data curation, I.-J.Y. and J.L.; Writing—original draft, S.-H.L. and I.-J.Y.; Writing—review & editing, S.-H.L.; Supervision, S.-W.S., J.L. and W.-H.K.; Project administration, J.L. and W.-H.K. All authors have read and agreed to the published version of the manuscript.

Funding: This work was partly supported by the Korea Institute of Energy Technology Evaluation and Planning (KETEP) grant funded by the Korean government (MOTIE) (20214000000060, Department of Next Generation Energy System Convergence Based on Techno-Economics—STEP). In part, this research was funded by the Energy Demand Management Core Technology Development Project (Development and Demonstration of Medium-sized Constant Speed High-Efficiency Electric Motors, No. RS-2023-00232767), funded by the Ministry of Trade, Industry, and Energy, and supported by the Korea Energy Technology Evaluation Institute (KETEP).

Data Availability Statement: The original contributions presented in the study are included in the article, further inquiries can be directed to the corresponding author.

Conflicts of Interest: The authors declare no conflicts of interest.

References

1. Song, S.-W.; Kim, W.-H.; Lee, J.; Jung, D.-H. Design to Reduce Cogging Torque and Irreversible Demagnetization in Traction Hybrid Motor Using Dy-free Magnet. *Machines* **2023**, *11*, 345. [CrossRef]

2. Song, P.; Li, W.; Li, Z.; Toulabi, M.S.; Kar, N.C. Noise and Vibration Prediction of a Six-Phase IPMSM in a Single Open-Phase Failure Under a Negative Sequence Current Compensated Fault Tolerant Control Mode. *IEEE Trans. Magn.* **2022**, *58*, 1–6. [CrossRef]
3. Lin, F.; Zuo, S.G.; Deng, W.Z.; Wu, S.L. Reduction of vibration and acoustic noise in permanent magnet synchronous motor by optimizing magnetic forces. *J. Sound Vibr.* **2018**, *429*, 193–205. [CrossRef]
4. Gieras, A.; Jacek, F. *Noise of Polyphase Electric Motors*; Taylor & Francis CRC Press: Boca Raton, FL, USA, 2006.
5. Hartani, K.; Miloud, Y.; Miloudi, A. Improved direct torque control of permanent magnet synchronous electrical vehicle motor with proportional-integral resistance estimator. *J. Electr. Eng. Technol.* **2010**, *5*, 451–461. [CrossRef]
6. Wang, S.; Li, H. Reduction of Electromagnetic Vibration and Noise in Permanent Magnet Motor for EVs by Optimizing Design of Rotor Based on GPR-PSO Model. *J. Electr. Eng. Technol.* **2020**, *15*, 1231–1243. [CrossRef]
7. Park, Y.-W.; Ko, J.-S.; Kim, D.-K. Optimal Design of Step-Sloping Notches for Cogging Torque Minimization of Single-Phase BLDC Motors. *Energies* **2021**, *14*, 7104. [CrossRef]
8. Jo, I.-H.; Lee, H.-W.; Jeong, G.; Ji, W.-Y.; Park, C.-B. A Study on the Reduction of Cogging Torque for the Skew of a Magnetic Geared Synchronous Motor. *IEEE Trans. Magn.* **2019**, *55*, 1–5. [CrossRef]
9. Du, Z.S.; Lipo, T.A. High Torque Density and Low Torque Ripple Shaped-Magnet Machines Using Sinusoidal Plus Third Harmonic Shaped Magnets. *IEEE Trans. Ind. Appl.* **2019**, *55*, 2601–2610. [CrossRef]
10. Adase, L.A.; Alsofyani, I.M.; Lee, K.-B. Predictive Torque Control with Simple Duty-Ratio Regulator of PMSM for Minimizing Torque and Flux Ripples. *IEEE Access* **2020**, *8*, 2373–2381. [CrossRef]
11. Zhao, J.; Quan, X.; Tong, X.; Lin, M. Cogging Torque Reduction in Double-Rotor Hybrid Excited Axial Switched-Flux Permanent Magnet Machine. *IEEE Trans. Appl. Supercond.* **2020**, *30*, 1–5. [CrossRef]
12. Liu, N.-W.; Hung, K.-Y.; Yang, S.-C.; Lee, F.-C.; Liu, C.-J. Design of High-Speed Permanent Magnet Motor Considering Rotor Radial Force and Motor Losses. *Energies* **2020**, *13*, 5872. [CrossRef]
13. Jang, D.; Chang, J. Effects of Flux Modulation Poles on the Radial Magnetic Forces in Surface-Mounted Permanent-Magnet Vernier Machines. *IEEE Trans. Magn.* **2017**, *53*, 1–4. [CrossRef]
14. Hong, J.; Zhao, K.; Gui, L.; Cao, J. A Comprehensive Investigation of Tangential Force and Radial Force in PM Motor. *IEEE Trans. Ind. Electron.* **2024**, *71*, 16286–16296. [CrossRef]
15. *ISO 2372*; Mechanical Vibration of Machines with Operating Speeds from 10 to 200 rev/s—Basis for Specifying Evaluation Standards. ISO: Geneva, Switzerland, 1974.
16. *ISO 10816-1*; Mechanical Vibration—Evaluation of Machine Vibration by Measurements on Non-Rotating Parts—Part 1: General Guidelines. ISO: Geneva, Switzerland, 1995.
17. Zhang, G.; Hou, P. Optimization Design of Cogging Torque for Electric Power Steering Motors. *Machines* **2024**, *12*, 517. [CrossRef]
18. Chen, W.; Ma, J.; Wu, G.; Fang, Y. Torque Ripple Reduction of a Salient-Pole Permanent Magnet Synchronous Machine with an Advanced Step-Skewed Rotor Design. *IEEE Access* **2020**, *8*, 118989–118999. [CrossRef]
19. Yao, L.; Shen, J.-X.; Wang, Y.; Shi, D. Fast Prediction and Staged Optimization of Electromagnetic Vibration in Permanent Magnet Synchronous Motor. *IEEE Trans. Ind. Electron.* **2024**, *71*, 115–125. [CrossRef]
20. Siddique, F.; Singh, B.; Shastri, S. RSM Assisted MOGA for SRM EV Drive Control Factors Optimization. *IEEE Trans. Ind. Appl.* **2024**, *60*, 3200–3209. [CrossRef]
21. Calderón-Ramírez, M.; Gómez-Náfate, J.A.; Rico-Martínez, R.; Rodríguez-Castro, R.; Bravo-Sánchez, M.G.; Alcaraz-Caracheo, L.A. Improvement of Ohmic Pasteurization in Mango Pulp Through CFD and RSM. *IEEE Access* **2022**, *10*, 81380–81389. [CrossRef]
22. Lu, Y.; Li, J.; Yang, K. A Hybrid Calculation Method of Electromagnetic Vibration for Electrical Machines Considering High-Frequency Current Harmonics. *IEEE Trans. Ind. Electron.* **2022**, *69*, 10385–10395. [CrossRef]
23. Liang, J.; Jiang, J.; Callegaro, A.; Bilgin, B.; Dong, J.; Reeves, D.; Emadi, A. Prediction of Acoustic Noise and Vibration of a 24/16 Traction Switched Reluctance Machine. IET Electr. *Syst. Transp.* **2020**, *10*, 35–43.
24. Dong, Q.; Liu, X.; Qi, H.; Zhou, Y. Vibro-acoustic prediction and evaluation of permanent magnet synchronous motors. *Proc. Inst. Mech. Eng. Part D J. Automob. Eng.* **2020**, *234*, 2783–2793. [CrossRef]
25. Ibrahim, I.; Silva, R.; Mohammadi, M.H.; Ghorbanian, V.; Lowther, D.A. Surrogate-Based Acoustic Noise Prediction of Electric Motors. *IEEE Trans. Magn.* **2020**, *56*, 1–4. [CrossRef]
26. Yang, S.J. *Low-Noise Electrical Motors*; Clarendon Press: Oxford, UK, 1981.
27. You, Y.-M. Multi-Objective Optimal Design of Permanent Magnet Synchronous Motor for Electric Vehicle Based on Deep Learning. *Appl. Sci.* **2020**, *10*, 482. [CrossRef]
28. Sun, X.; Shi, Z.; Lei, G.; Guo, Y.; Zhu, J. Multi-Objective Design Optimization of an IPMSM Based on Multilevel Strategy. *IEEE Trans. Ind. Electron.* **2021**, *68*, 139–148. [CrossRef]

energies

Review

Analysis and Compensation of Current Measurement Errors in Machine Drive Systems—A Review

Pingyue Song, Tao Wang *, Lijian Wu *, Hao Li, Xiang Meng and Cheng Li

College of Electrical Engineering, Zhejiang University, Hangzhou 310012, China; py_song@zju.edu.cn (P.S.); hale_lee@zju.edu.cn (H.L.); mengx@zju.edu.cn (X.M.); 22310107@zju.edu.cn (C.L.)
* Correspondence: wangtaoee@zju.edu.cn (T.W.); ljw@zju.edu.cn (L.W.)

Abstract: The rapid development of machine drive systems is closely tied to advancements in signal sampling technology. Accurate measurement results, especially current measurement results, are crucial for high-performance machine drive systems. However, current measurement errors (CME) caused by circuit parameter inconsistencies, aging, and temperature variations can significantly affect the control performance of drive systems, thus necessitating compensation. To better understand the mechanisms of CME in drive systems, this paper reviews existing research on CME analysis and compensation techniques. Based on the source of CME, this paper classifies CME into high-frequency and low-frequency components and discusses their causes and compensation methods, respectively. Additionally, the influence of different types of CME on other control strategies, such as sensorless and model predictive control, is discussed. The characteristics of CME compensation strategies for special current sensor installation positions are also explored. Finally, the paper summarizes the state of the art in CME analysis and compensation and discusses future trends in this field.

Keywords: current measurement error; machine drive system; current sensor; error compensation

1. Introduction

Due to the widespread development of various permanent magnet materials, permanent magnet synchronous machines (PMSM) have received increasing attention in research. Their significant advantages, such as simple structure, high power density, and high efficiency, have led to unprecedented applications in fields such as ship propulsion [1], electric vehicles [2], rail transportation [3], and wind turbines [4]. To meet the diverse demands of various applications, a wide range of PMSMs has been developed, including multi-phase [5], linear [6], and open-winding structures [7]. Meanwhile, diverse control strategies, such as sensorless control [8], model predictive control [9], and direct torque control [10], have been extensively developed.

Regardless of the control strategy used, precise feedback data sampling is essential. In a typical PMSM drive system, rotor position and phase current are the necessary feedback signals for achieving the desired control performance. The rotor position can be indirectly observed through phase current, enabling sensorless algorithms. Therefore, the precise measurement of phase currents is critical for high-performance motor control.

However, due to the existence of non-ideal factors in the current sampling paths, such as device aging, temperature drift, etc., current measurement error (CME) is inevitable [11]. CME refers to the deviation of the sampled current result from the actual values. The presence of CME affects the accuracy of current controllers, which in turn leads to various issues, such as electromagnetic torque pulsation, current ripple, and mechanical vibration. These undesired problems severely threaten the reliable and stable operation of PMSM. Hence, extensive studies have been conducted to clarify the causes, effects, and compensation methods of CME.

This paper reviews previous studies on CME. It should be noted that this paper only focuses on the influence and compensation of low- and high-frequency current harmonic disturbances in the entire current sampling path of the machine drive system. The analysis and improvement of current sensor performance are outside the scope of this paper. Different types of CME have varying effects in different topologies, and the compensation methods also differ. Finally, this article presents future development trends of the studies related to CME, including the influence of CME in different topologies, compensation methods for different types of CME, the interaction of CME with other control algorithms, and current sensor fault diagnosis methods.

The remaining parts of this article are arranged as shown in Figure 1: Section II analyzes the causes of CME and its impact on different drive systems. Section III classifies CME and reviews compensation methods for different types of CME. Section IV reviews the research conducted in the abovementioned areas and provides future trends related to CME. Finally, Section V concludes the article.

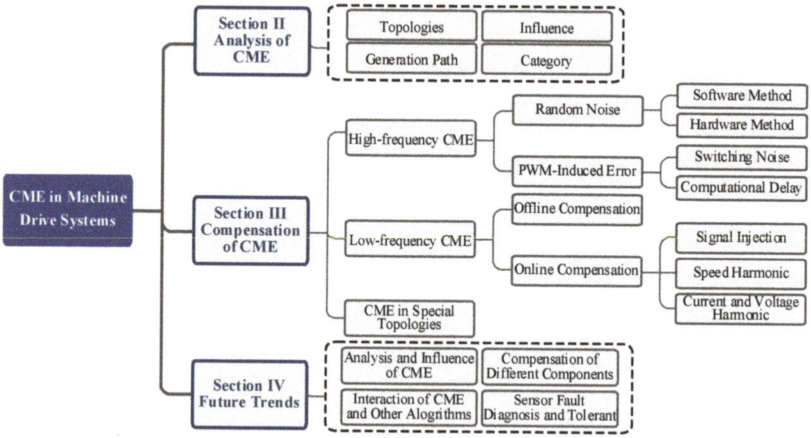

Figure 1. Arrangements for the subsequent sections of this paper.

2. Analysis of Current Measurement Error

Commonly utilized current sensors include sampling resistors [12], open-loop or closed-loop Hall sensors [13], fluxgate sensors [14], Rogowski coils [15], Giant Magneto Resistive (GMR) sensors [16], and current transformers [17]. Table 1 summarizes the performance characteristics of seven typical current sensors. These sensors exhibit notable variations in bandwidth, accuracy, size, cost, and power loss [17–19]. Table 1 provides different types of current sensors and their main parameters in [20–43]. When selecting current sensors, it is essential to consider the specific requirements of different applications, such as temperature tolerance, accuracy, bandwidth, and measurement range. For machine drive systems, shunt resistance (sampling resistance) and Hall sensors are among the most commonly used types.

Table 1. Performance comparison of typical current sensors.

Reference	Sensor Type	Application	PCB Integrated	Bandwidth	Tested Frequency	Accuracy	Range	Main Contribution
[20]	Shunt resistance	Electric vehicles	Y	3.7 MHz	162.8 Hz	0.12%	±1.6 A(rms)	Calibrate aging drift of resistance; low loss (43 mA/A);
[21]	Shunt resistance	Power module (GaN)	Y	150 MHz	Switching current	0.40%	<20 A	Low parasitic inductance (0.55 nH);
[22]	Shunt resistance	Power module (GaN)	Y	2.23 GHz	Switching current	0.90%	<80 A	High bandwidth; low parasitic inductance (0.12 nH);
[23]	Shunt resistance	In-wheel motor	Y	-	-	-	±100 A	Inverter leg current sensing;
[24]	MR + Rogowski coil	Power module	Y	30 MHz	Switching current	2%	±30 A	DC current measurable;
[25]	Rogowski coil	Power module	N	8 MHz	Switching current	0.50%	±60 A	Bandwidth improvement with segmented coils;
[26]	Rogowski coil	Resonant converter	Y	40–150 MHz	100 kHz	-	±27 A	PCB mounted structure;
[27]	Rogowski coil	Power module (GaN)	Y	30 MHz	Switching current	-	±20 A	PCB mounted structure;
[28]	Current transformer	Energy meter	Y	-	50 Hz	-	±10 A	Electromagnetic saturation compensation;
[29]	Current transformer	HVDC	N	-	50 Hz	-	±60 A	DC current measurement periodically with switch device;
[30]	Current transformer	STATCOM	N	-	50 Hz	-	±10 A	DC current suppression;
[31]	High-frequency current transformer	HVDC	N	-	900 kHz	-	±100 A	Low loss with core airgap profiles design (1.1–1.5 W);
[32]	High-frequency current transformer	PD detection	N	10 MHz	PD pulses	-	±75 A	Optimal air gap length calculation;
[33]	Current transformer	AC transmission lines	N	-	50 Hz	-	±2 A~1.25 kA	Wide range sensing;
[34]	Hall sensor	Electric vehicles	N	40 kHz	500 Hz–1.5 kHz	3.5% *	±40 A	Phase shift error compensation;
[35]	Hall sensor	-	Y	250 kHz	-	0.50%	±500 A	Mechanically flexible sensor array;
[36]	Fluxgate	Power converter	Y	-	20 kHz	0.03%	15 A	DC bias elimination;
[37]	Fluxgate + Rogowski coil	Power module	Y	50 MHz	Switching current	-	-	Hybrid current sensing; light-size;
[38]	Fluxgate	-	N	-	Switching current	-	±1.5 kA	Suppress magnetization distortion;
[39]	Fluxgate	Electric vehicles	N	-	-	0.50%	±500 A	Zero-point drift elimination;
[40]	Fluxgate	-	Y	1–512 Hz	-	-	-	Low-noise (SNR: 1.04.2 dB);
[41]	GMR	-	-	100 kHz	-	±0.7% **	±10 mA~20 A	Temperature drift suppression;
[42]	GMR	PD detection	N	1–10 MHz	PD pulses	-	±4.5 mA(rms)	Non-contacting PD detection;
[43]	GMR	Power module	Y	1 MHz	Switching current	1.25%	±80 A	Active temperature drift suppression circuit;

MR: Magnetoresistance; HVDC: high-voltage dc transmission; -: not mentioned; *: phase error at 1.5 kHz; PD: partial discharge; STATCOM: static synchronous compensator; **: linearity error; SNR: Signal-to-Noise Ratio.

Figure 2 shows a schematic diagram of the entire current sampling path, where the Hall sensor and sampling resistance are utilized for current sampling. The feedback signal from the sensor first passes through a low-pass filter circuit to eliminate high-frequency interference and noise. The filtered signal is then amplified by a differential operational amplifier circuit to convert the feedback signal range into the desired level range, for example, converting ±5 V to 0–3.3 V. The conditioned signal is then converted into a digital signal by the A/D conversion circuit. Finally, the digital signal is sent to the Digital Signal Processor (DSP) for closed-loop control [44].

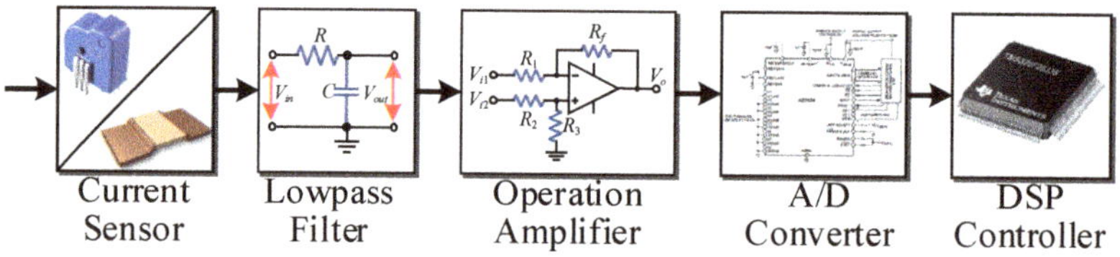

Figure 2. Schematic diagram of the current sampling path.

It should be noted that the CME discussed in this paper refers not only to the "current measurement error" resulting from the degradation of current sensor performance but also to the cumulative errors arising from the sensor, filtering circuits, operational amplifier circuits, and analog-to-digital (AD) conversion in the overall sampling path shown in Figure 2. Rather than focusing on the specific stages where errors occur, this paper concentrates on the practical influence of the sampling error, specifically the frequency characteristics of the additional disturbance signals. On this basis, this paper explores compensation methods for different types of disturbance signals. According to their harmonic characteristics, these errors can be classified into the following categories:

(1) High-frequency random error: Due to the resolution limitations of the A/D conversion circuit, the analog signals are quantized into 2^n digital signals, where n represents the resolution of the A/D conversion circuit. Additionally, the sampled results also involve measurement noise caused by environmental disturbances. Both quantization errors and measurement noise lead to high-frequency random errors in the measurement results [45].

(2) High-frequency PWM-induced error: Due to non-ideal factors such as EMI interference, current oscillation during the device commutation process, sampling, and PWM delay, the measured results may deviate from the average current value within one switching period. This deviation leads to a PWM-induced error between the sampled and desired data, which threatens the current control performance and the stable operation of current controllers [46].

(3) Low-frequency offset error: Due to the deviation of the resistance or other parameters from the pre-designed values in each phase conditioning circuit, a constant or slowly varying low-frequency offset error exists between the measured and desired data. For example, if the expected measured voltage range is 0–3.3 V, the actual measured voltage range could be 0.1–3.4 V [47].

(4) Low-frequency gain error: Due to parameter deviations in the conditioning circuit, the amplification factor in each phase may differ from the pre-designed slope, which leads to low-frequency gain error. For example, if the expected measured voltage slope is 0.1 V/A, the actual measured slope could be 0.09 V/A [48].

Figure 3 shows schematics of different types of CME. Figure 3a,b represent high-frequency random errors and one kind of PWM-induced error, respectively. The current harmonics generated by these high-frequency errors are typically distributed around the switching frequency. Figure 3c,d represent current measurement offset errors (CMOE) and current measurement gain errors (CMGE), respectively. The current harmonics generated by these low-frequency errors are typically DC or at the fundamental frequency. High-frequency errors cause the measured result to deviate from the average current value within this switching period, affecting the stable operation of current controllers [49]. According to the findings of [49], at a switching period of T_s, a one-cycle delay error results in a phase margin reduction of $2\pi f_c T_s$, where f_c denotes the cutoff frequency. However, low-frequency CMGE and CMOE generate additional torque ripples at the fundamental frequency and twice that frequency, resulting in additional loss and vibration [50]. According to reference [44], the typical sources of errors in electric powertrain systems are summarized, highlighting that CMEs can cause problems such as speed fluctuation, torque ripple, phase current mismatch, and additional losses.

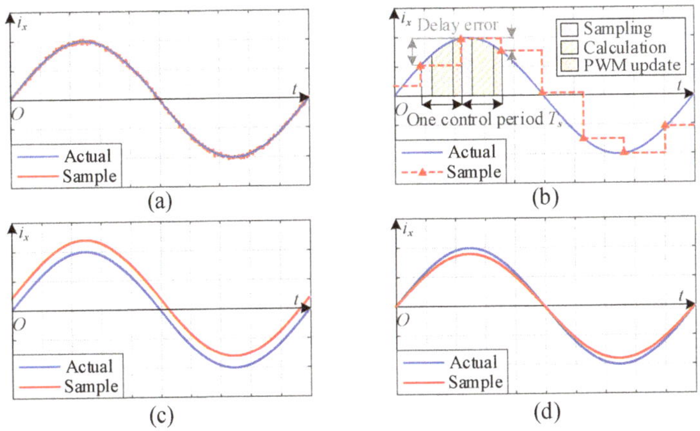

Figure 3. Schematic diagram of different current sampling errors. (**a**) High-frequency random error. (**b**) High frequency PWM-induced error caused by PWM delay. (**c**) Low-frequency offset error. (**d**) Low-frequency gain error.

Additionally, the influence of CME on the drive system depends on the sensor installation position. Figure 4 shows several installation positions for current sensors. In driver systems with the installation shown in Figure 4a, CMGE can lead to second torque and speed ripples. However, in driver systems with a single current sensor, as depicted in Figure 4b–d, CMGE does not exist. This is because the additional second torque ripple is caused by the difference in gain coefficients of each phase sensor. Therefore, the problem does not occur in systems that only use a single current sensor [51,52]. Nonetheless, CMGE still causes a deviation in the current limit, which can result in the output current either exceeding the limit or not being fully utilized [53]. This issue is present in drive systems that use either a single current sensor or multiple current sensors.

For all the topologies in Figure 4, Figure 4a is the most commonly used installation. Sensors are installed on the output branch in each phase or on the DC bus simultaneously. For a Y-connected three-phase PMSM, the sum of the three-phase currents is zero. Thus, only two-phase current sensors are needed to achieve closed-loop current control. However, in applications requiring high safety, such as elevators and cranes, three-phase current sensors are utilized to ensure fault tolerance during current sensor failures [48]. The current sensor installed on the DC bus is commonly implemented in electric vehicles and wind turbines. [47,54]. For low-cost machine drive systems, only one single current sensor is used, as shown in Figure 4b,c. In these cases, currents are sampled multiple times in one switching period, and all phase currents are reconstructed according to different space voltage vectors [51,52,55]. In addition to the installation described in Figure 4a–c, other signal sensor installation positions are studied to improve the performance of the current reconstruction algorithms at all modulation indices, as shown in Figure 4d [56–58].

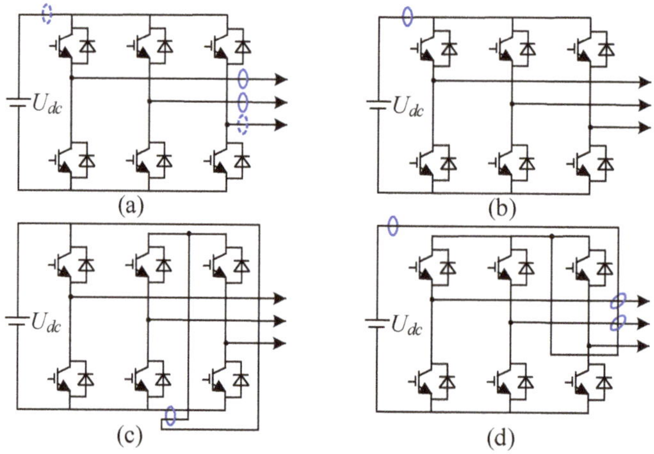

Figure 4. Schematic diagram of different installation positions for current sensors. (**a**) Sensors installed on the 2/3 phase output branch or the DC-bus. (**b**) Sensors installed only on the DC-bus. (**c**) Sensors installed on the phase-to-phase interleaved branch. (**d**) Sensors installed on the DC-bus and the interleaved branch between phase output and DC-bus.

The influence of CME is also related to the implemented control strategy. In drive systems that use disturbance rejection control algorithms, CMEs affect the accurate compensation of disturbances [59]. In multi-phase control systems, CMEs can cause an output power imbalance between different winding sets [60]. In open-wound control systems, CMOE may couple with the zero-sequence circulating current, which leads to non-zero circulating currents [61].

3. Compensation for Current Measurement Error

3.1. High-Frequency Current Measurement Error

According to Figure 3, CME can be divided into high- and low-frequency components. The different spectral distributions result in varying impacts and compensation methods for the two types of CME. In this section, two kinds of high-frequency CME, namely, random noise and PWM-induced error, and their compensation methods are briefly introduced.

(1) Random Noise: Random noise includes measurement noise and quantization noise. In drive systems that use proportional-integral (PI) controllers, reducing the bandwidth of the PI controller can suppress the influence of random noise; however, this will also reduce the dynamic characteristics of the current controller. In drive systems that use model predictive control and direct torque control, random noise directly affects the selection of voltage space vectors, significantly impacting the total harmonic distortion (THD) of the output currents.

(i) Software Method: The compensation of measurement noise can be achieved through software or hardware-based methods. For software compensation, low-pass filters (LPF), moving average filters, recursive least squares (RLS) [62], and Kalman filters [59,63] are effective techniques for pre-filtering the measured currents. In [64], a compensation method based on a Luenberger observer was proposed, which effectively suppresses the influence of measurement noise on the voltage vector selection of MPC. Similarly, a parallel cascaded extended observer was proposed in [65], which effectively suppresses high-frequency measurement noise in direct torque control.

(ii) Hardware Method: Due to the limitation of the switching period, hardware-based techniques are more effective solutions compared to software filtering. Using resistors and capacitors to form a first-order LPF circuit is a widely adopted method in engineering; however, there is a trade-off between noise suppression and transmission delay. A variable-gain ADC conditioning circuit is designed in [66] to achieve full ADC range utilization considering the different amplitudes of the input signals. The enhanced data acquisition system with this circuit significantly suppresses the random noise of the measured signal and enhances the accuracy of online machine state monitoring. In addition, oversampling is also an effective solution [67]. In sensorless drive systems, oversampling significantly improves position estimation and current measurement accuracy [68,69].

Many chips in machine control systems are equipped with oversampling capabilities. For instance, in the TI C2000 digital signal processor (DSP), oversampling can be achieved by modifying the registers of the ADC peripheral [70]. Similarly, in ADI's AD7606 chip, adjusting the voltage of the OS pin allows for the adjustment of the oversampling mode [71].

However, regardless of the chip used for implementation, the oversampling factor M_{os} can be utilized to characterize different oversampling levels. A higher M_{os} results in improved suppression of random noise. The suppression effect of oversampling on random noise can be evaluated using the signal-to-noise ratio (SNR), which can be defined as follows:

$$SNR = 10\lg\frac{P_s}{P_n} \tag{1}$$

where P_s and P_n are the signal and noise power, respectively. Oversampling technology allows ADC circuits to attain higher effective resolution and significantly decrease the SNR of measured results. Theoretically, with each four-fold increase in M_{os}, the SNR of the input signal is expected to rise by 6 dB, and the ADC's resolution will also increase by 1 bit [70]. Figure 5 shows the relationship between SNR improvement and M_{os} for the oversampling operation of the AD7606 chip [71]. It can be seen from Figure 5 that oversampling at different M_{os} levels improves the low-frequency SNR of the input signals. However, when M_{os} becomes too large, the improvement in SNR diminishes, and the high-frequency SNR may even decrease. Although oversampling can significantly improve SNR and equivalent resolution, it also entails some performance trade-offs. Higher M_{os} raises the execution time of the sampling process, thereby decreasing the maximum sampling speed of the ADC. According to the Nyquist sampling theorem, this operation limits the maximum bandwidth of captured signals.

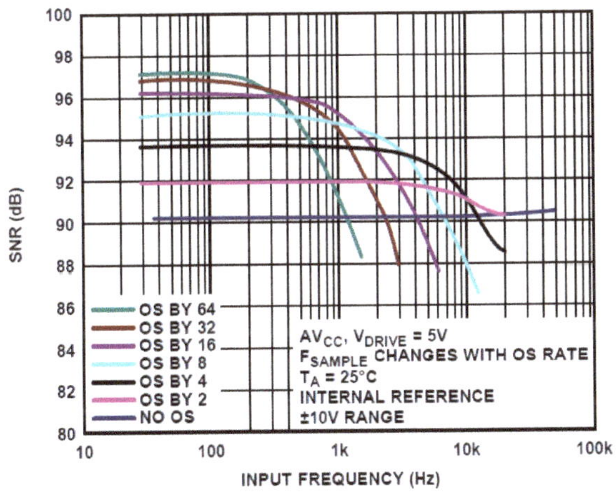

Figure 5. Signal-to-noise ratio of oversampling operation under different oversampling factor M_{os} [71].

Figure 6 shows the schematic of the oversampling operation when $M_{os} = 4$, where the blue dashed line and the black solid line represent the ideal and noisy signals, respectively, and the red triangles represent the measured points. If a 0–3 V signal is sampled with a 12-bit ADC, the sampled result for a 1.5 V signal should be 2048. However, noise causes a deviation in the sampled result. By using oversampling, four consecutive data points are collected, accumulated, and then down-sampled to obtain the equivalent oversampled result. Compared to single sampling, oversampling effectively suppresses the impact of random noise. The sampled results are closer to the actual values.

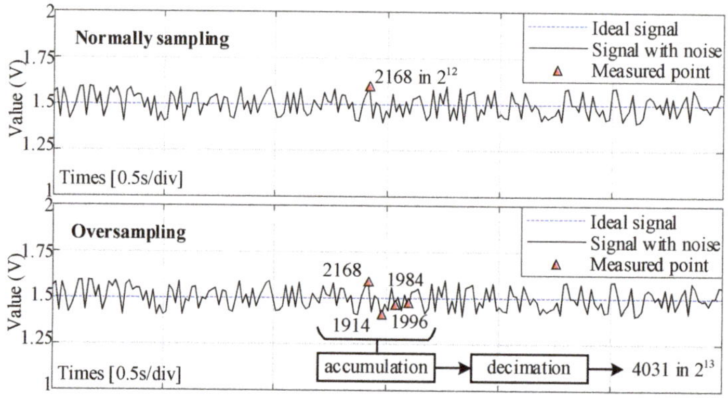

Figure 6. Schematic diagram of oversampling to suppress the random noise of measurement.

Apart from oversampling techniques, Σ-Δ ADC conversion technology is also an effective method for attenuating high-frequency random noise [72]. This technology was initially used for precise temperature measurements and audio processing. Compared to traditional successive approximation register (SAR) ADCs, Σ-Δ ADCs can achieve 16–18 bits of resolution even without special fine-tuning or calibration [73].

Figure 7 shows the block diagram of the Σ-Δ ADC [74]. The random noise in the input signal is initially mitigated through an oversampling operation, followed by digital filtering and a decimation structure to resample it with f_s. Subsequently, the noise is shaped using the Σ-Δ modulator, concentrating the noise energy in the high-frequency band. Finally, the shaped noise is attenuated again by the digital filter and the decimation structure to complete the Σ-Δ conversion process.

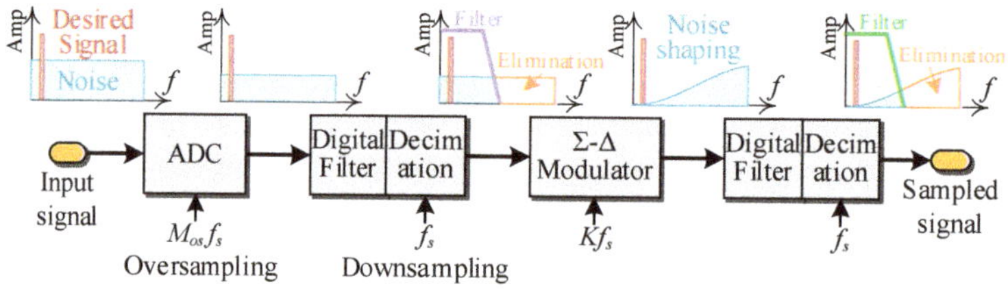

Figure 7. Block diagram of the Σ-Δ ADC techniques.

The Σ-Δ modulator serves as the central structure in all Σ-Δ conversion processes, facilitating noise shaping to significantly improve noise attenuation effects. Figure 8 shows the block diagram of a first-order Σ-Δ modulator. By employing closed-loop regulation of the integrator and comparator, sinusoidal signals are transformed into PWM signals with amplitudes of $\pm V_{ref}$. A study in [72] compared the noise suppression capabilities of oversampling operations and Σ-Δ-type ADCs. The analysis results validated the remarkable noise suppression capabilities of the Σ-Δ modulator. Compared to oversampling techniques, the Σ-Δ type ADC enhances the SNR by nearly 20 dB.

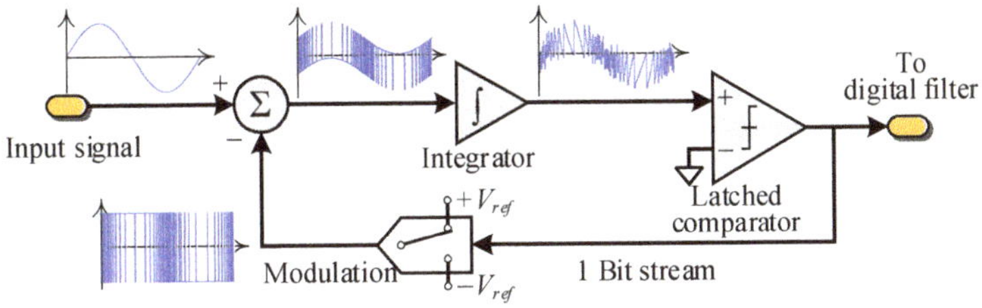

Figure 8. Block diagram of the Σ-Δ modulator [74].

(2) PWM-Induced Error: Compared to random high-frequency noise, the frequency of PWM-induced error typically aligns near the switching frequency, f_s. Notably, switch noise and computational delay are two typical PWM-induced errors.

(i) Switching Noise: Switch noise-induced errors are associated with the use of wide-band gap devices, such as SiC. While SiC devices shorten switching time, they also exacerbate current oscillations during the switching process. If the moment of current sampling closely coincides with the switching process, these current oscillations can lead to deviations in the sampled results from the average current within one switching cycle. Consequently, this phenomenon degrades the current closed-loop control performance and results in output current distortion [75]. Moreover, the rapid switching of SiC devices can result in serious electromagnetic interference (EMI) problems, which may cause the ADC circuits to produce error signals correlated with the PWM frequency. This phenomenon similarly degrades the performance of the drive system [76].

To address these issues, some device drive circuits use fiber optic interfaces instead of traditional isolation transformers to enhance electrical isolation [77]. Furthermore, in [78], both the driving and sampling signals are optically isolated and transmitted to the DSP. These optimized hardware designs effectively address EMI interference issues in the transmission lines. However, the additional fiber optic interfaces increase the cost of the drive system and complicate the design of the system's auxiliary power supply.

Another solution to PWM-induced error is the use of digital filters. A moving average filter (MAF) is a commonly used solution to suppress random white noise errors. However, for intermittent, non-stationary PWM-induced errors, especially when the number of erroneous sampling points is high compared to the total number of points in the sampling window, the performance of this MAF method is less effective. According to the analysis results in [79], any presence of switching noise can affect the averaged result. Thus, the MAF can only partially mitigate the effect of switching noise.

The median filter (MDF) was proposed to suppress switching noise-induced errors, combined with an amplitude band-pass filter to eliminate sampling points with instantaneous amplitude violations, effectively reducing EMI issues. The MDF can be expressed as follows:

$$i_{smd}[k] = \text{Median}(i_s[k], i_s[k-1], \ldots, i_s[k-(N-1)]) \quad (2)$$

where $i_s[k]$ represents the sampled stator current at time k, and N is the number of sample points for the MDF window. The "Median" function selects the median value from the sorted set of N samples, averaging the middle two if N is even. It is shown in [79] that if the influence of switching noise does not exceed half the peak-to-peak value of the current ripple, MDF can eliminate the undesired wrong sampling results by sorting the noise-affected points to both sides of the sampling window. When switching noise-induced CME exceeds half the current ripple's peak-to-peak value, [80] suggests adding a repetitive ripple removal (RR) algorithm to the MDF to solve this problem. The transfer function of the RR controller $G_{RR}(s)$ can be expressed as follows:

$$G_{RR}(s) = \frac{(1+R)\left(1-\left(z^{-N}-\frac{1}{N}\sum_{q=1}^{N} z^{-q}\right)\right)}{1-\left(z^{-N}-\frac{1}{N}\sum_{q=1}^{N} z^{-q}\right)+R} \quad (3)$$

where R = 0.125 is used to adjust the time constant of $G_{RR}(s)$ to approximately 10 T_s [81]. The basic principle of the RR controller is to estimate and suppress periodic disturbances using an internal model structure. Figure 9 shows the schematic diagram of the MDF and the combination of MDF and RR. The blue curve represents the real current, the yellow curve represents the average current, and the black discrete stair represents the sampled current. The black and red sampling points represent correct and noise-affected samples, respectively. As shown in Figure 9, when PWM-induced errors exceed half the peak-to-peak ripples, MDF cannot filter out the erroneous sampling points. However, with the help of RR, the bad points are identified and can still be sorted into the two sides of the sampling sequence after MDF, which demonstrates the effectiveness of the RR structure.

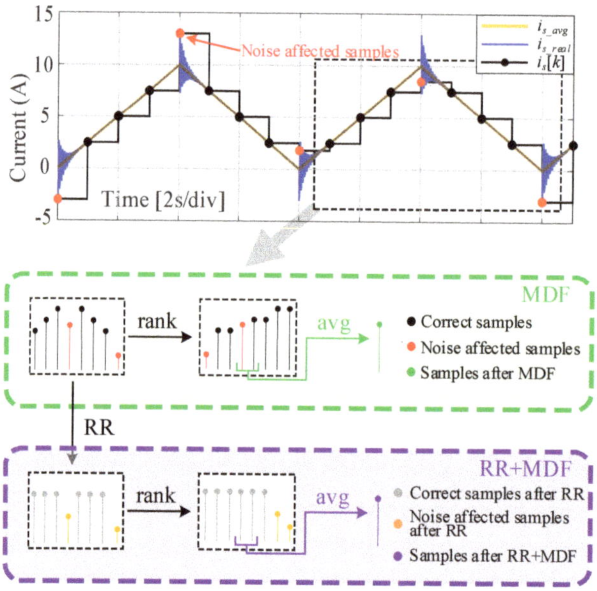

Figure 9. Schematic diagram of the median filter (MDF) and the MDF with a repetitive ripple removal structure (RR).

(ii) Computational Delays: The second type of PWM-induced CME is caused by computational delays. The delays in PWM updates and ADC sampling lead to the sampled current deviating from the average current during the switching cycle, which reduces the stable operating range of the current controller. Although the sampling process itself is not inherently problematic in this case, the sampled result deviates from the current expected by the controller. Therefore, this paper still reviews the related studies for the sake of generality. Based on the analysis in [82], the delay between the ADC and PWM update in a single-sample, single-update driving system is T_s.

To address this issue, one approach is to perform multiple PWM updates within one switching cycle. Figure 10a,b compare the control frameworks for single-sample single-update and single-sample dual-update, respectively, where the yellow triangular wave represents the carrier, the blue solid line is the modulation wave, and the green curve is the output PWM voltage. The red and blue arrows indicate the time points for triggering the ADC and PWM updates, respectively. The purple T_d denotes the delay time between ADC sampling and PWM update. Compared to the single-update algorithm in Figure 10a, the dual-update algorithm significantly reduces the delay from ADC sampling to the duty cycle update. A multi-sample, multi-update structure can further reduce this delay, but it demands higher bandwidth for current sampling and a faster processing speed from the processor. In some algorithms, an FPGA is also required, which increases the complexity of system design and hardware costs [83].

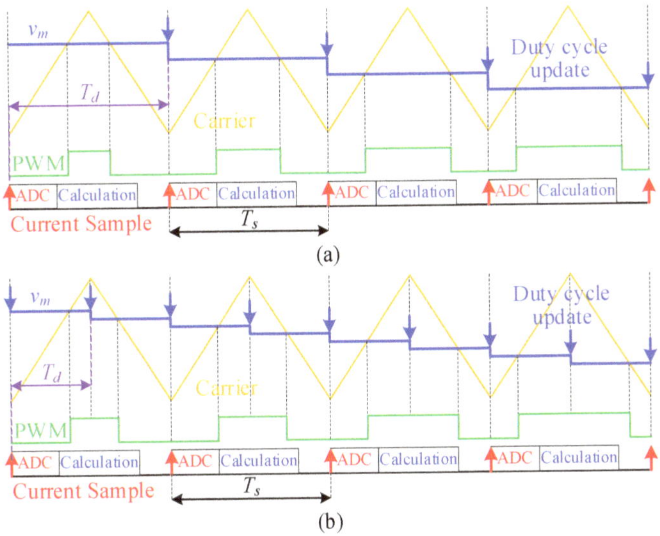

Figure 10. Comparison of the delay time between the single-sample single-update and single-sample dual-update control algorithms. (**a**) single-sample single-update control alogrithm; (**b**) single-sample dual-update control alogorithm.

In addition, adjusting the execution order of each controller task, such as ADC, control computation, and PWM updates, can help reduce the equivalent delay. A method for decreasing PWM updates is proposed in [84], which places time-consuming control computation, such as integration and resonance, outside the ADC sampling and PWM update processes, thereby minimizing the delay between them. Based on this, [49] proposed a control architecture to ensure minimized update delays for all driving systems that use any number of current controllers while maintaining anti-windup characteristics. This algorithm advances the outer loop control program and decomposes the inner loop control program into necessary pre-calculation and post-calculation parts. The addition of more current controllers only increases the post-calculation time and does not affect the pre-calculation part. As a result, this architecture effectively minimizes the delay from ADC sampling to duty cycle update.

3.2. Low-Frequency Current Measurement Error

In this section, two typical types of low-frequency CME, namely, current sampling offset error (CMOE) and current sampling gain error (CMGE), as well as their compensation methods, are briefly introduced. Compared to high-frequency CME, these two low-frequency CMEs lead to additional fundamental and twice-frequency ripples in the output torque. Since these two errors are usually considered and compensated together, the following studies are classified and introduced according to different compensation methods for clarity.

(1) Offline Compensation Method: Offline compensation of CME is the most common method in engineering. These methods are typically used during the initial debugging of the drive system or for pre-correction before factory delivery. In [85], an offline CMGE compensation method is proposed, in which the PMSM operates in standstill mode, and the currents are closed-loop controlled in the stationary frame, and the *dq*-axis currents, i_d and i_q, and the midpoint current are sampled while ensuring that $i_q = 0$. This method is simple and offers excellent compensation results; however, it requires precise resistance measurements and the use of three current sensors.

An offline CMGE compensation method is proposed that analyzes the compensation ability under resistance asymmetry conditions [48]. However, this method still requires three current sensors, and the rotor of the PMSM must be locked. In [53], an offline compensation method based on elliptical trajectories and sequence signal injection is proposed. By correcting the midpoint current, this algorithm can effectively suppress the influence of resistance offsets and inverter non-linearity on compensation performance, thereby improving the precision of offline compensation. Furthermore, both the deviation and average components of CMGE are analyzed and compensated.

(2) Online Compensation Method: Due to factors such as temperature drift and aging, the CME slowly changes during PMSM operation. These varying errors cannot be addressed by the offline compensation methods mentioned above. Since CMOE and CMGE cause additional fundamental and twice-frequency harmonics, several online compensation methods based on characteristic harmonics suppression have been extensively studied.

(i) Signal Injection-Based Method: By injecting fixed frequency signals into the regulated currents, the magnitude and phase of the feedback signals can be used to indirectly estimate CMOE and CMGE. Zhang et al. [86] proposed a compensation method in which additional CMOE and CMGE are injected. Then, CMEs are estimated by detecting the variation in the speed harmonic magnitude. However, this method is unsuitable for varying speed conditions, and the injected additional CMEs lead to more torque ripples.

Besides injecting additional CME, injecting high-frequency (HF) signals is also an effective means of compensating for CME. With the help of rotational transformation and the amplitude and phase information of high-frequency negative sequence harmonic currents, CME can be reconstructed [87]. Compared to the method in [86], this approach increases the frequency of the injected signal, thereby reducing the negative effect on torque ripples. Further, [60] proposed an HF injection-based CMGE online compensation method for a dual three-phase PMSM. The measured currents are decoupled into the fundamental and harmonic subspaces via vector space transformation (VSD). Since HF signals are injected only into the harmonic subspace, this algorithm has a minimal effect on torque ripple. Additionally, Lu et al. proposed a CME compensation method for electric vehicle drive systems [88]. This algorithm injects reverse voltage vectors to construct additional constraints. The calculation for reconstructing CME is significantly reduced. This method does not require continuous signal injection, further reducing the undesired ripples. However, it necessitates an additional DC-bus current sensor and is unsuitable for systems that only have phase current sensors.

(ii) Speed Harmonic-Based Method: Whether HF or low-frequency (LF) signals are injected, almost all of these methods affect the output torque and vibration noise. However, closed-loop compensation algorithms based on harmonic extraction do not have this issue.

In 1998, Chung et al. proposed a CME compensation method based on speed harmonics [50]. It reveals that CMOE and CMGE can be considered fundamental and as the second load disturbances. The speed harmonics are separated using spectral analysis and extracted through the proportional-integral (PI) controller, thereby enabling closed-loop compensation for CME. However, this algorithm requires the parameters of the mechanical system, such as damping and inertia. In [89–91], the authors proposed replacing the PI controller in the speed controller with a proportional-resonant (PR) or repetitive controller, thereby achieving indirect suppression of errors caused by CME. In [92,93], iterative learning or repetitive learning methods were proposed for CME compensation. However, both methods require real-time online learning or fitting, which leads to a significant computational burden. In [94], an improved adaptive selected harmonic elimination algorithm was introduced to compensate for CME. This algorithm extracts the speed harmonics and then achieves compensation for CME with the help of the phase relationship between the d-axis and q-axis harmonic currents.

However, these speed harmonic-based compensation methods in [50,89–94] only suppress the additional speed harmonic disturbances caused by CME but do not compensate for the CME itself. This means that these methods are not suitable for conditions with varying loads and speeds. Moreover, these algorithms always need to be executed during the operation of PMSM, which leads to a greater computational burden. Therefore, they also affect the performance of the original control algorithms under low-speed [91] and variable-speed conditions [89,90]. Additionally, the amplitude of torque and speed harmonics at different fundamental frequencies is dependent on the mechanical damping and inertia parameters, which can be expressed as follows [50]:

$$\Delta\omega_m = -\frac{s}{Js^2 + (K_{ps} + B)s + K_{is}}\Delta T_L \tag{4}$$

where J and B are damping and inertia parameters, respectively, and K_{ps} and K_{is} are the proportional and integral parameters of the speed controller, respectively. The speed harmonic magnitude first increases and then decreases as the fundamental frequency increases [50]. Considering that torque harmonics and current harmonics are approximately proportional, the compensation performance of the speed harmonic-based CME compensation method deteriorates at both low and high frequencies.

(iii) Current and Voltage Harmonic-Based Method: Compared to speed harmonic-based CME compensation methods, current and voltage harmonic-based methods are not affected by mechanical system parameters and have been widely studied.

Figure 11 shows the general block diagram of a CME compensation algorithm based on current and voltage harmonics. The blue part represents the filter used to extract harmonic components from the current and voltage, and in the red part, harmonic voltage and current information are used to reconstruct and compensate for CME.

A CME compensation method using Support Vector Regression (SVR) was proposed in [95], which considers both inverter nonlinearity and parameter uncertainty. The authors of [96] introduced a non-intrusive CMGE compensation method. However, this method requires the tested machine to be controlled to a short-circuit state. Additionally, prior knowledge of the short-circuit current characteristics of each machine is needed, which limits its general applicability.

Considering the machine control system under normal operating conditions with limited computational resources, a CMGE reconstruction approach based on the harmonic spatial effects of CMGE was proposed [97]. However, this method requires three current sensors. To improve applicability, a CME compensation method using two phase current sensors was proposed, in which the effect of the speed controller on CME compensation performance was analyzed [98]. However, this algorithm is not applicable when the current controller is saturated. To address this issue, S.-I. Kim et al. introduced an online compensation algorithm for CMOE, which enables its application under current saturation conditions [99], but the compensation of CMGE has not been considered. To compensate for the CMOE and CMGE simultaneously, a voltage harmonic-based compensation method was proposed [100], but it does not consider the impact of parameter mismatch, and the convergence of the estimated result is relatively slow.

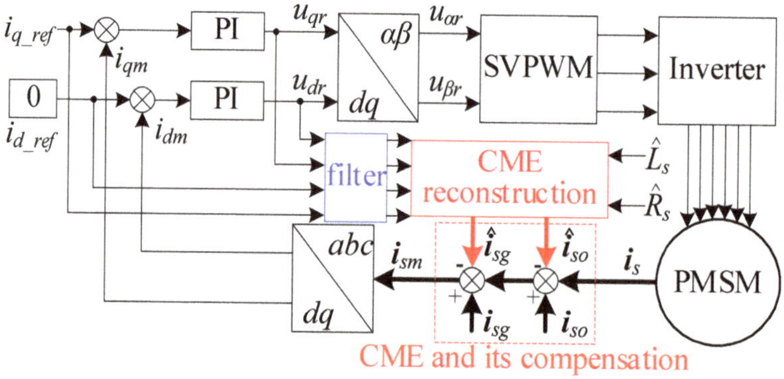

Figure 11. General block diagram of current sampling error compensation based on voltage and current harmonics.

A CME vector reconstruction technique was proposed in [101], in which the fundamental and harmonic components induced by CME are decoupled by a single-frequency suppressor. While this method enables rapid CME compensation, the overall compensation accuracy remains insufficient. The angle-domain definite integral approach is employed to extract current harmonics and reconstruct CME in [102,103]. This method achieves high-precision compensation; however, the performance under varying speed and load conditions has not been thoroughly examined. An online compensation method for CMOEs based on state observers is introduced in [104], which accounts for the influence of load and speed variations, as well as parameter deviations, on compensation performance. M. Hu et al. proposed an LPF-based compensation method for CMOE in [105], in which the effect of transients and saturation are considered. Filter parameters are also designed. However, compensation for CMGE is not provided.

An online CMOE compensation method based on an expanded state observer was introduced in [106], which effectively mitigates the influence of parameter deviations and ensures consistent compensation accuracy at different speeds. Kyung-Rae Cho et al. proposed a structure comprising a resonant observer and a residual compensator in [107], while a compensation method utilizing a virtual winding structure was proposed in [108]. Both methods enable simultaneous compensation for CMOE and CMGE without dependence on motor parameters and provide rapid compensation with high efficiency. Additionally, the modified complex coefficient filter can be used to compensate for CME. Beyond the advantages mentioned above, this method demonstrates satisfactory performance under varying speeds and loads [109].

(iv) Current and Voltage Harmonic-Based Method: Table 2 presents a comparison of the characteristics of various low-frequency CME compensation algorithms in machine drive systems. In addition, to further analyze the performance of different methods based on speed harmonics, high-frequency injection, voltage harmonics, and current harmonics, simulations of the methods in [87,94,100], and [109] were conducted.

Table 2. Performance comparison of low-frequency CME compensation methods.

Classification		[62]	[64]	[65]	[59]	[63]	[66]	[68]	[67]	[70]	[71]	[72]	[73]	[74]	[75]	[76]	[77]	[78]	[79]	[80]	[81]	[82]	[83]	[84]	[49]	[85]	[48]	[53]	[86]	[87]
Source of information	Speed	×	×	×	√	×	×	×	√	√	√	√	√	√	√	×	×	×	×	×	×	×	×	×	×	×	×	×	×	×
	Voltage	×	√	√	×	×	×	×	×	×	×	×	×	×	×	√	×	×	×	√	√	×	×	×	√	√	√	√	×	×
	Current	√	√	√	×	√	√	√	×	×	×	√	×	×	×	√	√	√	√	√	×	√	√	√	√	√	√	√	√	√
CME type	CMGE	√	√	√	√	√	√	√	√	√	√	√	√	√	√	√	√	√	×	×	√	√	√	√	×	×	×	√	√	√
	CMOE	×	×	√	√	×	×	√	√	√	√	√	√	√	√	√	×	×	√	√	√	√	√	√	√	√	√	√	√	√
Operating state	Running				√	√	√	√	√	√	√	√	√	√	√	√	√	√	√	√	√	√	√	√	√	√	√	√	√	√
	Standstill	√	√	√																										
Different conditions	Speed varying	-	-	-	×	×	√	×	×	√	√	√	×	×	×	×	×	×	√	×	√	√	×	×	√	√	√	√	×	√
	Load varying	-	-	-	×	×	√	×	×	√	√	√	×	×	×	×	×	×	√	×	√	√	×	×	√	√	√	×	×	√
Current sensor numbers		3	2	3	2	2	2 (1 set)	3 *	2	2	2	Not mentioned	2	Not mentioned	2	Not mentioned	2	3	2	2	2	2	2	2	2	2	2	2	2	2
Other issues		[63,66]: need HF injection; [68]: need reverse voltage vector injection; [77]: need short circuit test at coasting down condition.																												

* denotes 2 phase sensors and 1 idc + sensor. Other sensors are all phase sensors.

Figure 12 illustrates the simulation results of the CME compensation algorithm based on speed harmonics in [94]. The compensation process is divided into three stages: "Normal", "CME injection", and "CME compensation", which are represented by blue, red, and green regions, respectively. The reference values for CMOE are $o_{\alpha\_ref} = 0$ A and $o_{\beta\_ref} = 0.2$ A, while the reference value for CMGE is $g_{\alpha\beta\_ref} = 0.2$. The amplitude of the fundamental current is 2 A. It can be seen in Figure 12 that the estimated value of CMOE ($o_{\alpha\beta}$) effectively converges to its reference value after enabling CME compensation. However, the negative sequence currents $i_{\alpha\beta n}$, rather than CMGE, are compensated. Consequently, this algorithm is not suitable for compensating for the negative sequence disturbance under variable load conditions.

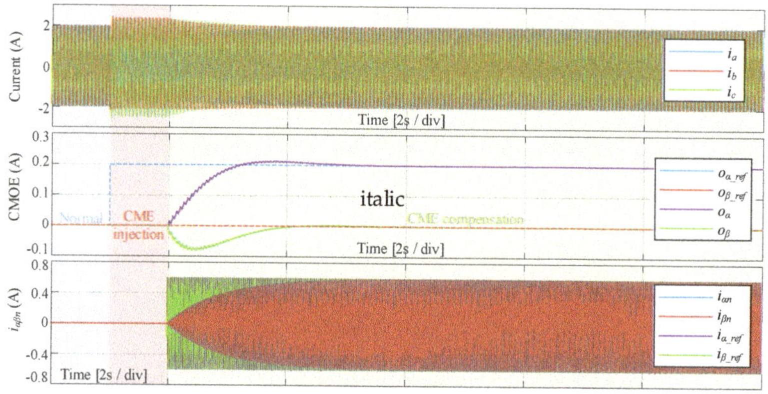

Figure 12. Simulation results of the CME compensation algorithm based on speed harmonics.

Figure 13 presents the simulation results of the CME compensation algorithm based on HF injection in [87]. The compensation process is divided into four stages: "Normal", "CME injection", "HF injection and compensation", and "Completion of CME compensation", represented by blue, red, purple, and green regions, respectively, where $o_{\alpha\_ref} = o_{\beta\_ref} = 0$ A and $g_{\alpha\beta\_ref} = 0.2$. It is worth noting that since HF signals do not affect CMOE, this algorithm is only suitable for compensating CMGE. It is evident in Figure 12 that after enabling HF injection, the estimated $g_{\alpha\beta}$ rapidly converges to the reference value. Although HF injection introduces additional losses, the estimated results can be locked after convergence, thus preventing the long-term operation of this compensation algorithm.

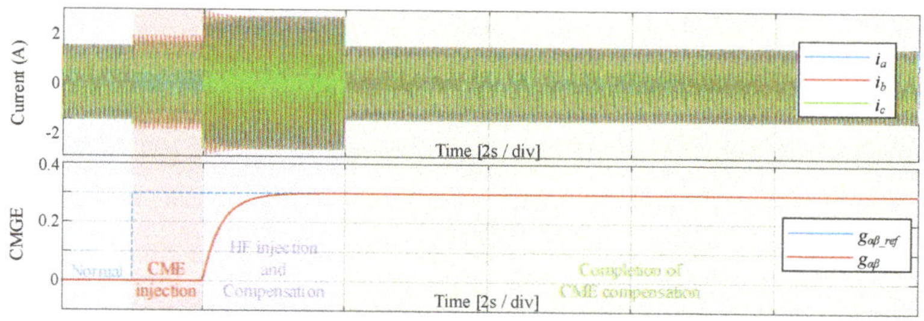

Figure 13. Simulation results of the CME compensation algorithm based on HF signal injection.

Figure 14 shows the simulation results of the CME compensation algorithm based on voltage harmonics in [100]. The experimental conditions are identical to those in Figure 12, and the compensation process can be divided into the same three stages. It can be seen in Figure 14 that although the estimated results for both CMOE and CMGE gradually converge to their reference values after 2 s, the overall convergence speed is relatively slow.

Additionally, there is a noticeable overshoot in the estimated result of CMOE. This indicates that the compensation algorithm, which relies solely on voltage harmonics, requires further improvement to enhance its response performance.

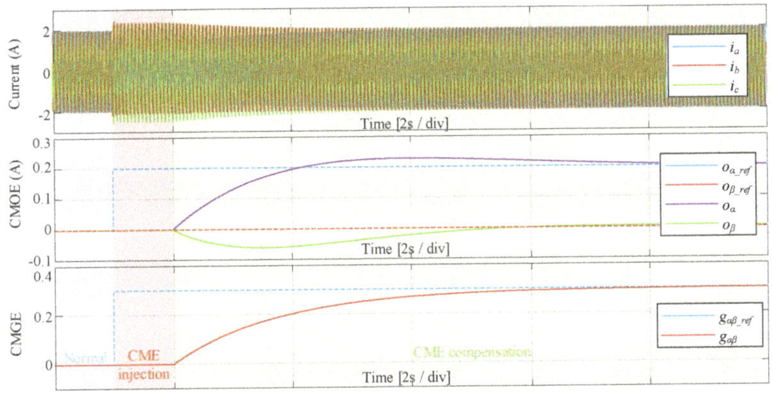

Figure 14. Simulation results of the CME compensation algorithm based on voltage harmonics.

Figure 15 shows the simulation results of the CME compensation algorithm based on current harmonics in [109]. The experimental conditions are the same as those in Figure 12, and the compensation process follows the same three stages. It can be seen that after enabling the compensation algorithm at 2 s, all estimated results rapidly converge to the reference value within 1 s. Moreover, the estimated results can be locked after stabilization to improve their applicability under varying operating conditions, which demonstrates the excellent performance of this current harmonic-based compensation algorithm.

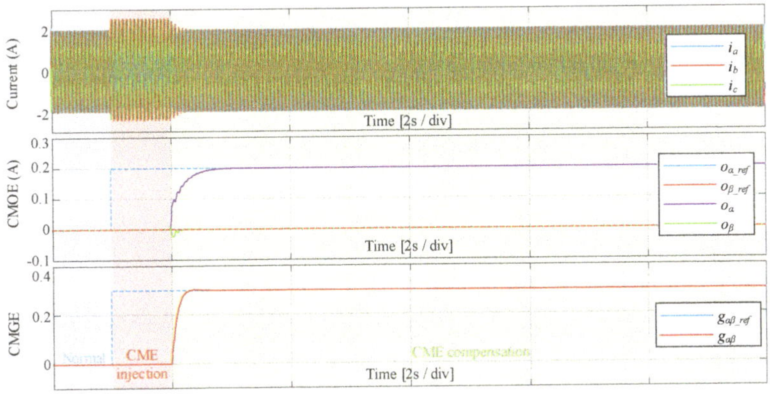

Figure 15. Simulation results of the CME compensation algorithm based on current harmonics.

3.3. CME in Special Topologies

In addition to the typical online compensation methods mentioned above, there are several specialized CME compensation algorithms with current sensor installations, as shown in Figure 4. For instance, in the three-phase, four-bridge-arm drive system, a current reconstruction technique based on fixed-point sampling was proposed in [56], in which the difference between the positive and negative bus currents is measured by a signal current sensor. By specifying the current during different switching states and predicting the current slope based on each state, compensation for the deviation of the sampled current from the average current within the current period can be achieved.

An all-current-survivable algorithm was proposed in [110] by adding three additional sensors that sample the sum of DC-bus and phase currents. This installation method ensures that errors or faults in any individual sensor do not impact the normal operation of the drive system. Additionally, a CME compensation method with a similar sensor installation was proposed in [57], which significantly reduces the computational burden by avoiding additional filters or observers. For a dual-PMSM-based system, fast calibration can be achieved through a consistency calibration method [111], which enables the simultaneous calibration of two sets of three-phase windings within 1 s. Furthermore, with the help of an additional DC-bus sensor, an orthogonal pulse modulation carrier strategy combined with fixed-point sampling was proposed in [54], which can reconstruct and compensate for CME with some simple algebraic calculations.

4. Future Trends

The high-performance demands in fields such as CNC machine tools and servo drives have posed significant challenges to both hardware and software design. Sensor errors, especially CME, have emerged as crucial factors influencing the control performance of drive systems. The analysis and compensation of CME have been widely studied and developed over the past two decades. Future research trends in this area can be categorized into the following aspects:

(1) Analysis of the Influence of CME under Different Topologies: The effects of CME vary across different topological structures and deserve further investigation. In drive systems utilizing wide-bandgap devices, parasitic capacitances between the inverter and the sensor can result in additional HF switching noise induced by CME [112]. The generation mechanism of this error was carried out by establishing a model of the parasitic parameters of the drive system. Additionally, the influence of this error on HF injection-based sensorless algorithms is analyzed. Similar LF or HF disturbances due to CME can also occur in other types of power electronic converters. For instance, in three-phase grid-connected inverters, errors from additional voltage sensors may couple with CME, thereby affecting fault detection and compensation performance [113–115].

Additionally, the compensation algorithms for CME in multi-phase machines [60] and low-carrier ratio scenarios [69] are worth further exploration. In circuit topologies with special current sensor placements, as illustrated in Figure 4, the compensation performance of CME may be improved. However, this also introduces some specific issues. For example, in drive systems with a single current sensor, there exists a region in the space voltage vector where current reconstruction cannot be achieved, leading to current distortion in closed-loop control [56]. Moreover, in systems utilizing position sensors such as resolvers and Hall sensors, measured position signals can also induce errors similar to CME. The combined effects of CME and position errors on the drive system should also be explored further [116].

(2) Compensation for Different Components of CME: CME includes both HF and LF components. For the HF component, existing manufacturing processes have successfully implemented oversampling techniques [70,71] and Σ-Δ ADC technology [73,74] in hardware that significantly enhance the SNR for random noise. Therefore, in applications with fewer cost constraints, suppressing random noise through software algorithms may no longer be a valuable area of research. However, with the use of wide-bandgap semiconductors that have faster switching speeds, the influence of PWM-induced non-random error has become increasingly significant [112]. Especially in cases with high environmental EMI [75] or in which the sensor range is not fully utilized [66], the causes of PWM-induced errors and their compensation strategies still require further investigation. For some special

applications, such as those with low inductance and a high carrier ratio, the influence of PWM-induced errors may become more significant.

In the compensation methods for the LF component, offline compensation methods cannot compensate for real-time varying errors caused by temperature drift and aging. Signal injection-based methods introduce additional losses and torque ripples. The performance of methods based on speed harmonic extraction is limited by load, damping, and the fundamental frequency. Therefore, compensation methods based on voltage and current harmonics are expected to be a key area for future development. Moreover, improving the CME compensation performance to achieve satisfactory accuracy at any fundamental frequency is crucial [106]. Furthermore, enhancing the applicability of compensation algorithms under startup, acceleration, deceleration, and load variation conditions is a critical direction for future research [104,105].

(3) Interaction between CME and Other Algorithms: For drive systems with multi-sampling and multi-update PWM, the ADC process is more susceptible to the influence of PWM switching noise, which in turn causes the measured results to deviate from the actual currents. Moreover, CME also depends on the magnitude of overshoot during the device commutation process. For sensorless control algorithms based on observers, the LF components of CME lead to additional negative sequence disturbances in the position estimation results [117]. For sensorless algorithms based on HF injection, PWM-induced errors cause regular deviations in position estimation values [112]. In drive systems using two-phase current sensors, the random noise in CME is significantly higher than in systems using three-phase current sensors, which reduces the SNR of the estimated position [118]. Finally, for fault diagnosis algorithms such as high-resistance connections, inter-turn short circuits, and machine eccentricity, LF CME also affects estimation accuracy and can even lead to fault misdiagnosis in severe cases. Identifying the differences between these faults and CME requires further investigation.

(4) Sensor Fault Diagnosis and Fault-Tolerant Control: In addition to generating HF and LF errors, current sensors can also produce other non-periodic errors due to line damage and power supply loss. Diagnosing different sensor faults and improving the fault tolerance capability of current sensors are crucial research directions. A diagnostic technique for common faults in IPMSM is proposed in [119]. The algorithm injects HF signals and combines current responses to construct multiple fault indicators for identifying inter-turn short circuits, CMOE, and CMGE. However, the application of this algorithm in SPMSM requires further investigation. In [120], for SPMSM, a method is proposed for diagnosing and locating current sensor faults, which can distinguish the number and location of faulty sensors. The authors of [121] present a fault-tolerant algorithm for current sensors in IPMSM, in which state observers are constructed to estimate current results in real time, thereby enabling speed closed-loop control, even with a single-phase sensor or no current sensor. However, mismatched machine parameters can affect control performance. The authors of [122] propose a current sensor fault diagnosis method based on a Multichannel Global Maximum Pooling CNN, which diagnoses not only CMOE and CMGE but also broken-line faults and stuck faults.

5. Conclusions

This paper provides an overview of the existing analysis and compensation methods for CME in PMSM. These studies include the current sampling path and classification of CME, the effects of CME on various topologies and control algorithms, and the different techniques for CME compensation. From a frequency perspective, CME consists of two main components: HF and LF. HF errors include random noise and periodic PWM-induced errors. Some hardware chips equipped with oversampling and Σ-Δ ADC functions can effectively address the issue of random noise. However, PWM-induced errors are related to current oscillations during the device commutation process and ADC delays, necessitating modifications to the control program. LF errors include CMOE and CMGE. Compensation methods based on signal injection introduce additional losses, vibrations, and noise. Methods based on speed harmonic extraction exhibit decreased performance at both low and high speeds. In contrast, voltage and current harmonic-based methods provide relatively superior performance. Finally, trends in the development of CME influence analysis, compensation, and sensor fault-tolerant strategies are discussed.

Author Contributions: Conceptualization, P.S. and T.W.; methodology, P.S.; software, P.S.; validation, P.S., C.L. and X.M.; formal analysis, H.L.; investigation, H.L.; resources, C.L.; data curation, X.M.; writing—original draft preparation, P.S.; writing—review and editing, L.W.; visualization, T.W.; supervision, L.W.; project administration, L.W.; funding acquisition, L.W. All authors have read and agreed to the published version of the manuscript.

Funding: This work was supported in part by the National Key Research and Development Program of China under Grant 2023YFB4202800; in part by the National Science Fund for Distinguished Young Scholars under Grant 52225703.

Conflicts of Interest: The authors declare no conflict of interest.

References

1. Sonandkar, S.; Selvaraj, R.; Chelliah, T.R. Fault Tolerant Capability of Battery Assisted Quasi-Z-Source Inverter Fed Five Phase PMSM Drive for Marine Propulsion Applications. In Proceedings of the 2020 IEEE International Conference on Power Electronics, Drives and Energy Systems (PEDES), Jaipur, India, 16–19 December 2020; pp. 1–6. [CrossRef]
2. Choo, K.-M.; Won, C.-Y. Design and Analysis of Electrical Braking Torque Limit Trajectory for Regenerative Braking in Electric Vehicles With PMSM Drive Systems. *IEEE Trans. Power Electron.* **2020**, *35*, 13308–13321. [CrossRef]
3. Bharatiraja, C.; Vinoth, J. A New Three Phase Twelve Switch Inverter for Three Wheel PMSM Drive Electric Transportation. In Proceedings of the 2024 IEEE 4th International Conference on Sustainable Energy and Future Electric Transportation (SEFET), Hyderabad, India, 31 July–3 August 2024; pp. 1–6. [CrossRef]
4. Wang, Y.; Meng, J.; Zhang, X.; Xu, L. Control of PMSG-Based Wind Turbines for System Inertial Response and Power Oscillation Damping. *IEEE Trans. Sustain. Energy* **2015**, *6*, 565–574. [CrossRef]
5. Miyama, Y.; Ishizuka, M.; Kometani, H.; Akatsu, K. Vibration Reduction by Applying Carrier Phase-Shift PWM on Dual Three-Phase Winding Permanent Magnet Synchronous Motor. *IEEE Trans. Ind. Appl.* **2018**, *54*, 5998–6004. [CrossRef]
6. Cheema, M.A.M.; Fletcher, J.E.; Xiao, D.; Rahman, M.F. A Linear Quadratic Regulator-Based Optimal Direct Thrust Force Control of Linear Permanent-Magnet Synchronous Motor. *IEEE Trans. Ind. Electron.* **2016**, *63*, 2722–2733. [CrossRef]
7. Jiang, C.; Liu, H.; Wheeler, P.; Wu, F.; Cai, Z.; Huo, J. A Novel Open-Circuit Fault Detection and Location for Open-End Winding PMSM Based on Differential-Mode Components. *IEEE Trans. Ind. Electron.* **2022**, *69*, 7776–7786. [CrossRef]
8. Lee, H.; Lee, J. Design of Iterative Sliding Mode Observer for Sensorless PMSM Control. *IEEE Trans. Control. Syst. Technol.* **2013**, *21*, 1394–1399. [CrossRef]
9. Gao, J.; Gong, C.; Li, W.; Liu, J. Novel Compensation Strategy for Calculation Delay of Finite Control Set Model Predictive Current Control in PMSM. *IEEE Trans. Ind. Electron.* **2020**, *67*, 5816–5819. [CrossRef]
10. Niu, F.; Wang, B.; Babel, A.S.; Li, K.; Strangas, E.G. Comparative Evaluation of Direct Torque Control Strategies for Permanent Magnet Synchronous Machines. *IEEE Trans. Power Electron.* **2016**, *31*, 1408–1424. [CrossRef]
11. Cho, K.-R.; Seok, J.-K. Correction on Current Measurement Errors for Accurate Flux Estimation of AC Drives at Low Stator Frequency. *IEEE Trans. Ind. Appl.* **2008**, *44*, 594–603. [CrossRef]
12. Bourns, Inc. Using Current Sense Resistors for Accurate Current Measurement, Jan. 2017. Available online: https://www.bourns.com/docs/technical-documents/technical-library/current-sense-pulse-power-high-power-resistors/application-notes/bourns_n1702_current_sense_accurate_measurement_appnote.pdf (accessed on 5 March 2025).
13. Crescentini, M.; Syeda, S.F.; Gibiino, G.P. Hall-Effect Current Sensors: Principles of Operation and Implementation Techniques. *IEEE Sensors J.* **2022**, *22*, 10137–10151. [CrossRef]
14. Li, J.; Ren, W.; Luo, Y.; Zhang, X.; Liu, X.; Zhang, X. Design of Fluxgate Current Sensor Based on Magnetization Residence Times and Neural Networks. *Sensors* **2024**, *24*, 3752. [CrossRef] [PubMed]

15. Mühlthaler, J.; Lehner, B.; Reeh, A. Detection of Inter-Turn Short-Circuits in Permanent Magnet Machines Based on Rogowski & Search Coil Based Monitoring. In Proceedings of the 2024 International Conference on Electrical Machines (ICEM), Torino, Italy, 1–4 September 2024; pp. 1–8. [CrossRef]
16. Singh, R.P.; Khambadkone, A.M. Giant Magneto Resistive (GMR) Effect Based Current Sensing Technique for Low Voltage/High Current Voltage Regulator Modules. *IEEE Trans. Power Electron.* **2008**, *23*, 915–925. [CrossRef]
17. Xiao, C.; Zhao, L.; Asada, T.; Odendaal, W.; Van Wyk, J. An overview of integratable current sensor technologies. In Proceedings of the 38th IAS Annual Meeting on Conference Record of the Industry Applications Conference, Salt Lake City, UT, USA, 12–16 October 2003; Volume 2. [CrossRef]
18. Patel, A. Current Measurement in Power Electronic and Motor Drive Applications—A Comprehensive Study. Master's Thesis, University of Missouri, Rolla, MO, USA, 2007. Available online: https://scholarsmine.mst.edu/masters_theses/4581 (accessed on 5 March 2025).
19. Ziegler, S.; Woodward, R.C.; Iu, H.H.-C.; Borle, L.J. Current Sensing Techniques: A Review. *IEEE Sensors J.* **2009**, *9*, 354–376. [CrossRef]
20. Kawahara, S.; Furuta, Y.; Wada, S.; Taniguchi, S.; Nezuka, T. A 30-nΩ Accuracy Low Power Two-Step Ratiometric Shunt Resistance Measurement System Using a Switching Regulator- Based Current Generator for Shunt-Based Current Sensors. *IEEE Sens. J.* **2024**, *24*, 40188–40195. [CrossRef]
21. Chen, Q.; Li, K.; Ahmed, R. A high-bandwidth current shunt resistor for testing GaN devices. *IET Conf. Proc.* **2024**, *2024*, 547–552. [CrossRef]
22. Zhang, W.; Zhang, Z.; Wang, F.; Brush, E.V.; Forcier, N. High-Bandwidth Low-Inductance Current Shunt for Wide-Bandgap Devices Dynamic Characterization. *IEEE Trans. Power Electron.* **2021**, *36*, 4522–4531. [CrossRef]
23. Moon, J.-S.; Kim, J.-Y.; Lee, J.-S.; Kim, S.-J.; Lee, W.-R. The control strategy of In-wheel motor drive system using inverter leg shunt resistors. In Proceedings of the 2021 24th International Conference on Electrical Machines and Systems (ICEMS), Gyeongju, Republic of Korea, 31 October–3 November 2021; pp. 1742–1746.
24. Guo, W.; Xiao, G.; Wang, L.; Gao, K. A Current Monitoring Method for Wire-bonding Power Modules based on Magnetoresistance-Planar Rogowski Coil. In Proceedings of the 2024 IEEE 10th International Power Electronics and Motion Control Conference (IPEMC2024-ECCE Asia), Chengdu, China, 17–20 May 2024; pp. 702–706.
25. Liu, X.; He, Y.; Huang, H.; Xu, Q.; Huang, T. A Method for Increasing the Bandwidth of Rogowski Coils Without Changing Their Size. *IEEE Trans. Instrum. Meas.* **2022**, *71*, 1–10. [CrossRef]
26. Krýsl, P.; Jára, M. Design of Current Measurement Circuit Using PCB Rogowski Coils for Resonant Converters. In Proceedings of the 2024 21st International Conference on Mechatronics-Mechatronika (ME), Brno, Czech Republic, 4–6 December 2024; pp. 1–4.
27. Moser, S.; Incurvati, M.; Schiestl, M.; Stärz, R. Non-invasive wide-bandwidth current sensor for wide-bandgap devices. In Proceedings of the 2023 25th European Conference on Power Electronics and Applications (EPE'23 ECCE Europe), Aalborg, Denmark, 4–8 September 2023; pp. 1–7.
28. Chen, B.; Zhang, Z.; Wu, Y.; Tian, C.; Chen, Y. A Novel Anti-DC Bias Energy Meter Based on Magnetic-Valve-Type Current Transformer. *IEEE Trans. Instrum. Meas.* **2022**, *71*, 1–9. [CrossRef]
29. Hu, Y.; Ottoboni, R.; Li, J. A Novel Method for DC Current Measurement Using Current Transformer. *IEEE Trans. Instrum. Meas.* **2024**, *73*, 1–9. [CrossRef]
30. Simonetti, F.; Mohamadian, S.; Buccella, C.; Cecati, C. DC Current Suppression in CHB-STATCOM With Model Predictive Control Employing Current Transformers. *IEEE Trans. Ind. Electron.* **2024**, *71*, 13135–13145. [CrossRef]
31. Bashir, S.; Paul, S.; Chang, J. Novel Core Airgap Profiles Design Scheme for Winding and Thermal Loss Reduction in High-Frequency Current Transformer Sensors. *IEEE Sensors J.* **2020**, *20*, 892–898. [CrossRef]
32. Fritsch, M.; Wolter, M. Saturation of High-Frequency Current Transformers: Challenges and Solutions. *IEEE Trans. Instrum. Meas.* **2023**, *72*, 1–10. [CrossRef]
33. Wang, R.; Du, G.; Xiao, W.; Zhang, B.; Qiu, D. Wide Range Energy Harvesting Technique for Current Transformer Based on Coil Adaptive Switching. *IEEE Sensors J.* **2024**, *24*, 41875–41884. [CrossRef]
34. Kim, K.; Kim, Y.; Han, J.-H.; Hong, S.-K. Analysis and Compensation of Phase Shift Errors of an Open-Loop Current Transducer Considering Eddy Current. *IEEE Access* **2024**, *12*, 62730–62737. [CrossRef]
35. Zapf, F.; Weiss, R.; Itzke, A.; Gordon, R.; Weigel, R. Mechanically Flexible Sensor Array for Current Measurement. *IEEE Trans. Instrum. Meas.* **2020**, *69*, 8554–8561. [CrossRef]
36. Qiu, G.; Ran, L.; Feng, H.; Jiang, H.; Long, T.; Forsyth, A.J.; Shao, W.; Hou, X. A Fluxgate-Based Current Sensor for DC Bias Elimination in a Dual Active Bridge Converter. *IEEE Trans. Power Electron.* **2022**, *37*, 3233–3246. [CrossRef]
37. Sirat, A.P.; Niakan, H.; Gafford, J.; Parkhideh, B. Design and Development of Hybrid Current Sensors for Wide-Bandgap Power Electronics Applications. In Proceedings of the 2024 IEEE Applied Power Electronics Conference and Exposition (APEC), Long Beach, CA, USA, 25–29 February 2024; pp. 260–266.

38. Wei, Y.; Li, C.; Zhao, W.; Xue, M.; Cao, B.; Chu, X.; Ye, C. Electrical Compensation for Magnetization Distortion of Magnetic Fluxgate Current Sensor. *IEEE Trans. Instrum. Meas.* **2022**, *71*, 1–9. [CrossRef]
39. Tian, X.; Qian, Q.; Fu, W. Fluxgate Current Sensor Based on H-Bridge. In Proceedings of the 2023 6th International Conference on Electronics Technology (ICET), Chengdu, China, 12–15 May 2023; pp. 371–376.
40. Scherzer, M.; Auer, M.; Valavanoglou, A.; Magnes, W. Implementation of a Fully Differential Low Noise Current Source for Fluxgate Sensors. In Proceedings of the 2022 IEEE 13th Latin America Symposium on Circuits and System (LASCAS), Puerto Varas, Chile, 1–4 March 2022; pp. 1–4.
41. Yang, X.; Liu, H.; Wang, Y.; Wang, Y.; Dong, G.; Zhao, Z. A Giant Magneto Resistive (GMR) Effect Based Current Sensor with a Toroidal Magnetic Core as Flux Concentrator and Closed-Loop Configuration. *IEEE Trans. Appl. Supercond.* **2014**, *24*, 1–5. [CrossRef]
42. Chen, Y.; Heredia, L.C.C.; Smit, J.J.; Niasar, M.G.; Ross, R. Giant Magneto-Resistive (GMR) Sensors for Non-Contacting Partial Discharge Detection. *IEEE Trans. Instrum. Meas.* **2023**, *72*, 1–11. [CrossRef]
43. Kim, W.; Luo, S.; Lu, G.-Q.; Ngo, K.D.T. Integrated current sensor using giant magneto resistive (GMR) field detector for planar power module. In Proceedings of the 2013 Twenty-Eighth Annual IEEE Applied Power Electronics Conference and Exposition (APEC), Long Beach, CA, USA, 17–21 March 2013; pp. 2498–2505.
44. Wireko-Brobby, A.; Hu, Y.; Wang, G.; Gong, C.; Lang, W.; Zhang, Z. Analysis of the Sources of Error Within PMSM-Based Electric Powertrains—A Review. *IEEE Trans. Transp. Electrif.* **2024**, *10*, 6370–6406. [CrossRef]
45. Babayomi, O.; Zhang, Z.; Li, Z. Model-Free Predictive Control of DC–DC Boost Converters: Sensor Noise Suppression with Hybrid Extended State Observers. *IEEE Trans. Power Electron.* **2024**, *39*, 245–259. [CrossRef]
46. Yang, D.; Ruan, X.; Wu, H. A Real-Time Computation Method With Dual Sampling Mode to Improve the Current Control Performance of the *LCL*-Type Grid-Connected Inverter. *IEEE Trans. Ind. Electron.* **2015**, *62*, 4563–4572. [CrossRef]
47. Ding, X.; Zhang, Y.; Ye, Z. Current Sensors Offset Fault Online Estimation in Permanent Magnet Synchronous Generator (PMSG) Drives for Offshore Wind Turbines. *IEEE Access* **2021**, *9*, 135996–136003. [CrossRef]
48. Yoo, M.-S.; Park, S.W.; Choi, Y.-Y.; Han, S.-H.; Yoon, Y.-D. Current-Scaling Gain Compensation of Motor Drives Under Locked-Rotor Condition Considering Inequality of Phase Resistances. *IEEE Trans. Ind. Appl.* **2020**, *56*, 4915–4923. [CrossRef]
49. Hu, M.; Hua, W.; Xiao, H.; Wang, Z.; Liu, K.; Cai, K.; Wang, Y. Fast Current Control Without Computational Delay by Minimizing Update Latency. *IEEE Trans. Power Electron.* **2021**, *36*, 12207–12212. [CrossRef]
50. Chung, D.-W.; Sul, S.-K. Analysis and compensation of current measurement error in vector-controlled AC motor drives. *IEEE Trans. Ind. Appl.* **1998**, *34*, 340–345. [CrossRef]
51. Song, Y.; Lu, J.; Hu, Y.; Zhang, W.; Su, Y.; Wu, X.; Liu, J. Expanding Limit of Minimum Sampling Time Using Auxiliary Vectors for PMSM Drives With Single DC-Link Current Sensor. *IEEE Trans. Ind. Electron.* **2023**, *70*, 3437–3448. [CrossRef]
52. Lu, J.; Zhang, X.; Hu, Y.; Liu, J.; Gan, C.; Wang, Z. Independent Phase Current Reconstruction Strategy for IPMSM Sensorless Control Without Using Null Switching States. *IEEE Trans. Ind. Electron.* **2018**, *65*, 4492–4502. [CrossRef]
53. Retianza, D.V.; Arrozy, J.; Van Duivenbode, J.; Huisman, H.; Vermulst, B. Sequential Error Disentanglement of Three-Phase Current Sensor for AC Machine in Standstill Conditions. *IEEE Access* **2022**, *10*, 113902–113914. [CrossRef]
54. Lu, J.; Hu, Y.; Wang, J.; Song, Y.; Su, Y.; Liu, J. Synergistic Correction of Current Sampling Errors in Dual-Motor Powered Electric Powertrain for High Power Electric Vehicles. *IEEE Trans. Ind. Electron.* **2022**, *69*, 225–235. [CrossRef]
55. Xu, Y.; Yan, H.; Zou, J.; Wang, B.; Li, Y. Zero Voltage Vector Sampling Method for PMSM Three-Phase Current Reconstruction Using Single Current Sensor. *IEEE Trans. Power Electron.* **2016**, *32*, 3797–3807. [CrossRef]
56. Lu, J.; Hu, Y.; Liu, J. Analysis and Compensation of Sampling Errors in TPFS IPMSM Drives With Single Current Sensor. *IEEE Trans. Ind. Electron.* **2019**, *66*, 3852–3855. [CrossRef]
57. Lu, J.; Hu, Y.; Liu, J.; Wen, H. Self-Calibration of Phase Current Sensors With Sampling Errors by Multipoint Sampling of Current Values in a Single PWM Cycle. *IEEE Trans. Ind. Electron.* **2021**, *68*, 2942–2951. [CrossRef]
58. Lu, J.; Hu, Y.; Liu, J.; Wang, J.; Li, P. Fixed-Point Sampling Strategy for Estimation on Current Measurement Errors in IPMSM Drives. *IEEE Trans. Power Electron.* **2021**, *36*, 5748–5759. [CrossRef]
59. Gao, F.; Yin, Z.; Li, L.; Li, T.; Liu, J. Gaussian Noise Suppression in Deadbeat Predictive Current Control of Permanent Magnet Synchronous Motors Based on Augmented Fading Kalman Filter. *IEEE Trans. Energy Convers.* **2023**, *38*, 1410–1420. [CrossRef]
60. Yan, L.; Zhu, Z.Q.; Shao, B. Current Measurement Gain Compensation Using High-Frequency Signal Injection in Dual Three-Phase PMSM Systems. *IEEE Trans. Ind. Electron.* **2024**, *71*, 3472–3482. [CrossRef]
61. Song, Z.; Ma, X.; Yu, Y. Design of Zero-Sequence Current Controller for Open-End Winding PMSMs Considering Current Measurement Errors. *IEEE Trans. Power Electron.* **2020**, *35*, 6127–6139. [CrossRef]
62. Sun, P.; Wu, X.; Ni, R.; Nie, S. A Study of RLS Adaptive Filter Algorithm in Current Loop Noise Suppression of PMSM Vector Control System. In Proceedings of the 2019 22nd International Conference on Electrical Machines and Systems (ICEMS), Harbin, China, 11–14 August 2019; pp. 1–4.

63. Zhu, H.; Fujimoto, H. Suppression of Current Quantization Effects for Precise Current Control of SPMSM Using Dithering Techniques and Kalman Filter. *IEEE Trans. Ind. Inform.* **2014**, *10*, 1361–1371. [CrossRef]
64. Liu, D.; Wang, Q.; Li, Y.; Gao, J.; Dong, G.; Zhi, W.; Sun, Y.; Wang, Y. An Improved Model Predictive Control method with Luenberger Observer for Measurement Noise Suppression and Reduction of Current Ripples. In Proceedings of the 2023 3rd International Conference on Intelligent Power and Systems (ICIPS), Shenzhen, China, 20–22 October 2023; pp. 326–330.
65. Babayomi, O.; Li, Z.; Wang, Y.; Li, J.; Zhang, Z. Model-Free Predictive Torque Control of PMSM Drives with Measurement Noise Suppression. In Proceedings of the 2023 IEEE 2nd International Power Electronics and Application Symposium (PEAS), Guangzhou, China, 10–13 November 2023; pp. 912–917.
66. Dutta, R.; Badajena, H.; Chakraborty, B.; Routray, A.; Jenamani, M. Enhanced Data Acquisition System of Current Signature for Diagnosis of Variable Speed Induction Motor Drives with Full ADC Range Utilization and Noise Cancellation. In Proceedings of the 2023 IEEE 2nd Industrial Electronics Society Annual OnLine Conference (ONCON), Virtual, 8–10 December 2023; pp. 1–6.
67. Weber, B.; Wiedmann, K.; Mertens, A. Increased signal-to-noise ratio of sensorless control using current oversampling. In Proceedings of the 2015 9th International Conference on Power Electronics and ECCE Asia (ICPE 2015-ECCE Asia), Seoul, Republic of Korea, 1–5 June 2015; pp. 1129–1134.
68. Landsmann, P.; Paulus, D.; Dotlinger, A.; Kennel, R. Silent injection for saliency based sensorless control by means of current oversampling. In Proceedings of the 2013 IEEE International Conference on Industrial Technology (ICIT 2013), Cape Town, South Africa, 25–28 February 2013; pp. 398–403.
69. Zhang, H.; Lei, Y.; Zhang, Y. Harmonics Influence Analysis on Current-Oversampling-based Position Estimation under Multi-mode PWM for IPMSM Sensorless Drives in Rail Transit Applications. In Proceedings of the 2024 IEEE 10th International Power Electronics and Motion Control Conference (IPEMC2024-ECCE Asia), Chengdu, China, 17–20 May 2024; pp. 4227–4232.
70. Texas Instruments, Inc. *Application Note: ADC Oversampling, Susmitha Bumadi, August, 2024*. Available online: https://www.ti.com/lit/an/sprad55a/sprad55a.pdf (accessed on 5 March 2025).
71. Analog Devices, Inc. Datasheet: 8-/6-/4-Channel DAS with 16-Bit, Bipolar Input, Simultaneous Sampling ADC AD7606/AD7606-6/AD7606-4. Available online: https://www.analog.com/media/en/technical-documentation/data-sheets/AD7606_7606-6_7606-4.pdf (accessed on 5 March 2025).
72. Li, D.; Liu, Z.; Gui, X. Noise Analysis and Suppression for High Accuracy Current Sampling in Motor Control. In Proceedings of the 2024 IEEE 7th International Electrical and Energy Conference (CIEEC), Harbin, China, 10–12 May 2024; pp. 1123–1128.
73. Analog Devices, Inc. Technical Articles: Understanding SAR ADCs: Their Architecture and Comparison with Other ADCs, Oct, 2001. Available online: https://www.analog.com/en/resources/technical-articles/successive-approximation-registers-sar-and-flash-adcs.html (accessed on 5 March 2025).
74. Analog Devices, Inc. Technical Articles: Fundamental Principles Behind the Sigma-Delta ADC Topology: Part 1, Michael Clifford, Jan, 2016. Available online: https://www.analog.com/en/resources/technical-articles/behind-the-sigma-delta-adc-topology.html (accessed on 5 March 2025).
75. Wang, L.; Shi, Y.; Li, H. Anti-EMI Noise Digital Filter Design for a 60-kW Five-Level SiC Inverter Without Fiber Isolation. *IEEE Trans. Power Electron.* **2018**, *33*, 13–17. [CrossRef]
76. Oswald, N.; Anthony, P.; McNeill, N.; Stark, B.H. An Experimental Investigation of the Tradeoff between Switching Losses and EMI Generation With Hard-Switched All-Si, Si-SiC, and All-SiC Device Combinations. *IEEE Trans. Power Electron.* **2014**, *29*, 2393–2407. [CrossRef]
77. Zhang, X.; Li, H.; Brothers, J.A.; Fu, L.; Perales, M.; Wu, J.; Wang, J. A Gate Drive With Power Over Fiber-Based Isolated Power Supply and Comprehensive Protection Functions for 15-kV SiC MOSFET. *IEEE J. Emerg. Sel. Top. Power Electron.* **2016**, *4*, 946–955. [CrossRef]
78. Boroyevich, D. Use of SiC devices in medium-voltage converters. Proc. CPES Workshop WBG Power Electron. Adv. Distrib. Grids, Apr. 2016. Available online: https://www.nist.gov/system/files/documents/pml/high_megawatt/Boroyevich_db-Workshop-on-MV-WBG-PE-for-ADG-2016.pdf (accessed on 5 March 2025).
79. Cvetanovic, R.; Petric, I.; Mattavelli, P.; Buso, S. Median Filters for Switching Noise Mitigation in Oversampled Power Electronics Control Systems. In Proceedings of the 2023 IEEE Energy Conversion Congress and Exposition (ECCE), Nashville, TN, USA, 29 October–2 November 2023; pp. 2725–2731.
80. Cvetanovic, R.; Petric, I.Z.; Mattavelli, P.; Buso, S. Switching Noise Propagation and Suppression in Multisampled Power Electronics Control Systems. *IEEE Trans. Power Electron.* **2024**, *39*, 149–163. [CrossRef]
81. Corradini, L.; Mattavelli, P.; Tedeschi, E.; Trevisan, D. High-Bandwidth Multisampled Digitally Controlled DC–DC Converters Using Ripple Compensation. *IEEE Trans. Ind. Electron.* **2008**, *55*, 1501–1508. [CrossRef]
82. Kim, H.; Degner, M.W.; Guerrero, J.M.; Briz, F.; Lorenz, R.D. Discrete-Time Current Regulator Design for AC Machine Drives. *IEEE Trans. Ind. Appl.* **2010**, *46*, 1425–1435. [CrossRef]
83. Naouar, M.-W.; Monmasson, E.; Naassani, A.A.; Slama-Belkhodja, I.; Patin, N. FPGA-Based Current Controllers for AC Machine Drives—A Review. *IEEE Trans. Ind. Electron.* **2007**, *54*, 1907–1925. [CrossRef]

84. Ke, X.; Buchman, B. Current Control for Passivity of a Power Converter. International Patent Application WO2020048579A1, 12 March 2020.
85. Yoo, M.-S.; Park, S.-W.; Lee, H.-J.; Yoon, Y.-D. Offline Compensation Method for Current Scaling Gains in AC Motor Drive Systems With Three-Phase Current Sensors. *IEEE Trans. Ind. Electron.* **2021**, *68*, 4760–4768. [CrossRef]
86. Zhang, Q.; Guo, H.; Liu, Y.; Guo, C.; Zhang, F.; Zhang, Z.; Li, G. A Novel Error-Injected Solution for Compensation of Current Measurement Errors in PMSM Drive. *IEEE Trans. Ind. Electron.* **2023**, *70*, 4608–4619. [CrossRef]
87. Harke, M.C.; Guerrero, J.M.; Degner, M.W.; Briz, F.; Lorenz, R.D. Current Measurement Gain Tuning Using High-Frequency Signal Injection. *IEEE Trans. Ind. Appl.* **2008**, *44*, 1578–1586. [CrossRef]
88. Lu, J.; Hu, Y.; Chen, G.; Wang, Z.; Liu, J. Mutual Calibration of Multiple Current Sensors With Accuracy Uncertainties in IPMSM Drives for Electric Vehicles. *IEEE Trans. Ind. Electron.* **2019**, *67*, 69–79. [CrossRef]
89. Zhang, Q.; Guo, H.; Guo, C.; Liu, Y.; Wang, D.; Lu, K.; Zhang, Z.; Zhuang, X.; Chen, D. An adaptive proportional-integral-resonant controller for speed ripple suppression of PMSM drive due to current measurement error. *Int. J. Electr. Power Energy Syst.* **2021**, *129*, 106866. [CrossRef]
90. Zhang, Q.; Guo, H.; Liu, Y.; Guo, C.; Lu, K.; Wang, D.; Zhang, Z.; Sun, J. Robust plug-in repetitive control for speed smoothness of cascaded-PI PMSM drive. *Mech. Syst. Signal Process.* **2022**, *163*, 108090. [CrossRef]
91. Xia, C.; Ji, B.; Yan, Y. Smooth Speed Control for Low-Speed High-Torque Permanent-Magnet Synchronous Motor Using Proportional–Integral–Resonant Controller. *IEEE Trans. Ind. Electron.* **2015**, *62*, 2123–2134. [CrossRef]
92. Qian, W.; Panda, S.; Xu, J. Speed Ripple Minimization in PM Synchronous Motor Using Iterative Learning Control. *IEEE Trans. Energy Convers.* **2005**, *20*, 53–61. [CrossRef]
93. Qian, W.; Panda, S.; Xu, J. Periodic speed ripples minimization in PM synchronous motors using repetitive learning variable structure control. *ISA Trans.* **2003**, *42*, 605–613. [CrossRef] [PubMed]
94. Zhang, K.; Fan, M.; Yang, Y.; Zhu, Z.; Garcia, C.; Rodriguez, J. An Improved Adaptive Selected Harmonic Elimination Algorithm for Current Measurement Error Correction of PMSMs. *IEEE Trans. Power Electron.* **2021**, *36*, 13128–13138. [CrossRef]
95. Choi, C.-H.; Cho, K.-R.; Seok, J.-K. Inverter Nonlinearity Compensation in the Presence of Current Measurement Errors and Switching Device Parameter Uncertainties. *IEEE Trans. Power Electron.* **2007**, *22*, 576–583. [CrossRef]
96. Li, X.; Cheng, S.; Wang, D.; Ji, Z.; Hu, Y.; Lv, Y. A Nonintrusive Current Sensor Gain Tuning Method for Interior Permanent Magnet Synchronous Motor Drives Using Controlled Short-Circuit Tests. *IEEE Trans. Transp. Electrif.* **2022**, *8*, 1278–1288. [CrossRef]
97. Harke, M.C.; Lorenz, R.D. The Spatial Effect and Compensation ofCurrent Sensor Differential Gains for Three-Phase Three-Wire Systems. *IEEE Trans. Ind. Appl.* **2008**, *44*, 1181–1189. [CrossRef]
98. Lee, K.-W.; Kim, S.-I. Dynamic Performance Improvement of a Current Offset Error Compensator in Current Vector-Controlled SPMSM Drives. *IEEE Trans. Ind. Electron.* **2019**, *66*, 6727–6736. [CrossRef]
99. Kim, S.-I.; Kim, J.-Y.; Lee, K.-W. Current Measurement Offset Error Compensation Scheme Considering Saturation of Current Controller in SPMSM Drives. *IEEE Access* **2023**, *11*, 17233–17240. [CrossRef]
100. Kim, M.; Sul, S.-K.; Lee, J. Compensation of Current Measurement Error for Current-Controlled PMSM Drives. *IEEE Trans. Ind. Appl.* **2014**, *50*, 3365–3373. [CrossRef]
101. Bai, Y.; Li, B.; Wang, Q.; Ding, D.; Zhang, G.; Wang, G.; Xu, D. An Adaptive-Frequency Harmonic Suppression Strategy Based on Vector Reconstruction for Current Measurement Error of PMSM Drives. *IEEE Trans. Power Electron.* **2023**, *38*, 34–40. [CrossRef]
102. Jung, H.-S.; Hwang, S.-H.; Kim, J.-M.; Kim, C.-U.; Choi, C. Diminution of Current-Measurement Error for Vector-Controlled AC Motor Drives. *IEEE Trans. Ind. Appl.* **2006**, *42*, 1249–1256. [CrossRef]
103. Park, G.-G.; Hwang, S.-H.; Kim, J.-M.; Lee, K.-B.; Lee, D.-C. Reduction of Current Ripples due to Current Measurement Errors in a Doubly Fed Induction Generator. *J. Power Electron.* **2010**, *10*, 313–319. [CrossRef]
104. Lee, S.; Kim, H.; Lee, K. Current Measurement Offset Error Compensation in Vector-Controlled SPMSM Drive Systems. *IEEE J. Emerg. Sel. Top. Power Electron.* **2022**, *10*, 2619–2628. [CrossRef]
105. Hu, M.; Hua, W.; Wu, Z.; Dai, N.; Xiao, H.; Wang, W. Compensation of Current Measurement Offset Error for Permanent Magnet Synchronous Machines. *IEEE Trans. Power Electron.* **2020**, *35*, 11119–11128. [CrossRef]
106. Zuo, Y.; Wang, H.; Ge, X.; Zuo, Y.; Woldegiorgis, A.T.; Feng, X.; Lee, C.H.T. A Novel Current Measurement Offset Error Compensation Method Based on the Adaptive Extended State Observer for IPMSM Drives. *IEEE Trans. Ind. Electron.* **2024**, *71*, 3371–3382. [CrossRef]
107. Cho, K.-R.; Seok, J.-K. Pure-Integration-Based Flux Acquisition with Drift and Residual Error Compensation at a Low Stator Frequency. *IEEE Trans. Ind. Appl.* **2009**, *45*, 1276–1285. [CrossRef]
108. Song, P.; Ma, E.; Wang, Y.; Wang, T.; Li, H.; Wu, L. Compensation of Current Measurement Error in PMSM Drives Based on Virtual Windings. In Proceedings of the 2024 International Conference on Electrical Machines (ICEM), Torino, Italy, 1–4 September 2024; pp. 1–7.
109. Song, P.; Wang, Y.; Ma, E.; Wu, L.; Wang, T.; Li, H. Current Measurement Error Compensation Based on Modified Multiple Complex-Coefficient Filters in PMSM Drives. *IEEE Trans. Power Electron.* **2025**, *40*, 4067–4079. [CrossRef]

110. Lu, J.; Hu, Y.; Liu, J.; Wang, Z. All Current Sensor Survivable IPMSM Drive With Reconfigurable Inverter. *IEEE Trans. Ind. Electron.* **2020**, *67*, 6331–6341. [CrossRef]
111. Song, Y.; Lu, J.; Hu, Y.; Wu, X.; Wang, G. Fast Calibration With Raw Data Verification for Current Measurement of Dual-PMSM Drives. *IEEE Trans. Ind. Electron.* **2024**, *71*, 6875–6885. [CrossRef]
112. Lee, Y.-R.; Yoo, J.; Hwang, I.; Sul, S.-K. Analysis of Position Estimation Error in Signal-Injection Sensorless Control Induced by Inverter dv/dt-Based Current Measurement Noise. *IEEE Trans. Power Electron.* **2023**, *38*, 839–851. [CrossRef]
113. Trinh, Q.-N.; Choo, F.H.; Wang, P. Control Strategy to Eliminate Impact of Voltage Measurement Errors on Grid Current Performance of Three-Phase Grid-Connected Inverters. *IEEE Trans. Ind. Electron.* **2017**, *64*, 7508–7519. [CrossRef]
114. Trinh, Q.N.; Wang, P.; Tang, Y.; Koh, L.H.; Choo, F.H. Compensation of DC Offset and Scaling Errors in Voltage and Current Measurements of Three-Phase AC/DC Converters. *IEEE Trans. Power Electron.* **2018**, *33*, 5401–5414. [CrossRef]
115. Trinh, Q.N.; Choo, F.H.; Tang, Y.; Wang, P. Control Strategy to Compensate for Current and Voltage Measurement Errors in Three-Phase PWM Rectifiers. *IEEE Trans. Ind. Appl.* **2019**, *55*, 2879–2889. [CrossRef]
116. Pramod, P. Synchronous Frame Current Estimation Inaccuracies in Permanent Magnet Synchronous Motor Drives. In Proceedings of the 2020 IEEE Energy Conversion Congress and Exposition (ECCE), Detroit, MI, USA, 11–15 October 2020; pp. 2379–2386.
117. Han, J.; Kim, B.-H.; Sul, S.-K. Effect of current measurement error in angle estimation of permanent magnet AC motor sensorless control. In Proceedings of the 2017 IEEE 3rd International Future Energy Electronics Conference and ECCE Asia (IFEEC 2017-ECCE Asia), Kaohsiung, Taiwan, 3–7 June 2017; pp. 2171–2176.
118. Kang, Y.G.; Reigosa, D.D. Dq-Transformed Error and Current Sensing Error Effects on Self-Sensing Control. *IEEE J. Emerg. Sel. Top. Power Electron.* **2022**, *10*, 1935–1945. [CrossRef]
119. Feng, X.; Wang, B.; Wang, Z.; Hua, W.; Cheng, M. Diagnosis and Identification of Common Electrical Faults in PM Machine Drives. *IEEE Trans. Power Electron.* **2024**, *39*, 13686–13695. [CrossRef]
120. Feng, X.; Wang, B.; Wang, Z.; Hua, W. Investigation and Diagnosis of Current Sensor Fault in Permanent Magnet Machine Drives. *IEEE Trans. Ind. Electron.* **2024**, *72*, 1261–1270. [CrossRef]
121. Jeong, Y.-S.; Sul, S.-K.; Schulz, S.; Patel, N. Fault detection and fault-tolerant control of interior permanent-magnet motor drive system for electric vehicle. *IEEE Trans. Ind. Appl.* **2005**, *41*, 46–51. [CrossRef]
122. Wu, S.; Ma, G.; Yao, C.; Sun, Z.; Xu, S. Current Sensor Fault Detection and Identification for PMSM Drives Using Multichannel Global Maximum Pooling CNN. *IEEE Trans. Power Electron.* **2024**, *39*, 10311–10325. [CrossRef]

energies

MDPI

Article

Improved Discrete-Time Active Disturbance Rejection Control for Enhancing Dynamics of Current Loop in LC-Filtered SPMSM Drive System

Zibo Li, Haitao Yang, Jin Wang *, Yali Wang and Libing Zhou

State Key Laboratory of Advanced Electromagnetic Technology, School of Electrical and Electronic Engineering, Huazhong University of Science and Technology, Wuhan 430074, China; lizibo@hust.edu.cn (Z.L.); haitaoyang@hust.edu.cn (H.Y.); yaliwang@hust.edu.cn (Y.W.); zlb@mail.hust.edu.cn (L.Z.)

* Correspondence: jinwang@hust.edu.cn

Abstract: Active disturbance rejection control is implemented in a LC-filtered surface-mounted permanent magnet synchronous motor (SPMSM) drive system to enhance current control dynamics. However, the combined effects of computation one-beat delay and the pulse-width modulation zero-order hold (ZOH) effect significantly degrade system stability and dynamic performance. To address these limitations, an improved predictive extended state observer (ESO) with an accurate ZOH discretization method is proposed to ensure fast and robust dynamic performance. The predictive ESO predicts one beat to compensate for the delay effect, while the ZOH discretization yields a more precise discrete dynamic model of the system. These combined improvements substantially enhance the system's phase and gain margins, leading to superior dynamic performance. Furthermore, a discrete-domain transfer function of the control system is analytically derived, with the control parameters systematically designed using frequency-domain analysis to guarantee robust performance. Experimental validation on a LC-filtered SPMSM drive system demonstrates remarkable enhancement in current control dynamics while maintaining sufficient robustness.

Keywords: LC filter; active disturbance rejection control (ADRC); computation delay; zero-order hold (ZOH); predictive extended state observer (ESO)

1. Introduction

In pursuit of reliable operation and high torque density, a surface-mounted permanent magnet synchronous motor (SPMSM) is frequently applied to a system with voltage-source inverter (VSI) connected [1]. For applications such as long-cable fed drive [2], wide-bandgap semiconductor-based VSI [3] and high-speed motor systems [4], an LC filter is needed to suppress the negative effects of high dv/dt in PWM waveform on the stator insulation of SPMSM. The integration of an LC filter in the system may introduce resonance issues, significantly degrading the dynamic performance of motor current control, thereby limiting its applicability in high-performance applications [5]. Therefore, developing a control strategy with enhanced dynamic response becomes crucial.

Extensive research efforts have been devoted to developing advanced control strategies for LC-filtered SPMSM current regulation, as evidenced in [6–17]. These approaches encompass model predictive control (MPC) [6–8], active damping techniques (AD) [9–11], and triple-cascaded proportional-integral control (TCPI) [12,13]. These methodologies demonstrate substantial improvements in resonance attenuation and dynamic response compared to conventional PI controllers [14]. However, they are intrinsically limited by the fundamental constraints of digital control systems. Specifically, the inevitable one-beat computational delay and zero-order hold (ZOH) effects pose significant challenges. These inherent limitations present significant challenges in achieving further enhancements in transient response performance.

Digital control, with its high programmability and widespread industrial use, is limited by one-beat delay and ZOH effects. In traditional continuous-domain controller design, the one-beat delay and ZOH effect can be approximated as a delay of one and a half sampling periods [15,16]. In motor control, this delay is represented as a lagging angle in the rotating dq-axis, providing an equivalent spatial interpretation of digital computation delay [17]. Such delays can generate output voltages with phase retardation in both d-axis and q-axis, resulting in reduced stability margins and degraded dynamic performance [18].

To deal with the one-beat delay and ZOH effect in SPMSM control without LC filter, there are strategies, including the phase-lead corrector [19,20], deadbeat prediction [21,22], and discrete-domain pole-zero assignment [23,24]. The phase-lead correction method proposed in [19] is popularly used in motor control applications. In this method, the time-delay effect is compensated by introducing a phase-lead correction prior to the inverse Park transformation. This correction is achieved by calculating the hysteretic rotation angle through the product of the measured angular velocity and one and a half sampling periods [20]. While effective, this phase-lead correction method exhibits limited dynamic performance [25] and may lead to system instability under high-speed operating conditions [26]. Deadbeat prediction has emerged as a prevalent delay compensation technique in digital control systems. This method employs system model-based one-beat prediction to counteract the computational delay inherent in digital processors, thereby significantly enhancing transient response characteristics [21,22]. Nevertheless, its performance is critically dependent on precise parameter matching. Practical implementations reveal that parameter deviations can induce oscillatory instability and steady-state tracking errors, severely degrading system robustness. In comparison, discrete-domain pole-zero assignment demonstrates superior parameter robustness through explicit specification of closed-loop pole locations [23]. By incorporating the time delay as an inherent component of the discrete system model, this approach achieves effective delay compensation while maintaining stability margins [24]. However, the absence of predictive compensation necessitates conservative phase margin allocation. Therefore, the advantages of deadbeat prediction and discrete-domain pole-zero assignment can be combined to generate one-beat prediction while ensuring system robustness through pole placement.

Active disturbance rejection control (ADRC), a derivative of the classic PID controller, retains the strong practical applicability and ease of parameter tuning characteristic of PID controllers. By employing a tracking differentiator (TD), extended state observer (ESO), and state error feedback (SEF), ADRC significantly improves both the dynamic response and robustness of the system [27,28]. An observer-based control method has been extensively applied in SPMSM drive systems, including address parameter uncertainties [29], current control [30], speed regulation [31], and sensorless control strategies [32]. In the field of LC-filtered SPMSM control, current control systems face heightened susceptibility to multiple disturbance sources. These include pronounced resonance phenomena induced by LC filter dynamics and cross-coupling effects between dq-axes, which can collectively drive the system towards instability if not properly compensated. Owing to its superior disturbance rejection capabilities, ADRC has witnessed expanding industrial adoption, particularly in applications requiring robust resonance attenuation and rapid dynamic compensation [33–36]. The application of ADRC has been further extended to LCL-type grid-connected inverters, where it effectively mitigates resonance phenomena and reduces current harmonics. Building on these developments, a third-order ADRC architecture was proposed in [34], which introduces the estimated first derivative of current as an additional damping term for resonance suppression. Alternative approaches include the first-order system approximation method in [35], which leverages capacitor current feedback to provide inherent damping characteristics. The resonant ESO technique in [36] estimates resonance information for adaptive controller design. However, despite these significant contributions, existing methodologies have not adequately addressed the critical challenges posed by computational delays and discretization effects in digital control systems.

Various improved discretization methods have been proposed for its practical implementation. In the transition from continuous to discrete domain, a current-observer-based observer structure was developed in [37], while [38] employed the Euler method to derive the discrete observer matrix. Predictive schemes have been effectively incorporated into ADRC to mitigate delay effects [39,40]. The study in [39] cascades a Smith predictor with the ADRC observer, demonstrating superior uncertainty tolerance and delay compensation compared to conventional prediction methods. Furthermore, Ref. [40] introduces a predictive ADRC design utilizing fixed-time prediction instead of the Smith predictor, achieving enhanced robustness and faster dynamic response. However, these ADRC discretization approaches fail to specifically address the one-beat computation delay, and the ZOH effect remains unmitigated. Consequently, the dynamic performance of ADRC-based current control in LC-filtered SPMSM systems cannot be guaranteed.

Although Euler discretization and current observers offer structural simplicity and implementation ease, their direct applications in LC-filtered SPMSM current control demonstrate limited performance. This limitation stems from two fundamental issues: Euler discretization inadequately represents the discrete dynamics of PWM voltage inputs, while current observers fail to compensate for the inherent one-step computation delay in digital control. These deficiencies amplify resonance issues in LC-filtered SPMSM systems, significantly degrading both phase and gain margins. Consequently, the dynamic performance of ADRC-based current control is substantially compromised, restricting the application of LC-filtered SPMSM in high-performance drive systems requiring fast dynamic response.

To address these limitations, this paper proposes an enhanced ADRC discretization method. The proposed method improves the current control loop by incorporating ZOH discretization instead of Euler discretization and introducing an improved predictive observer to replace the conventional current observer, significantly enhancing both dynamic performance and system stability. The main contributions of this work are threefold:

(1) A comprehensive comparison demonstrating the superior performance of ZOH discretization in accurately modeling discrete system dynamics and the enhanced delay compensation capability of the proposed predictive observer over conventional approaches, establishing the necessity of the proposed method for high-performance applications.
(2) Development of an accurate discrete-domain transfer function for the control system, accompanied by a systematic discrete-domain parameter design methodology based on frequency-domain analysis, ensuring robust stability and consistent dynamic performance under parameter variations.
(3) Experimental validation showing a substantial improvement in current response speed, achieving over 72% reduction in settling time compared to existing methods [7,9,13].

The rest of this paper is organized as follows. Section 2 presents the system model and outlines the limitations of traditional ADRC discretization methods. Section 3 details the ZOH discretization and the improved predictive observer. In Section 4, the impulse transfer function of the discrete ADRC is derived, along with the corresponding parameter design method. Section 5 provides experimental results to validate the effectiveness of the proposed method. Finally, Section 6 concludes the paper.

2. Modeling of LC-Filtered SPMSM Drive System and Traditional ADRC Design

2.1. Mathematical Modeling of System

The block diagram of the LC-filtered SPMSM drive system in the *dq* synchronous reference frame is shown in Figure 1, which includes the VSI, LC filter, and SPMSM. The output voltage and output current of VSI are defined as $\boldsymbol{u}_i = [u_{id}, u_{iq}]$ and $\boldsymbol{i}_i = [i_{id}, i_{iq}]$, respectively. The output voltage and output current of the LC filter are defined as $\boldsymbol{u}_f = [u_{fd}, u_{fq}]$ and $\boldsymbol{i}_f = [i_{fd}, i_{fq}]$, respectively. The LC filter is composed of the filter inductance L_f, the filter resistance R_f, and the filter capacitor C_f. The SPMSM parameters include stator inductance L_s, stator resistance R_s, electrical angular velocity ω_e, pole pairs n_p and permanent magnet flux linkage ψ_m.

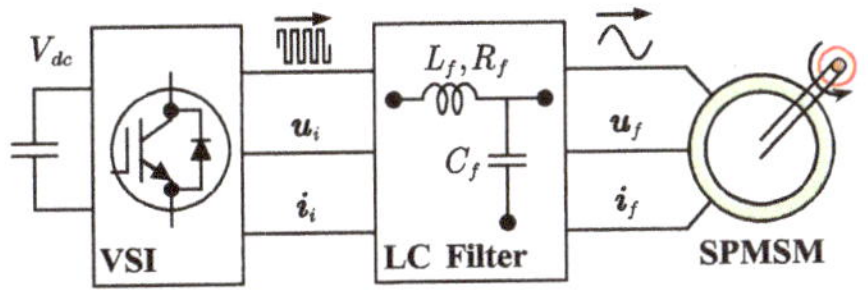

Figure 1. LC-filtered SPMSM drive system.

Modeling the LC-filtered SPMSM drive system in the *dq* synchronous reference frame gives [13]

$$\begin{cases} u_{id} = L_f(di_{id}/dt) + R_f i_{id} - \omega_e L_f i_{iq} + u_{fd} \\ u_{iq} = L_f(di_{iq}/dt) + R_f i_{iq} + \omega_e L_f i_{id} + u_{fq} \end{cases} \tag{1}$$

$$\begin{cases} i_{id} = C_f\left(du_{fd}/dt\right) - \omega_e C_f u_{fq} + i_{fd} \\ i_{iq} = C_f\left(du_{fq}/dt\right) + \omega_e C_f u_{fd} + i_{fq} \end{cases} \tag{2}$$

$$\begin{cases} u_{fd} = L_s\left(di_{fd}/dt\right) + R_s i_{fd} - \omega_e L_s i_{fq} \\ u_{fq} = L_s\left(di_{fq}/dt\right) + R_s i_{fq} + \omega_e L_s i_{fd} + \omega_e \psi_m \end{cases} \tag{3}$$

2.2. Design of the Third-Order ADRC in Continues Domain

The control block diagram of the LC-filtered SPMSM drive system is shown in Figure 2. The speed loop is based on the classical PI control, and the current loop is based on the third-order ADRC.

The design of the third-order ADRC in continuous domain includes extended state observer (ESO), tracking differentiator (TD) and state error feedback (SEF). ESO plays a very important role, relying on system model information for state variable estimation and disturbance rejection. TD is used to generate the reference signal traces. SEF is used to generate control signals with desired controller bandwidth.

Taking *q* axis as an example, substituting the differential of (1) and (2) into (3) yields the third-order expression of the *q*-axis LC filter output current as

$$\dddot{y}(t) = -a_2\ddot{y}(t) - a_1\dot{y}(t) - a_0 y(t) + b_0 u_p(t) + f_l(t) \tag{4}$$

where

$$a_0 = (R_s + R_f)/(C_f L_f L_s); a_1 = (L_f + L_s)/(C_f L_f L_s)$$
$$a_2 = (C_f L_f R_s + C_f L_s R_f)/(C_f L_f L_s); b_0 = 1/(C_f L_f L_s).$$

$y = i_{fq}$ is the q-axis current, $u_p = u_{iq}$ is the q-axis input voltage, and the number of dots above a letter represents the order of its time derivative. $f_l(t)$ represents the lumped disturbance of the system, which includes the back electromotive force, the dq-axis coupling, and the unmodeled dynamics.

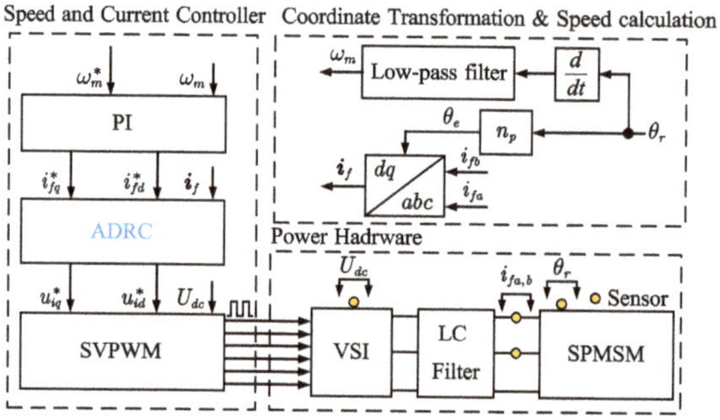

Figure 2. Control Block diagram of LC-filtered SPMSM drive system. The superscript * means the reference value.

An extended state is added to describe the dynamic of lumped disturbance in the system (4), thus the state-space expression of the extended system is noted as

$$\begin{cases} \dot{x}(t) = A_p x(t) + B_p u_p(t) + E h(t) \\ y(t) = C x(t) \end{cases} \tag{5}$$

where

$$x(t) = \begin{bmatrix} x_1(t) \\ x_2(t) \\ x_3(t) \\ x_4(t) \end{bmatrix}; A_p = \begin{bmatrix} 0 & 1 & 0 & 0 \\ 0 & 0 & 1 & 0 \\ 0 & 0 & 0 & 1 \\ 0 & -a_0 & -a_1 & -a_2 \end{bmatrix}; B_p = \begin{bmatrix} 0 \\ 0 \\ b_0 \\ -a_2 b_0 \end{bmatrix}; E = \begin{bmatrix} 0 \\ 0 \\ 0 \\ 1 \end{bmatrix}; C = \begin{bmatrix} 1 \\ 0 \\ 0 \\ 0 \end{bmatrix}^T.$$

$y(t) = i_{fq}(t)$ is the q-axis output current of LC filter, $u_p(t)$ is the system plant input, and $h(t) = \dot{f}_l(t)$ is the derivative of the lumped disturbance.

The design of ESO is well-established in the continuous domain based on the extended system (5). With the observer gain matrix to correct the estimated states, the state-space expression of ESO is constructed as

$$\dot{\hat{x}}(t) = A_e \hat{x}(t) + B_e u_e(t) + L_c(y(t) - C\hat{x}(t)) \tag{6}$$

with $\hat{x}(t) = \begin{bmatrix} \hat{x}_1(t) & \hat{x}_2(t) & \hat{x}_3(t) & \hat{x}_4(t) \end{bmatrix}^T$, $A_e = A_p$, $B_e = B_p$. L_c is the observer gain matrix in continuous domain, u_e is the input of ESO.

TD can be designed as

$$\dot{v}(t) = A_t v(t) + B_t r(t) \tag{7}$$

where

$$v(t) = \begin{bmatrix} v_1(t) \\ v_2(t) \\ v_3(t) \end{bmatrix}; A_t = \begin{bmatrix} 0 & 1 & 0 \\ 0 & 0 & 1 \\ -\omega_t^3 & -3\omega_t^2 & -3\omega_t \end{bmatrix}; B_t = \begin{bmatrix} 0 \\ 0 \\ \omega_t^3 \end{bmatrix}.$$

$v(t)$ is the reference signal traces matrix, ω_t is the tracking bandwidth, and $r(t)$ is the current-loop reference.

SEF is designed as

$$u_c(t) = K_v v(t) - K_x \hat{x}(t) \tag{8}$$

where control gains K_v and K_x are determined by bandwidth method commonly adopted in [32] as

$$K_v = \begin{bmatrix} \omega_c^3 & 3\omega_c^2 & 3\omega_c \end{bmatrix} / b_0; K_x = \begin{bmatrix} \omega_c^3 & 3\omega_c^2 & 3\omega_c & 1 \end{bmatrix} / b_0. \tag{9}$$

2.3. Limitations of Euler Discretion and Current Observer

The control block diagram is shown in Figure 3. Based on the work of [37] on discrete ESO, the current observer form ESO is adopted as

$$\begin{cases} \overline{x}(k) = \overline{\Phi}_e \hat{x}(k-1) + \overline{\Gamma}_e u_p(k-1) \\ \hat{x}(k) = \overline{x}(k) + \overline{L}_d(y(k) - C\overline{x}(k)) \end{cases} \tag{10}$$

where k denotes the value at the time $t = kT$. $\overline{x}(k)$ is the predicted value obtained from the assumed plant dynamics. $\overline{\Phi}_e$ and $\overline{\Gamma}_e$ are the discrete dynamic matrices of the assumed plant. $u_p(k-1)$ is the input to the ESO, which is equal to $u_c(k-2)$. $\hat{x}(k)$ is the corrected value derived from $\overline{x}(k)$ and the plant output $y(k)$. $\overline{L}_d$ is the observer gain matrix.

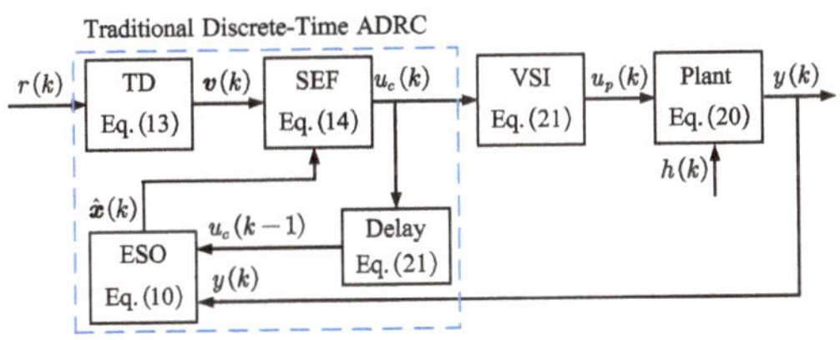

Figure 3. Block diagram of discrete-time ADRC with current observer and Euler discretization.

Using Euler discretization [38], the values of $\overline{\Phi}_e$ and $\overline{\Gamma}_e$ are derived as

$$\overline{\Phi}_e = I_4 + TA_e; \overline{\Gamma}_e = TB_e. \tag{11}$$

The poles are then configured in the discrete domain to determine $\overline{L}_d$ as

$$\left| zI_4 - (\overline{\Phi}_e - \overline{L}_d C \overline{\Phi}_e) \right| = (z - z_o)^4 \tag{12}$$

where I_4 is a fourth-order identity matrix, z is a Z-transform operator, $z_o = \exp(-\omega_o T_s)$ is the poles of the observer in discrete domain.

TD (7) can be discretized by Euler method as

$$\hat{v}(k) = \overline{\Phi}_t \hat{v}(k-1) + \overline{\Gamma}_t r(k-1) \tag{13}$$

where $\overline{\Phi}_t = I_3 + TA_t$, $\overline{\Gamma}_t = TB_t$, I_3 is a third-order identity matrix, and $r(k)$ is the reference signal of current loop.

SEF (8) can be discretized naturally as

$$u_c(t) = \boldsymbol{K}_v \boldsymbol{v}(k) - \boldsymbol{K}_x \hat{\boldsymbol{x}}(k). \tag{14}$$

The traditional discrete methods mentioned above mainly suffer from two issues: discretization inaccuracies and the impact of one-beat time delay.

Backward Euler discretization introduces errors between the real system model and the ESO. These errors increase as the switching frequency decreases, potentially leading to resonance peaks and additional phase lag.

The one-beat computation delay in digital control systems significantly impacts system stability by reducing the phase margin. In the frequency domain, this delay introduces frequency-dependent phase lag that increases proportionally with frequency.

Consequently, the combined effects of discretization inaccuracies and the one-beat time delay diminish the stability margin of ADRC implemented with Euler discretization and the current-observer method. Therefore, when attempting to improve dynamic performance, the system can easily become unstable. This can be observed in the variation of the closed-loop poles shown in Figure 4. The range of control bandwidth ω_c varies from 50 Hz to 500 Hz, with a step size of 50 Hz, while ω_o is fixed at 1500 Hz. The system parameters used in this analysis are consistent with those listed in Table 3 of Section 5. As the control bandwidth ω_c increases, the poles in Figure 3 gradually shift from inside the stable unit circle to outside the unstable unit circle. The system becomes unstable around at ω_c = 250 Hz. The detailed derivation is conducted in Section 4.

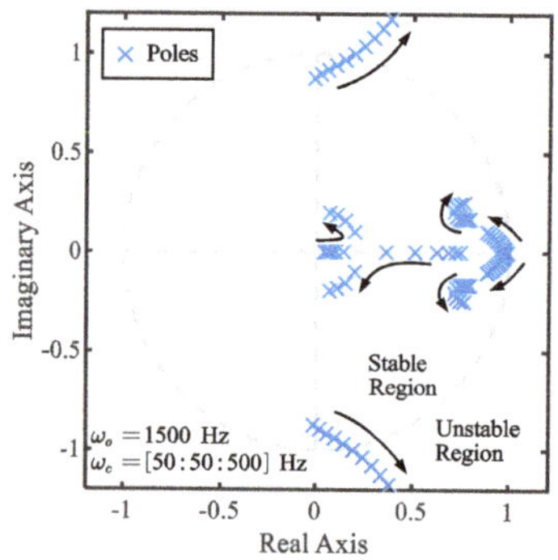

Figure 4. Closed-loop pole trajectories of ADRC with current observer and Euler discretization.

3. Proposed Predictive ESO for Delay Compensation and ZOH Accurate Discretization

Discretization inaccuracies and one-beat time delays reduce controller bandwidth and degrade dynamic performance. In this section, an accurate discretization and delay-compensated method are proposed to address this issue.

3.1. Accurate Plant Discretization via ZOH Method

The discrete observer (10) can be divided into two parts:

one part is the observer dynamics item which simulates the system dynamics, represented as $\overline{\boldsymbol{\Phi}}_e \hat{\boldsymbol{x}}(k-1) + \overline{\boldsymbol{\Gamma}}_e u_p(k-1)$. When the $\overline{\boldsymbol{\Phi}}_e$ and $\overline{\boldsymbol{\Gamma}}_e$ matrices of the observer closely match those of the real plant, the system's stability margin is higher, allowing for greater dynamic responce.

The other part is the correction item based on the feedback value, represented as $\overline{L}_d(y(k)-C\overline{x}(k))$.The selection and impact of $\overline{L}_d$ will be analyzed in the next subsection.

Compared to the Euler discretization method, the ZOH discretization method can more accurately obtain the plant's dynamic matrices under PWM voltage input. The core concept of ZOH discretization is to assume that the plant's input remains constant during each sampling interval. Simultaneously, the PWM output is approximated as a constant value within each interval, following the volt-second balance principle [41]. The following is the derivation process.

According to the solution of state-space Equation (5), state variables at instant $t = kT$ and $t = (k+1)T$ are obtained as

$$x(k) = e^{A_p kT} x(0) + e^{A_p kT} \int_0^{kT} e^{-A_p \tau} (B_p u_p(\tau) + Eh(\tau)) d\tau \tag{15}$$

$$x(k+1) = e^{A_p (k+1)T} x(0) + e^{A_p (k+1)T} \int_0^{(k+1)T} e^{-A_p \tau} (B_p u_p(\tau) + Eh(\tau)) d\tau \tag{16}$$

Multiply (15) by $\exp(A_p T)$, and then subtract Equation (16) to obtain:

$$x(k+1) = e^{A_p T} x(k) + e^{A_p (k+1)T} \int_{kT}^{(k+1)T} e^{-A_p \tau} (B_p u_p(\tau) + Eh(\tau)) d\tau \tag{17}$$

The PWM output voltage $u_p(t)$ between kT and $(k+1)T$ can be approximated as a zero-order holder, which can be expressed as

$$u_p(t) = u_p(k), h(t) = h(k)(kT \leqslant t < kT + T). \tag{18}$$

According to the ZOH characteristics, (17) can be simplified as

$$\begin{aligned} x(k+1) = e^{A_p T} \quad & x(k) + e^{A_p T} \textstyle\int_0^T e^{-A_p t} (B_p u_p(k) + Eh(k)) dt \\ & = e^{A_p T} x(k) + \textstyle\int_0^T e^{A_p \lambda} (B_p u_p(k) + Eh(k)) d\lambda \end{aligned} \tag{19}$$

where $\lambda = T - t$.

The discrete state equation of the plant based ZOH method is listed as

$$\begin{cases} x(k+1) = \boldsymbol{\Phi}_p x(k) + \boldsymbol{\Gamma}_p u_p(k) + \Delta_p h(k) \\ y(k) = Cx(k) \end{cases} \tag{20}$$

where $\boldsymbol{\Phi}_p = e^{A_p T}; \boldsymbol{\Gamma}_p = \left(\int_0^T e^{A_p \tau} d\tau\right) B_p; \Delta_p = \left(\int_0^T e^{A_p \tau} d\tau\right) E$.

The plant's precise discrete model, obtained using the ZOH method, as given in (20), serves as the basis for designing the discrete ESO. The dynamic process of the discrete ESO designed based on (20) closely mimics the dynamics of the plant, thereby improving the system's stability margin.

3.2. Improved Predictive ESO with One-Beat Lead

In addition to obtaining the plant's accurate discrete dynamic model through the ZOH method, the ESO also needs to be improved to account for the one-beat time delay introduced by digital computation.

Time delay is noted in discrete domain as

$$G_d(z) = z^{-1} \tag{21}$$

The current observer, as shown in (10), provides the estimate at the current time $t = kT$, and it does not compensate for the one-beat time delay. Similarly, the traditional predictive observer, as illustrated by the dashed line in Figure 5, outputs the state estimate to the SEF, which is still based on the current time $t = kT$. In addition, since the traditional predictive observer uses $u_p(k-1)$ and $y(k-1)$ to calculate $\hat{x}(k)$, this method theoretically introduces a greater delay compared to the current observer.

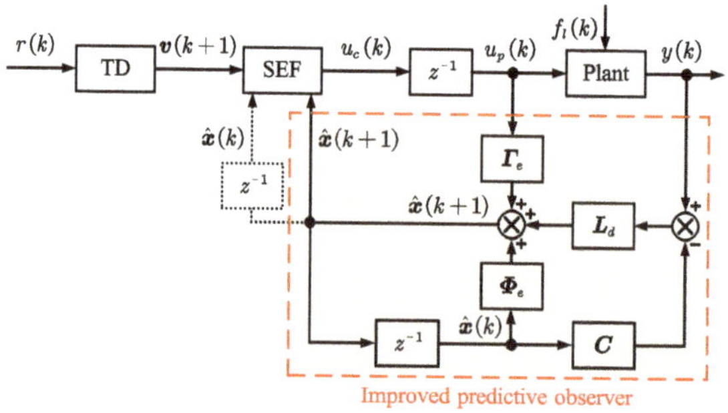

Figure 5. Block diagram of discrete-time ADRC with proposed improved predictive observer or traditional predictive observer.

This paper proposes an improved predictive observer that provides a one-step lead prediction, outputting the predicted value at time $t = (k+1)T$, which helps to counteract the one-beat delay in the forward channel. The proposed control method is shown in Figure 5 based on ADRC using the improved predictive ESO and accurate ZOH discretization.

The proposed predictive ESO based on the ZOH discretization method is given by

$$\hat{x}(k+1) = \boldsymbol{\Phi}_e\hat{x}(k) + \boldsymbol{\Gamma}_e u_e(k) + \boldsymbol{L}_d(y(k) - \boldsymbol{C}\hat{x}(k)) \quad (22)$$

where $\boldsymbol{\Phi}_e = \boldsymbol{\Phi}_p$; $\boldsymbol{\Gamma}_e = \boldsymbol{\Gamma}_p$.

The SEF based on the proposed predictive ESO is given by

$$u_c(k) = \boldsymbol{K}_v \boldsymbol{v}(k+1) - \boldsymbol{K}_x\hat{x}(k+1) \quad (23)$$

Subtracting the observer equation from the state equation with $\boldsymbol{e}(k) = \boldsymbol{x}(k) - \hat{x}(k)$ yields

$$\boldsymbol{e}(k+1) = (\boldsymbol{\Phi}_e - \boldsymbol{L}_d\boldsymbol{C})\boldsymbol{e}(k) \quad (24)$$

Based on the above equation, the dynamics of the error can be designed in the discrete domain. The observer feedback gain matrix $\boldsymbol{L}_d$ is derived as

$$\det(z\boldsymbol{I}_4 - (\boldsymbol{\Phi}_e - \boldsymbol{L}_d\boldsymbol{C})) = (z - z_o)^4 \quad (25)$$

where $z_o = e^{-\omega_o T}$ is parameter related to observer bandwidth, and z_o is the mapping point of ω_o in z domain.

To verify the improvement in dynamic performance of the ADRC based on the ZOH discretization method and the improved predictive observer, the closed-loop pole trajectories are illustrated in Figure 6. The control bandwidth ω_c varies from 350 Hz to 800 Hz, with a step size of 50 Hz, while the observer bandwidth ω_o is fixed at 1500 Hz. It can be seen that when the controller bandwidth reaches 800 Hz, the system's closed-loop poles remain inside the unit circle. In comparison to the ADRC based on the Euler discretization method and the conventional observer shown in Figure 4, the controller bandwidth can be significantly increased without losing stability.

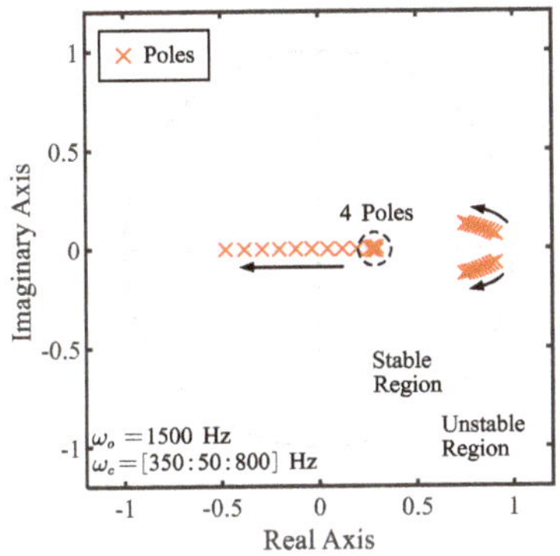

Figure 6. Closed-loop pole trajectories of the proposed discrete-time ADRC with ZOH discretization and improved predictive observer.

4. Robust Control Parameter Design Based on Open-Loop Transfer Function Margin Metrics

Predictive control strategies typically require rigorous stability analysis, and the proposed discrete ESO is no exception. This section presents a systematic approach for control parameter design based on margin metrics, aiming to achieve a trade-off between robustness and dynamic.

4.1. Open-Loop Transfer Function Derivation

The equivalent control block of the control system is derived as shown in Figure 7. $G_t(z)$ is the transfer function of TD derived in Section 4.3.

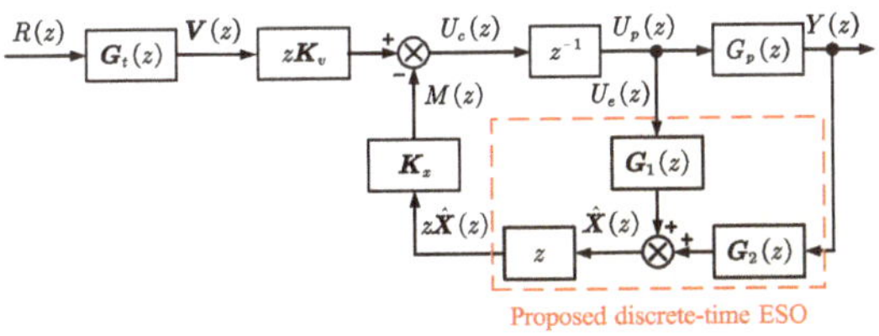

Figure 7. The equivalent control block of proposed discrete-time ADRC with ZOH discretization and improved predictive observer.

Based on the discrete expression of ESO in (22), the relation between the estimated state vector $\hat{X}(z)$ and the input of ESO $U_e(z)$ and $Y(z)$ is obtained as

$$\hat{X}(z) = G_1(z)U_e(z) + G_2(z)Y(z) \tag{26}$$

where uppercase letters denote variables in the Z-domain; for example, $Y(z)$ is the Z-transform of $y(k)$. The transfer function is defined as follows:

$$G_1(z) = (zI_4 - \Phi_e + L_dC)^{-1}\Gamma_e;\ G_2(z) = (zI_4 - \Phi_e + L_dC)^{-1}L_d.$$

The Z-transform of the ESO output $\hat{x}(k+1)$ is

$$\mathcal{Z}[\hat{x}(k+1)] = z\hat{X}(z) \tag{27}$$

where $\mathcal{Z}$ represents the Z-transform.

The modeling of plant in discrete domain with ZOH input according to (20) is obtained as

$$G_p(z) = Y(z)/U_p(z) = C\left(zI_4 - \Phi_p\right)^{-1}\Gamma_p. \tag{28}$$

The open-loop transfer function of the control system is obtained as

$$G_{op}(z) = U_c(z)/M(z) = K_x\left(G_1(z) + G_p(z)G_2(z)\right). \tag{29}$$

4.2. Control Parameters Design and Margins Improvement

Increased control bandwidth ω_c enhances the system's dynamic response performance and disturbance rejection capability. However, excessively high bandwidth ω_c may compromise system robustness, creating an inherent trade-off between dynamic response and stability.

Based on frequency-domain design principles, the open-loop transfer function must satisfy the following stability criteria to ensure both dynamic performance and robustness:

1. A minimum phase margin of 40°.
2. A minimum gain margin of 6 dB.

Following these design guidelines, the proposed improved discrete ADRC parameters are selected as

$$\omega_c = 500\text{ Hz};\ \omega_o = 3\omega_c = 1500\text{ Hz} \tag{30}$$

Figure 8 shows the Bode plots of the open-loop transfer functions for the four discretization methods of "Euler discretization + Current observer" ("Eul. + Cur."), "ZOH discretization + improved predictive observer" ("ZOH + Pre."), "Euler discretization + improved predictive observer" ("Eul. + Pre."), and "ZOH discretization + Current observer" ("ZOH + Cur."). The control bandwidth is set as ω_c = 500 Hz, and the ESO bandwidth is set as ω_o = 1500 Hz.

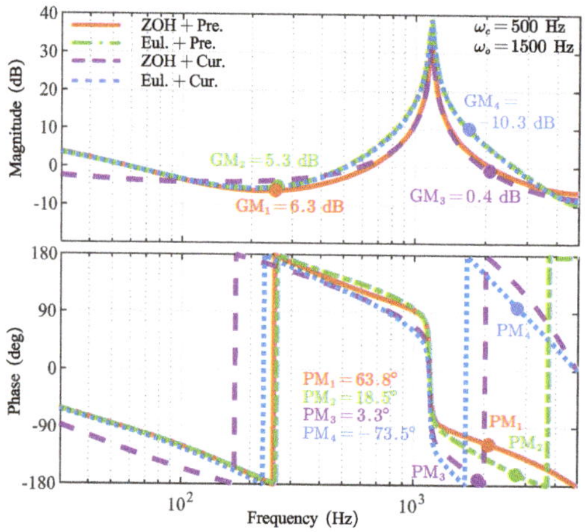

Figure 8. Bode plots of open-loop transfer function of the discrete ESO using four discretization methods. GM = gain margin, PM = phase margin.

As illustrated in Figure 8, the proposed "ZOH + Pre." method achieves a gain margin of 6.3 dB and a phase margin of 63.8°, demonstrating high stability margins and meeting robustness requirements. In contrast, the "Eul. + Pre." method yields a gain margin of 5.3 dB and a phase margin of 18.5°, indicating relatively lower stability margins and insufficient robustness. The "ZOH + Cur." method further degrades, with a gain margin of 0.4 dB and a phase margin of 3.3°, approaching the stability boundary. Notably, the "Eul. + Cur." method fails to maintain stability, exhibiting a gain margin of −10.3 dB and a phase margin of −73.5°. The proposed "ZOH + Pre." method significantly enhances system stability, improving the gain margin from −10.3 dB to 6.3 dB (a 16.6 dB increase) and the phase margin from −73.5° to 63.8° (a 137.3° increase).

It can be concluded that, under the same bandwidth, both the improved predictive observer method and the ZOH discretization can improve the gain margin and phase margin, compared to the current observer method and the Euler discretization. As a result, it offers a significant advantage when increasing the controller bandwidth to improve system dynamic performance. This alleviates the trade-off between system dynamic performance and robustness, ensuring stability at higher control bandwidths.

4.3. Tracking Differentiator and Step Response

While the open-loop transfer function determines the stability margin of the control system, it is the closed-loop transfer function that critically defines the system's tracking performance.

TD (7) can be discretized by ZOH method as

$$\hat{v}(k) = \boldsymbol{\Phi}_t \hat{v}(k-1) + \boldsymbol{\Gamma}_t r(k-1) \tag{31}$$

where $\boldsymbol{\Phi}_t = e^{A_t T}; \boldsymbol{\Gamma}_t = \left(\int_0^T e^{A_t \tau} d\tau\right) \boldsymbol{B}_t$. The transfer function of the TD in the discrete domain is denoted as

$$\boldsymbol{G}_t(z) = (z\boldsymbol{I}_3 - \boldsymbol{\Phi}_t)^{-1} \boldsymbol{\Gamma}_t. \tag{32}$$

Design of bandwidth parameter for TD is based on the engineering configuration experience as

$$\omega_t = 2\omega_c. \tag{33}$$

Based on Figure 7, the closed-loop transfer function of the control system is hence derived as

$$G_{cl}(z) = \frac{R(z)}{Y(z)} = \frac{\boldsymbol{K}_v \boldsymbol{G}_t(z) G_p(z)}{1 + G_{op}(z)}. \tag{34}$$

The selection of controller and observer bandwidth parameters must consider the phase margin of the system's open-loop transfer function. Insufficient phase margin (typically < 40°) indicates proximity to instability, particularly under parameter uncertainties. Practical control methods therefore require adequate phase margin.

As Figure 8 demonstrates, the "Eul. + Pre.", "Eul. + Cur.", and "ZOH + Cur." methods exhibit critically low phase margins (<20°) at $\omega_c = 500$ Hz and $\omega_o = 1500$ Hz, rendering them unsuitable for direct implementation. Mitigation involves reducing controller and observer bandwidths to enhance phase margins. Table 1 quantifies the improved phase margins with reduced controller and observer bandwidths.

Table 1. Phase margins with reduced controller and observer bandwidth.

Methods	Control Bandwidth and Observer Bandwidth	Phase Margin	Reduced Bandwidth	Phase Margin
"ZOH + Pre."	500 Hz; 1500 Hz	63.8°	No need to reduce bandwidth	
"Eul. + Pre."	500 Hz; 1500 Hz	18.5°	300 Hz; 600 Hz	41.2°
"ZOH + Cur."	500 Hz; 1500 Hz	3.3°	300 Hz; 600 Hz	40.1°
"Eul. + Cur."	500 Hz; 1500 Hz	−73.5°	150 Hz; 600 Hz	34.5°

For TD bandwidth selection, empirical practice suggests setting it to twice the controller bandwidth in (33) to accelerate closed-loop response.

The resulting parameter configurations for different discretization methods are detailed in Table 2, with numerical calculation of step response shown in Figure 9. For "ZOH + Pre." method, the settling time is 1.0 ms with an overshoot of 1%. "Eul. + Pre." method shows a settling time of 2.6 ms and an overshoot of 13%. For "Eul. + Cur." method, the settling time is 6.2 ms with an overshoot of 11%, while "ZOH + Cur." method has a settling time of 9.6 ms and an overshoot of 4%.

Table 2. Selection of bandwidth parameters for step response.

Methods	Control Bandwidth ω_c	Observer Bandwidth ω_o	TD Bandwidth ω_t
"ZOH + Pre."	500 Hz	1500 Hz	1000 Hz
"Eul. + Pre."	300 Hz	600 Hz	600 Hz
"ZOH + Cur."	300 Hz	600 Hz	600 Hz
"Eul. + Cur."	150 Hz	600 Hz	300 Hz

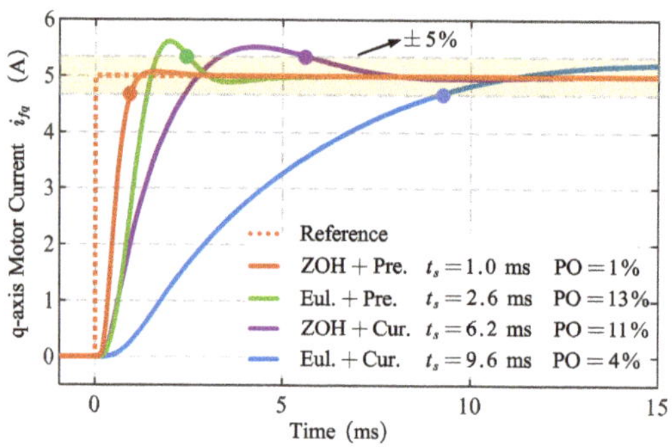

Figure 9. Numerical calculation: step response of the current under the four discretization methods. PO = percentage of overshoot. t_s = settling time.

With the inclusion of accurate discretization and delay compensation, both the settling time and overshoot improve in current tracking performance.

5. Experimental Verification

The effectiveness of the proposed method was validated experimentally. All control parameters used in the experimental verification are listed in Table 2. The experimental setup of the drive system is shown in Figure 10. The basis for the design and selection of LC filters can be seen in [4]. The experiments were conducted using a dSPACE MicroLabBox (ds1202) real-time platform along with the hardware prototype. The system parameters are listed in Table 3. The system is powered by a 311 V DC source, making it suitable for industrial applications. The sampling rate and control frequency are both set to 10 kHz, with the inverter dead-time set to 1 μs. The LC filter consists of three-phase inductors and capacitors. The SPMSM is connected to a magnetic powder brake, which serves as the load via a shaft.

The output voltage of the LC filter and the DC bus voltage were measured by Hall-effect voltage sensors (LEM LV 25-P). The SPMSM are configured with the parameters specified in Table 3. The output current of the LC filter was measured by Hall-effect current sensors (LEM LA 25-P). The SPMSM is equipped with a high-resolution incremental encoder with 2500 pulses per revolution to ensure accurate position feedback. These measurement signals are fed into the dSPACE system, which generates PWM signals for driving the inverter through the control strategy.

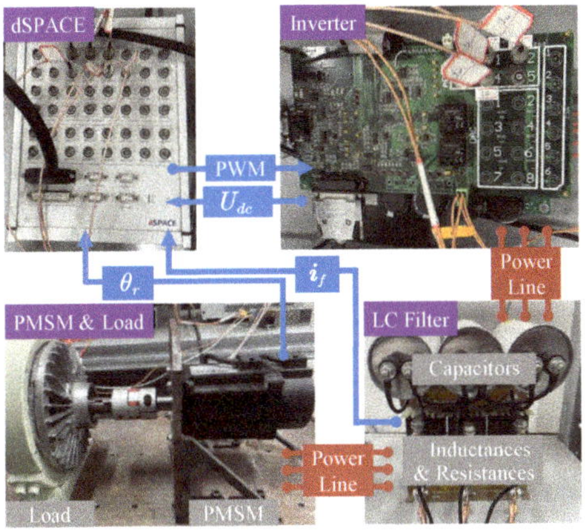

Figure 10. Experimental setup.

Table 3. Parameters of system.

Symbol	Parameter	Value
L_f	Inductance of filter	2.2 mH
R_f	Resistance of filter	0.5 Ω
C_f	Capacitor of filter	11 μF
L_s	Inductance of SPMSM	6.5 mH
R_s	Resistance of stator	1 Ω
n_p	Number of pole pairs	4
ψ_m	Motor flux linkage	0.086 Wb
I_n	Rated motor current	3.5 A
P_n	Rated motor power	750 W
n_n	Rated rotor speed	1500 r/min
J	System moment of inertia (with load)	0.005 kg·m^2
U_{dc}	DC-link voltage	311 V
f_s	Sampling and Switching frequency	10 kHz

5.1. Speed and Current Response During Startup

Figure 11 shows the rotor speed, and motor current waveforms during the startup process of the SPMSM. Both the "ZOH + Pre." and "Eul. + Cur." discretization methods were applied to discretize the ADRC in the current controller. Control parameters are selected from Table 2. Both speed controllers employ conventional PI control with a bandwidth set to 5 Hz. An enlarged version of the current waveforms is also provided to observe more details of the dynamic process during the current step response.

Figure 11a shows the motor speed waveform during startup, with the speed reference value stepping from 0 to 1000 r/min at 0 s. The rotor speed settling time (5%) for both "ZOH + Pre." and "Eul. + Cur." discretization methods are close to 210 ms.

Figure 11b shows the current waveform during the startup process. The steady-state current fluctuation amplitude for the "ZOH + Pre." discretization method is 0.5 A (10% of reference current), while for the "ZOH + Pre." discretization method, it is 2.2 A (44% of reference current).

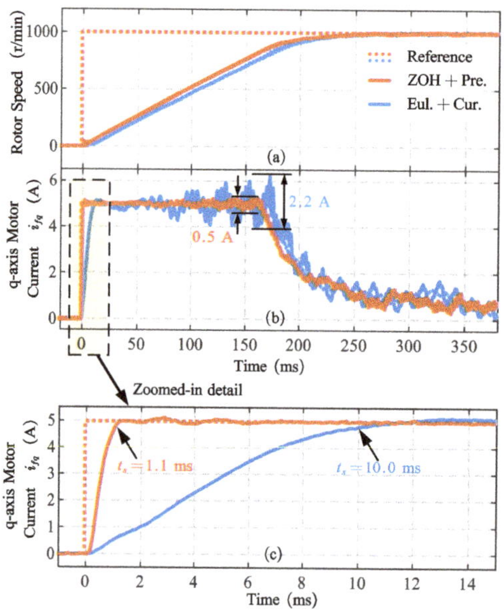

Figure 11. Experimental results: Acceleration from standstill to 1000 r/min using "ZOH + Pre." method and "Eul. + Cur." method. (**a**) Rotor speed. (**b**) Motor current. (**c**) Zoomed-in motor current.

Figure 11c shows more details of the current step response at zero speed. The current settling time (5%) of the "ZOH + Pre." discretization method is 1.1 ms with no overshoot, compared to 10.0 ms with a 4% overshoot for the "Eul. + Cur." discretization method. The experimental current step response waveforms closely match the numerical calculation results shown in Figure 9.

Therefore, the proposed "ZOH + Pre." discretization method significantly outperforms the "Eul. + Cur." discretization method in terms of current settling time, highlighting the effectiveness and necessity of accurate discretization and delay compensation techniques in applications demanding fast dynamic responses.

5.2. Current Step Responses of Proposed Discretization Method at Different Rotor Speeds

Figure 12a–d show the motor current step response of the proposed discretization ADRC at different rotor speeds: 375 r/min (25 Hz), 750 r/min (50 Hz), 1125 r/min (75 Hz), and 1500 r/min (100 Hz). It is noteworthy that the current response shown in Figure 11c can be approximately considered at a rotor speed of 0 r/min (0 Hz). At different rotor speed, a 5A (rated current) step signal is applied to the q-axis current reference of the SPMSM at 0 s.

The integral of the time multiplied by the absolute error (ITAE) index is commonly used to evaluate the tracking dynamic performance, and it is defined as

$$\text{ITAE}\left(y^*,y\right) = \sqrt{\sum_{k=1}^{m}(kT|y^*(k) - y(k)|)} \tag{35}$$

where y^* is the reference value, y is the actual value, m is the number of points in the selected interval, T is the sampling time, $y^*(k)$ and $y(k)$ are the value of y^* and y, respectively, at time k.

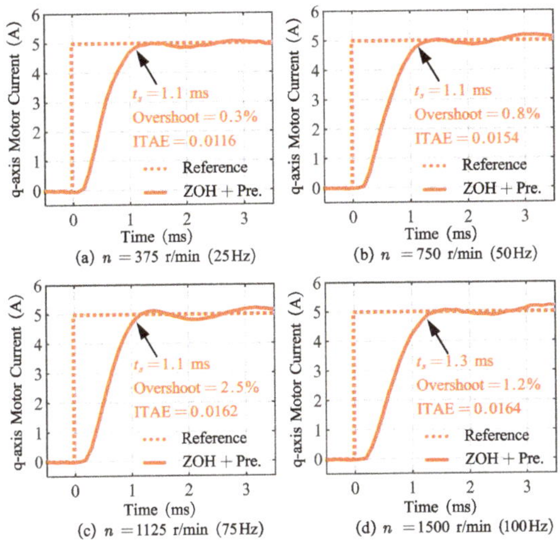

Figure 12. Experimental results: Step responses of motor current using "ZOH + Pre." discretization method. (**a**) 375 r/min (25 Hz). (**b**) 750 r/min (500 Hz). (**c**) 1125 r/min (75 Hz). (**d**) 1500 r/min (100 Hz).

Based on Figures 12 and 11c, the settling time of motor current with the proposed discretization method is 1.1 ms for frequencies ranging from 0 Hz up to 75 Hz and 1.3 ms at 100 Hz. The overshoot increases slightly with rotor speed, rising from 0.3% to 2.5%. In the interval [0 ms, 3.5 ms], the ITAE increases slightly by 0.005 for frequencies exceeding 25 Hz, while maintaining a baseline value of 0.0109 at zero speed. These results indicate that the current dynamic performance remains similar to the zero-speed condition when the rotational speed operates below the fundamental frequency. The increased settling time and ITAE observed at higher speeds during dynamic processes are likely attributable to the increased back electromotive force and dq-axis coupling.

5.3. Comparison of Closed-Loop Performance with Different Discretization Methods

This section validates the improvement in dynamic performance of the current loop achieved by the proposed precise discretization and delay compensation method, compared to traditional discretization methods, through both frequency-domain Bode plots and time-domain step response analysis.

(a) Frequency-domain: closed-loop frequency sweep with different discretization methods

Figure 13 presents the closed-loop Bode plots in the frequency domain for different discretization methods, with the speed condition set to 0 r/min. The frequency sweep is conducted using a chirp signal in the range of 10–1000 Hz as the reference signal. From this, the values of reference and feedback are obtained to derive the Bode plot of the closed-loop response. Bandwidth is commonly used as a metric to assess frequency-domain performance. The magnitude bandwidth is defined as the frequency at which the magnitude drops to −3 dB, while the phase bandwidth is the frequency at which the phase decreases to −45°.

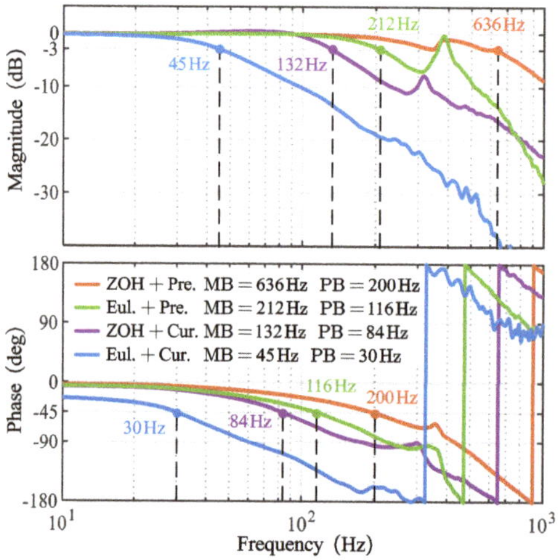

Figure 13. Experimental results: Closed-loop frequency sweep with different discretization methods. MB = Magnitude bandwidth. PB = Phase bandwidth.

From Figure 13, it can be seen that for "ZOH + Pre." method, the magnitude bandwidth is 636 Hz, and the phase bandwidth is 200 Hz. For the "Eul. + Pre." method, the magnitude bandwidth is 212 Hz, and the phase bandwidth is 116 Hz. The "Eul. + Cur." method shows the magnitude bandwidth at 132 Hz and the phase bandwidth at 84 Hz, while the "ZOH + Cur." method has the magnitude bandwidth at 45 Hz and the phase bandwidth at 30 Hz.

These results indicate that with precise discretization and delay compensation, the system's bandwidth for tracking performance improves significantly. Specifically, the magnitude bandwidth is increased by a factor of 14.1, and the phase bandwidth is increased by a factor of 6.7.

(b) Time-domain: step response with different discretization methods

Figure 14 illustrates the current step response in the time-domain for different discretization methods, with the speed condition again set to 0 r/min. For the "ZOH + Pre." method, the settling time is 1.1 ms with almost no overshoot. The "Eul. + Pre." method shows a settling time of 2.7 ms and an overshoot of 18%. For the "Eul. + Cur." method, the settling time is 5.8 ms with an overshoot of 15%, while the "ZOH + Cur." method has a settling time of 9.7 ms and an overshoot of 3%.

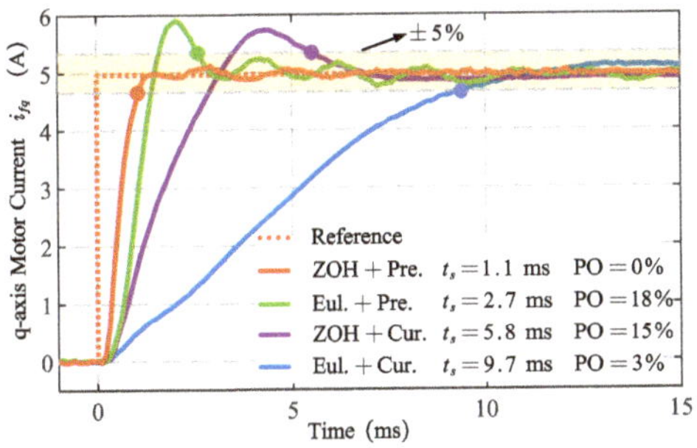

Figure 14. Experimental results: Closed-loop step response with different discretization methods. t_s = settling time. PO = percentage of overshoot.

It is clear that, with the inclusion of precise discretization and delay compensation, both the settling time and overshoot improve in current tracking performance. Specifically, the settling time is reduced by 88% compared to traditional discrete methods. Furthermore, comparing the numerical calculations from Figure 9 with the experimental results in Figure 14, it is evident that the experimental outcomes are in good agreement with the theoretical predictions.

From both the frequency-domain and time-domain results, it is evident that the proposed discretization method significantly enhances dynamic performance and outperforms existing discretization methods.

5.4. Comparison with Existing Methods

To facilitate comparison with settling times obtained at different switching frequencies in different papers, the unit of settling time has been changed from milliseconds to the number of control periods in this subsection. This means that the settling periods are now expressed as the number of control periods required for the step response to reach a steady-state error of less than 5%. A lower settling period value indicates a quicker system response, making it a useful metric for evaluating the performance of control strategies with different switching periods.

The experimental data for MPC [7], AD [9], and TCPI [13] are extracted from their respective references. The experimental results of the proposed method are provided in Figure 14. Table 4 presents a comparison between the proposed method and existing methods in terms of settling periods, overshoot, and number of voltage and current sensors.

Table 4. Step response and sensors comparison with existing methods.

Methods	Settling Periods	Percentage of Overshoot	Voltage and Current Sensors
Proposed	11	Nearly 0%	Filter output current *2
MPC [7]	52	24%	Filter output current *2 Filter capacitor current *2 Inverter output current *2
AD [9]	40	30%	Filter output current *2
TCPI [13]	41	8%	Filter output current *2 Filter output voltage *2 Inverter output current *2
ADRC [1]	97	2%	Filter output current *2

[1] The ADRC is based on the Euler discretization and current observer method in this paper. *2 means two sensors.

The proposed method demonstrates a significant advantage in current loop dynamic performance, with a settling time of 11 cycles and minimal overshoot. This method requires only two filter-output current sensors in addition to the DC-link voltage sensor. In contrast, the MPC [7] method exhibits a settling time of 52 cycles, an overshoot of 24%, and requires six sensors. The AD [9] method has a settling time of 40 cycles, an overshoot of 30%, and requires two sensors. The TCPI [13] method shows a settling time of 41 cycles, an overshoot of 8%, and requires six sensors. The Euler discretization and current observer method ADRC shows a settling time of 97 cycles, an overshoot of 3%, and requires two sensors.

The comparison with existing methods is summarized in Table 4. The proposed method reduces the settling time by over 72%, eliminates overshoot entirely, and requires the fewest sensors. These results demonstrate that the improved discrete-time ADRC method offers a remarkably fast dynamic response, and reduced hardware complexity.

5.5. Robustness Validation via Open-Loop Margins and Step Response Under Parameter Mismatches

This section validates the robustness of the proposed precise discretization and delay compensation method through both the open-loop transfer function and the step response under parameter mismatches.

(a) Frequency-domain: open-loop transfer function frequency sweep

Figure 15 shows the frequency sweep results of the system's open-loop transfer function. The sweep is conducted using a chirp signal in the range of 10–4000 Hz as the reference signal, from which the Bode plot of the open-loop transfer function in the frequency domain is derived. In Figure 15, the open-loop gain margin is 6.6 dB and the phase margin is 58.7°, which closely matches the theoretical values shown in Figure 8: 6.3 dB for the gain margin and 63.8° for the phase margin, with an error of less than 10%.

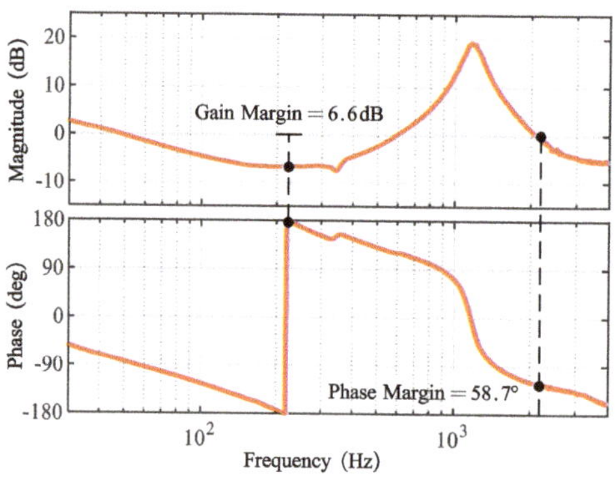

Figure 15. Experimental results: Open-loop frequency sweep using "ZOH + Pre." discretization methods.

(b) Time-domain: step response under parameter mismatches

Inductance may vary due to magnetic saturation, capacitance may deviate due to measurement errors, and resistance may change with temperature fluctuations. Therefore, this subsection involves the following parameter mismatches: filter inductance L_f, filter capacitance C_f, stator inductance L_s, and the total single-phase resistance R (where $R = R_f + R_s$). The controller uses the values marked with a hat, while the unmarked values represent the actual system parameters.

Figure 16(a1,a2) shows the step response of the current under filter inductance mismatch. When $\hat{L}_f = 0.7L_f$, the current settling time is 2.1 ms, with an overshoot of 4.8%, and ITAE = 0.0164. When $\hat{L}_f = 1.3L_f$, the settling time is 2.5 ms, with an overshoot of 4.7%, and ITAE = 0.0177.

Figure 16(b1,b2) present the step response under filter capacitance mismatch. When $\hat{C}_f = 0.7C_f$, the settling time is 3.3 ms, with an overshoot of 5.5%, and ITAE = 0.0175. When $\hat{C}_f = 1.3C_f$, the settling time is 2.0 ms, with an overshoot of 3.5%, and ITAE = 0.0185.

Figure 16(c1,c2) illustrate the step response under stator inductance mismatch. When $\hat{L}_s = 0.7L_s$, the settling time is 1.8 ms, with an overshoot of 3.8%, and ITAE = 0.0228. When $\hat{L}_f = 1.3L_f$, the settling time is 2.4 ms, with an overshoot of 5.7%, and ITAE = 0.0192.

Figure 16(d1,d2) show the step response under total single-phase resistance mismatch. When $\hat{R} = 0.5R$, the settling time is 1.0 ms, with an overshoot of 4.4%, and ITAE = 0.0140. When $\hat{R} = 2R$, the settling time is 1.1 ms, with an overshoot of 4.9%, and ITAE = 0.0179.

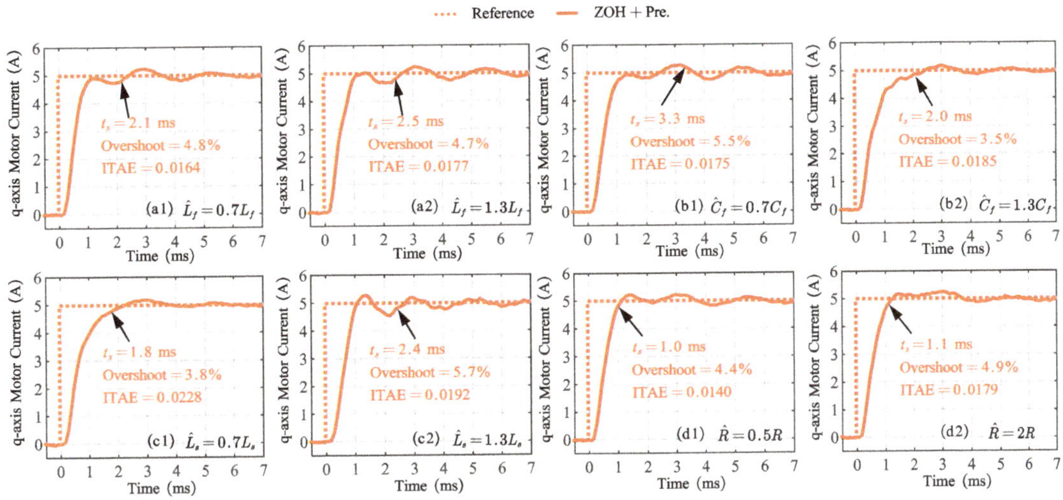

Figure 16. Experimental results: Parameter mismatches under the "ZOH + Pre." discretization method through step response evaluations (**a1**,**a2**) Decrease and increase in the filter inductance. (**b1**,**b2**) Decrease and increase in the filter capacitance. (**c1**,**c2**) Decrease and increase in the stator inductance. (**d1**,**d2**) Decrease and increase in all resistances. t_s = settling time. ITAE = The integral of the time multiplied by the absolute error.

When parameter mismatches of ±30% occur, the system remains stable. The maximum settling time is 3.1 ms, and the maximum overshoot is 5.5%, occurring when $\hat{C}_f = 0.7C_f$. The maximum ITAE occurs when $\hat{L}_s = 0.7L_s$, with a value of 0.0228. The method is not sensitive to variations in resistance, and the settling time remains nearly unchanged when resistance varies. To achieve the best dynamic performance, it is recommended to accurately measure the filter capacitance parameters and use look-up tables to handle the saturation characteristics of the inductance, while more relaxed tolerances can be applied to resistance values.

Thus, the control parameters designed based on the proposed frequency-domain margin metrics ensure good parameter robustness and maintain excellent dynamic performance, even under parameter mismatches.

5.6. Increasing and Reducing Load Torque

Figure 17 illustrates the speed and current variations during load torque application and shedding. Figure 17(a1,a2) depict the speed waveform and current profile when applying a 2 Nm load torque (80% of rated torque). Figure 17(b1,b2) show the corresponding responses during torque shedding from 2 N·m to 0. The waveforms validate that the proposed control method enables rapid tracking of the q-axis current reference during both loading and unloading transients.

5.7. Voltage and Current Characteristics at Inverter Output and LC Filter Output

Figure 18 demonstrates the voltage and current waveforms at inverter output and LC filter output under 50 Hz electrical frequency (750 r/min) at 80% rated load. Specifically, Figure 18a shows the inverter and filter output line voltage, while Figure 18b displays the inverter and filter output phase current. As shown in Figure 18a, the inverter output voltage exhibits a PWM waveform, while the filtered output voltage demonstrates a quasi-sinusoidal profile. Figure 18b reveals that the filter output current contains significantly reduced high-frequency components compared to the inverter output current, confirming the effective harmonic suppression capability of the LC filter. This results in motor terminal voltage and current waveforms closely approximating ideal sinusoidal characteristics.

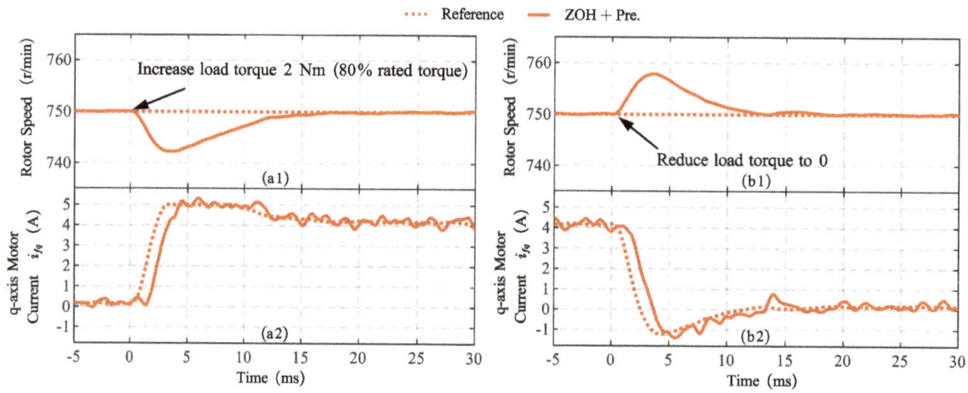

Figure 17. Experimental results: Increasing torque and reducing torque application experiments conducted at rated speed.

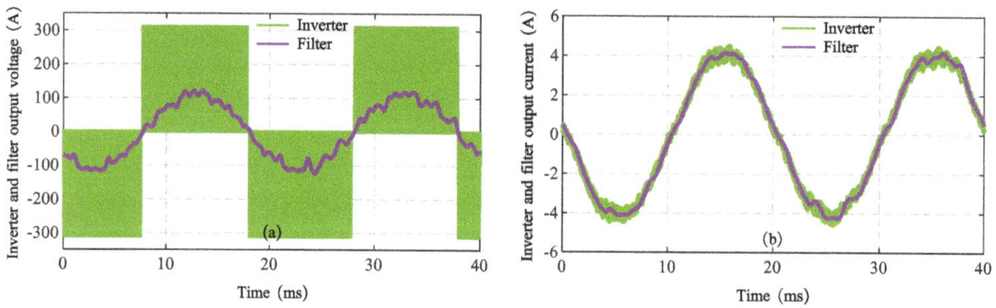

Figure 18. Experimental results: Voltage and current waveforms at inverter output and LC filter output. (**a**) Inverter and filter output line voltage; (**b**) Inverter and filter output phase current.

6. Conclusions

This paper proposes an improved discrete-time ADRC strategy for fast and robust current control in LC-filtered SPMSM drive systems. The proposed strategy incorporates an accurate zero-order hold discretization method and an improved predictive observer, which significantly enhances the control system's amplitude and phase margins over traditional Euler discretization and current observer. The control parameters are meticulously designed based on discrete-time frequency-domain metrics to ensure enhanced robustness. Experimental results demonstrate that the improved discrete-time ADRC strategy significantly improves the dynamic performance of motor current. It reduces the settling time by over 72% compared to existing methods, while also having enough robustness to handle common parameter mismatches.

The findings from this paper underscore the efficacy of accurate discretization and delay compensation, which could be instrumental in advancing ADRC for applications demanding high dynamic performance.

Author Contributions: Conceptualization, methodology and software, Z.L.; validation and writingoriginal draft preparation, Z.L. and H.Y.; writing—review and editing, J.W., L.Z. and Y.W.; supervision, L.Z. All authors have read and agreed to the published version of the manuscript.

Funding: This research received no external funding.

Data Availability Statement: The original contributions presented in this study are included in the article. Further inquiries can be directed to the corresponding author.

Conflicts of Interest: The authors declare no conflicts of interest.

Abbreviations

The following abbreviations are used in this manuscript:

SPMSM	Surface-mounted permanent magnet synchronous motor
ZOH	Zero-order hold
Eul.	Euler discretization
Cur.	Current observer
Pre.	Improved predictive observer
GM, PM	Gain margin, phase margin
MB, PB	Magnitude bandwidth, phase bandwidth
PO	Percentage of overshoot
t_s	Settling time
ITAE	Integral of the time multiplied by the absolute error

References

1. Yang, J.; Chen, W.H.; Li, S.; Guo, L.; Yan, Y. Disturbance/ uncertainty estimation and attenuation techniques in PMSM drives—A survey. *IEEE Trans. Ind. Electron.* **2017**, *64*, 3273–3285. [CrossRef]
2. Jiang, Y.; Wu, W.; He, Y.; Chung, H.S.; Blaabjerg, F. New passive filter design method for overvoltage suppression and bearing currents mitigation in a long cable based PWM inverter-fed motor drive system. *IEEE Trans. Power Electron.* **2017**, *32*, 7882–7893. [CrossRef]
3. Velander, E.; Bohlin, G.; Sandberg, A.; Wiik, T.; Botling, F.; Lindahl, M.; Zanuso, G.; Nee, H.-P. An ultralow loss inductorless dv/dt filter concept for medium-power voltage source motor drive converters with SiC devices. *IEEE Trans. Power Electron.* **2018**, *33*, 6072–6081. [CrossRef]
4. Mishra, P.; Maheshwari, R. Design analysis and impacts of sinusoidal LC filter on pulsewidth modulated inverter fed-induction motor drive. *IEEE Trans. Ind. Electron.* **2020**, *67*, 2678–2688. [CrossRef]
5. Li, Z.; Zhang, Q.; Luo, H.; Wang, H.; Wang, J.; Han, F.; Wang, A.; Liu, X.; Yu, X.; Zhou, L. Sensorless starting control of permanent magnet synchro-nous motors with Step-up transformer for downhole electric drilling. In Proceedings of the 44th Annual Conference of the IEEE Industrial Electronics Society, Washington, DC, USA, 21–23 October 2018; pp. 689–694.
6. Zheng, C.; Xie, M.; Dong, X.; Gong, Z.; Dragičević, T.; Rodriguez, J. Design and analysis of predictive inductor current control for PMSM drives with LC filter. *IEEE Trans. Power Electron.* **2025**, *40*, 1872–1885. [CrossRef]
7. Zheng, C.; Xie, M.; Dong, X.; Gong, Z.; Dragičević, T.; Rodriguez, J. Modulated model predictive current control with active damping for LC-filtered PMSM drives. *IEEE J. Emerg. Sel. Top. Power Electron.* **2024**, *12*, 3848–3861. [CrossRef]

8. Xue, C.; Zhou, D.; Li, Y. Finite-control-set model predictive control for three-level NPC inverter-fed PMSM drives with LC filter. *IEEE Trans. Ind. Electron.* **2021**, *68*, 11980–11991. [CrossRef]
9. Zhou, J.; Yao, Y.; Huang, Y.; Peng, F. Motor Current feedback-only active damping controller with high robustness for LCL-equipped high-speed PMSM. *IEEE Trans. Power Electron.* **2023**, *38*, 8707–8718. [CrossRef]
10. Vaishnav, N.; Bajjuri, N.K.; Jain, A.K. Inductor selection, improved active damping, and speed sensorless operation without voltage sensors in IM drive with LC filter. *IEEE Trans. Power Electron.* **2022**, *37*, 15272–15282. [CrossRef]
11. Yao, Y.; Huang, Y.; Peng, F.; Dong, J.; Zhu, Z. Dynamic-decoupled active damping control method for improving current transient behavior of LCL-equipped high-speed PMSMs. *IEEE Trans. Power Electron.* **2022**, *37*, 3259–3271. [CrossRef]
12. Lyu, Z.; Wu, L. Current control scheme for LC-equipped PMSM drive considering decoupling and resonance suppression in synchronous complex-vector frame. *IEEE J. Emerg. Sel. Top. Power Electron.* **2023**, *11*, 2061–2073. [CrossRef]
13. Mukherjee, S.; Poddar, G. Fast control of filter for sensorless vector control SQIM drive with sinusoidal motor voltage. *IEEE Trans. Ind. Electron.* **2007**, *54*, 2435–2442. [CrossRef]
14. Vaishnav, N.; Jain, A.K. Limitations in inner-loop PI controllers in IM drive with an LC filter and its remedies. *IEEE Trans. Ind. Appl.* **2022**, *58*, 3602–3612. [CrossRef]
15. Li, S.; Lin, H. A capacitor-current-feedback positive active damping control strategy for LCL-type grid-connected inverter to achieve high robustness. *IEEE Trans. Power Electron.* **2022**, *37*, 6462–6474. [CrossRef]
16. Yang, S.; Yin, Z.; Tong, C.; Sui, Y.; Zheng, P. Active damping current control for current-source inverter-based PMSM drives. *IEEE Trans. Ind. Electron.* **2023**, *70*, 3549–3560. [CrossRef]
17. Li, Q.; Wang, D.; Liang, X.; Guo, P.; Ma, C. Complex vector-based stability analysis of the high-speed electric motor emulator. *IEEE Trans. Power Electron.* **2024**, *39*, 346–360. [CrossRef]
18. Yepes, A.G.; Freijedo, F.D.; Doval-Gandoy, J.; López, Ó.; Malvar, J.; Fernandez-Comesaña, P. Effects of discretization methods on the performance of resonant controllers. *IEEE Trans. Power Electron.* **2010**, *25*, 1692–1712. [CrossRef]
19. Bae, B.-H.; Sul, S.-K. A compensation method for time delay of full-digital synchronous frame current regulator of PWM AC drives. *IEEE Trans. Ind. Appl.* **2003**, *39*, 802–810. [CrossRef]
20. Wang, M.; Buticchi, G.; Li, J.; Gu, C.; Gerada, D.; Degano, M.; Xu, L.; Li, Y.; Zhang, H.; Geradab, C. Decoupled discrete current control for ac drives at low sampling-to-fundamental frequency ratios. *IEEE J. Emerg. Sel. Top. Power Electron.* **2023**, *11*, 1358–1369. [CrossRef]
21. Pei, G.; Liu, J.; Gao, X.; Tian, W.; Li, L.; Kennel, R. Deadbeat predictive current control for SPMSM at low switching frequency with moving horizon estimator. *IEEE J. Emerg. Sel. Top. Power Electron.* **2021**, *9*, 345–353. [CrossRef]
22. Dai, S.; Wang, J.; Sun, Z.; Chong, E. Deadbeat predictive current control for high-speed permanent magnet synchronous machine drives with low switching-to-fundamental frequency ratios. *IEEE Trans. Ind. Electron.* **2022**, *69*, 4510–4521. [CrossRef]
23. Kim, H.-S.; Jung, H.-S.; Sul, S.-K. Discrete-time voltage controller for voltage source converters with LC filter based on state-space models. *IEEE Trans. Ind. Appl.* **2019**, *55*, 529–540. [CrossRef]
24. Pérez-Estévez, D.; Doval-Gandoy, J.; Yepes, A.G.; López, Ó. Positive- and negative-sequence current controller with direct discrete-time pole placement for grid-tied converters with LCL filter. *IEEE Trans. Power Electron.* **2017**, *32*, 7207–7221. [CrossRef]
25. Huang, K.; Zhou, J.; Zhao, H.; Lv, W.; Huang, S. Novel compensation method of digital delay for high-speed permanent magnet synchronous motor under low carrier ratio. In Proceedings of the IEEE Energy Conversion Congress and Exposition (ECCE), Detroit, MI, USA, 11–15 October 2020; pp. 3854–3861.
26. Yim, J.-S.; Sul, S.-K.; Bae, B.-H.; Patel, N.R.; Hiti, S. Modified current control schemes for high-performance permanent-magnet ac drives with low sampling to operating frequency ratio. *IEEE Trans. Ind. Appl.* **2009**, *45*, 763–771. [CrossRef]
27. Han, J. From PID to active disturbance rejection control. *IEEE Trans. Ind. Electron.* **2009**, *56*, 900–906. [CrossRef]
28. Gao, Z. Scaling and bandwidth-parameterization based controller tuning. In Proceedings of the American Control Conference, Denver, CO, USA, 4–6 June 2003; pp. 4989–4996.
29. Lin, X.; Wu, C.; Yao, W.; Liu, Z.; Shen, X.; Xu, R.; Sun, G.; Liu, J. Observer-Based Fixed-Time Control for Permanent-Magnet Synchronous Motors With Parameter Uncertainties. *IEEE Trans. Power Electron.* **2023**, *38*, 4335–4344. [CrossRef]
30. Lin, X.; Liu, J.; Liu, Z.; Gao, Y.; Peretti, L.; Wu, L. Model-Free Current Predictive Control for PMSMs With Ultra-Local Model Employing Fixed-Time Observer and Extremum-Seeking Method. *IEEE Trans. Power Electron.* **2025**, *40*, 10682–10693. [CrossRef]
31. Lin, X.; Xu, R.; Yao, W.; Gao, Y.; Sun, G.; Liu, J.; Peretti, L.; Wu, L. Observer-Based Prescribed Performance Speed Control for PMSMs: A Data-Driven RBF Neural Network Approach. *IEEE Trans. Ind. Inform.* **2024**, *20*, 7502–7512. [CrossRef]
32. Li, P.; Bazzi, A.; Lin, H.; Gultekin, M.A. Seamless Start-up of Loaded PMSM Drive from Open-Loop Id/F to Sensorless FOC. In Proceedings of the 2023 IEEE International Electric Machines & Drives Conference (IEMDC), San Francisco, CA, USA, 15–18 May 2023; pp. 1–5.
33. Li, Z.; Wang, J.; Yang, H.; Zhou, L. Fast and robust current controller for long-cable-fed PMSM drive using cascaded model-assisted active disturbance rejection control. *IEEE J. Emerg. Sel. Top. Power Electron.* **2025**, *13*, 480–491. [CrossRef]

34. Ma, W.; Guan, Y.; Zhang, B.; Wu, L. Active disturbance rejection control based single current feedback resonance damping strategy for LCL-type grid-connected inverter. *IEEE Trans. Energy Conver.* **2021**, *36*, 48–62. [CrossRef]
35. Cai, Y.; He, Y.; Zhou, H.; Liu, J. Active-damping disturbance rejection control strategy of LCL grid-connected inverter based on inverter-side current feedback. *IEEE J. Emerg. Sel. Top. Power Electron.* **2021**, *9*, 7183–7198. [CrossRef]
36. Tran, T.V.; Kim, K.H.; Lai, J.S. Optimized active disturbance rejection control with resonant extended state observer for grid voltage sensorless LCL-filtered inverter. *IEEE Trans. Power Electron.* **2021**, *36*, 13317–13331. [CrossRef]
37. Miklosovic, R.; Radke, A.; Gao, Z. Discrete implementation and generalization of the extended state observer. In Proceedings of the American Control Conference, Minneapolis, MN, USA, 14–16 June 2006; pp. 1–6.
38. Du, B.; Wu, S.; Han, S.; Cui, S. Application of linear active disturbance rejection controller for sensorless control of internal permanent-magnet synchronous motor. *IEEE Trans. Ind. Electron.* **2016**, *63*, 3019–3027. [CrossRef]
39. Zheng, Q.; Gao, Z. Predictive active disturbance rejection control for processes with delay. In Proceedings of the 32nd Chinese Control Conference, Xi'an, China, 26–28 July 2013; pp. 4108–4113.
40. Wu, J.-A.; Tian, C.; Yan, P. A predictor-based ADRC for input delay systems subject to unknown disturbances. In Proceedings of the 2018 Chinese Automation Congress (CAC), Xi'an, China, 2 November–2 December 2018; pp. 231–236.
41. Yan, Q.; Wu, X.; Yuan, X.; Geng, Y. An improved grid-voltage feedforward strategy for high-power three-phase grid-connected inverters based on the simplified repetitive predictor. *IEEE Trans. Power Electron.* **2016**, *31*, 3880–3897. [CrossRef]

energies

MDPI

Review

Review on Research and Development of Magnetic Bearings

Yuanhao Du, Gan Zhang * and Wei Hua

School of Electrical Engineering, Southeast University, Nanjing 210096, China; 230228762@seu.edu.cn (Y.D.); huawei1978@seu.edu.cn (W.H.)
* Correspondence: zhanggan@seu.edu.cn; Tel.: +86-1570-518-1487

Abstract

This paper reviews the research advancements and development in magnetic bearings. Firstly, from the technical principle, the design differences and application areas of active magnetic bearings, permanent magnetic bearings and hybrid structures are clarified. At the key technology level, focusing on electromagnetic design optimization, control strategy innovation and power-driven energy management, the breakthrough points of multi-physics coupling modeling, vibration suppression and energy efficiency improvement are revealed. Through the analysis of its engineering cases in the fields of high-speed motors, flywheel energy storage, aerospace and so on, the feasibility and economy of the technical scheme are verified. Further, the technical bottlenecks that need to be broken through are pointed out. For the future trend, this paper suggests that integration of interdisciplinary high-precision modeling, intelligent control algorithm and miniaturized integrated design should be deeply integrated to promote the large-scale application of magnetic bearing in frontier fields. This paper provides theoretical reference and engineering practice guidance for the technology iteration and cross-field integration of magnetic bearings.

Keywords: magnetic bearings; topology; control method; multi-field coupled modeling; optimization design

1. Introduction

The traditional bearings mainly include rolling bearings, sliding bearings and oil film bearings. With the development of industrial technology towards high speed, high precision and high reliability, the limitations of traditional bearings under special working conditions have become increasingly prominent: friction and wear, speed limit, lubrication, vibration and noise, and maintenance costs. These limitations have prompted people to seek new non-contact support technologies, and magnetic bearing is a breakthrough solution that emerged in this context [1–6].

Traditional bearings continue to have cost advantages under conventional operating conditions, but their technical limitations are becoming more apparent in specialized environments such as high-speed, vacuum, and ultra-clean conditions. This highlights the clear need for the development of magnetic bearings in such applications. Magnetic bearings represent an advanced support technology that utilizes magnetic forces to achieve non-contact suspension of the rotor. The basic principle involves balancing the rotor's weight and other loads through the magnetic force generated by electromagnets or permanent magnets (PM), thereby enabling frictionless suspension between the rotor and stator [7–9]. Compared with traditional bearings, magnetic bearings have the following advantages: no contact and no wear, high speed, energy saving and efficiency, clean and environmental protection, as well as strong controllability [10,11].

Since its inception in the 19th century, maglev technology has evolved from theoretical exploration to practical engineering applications. Its development can be broadly categorized into three key stages:

1. Concept Exploration Period (1842–1960):
   In 1842, Earnshaw introduced the concept of magnetic levitation and demonstrated that stable levitation of ferromagnetic materials in all degrees of freedom cannot be achieved using only permanent magnets, thereby establishing the mathematical foundation of magnetic levitation theory [12]. In 1937, Kemper filed the first patent related to magnetic levitation technology and proposed the use of actively controlled electromagnets for achieving stable levitation—a milestone in the development of magnetic bearings and maglev transportation systems [13,14].
2. Technology Development Period (1960–1990):
   Between the 1950s and 1960s, Beams successfully applied magnetic levitation to ultra-high-speed centrifuges, marking the first implementation of magnetic support for rotating bodies [15]. In 1972, the LRBA Laboratory pioneered the application of magnetic bearings in satellite reaction wheels, setting a precedent for engineering applications of the technology [16].
3. Engineering Application Period (1990–Present):
   In 1983, the United States employed a magnetic bearing vacuum pump aboard the Space Shuttle, demonstrating its reliability under extreme conditions [17]. By 1997, Draper Laboratory reported a series of advancements in high-temperature magnetic bearings capable of operating at 510 °C for aerospace engine applications [18]. As of 2021, magnetic levitation technology has been widely adopted in molecular pumps, blowers, compressors, and other industrial systems [19–21]. According to industry reports [22–25], the global magnetic bearing market size was about $2 billion in 2023 and is expected to grow to about $3.34 billion by 2032, with a compound annual growth rate (CAGR) of about 5.96% from 2024 to 2032. The top five manufacturers include SKF, Schaeffler, Siemens, GE and NSK, which together occupy more than 60% of the market share.

Table 1 summarizes representative applications and key performance parameters of magnetic bearings [26–28]. Despite remarkable progress, magnetic bearing technology still faces several challenges, including high cost (approximately 3–5 times that of conventional bearings), complex control requirements, and limited reliability. In particular, applications involving heavy loads (e.g., large-scale generator sets) and extreme environments (e.g., ultra-high temperatures) continue to demand critical technological breakthroughs. Current research is shifting from single-objective performance optimization toward integrated design approaches that emphasize high speed, high load capacity, and enhanced reliability.

Table 1. Magnetic bearing application and index [28].

Applications	Power (kW)	Speed (r/min)	Characteristic
Power Generation	2~150	35,000~220,000	Wide power range, super high speed
Flywheel Energy Storage System	120	40,000	High power, high speed
High-Speed Spindles	1~24	9000~180,000	Low power, wide speed range
Turbo Molecular Pumps	Few hundred Watt	100,000	Micropower, ultra-high speed
Gas Compressors	10,000	20,000	Industrial high power, medium speed
Air Compressors	100~150	80~15,000	High power, wide speed range
Micro Turbines	50	80,000	Compact high power
Turbo Generators	30	60,000	High efficiency, high speed

This review aims to provide a systematic overview of the development of magnetic bearing technology, outlining major technical challenges and future research directions. It is intended as a comprehensive reference for both academic researchers and industrial practitioners. The structure of the review is as follows:

- Section 2 discusses the classification and operating principles of magnetic bearings, including the technical characteristics and comparisons of active, passive, and hybrid types.
- Section 3 focuses on core technologies, including structural design, control algorithms, and drive systems.
- Section 4 presents typical application scenarios in energy, transportation, industrial systems, and other domains.
- Section 5 addresses current technical challenges and explores future development trends.
- Section 6 concludes the review with final remarks.

2. Classification and Comparison of Magnetic Bearing

Magnetic bearings can be classified in various ways depending on different criteria:

1. Type of Magnetic Force: Attractive or repulsive forces.
2. Suspension Mode: Active, passive, or hybrid systems.
3. Magnet Type: Superconducting, permanent magnet, or electromagnetic.
4. Structural Configuration: Horizontal or vertical orientation; internal or external rotor.
5. Degree of Contact: Fully non-contact or partially contact-based systems.
6. Control Current Type: Alternating current (AC) or direct current (DC).
7. Magnetic Pole Arrangement: Heteropolar or homopolar configurations.
8. Degrees of Freedom (DOF): Axial (1 DOF), radial (2 DOF), combined radial-axial (3 DOF), or extended to 4 and 5 DOF systems.

Given the diversity of magnetic bearing types, it is challenging to encompass all variants from a single classification perspective. Therefore, it is reasonable to select a broad classification method that includes most of the styles according to the suspension mode and then subdivides them. At present, the most widely used is classified by suspension mode, that is, active, passive and hybrid magnetic bearings [29,30].

2.1. Classification of Magnetic Bearing

2.1.1. Active Magnetic Bearings

Active Magnetic Bearings (AMB) achieve stable rotor suspension through controllable electromagnetic forces [31,32]. The materials of its stator and rotor are mostly silicon steel sheets or electric pure iron, and the coils are wound with copper wires. The stiffness and damping characteristics of the electromagnetic force can be dynamically adjusted via control algorithms, enabling precise regulation of bearing performance. Due to this controllability, AMBs offer a broader range of applications compared to other types of magnetic bearings. A typical configuration of AMBs is the radial magnetic bearing (RMB). Common structural designs include the 4-pole and 8-pole radial AMB configurations, as illustrated in Figures 1 and 2 [28,33,34].

The 4-pole RMB represents the simplest structural form of radial magnetic bearings. However, in many industrial applications, the 8-pole RMB is more widely adopted due to its superior performance. In an 8-pole configuration, adjacent magnetic poles pair to form four independent magnetic circuits, effectively decoupling the magnetic forces along four radial directions. This structural advantage simplifies control and improves dynamic stability. To regulate the currents in the four coil pairs, either two bipolar power amplifiers or four unipolar power amplifiers can be employed. Position sensors continuously monitor rotor displacement and provide feedback to the control system, which adjusts the coil currents in real time to maintain the rotor at the center position [35], as shown in Figure 3.

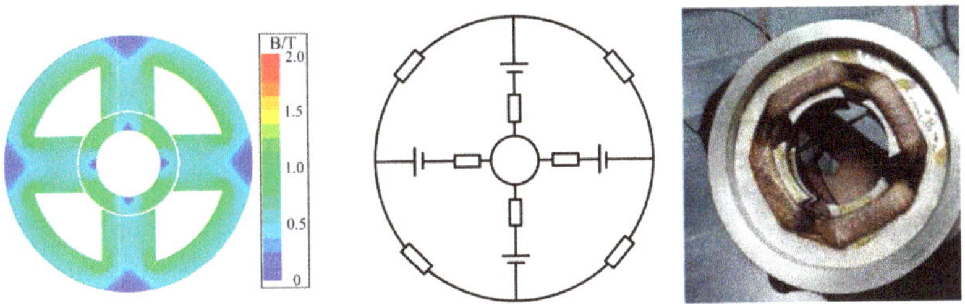

Figure 1. Magnetic circuit (flux lines) and prototype of 4-pole RMB [28,33].

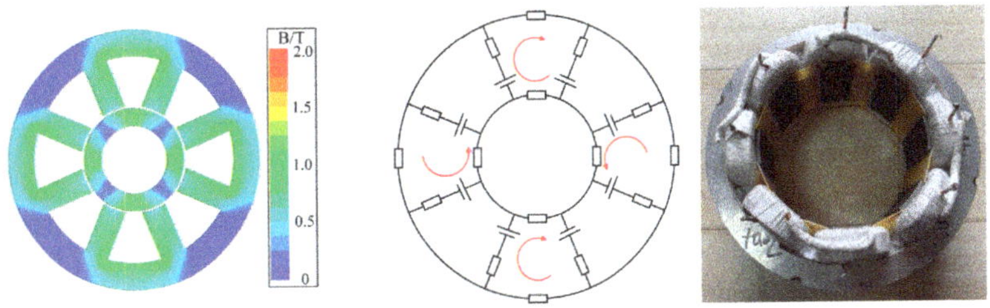

Figure 2. Magnetic circuit (flux lines) and prototype of 8-pole RMB [34].

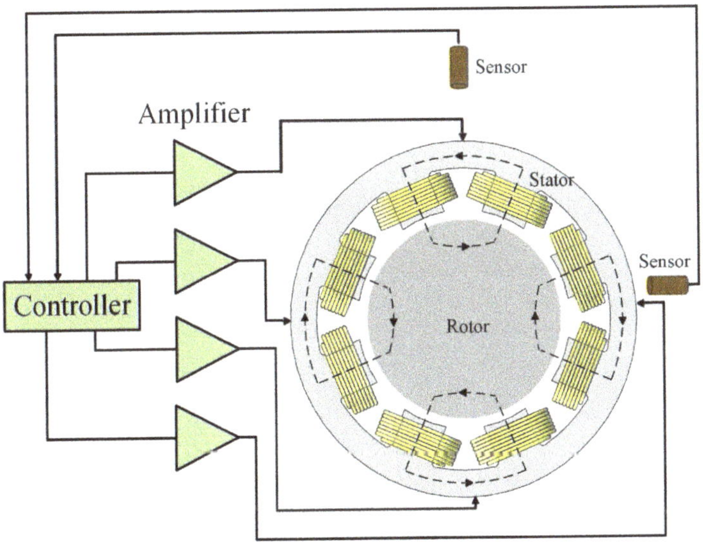

Figure 3. The operational concept of an RMB.

2.1.2. Passive Magnetic Bearings

Passive Magnetic Bearings (PMBs) achieve rotor suspension through magnetic fields generated by their own permanent or induced magnetism. A key advantage of PMBs is their simple structure, which requires no active control systems or external power, resulting in zero power consumption. However, this simplicity also means that the stiffness and damping of the system cannot be actively adjusted, which limits their versatility compared to active systems.

PMBs can be classified into three categories based on their suspension mechanisms:

1. Superconducting Magnetic Bearings,
2. Diamagnetic Magnetic Bearings and
3. Permanent Magnetic Bearings.

Among these, superconducting and diamagnetic bearings have limited practical use due to the need for cryogenic environments and inherently low stiffness, respectively. In contrast, permanent magnetic bearings are more commonly employed in practical applications [36,37]. The material of magnetic rings mostly adopts NdFeB with better performance. An example of a permanent magnetic bearing is illustrated in Figure 4.

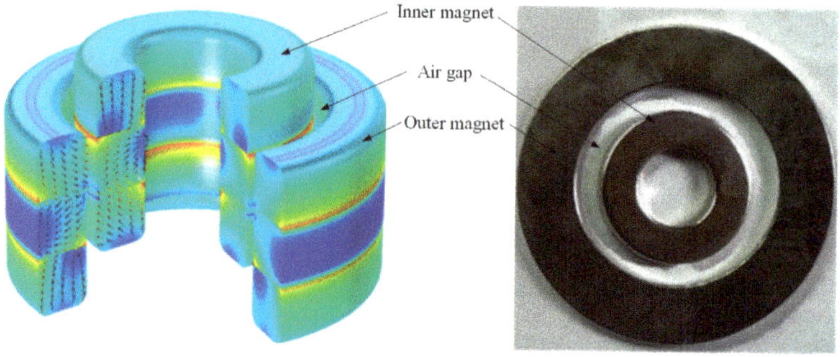

Figure 4. The 3-D constructor and prototype of a PMB.

Compared with AMB, the support force and stiffness provided by PMBs are relatively lower [38,39]. Therefore, PMBs generally adopt a stacked structure of multiple pairs of magnetic rings to achieve higher support stiffness. The basic design consists of two PM rings. Based on polarity and the layout direction, there are four basic configurations [40], namely radially attractive and repulsive types and axially attractive and repulsive types, as shown in Figures 5 and 6. In Figure 5a, when the two rings are offset in the radial direction, the attractive force will pull them back to the middle position, while the repulsive force will push them back to the middle position in Figure 5c.

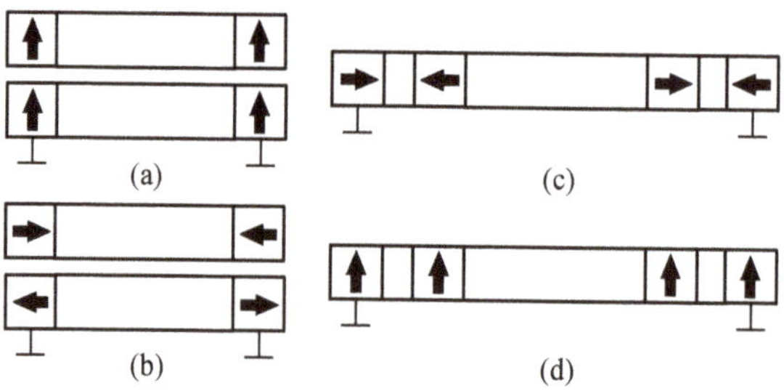

Figure 5. Radial PMB structure. (**a**) Attractive type; (**b**) Attractive type; (**c**) Repulsive types; (**d**) Repulsive types.

The attractive-type PMB is an unstable system when compared to the repulsive-type. For instance, in radial PMB (a), if the control failure occurs in the axial direction, the suspended rotor may drop or collide upward, which will damage the brittle magnetic ring. In contrast, the repulsive-type PMB overcomes the shortcomings of the attractive-type PMB. In the case of axial control failure, it can prevent the impact force caused by attraction and avoid damage to the bearing. However, repulsive PMB requires higher PM coercivity to resist demagnetization.

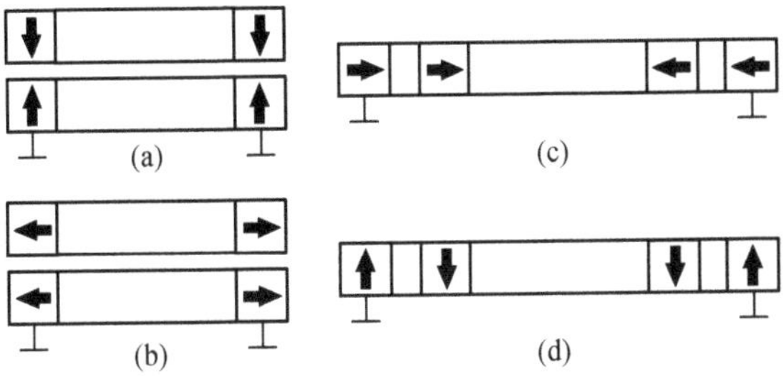

Figure 6. Axial PMB structure. (**a**) Repulsive types; (**b**) Repulsive types; (**c**) Attractive type; (**d**) Attractive type.

To address the low bearing capacity and stiffness associated with a single pair of magnetic rings, stacking is often used to improve their performance in engineering applications. Eight different topologies can be obtained from two stacking ways and four magnetization combinations [41–43]. The various combinations of stacking and magnetization are presented in Table 2.

Table 2. Stacking structure under different magnetization.

	Axial Magnetized	Radial Magnetized	Vertical Magnetized	Halbach Magnetized
Axial stacking	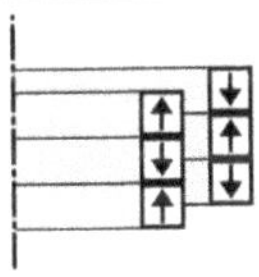	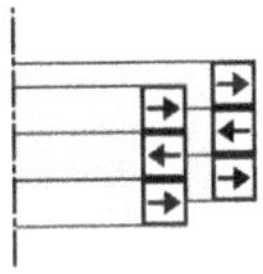	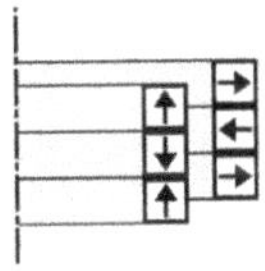	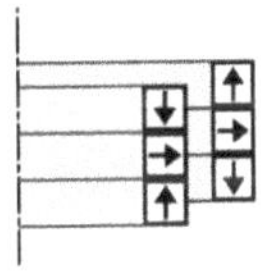
Radial stacking	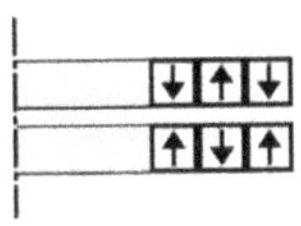	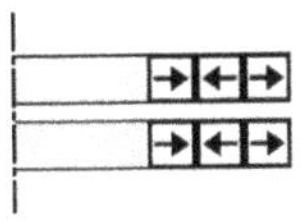	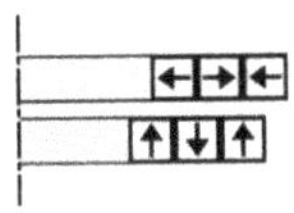	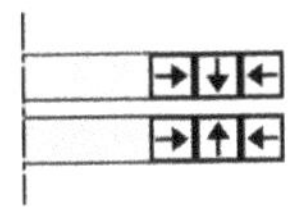

2.1.3. Hybrid Magnetic Bearings

Hybrid Magnetic Bearings (HMB) are an extension of the AMB design, incorporating permanent magnets (PMs) embedded within the tooth or yoke to generate a static bias magnetic field. This integration allows for the omission of the coil that typically generates the bias field, resulting in a more compact structure, reduced power consumption, and improved power density [44,45]. However, this configuration introduces greater complexity compared to traditional AMBs. For instance, in Figure 7, PMs are embedded in three alternating magnetic poles [46,47]. These permanent magnets create a permanent bias magnetic circuit in conjunction with the adjacent magnetic poles. The control current is then introduced into the coils to adjust the magnetic force. Compared to AMBs, which rely on coils to generate both bias and control magnetic fields, the current required in the coils of HMBs is inherently lower.

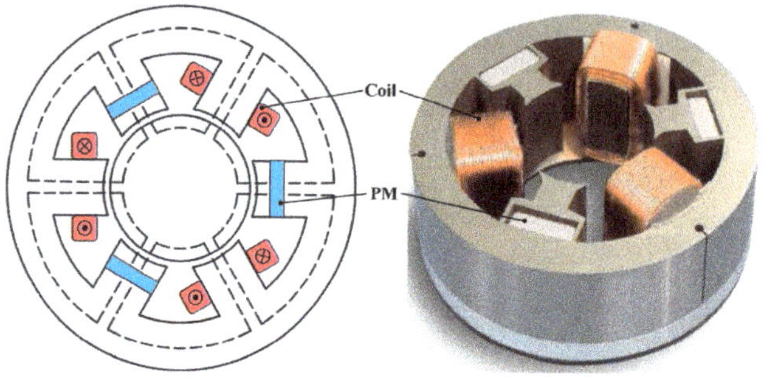

Figure 7. The magnetic circuit (flux lines) and prototype of a 6-pole HMB [46].

2.2. Comparison of Magnetic Bearing

The previous section provides a summary of the classification and fundamental principles of magnetic bearings. This classification framework serves as the foundation for understanding the diversity of magnetic bearing technologies and for selecting the most appropriate technical approaches. Table 3 presents a comparative analysis of the basic characteristics of the three main types of magnetic bearings.

Table 3. Comparison of basic characteristics of three types of magnetic bearings.

Characteristic	AMB	PMB	HMB
Source of force	Electromagnet [33,34]	PM [40–43]	PM + Electromagnet [46,47]
Control system	sensor and active control	Non-active control	Partial active control
Power consumption	High (continuous power supply)	0	Medium (adjustment only)
Stability	Full degree of freedom and stability	Required auxiliary stabilization	Full degree of freedom and stability
Typical application	High-precision machine tools [48], aircraft engines	Instrument, simple suspension device [43]	Compressor, flywheel energy storage [49]
Nonlinearity	Strong	Medium	Linearization around the operating point
Force density	Medium	Low	High
Speed range (r/min)	10^4~10^6	10^3~10^4	10^4~10^5
Reliability	High (electronic devices limit)	Extremely high (no active parts)	Very high

The diverse technical options available for magnetic bearings offer a wide range of choices for various industrial applications. Table 4 outlines the current application status and typical use cases of active, permanent magnet, and hybrid magnetic bearings across different industries, highlighting the relationship between their technical characteristics and application requirements.

Magnetic bearings offer numerous advantages; however, each type presents distinct technical challenges and application limitations that directly influence their adoption across various industries. Table 5 summarizes and compares the primary technical limitations associated with each type of magnetic bearing. These limitations not only act as barriers to the widespread use and implementation of magnetic bearings but also highlight potential avenues for future technological advancements.

Table 4. Comparison of the main application fields of three types of magnetic bearings.

Applications	AMB	PMB	HMB	Technical Selection Basis
High-speed precision machining	CNC machine tool spindle [48]	-	Mid-end machine tool [47]	Accuracy and response speed
Fluid machinery	Special operating compressor [50]	Micro-flow pump [42]	Centrifugal compressor, blower [51]	Balance of power and efficiency
Energy equipment	Flywheel energy storage [52]	-	Wind turbine, fuel cell compressor [45]	Reliability and maintenance
Transportation	Maglev train [53]	-	-	Large-scale accurate control
Aerospace	Satellite flywheel [54], APU	Simple suspension device [40]	Space actuators [55]	Environmental adaptability
Medical device	-	Artificial heart pump [56]	-	Zero power consumption and biological compatibility

Table 5. Comparison of the main technical limitations of three types of magnetic bearings.

Technical Limitation	AMB	PMB	HMB	Solutions
Cost	Very high (electronic system)	Low-medium	Medium-high	Scale and integration
Temperature limit	Medium (electronic devices)	High (PM limit)	Medium	high-temperature materials
Stability risk	Control failure	Earnshaw theorem	Medium	Improved protection bearing
Force density	Medium	Low	High	Optimize magnetic circuit
Standardization	High	Low	Low	Promoting industry standards
Typical failure mode	Electronic system failure	PM demagnetization	Electromagnetic under-regulation	Redundancy design

3. Research Progress on Key Technologies of Magnetic Bearing

Although magnetic bearings have been successfully applied across various fields, evolving demands require further enhancements in their performance. To fully harness their potential in high-speed and non-contact applications, the development and integration of several key technologies are crucial.

3.1. Topology and Modeling

3.1.1. Bearing Topologies

Various new configurations have been developed based on the conventional 4-pole and 8-pole active RMBs to enhance performance and reduce losses. The 4-pole RMB offers a simple coil layout and structural symmetry, making it ideal for small to medium-sized rotors. However, the magnetic circuits along the x and y axes are coupled. In contrast, the 8-pole RMB provides higher magnetic flux concentration and better control. However, with all eight poles arranged in a single plane and alternating north-south polarity, the magnetic field completes two full cycles per rotor revolution, leading to significant hysteresis and eddy current losses. To overcome these challenges, the 3-pole RMB presents a promising, compact alternative, particularly suited for rotors with small shaft diameters [48,57–60]. Additionally, using a three-phase inverter for control significantly reduces both system cost and complexity [61–65], as shown in Figure 8.

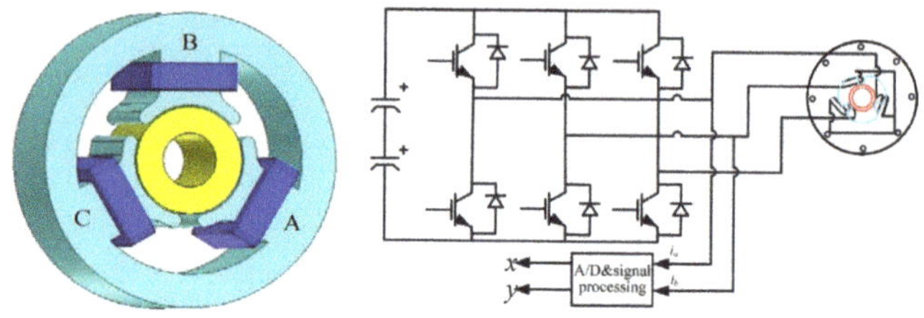

Figure 8. Prototype of 3-pole AMB and its control topology [57]. (**a**) Prototype; (**b**) Control topology.

In the 4-pole and 8-pole RMBs, the displacement in one axis is decoupled from the current and position in the orthogonal axis. However, the structural asymmetry of the 3-pole RMB introduces cross-coupling between electromagnetic forces along the x and y axes, thus increasing control complexity and susceptibility to instability. As a result, decoupling strategies must be integrated into the control algorithm to ensure system stability and precision [48,57–67]. To address this coupling issue, a 6-pole RMB topology has been proposed, as illustrated in Figure 9 [68]. Decoupling among the three pairs of electromagnets can be realized either by connecting coils on diametrically opposed poles or by linking coils on adjacent poles. These configurations enable a reduction in the number of drivers and power amplifiers, thereby lowering system cost and power consumption [47,69–72]. However, a notable limitation of this design is its relatively low resultant force coefficient due to the 120° spatial separation between electromagnets.

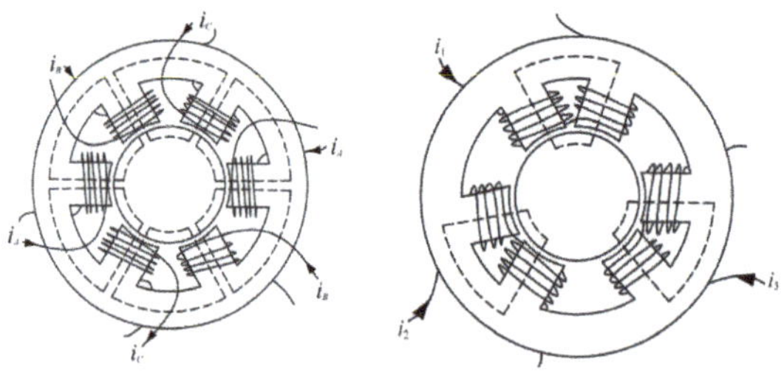

Figure 9. 6-poles AMB [68]. (**a**) Opposite poles connected; (**b**) Adjacent poles connected.

As the rotor diameter increases, the number of poles in magnetic bearings must also be increased to ensure adequate load capacity [73,74]. For instance, 12-pole and 16-pole RMBs, as shown in Figure 10a,b, provide comparable load-bearing capabilities to conventional 8-pole designs for medium- and large-sized rotors while offering reduced magnetic coupling effects and improved magnetic field uniformity. High-pole-count magnetic bearings are well-suited for applications such as industrial compressor blowers and heavy-duty centrifuges. However, these configurations require more power amplifiers and higher driving voltages, posing additional challenges in system design and integration. For magnetic bearings with a high number of poles used in very large rotors (as shown in Figure 10c, the marginal improvement in load capacity diminishes with increasing pole count, though the magnetic field becomes more uniformly distributed [75]. A modular pole design can enhance overall stiffness and mechanical robustness; however, it introduces greater assembly complexity, necessitates precise current control in each coil, and leads to elevated eddy current losses. Table 6 provides a comparative analysis of magnetic bearing technologies across different pole configurations.

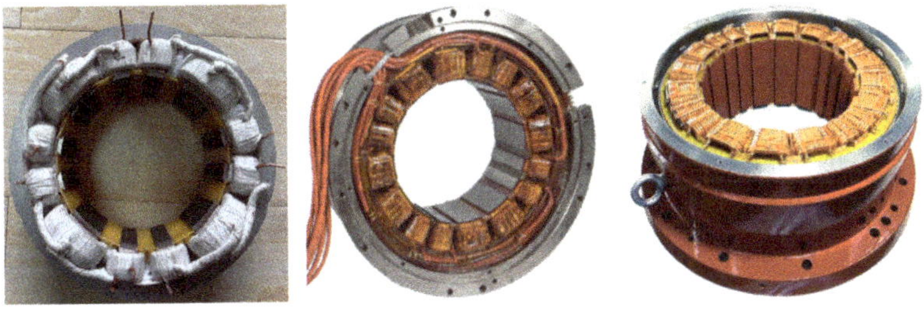

(**a**) (**b**) (**c**)

Figure 10. 12-poles, 16-poles and 20-poles AMB [73–75]. (**a**) 12-poles; (**b**) 16-poles; (**c**) 20-poles.

Table 6. Comparison of magnetic bearings with different poles.

Poles	Bearing Capacity	Stiffness	Power	Control Complexity	Typical Application
3 poles [57]	Low (micro-rotor)	Low	Medium-high	High	Microsensors, laboratory equipment
4 poles [33]	Medium (medium-small rotor)	Medium	Medium	Medium	Medical equipment, small compressor
6 poles [68]	High (medium rotor)	High	Medium-low	Higher	High-speed centrifuge, CNC machine tool
8 poles [34]	High (medium rotor)	Medium-low	Low	Low	High-speed motor, small flywheel energy storage
12 poles [73]	Medium (medium rotor)	Higher	Medium	Medium	Industrial compressors, medium blowers
16 poles [74]	High (large rotor)	High	High	High	Wind turbine, heavy centrifuge
24/32 poles	Stabilizing (oversized rotor)	Very high	Very high	Very high	Maglev train, nuclear main pump

In AMB, the magnetic field is generated entirely by electric coils, which results in significant heat generation. To reduce power consumption, HMB replaces the bias magnetic field coils with permanent magnets (PMs). HMBs are generally classified into homopolar and heteropolar types based on the polarity arrangement of the magnetic poles [47,76–78]. In homopolar HMBs, the bias and control magnetic fields are located in different planes, resulting in weak magnetic coupling. These systems typically use a split stator design, with PMs configured as axial rings or multiple rectangular blocks, leading to a relatively longer axial length. In contrast, heteropolar HMBs feature alternating pole polarities, with the bias and control magnetic fields in the same plane, resulting in stronger magnetic coupling but also a higher risk of PM demagnetization. In this configuration, PMs are usually embedded within the stator, resulting in a more compact structure and shorter axial length.

As a result, homopolar HMBs generally exhibit lower rotor eddy current losses, while heteropolar HMBs offer a higher critical speed [79,80]. Figures 11 and 12 illustrate typical configurations of both designs.

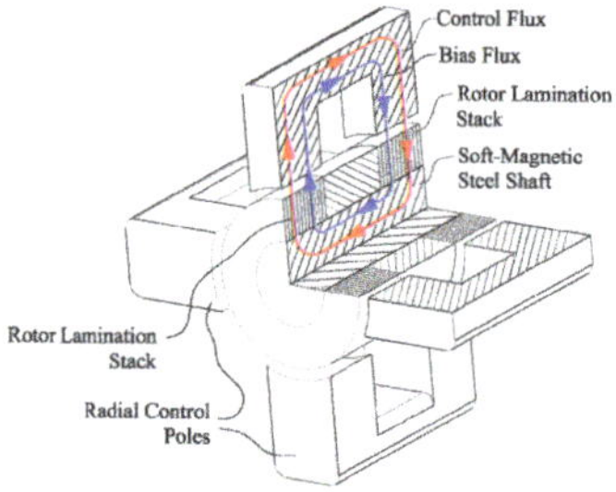

Figure 11. Model and prototype of 4-pole homopolar HMB [76].

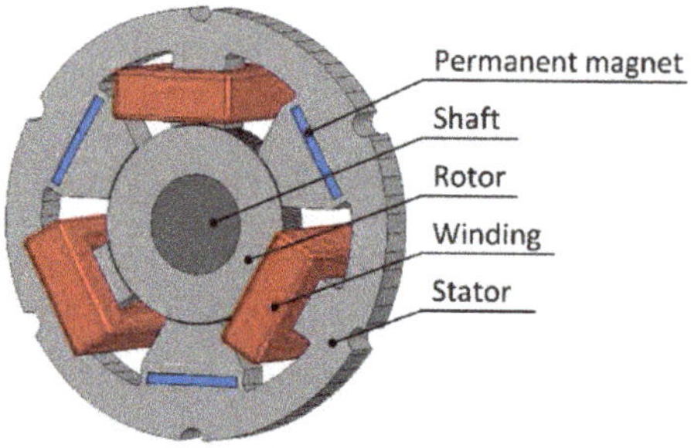

Figure 12. Model and prototype of 6-pole heteropolar HMB [47].

3.1.2. Modeling of Electromagnetic Force

The electromagnetic force modeling of magnetic bearing is the core foundation of system design and control, and its accuracy directly affects the bearing characteristics, dynamic response and stability. The current mainstream modeling methods can be divided into two major categories: analytical methods and numerical methods. In the following, the way of electromagnetic force modeling of magnetic suspension bearing is classified and elaborated.

1. Equivalent Magnetic Circuit Method
   The equivalent magnetic circuit method (EMC) is the most classical modeling method for the electromagnetic force of magnetic bearings, which is widely used in AMB and HMB [44,77,81–88]. EMC simplifies the topological relationship of the magnetic circuit and decomposes the magnetic circuit into series or parallel reluctances for calculation. It is suitable for magnetic bearing structures with relatively regular magnetic field distribution. For ease of calculation, only the working air-gap reluctance is generally considered, while the leakage reluctance and the core reluctance are ignored. The advantage of this method is its high calculation efficiency (more than 10 times faster than FEM), but the leakage effect needs to be corrected through the leakage coefficient compensation, with an error of approximately 5% to 10%.
2. Maxwell Tensor Method
   Based on Maxwell's equations, the electromagnetic force is calculated by integrating the air-gap magnetic field tensor, which is especially suitable for modeling the radial force of AC magnetic bearing [49,89–96]. In [91], an accurate radial force model is established for bearingless motors and quantifies the suspension force through the air-gap flux density component. The stiffness of the magnetic bearing is analyzed and quantified by the Maxwell tensor method in [49], and the error can be within 3%.

3. Equivalent Magnetic Charge Method
   The general magnetic circuit model proposed by Yonnet assumes that the PM is infinite in length, and the magnetic force analytical formula of PMB is derived by combining the equivalent magnetic charge method [97]. This method equivalents the permanent magnet as a virtual magnetic charge distribution and calculates the magnetic force in combination with Coulomb's law. According to the force relationship between the magnetic charges at two points, the numerical integration model of axial, radial, and Halback magnetized PMB can be established [98–101]. However, the numerical calculation methods of the model are generally complicated.
4. Magnetic Network Method
   Based on EMC, the magnetic network (MN) method constructs nonlinear equations by further discretizing the magnetic circuit nodes, which takes into account both computational efficiency and accuracy. For spherical magnetic bearings, the magnetic field is segmented accurately based on the flux-tube principle, and the edge flux and flux leakage are calculated accurately [11]. A network model in a spherical coordinate system is established to quantify the multi-degree freedom coupling effect, and more accurate calculation results are obtained.
5. Subdomain Method
   The subdomain method divides the magnetic field region into linear subdomains (such as air gaps, cores, and PMs) and solves them by the magnetic field boundary conditions. The magnetic field problem in sub-regions can be solved by using the vector magnetic potential within each sub-region. In [102–104], the radial AMB is divided into air-gap domain and slot domain. The zero-order equation and first-order equation of the magnetic field are calculated in polar coordinates, and the distribution of the magnetic field is obtained by using the method of variable separation.
6. Finite element method
   The finite element method (FEM) is the preferred tool for complex geometry and nonlinear material modeling, which has the highest accuracy and can construct multi-physics coupled models. However, the preprocessing and calculation time of the model are long, which limits the efficiency of design optimization and the speed of system-level simulation.

Table 7 compares and summarizes the electromagnetic force modeling methods mentioned above. For magnetic bearings with different structures, appropriate modeling methods should be selected according to requirements.

Table 7. Comparison of electromagnetic force modeling methods.

Modeling Methods	Advantage	Disadvantages	Applicable Scene
Equivalent Magnetic Circuit Method [44,77,81–87]	Computationally efficient, suitable for fast iterative	Ignore flux leakage and nonlinear effects, which require empirical correction	Magnetic bearing with regular magnetic field distribution, preliminary design stage.
Maxwell Tensor Method [49,89–96]	High precision, quantifiable current stiffness and eddy current loss.	High computational complexity	AC magnetic bearing, solid rotor
Equivalent Magnetic Charge Method [97–101]	Suitable for different magnetization directions, High theoretical accuracy	complex numerical integration model	Analysis of static characteristics of PMB

Table 7. *Cont.*

Modeling Methods	Advantage	Disadvantages	Applicable Scene
Magnetic Network Method [11]	Balance Efficiency and precision, Support 3D modeling	The error increases when the network is partitioned coarsely	Complex geometric structure, multi-DOF coupling analysis
Subdomain Method [102–104]	High analytical accuracy	Ignore core saturation	5-DOF magnetic bearing, multi-pole structure
FEM	Highest precision support for multi-physics field coupling	time-consuming, the hardware requirement	Analysis of complex physical field distribution

3.2. Control Strategy

3.2.1. Classical PID Control

The fundamental distinction between AMB systems and conventional control systems lies in the inherent open-loop instability and negative stiffness characteristics of both AMB and HMB systems [105,106]. These characteristics necessitate the use of closed-loop control to ensure stable rotor suspension and operation [35,107]. In modern industrial applications, the most widely adopted control strategy remains the Proportional-Integral-Derivative (PID) controller due to its simple structure, intuitive parameter tuning, and independence from the physical model of the controlled system [108–110]. For AMB systems with multiple DOFs and rigid rotors, each DOF can be considered independently and regulated using decentralized PID control. This approach simplifies the control architecture while maintaining effective dynamic performance [111,112]. Most existing magnetic bearing systems employ decentralized PID controllers or their variants [113–118], such as fractional order PID [116,119], segmented PID, Neural network PID [52,120,121], fast optimal PID, expert PID, and fuzzy PID [122–124] control, among others.

It should be emphasized that the AMB is a nonlinear and strongly coupled dynamic system. The widely used decentralized PID control strategy, while simple and effective in many cases, does not account for inter-axis coupling, which can significantly affect performance in high-speed and high-precision applications. As industrial demands for higher rotational speed, greater stability, and improved dynamic response continue to rise, traditional decentralized PID control increasingly falls short of meeting modern engineering requirements. To address the multi-input multi-output (MIMO) coupling present in AMB-rigid rotor systems, considerable research has been devoted to decoupling strategies. These aim to transform the coupled system into a set of independent single-input, single-output (SISO) subsystems under specific conditions. This approach preserves the simplicity and tuning advantages of PID controllers while incorporating anti-gyroscopic coupling characteristics from modern control theory, thus enhancing both system stability and control precision. Currently, several decoupling techniques have been applied to AMB systems, including linear state feedback decoupling [125], cross-feedback decoupling control [126,127], and modal decoupling control [128], among others.

3.2.2. Advanced Control Algorithm

Considering the significant nonlinearity and coupling phenomena of the rotor during high-speed rotation, researchers are extending intelligent control methods beyond traditional PID. With the rise of intelligent algorithms and high-performance processors, several more advanced control algorithms have been applied to magnetic bearing control, such as robust control [129–131], predictive control [132–134], sliding-mode control [135–138], neural network control [139,140], and adaptive control [54,141,142], fuzzy control [143–145], active disturbance rejection control [50,146,147]. These methods and their variants bring solutions to strong nonlinearity, parameter uncertainty, and external disturbances, greatly enhancing stability and accuracy and improving overall static and transient performance.

1. Robust control
   Robust control ensures the system stability under parameter variations or external disturbances by designing controllers that are insensitive to uncertainties, parameters in the model and disturbances. In the magnetic bearings, the robust controller based on μ analysis effectively deals with the high-frequency vibration and modal coupling problems in the rotor dynamics through frequency domain analysis and pole configuration and improves the stability margin of the system. Plus, H∞ control is applied to solve the mode control problem of the magnetic bearing rotor system. The system identification model is established by the orthogonal polynomial fitting method, and the stable suspension and mode suppression of the rotor are realized.
2. Model Predictive Control (MPC)
   MPC predicts future system states and optimizes control inputs based on a system model, making it well-suited for multivariable coupled control in magnetic bearing systems. MPC based on the radial 4-DOF global control model can achieve high-precision tracking of the rotor position and transient response optimization using the state-space model and optimal controller design. This approach demonstrates excellent performance in high-speed rotating machinery. However, it also entails high computational complexity and depends heavily on high-performance processors for real-time implementation.
3. Sliding-mode control (SMC)
   SMC is extensively applied in the nonlinear control of magnetic bearings due to its strong robustness against parameter variations and external disturbances. By appropriately designing the sliding surface and control law, SMC ensures fast convergence and effectively suppresses rotor displacement fluctuations. In applications such as magnetic bearings for wind turbines, the integration of SMC with PID control has been shown to reduce overshoot and enhance response speed. However, the well-known chattering phenomenon associated with SMC remains a challenge and necessitates further improvement through techniques such as adaptive boundary layer design and higher-order sliding mode methods.
4. Neural network control
   Neural networks can effectively model the complex behaviors of magnetic bearings by learning and approximating their nonlinear dynamics. A hybrid control architecture, integrating deep learning with PID feedback, is employed to design a compensation controller, significantly enhancing the suppression of unbalanced vibrations. Additionally, convolutional neural networks (CNN) and gated recurrent units (GRU) are utilized for fault prediction through the analysis of current waveforms and vibration spectra. However, neural networks face challenges, including a reliance on large volumes of training data and the need for optimization to meet real-time performance requirements.
5. Adaptive control
   Adaptive control adjusts system parameters online to accommodate changes, with methods like Model Reference Adaptive Control (MRAC) and Active Disturbance Rejection Control (ADRC). In magnetic bearing control systems, the ADRC algorithm, utilizing an extended state observer, estimates and compensates in real time, thereby minimizing manual parameter adjustments. Intelligent optimization techniques, such as Beetle Antennae Search (BAS), are employed to fine-tune PID parameters, achieving rapid convergence and low energy consumption in multi-DOF magnetic bearings.

6. Fuzzy control
   Fuzzy control does not require the precise mathematical model of the plant or the detailed system dynamics. It is particularly suitable for dealing with the nonlinearity and uncertainty of the magnetic bearing system. It relies on the relationship between error, error rate and output and uses fuzzy reasoning based on control rules to adjust control decisions according to specific system conditions to meet requirements. It overcomes the limitation of traditional PID control that cannot be adjusted in real-time. It also saves the time required for manual control parameter debugging.
7. Active disturbance rejection control
   As an advanced Control method that does not rely on accurate models, Active Disturbance rejection control (ADRC) has been widely used in magnetic bearings in recent years. The core idea is based on real-time estimation and compensation of disturbance. The multi-DOF coupling effect of the magnetic bearing is estimated in real-time through the Extended State Observer (ESO), and the nonlinear feedback control law is combined to realize decoupling and disturbance rejection. ADRC only needs to design the controller based on the input and output data. This property is especially suitable for scenarios where the nonlinear dynamics of magnetic bearings are difficult to model accurately.

A single control algorithm often cannot simultaneously optimize both transient performance and system robustness. Therefore, combining multiple control strategies can yield superior results. For instance, the integration of fuzzy control with PID combines the rule-based flexibility of fuzzy control with the steady-state accuracy of PID, striking a balance between transient response and disturbance rejection in HMB systems [143,145]. Another example is the fusion of adaptive control with robust control to address the strong coupling issues in magnetic bearings. Simulation results demonstrate that this approach reduces errors by over 70% compared to traditional methods [53,54]. Table 8 compares the aforementioned control methods.

Table 8. Comparative analysis of different control strategies.

Control Algorithm	Advantages	Limitations	Typical Application
PID control [111]	Simple structure, clear parameters, easy to implement	Difficult to adapt to nonlinearity/time variation, Limited anti-interference	Low-complexity systems (e.g., laboratory demonstration)
Variable parameter PID [112]	Adjust parameters dynamically, balancing dynamic response with accuracy.	Complex adjustment rules real-time performance is limited	Startup phase or set point changes frequently
Robust control [129]	Parameters and perturbations insensitive, multivariate coupling.	Relying on accurate models, high computational complexity, dynamic performance is limited	High-precision machine tools, aerospace flywheels, and strong interference environments.
Predictive control [132]	Predict the future state based on the model and adapt to nonlinear/time-varying systems.	Large computation, error sensitive, update frequently	Dynamic, high-speed response, multi-target coordination

Table 8. *Cont.*

Control Algorithm	Advantages	Limitations	Typical Application
Sliding-mode control [135]	Strong robustness, quick response, parameter insensitive	Chattering requiring higher-order precise sliding mode design or adaptive control	High anti-interference (e.g., flywheel energy storage)
Neural network control [139]	No need for an accurate mathematical model; online learning	Amount of training data, computationally intensive, poor interpretability	Adaptive adjustment of complex operating (e.g., variable load)
Adaptive control [54]	Automatically adjust parameters to improve robustness.	Slow convergence speed adaptability to fast time-varying systems is limited.	System with parameters changing slowly (e.g., compressor)
Fuzzy control [53]	Dynamically adjust parameters, no need for accurate models, anti-interference.	Rule design relies on experience; parameter tuning is complex, and high-frequency accuracy is insufficient.	Strong nonlinearity and unclear model (e.g., medical devices)

3.3. *Power Drive System and Controller of Bearings*

3.3.1. Driver Topology Design

An efficient and stable power amplification system is essential for achieving a magnetic bearing system with high sensitivity, rapid response, and low cost. Traditional linear power amplifiers have been replaced by power electronic converters due to their high loss and large size. One of the earliest solutions is the full-bridge topology shown in Figure 13a [148], which controls the coil current through four switches, offering a three-level output and good transient performance. However, multi-DOF magnetic bearing systems require a large number of devices, leading to high cost and volume. Improved topologies, such as the unipolar half-bridge structure in Figure 13b, reduce the number of devices by half and are ideal for active magnetic bearings (AMBs) that require only unidirectional current [149,150].

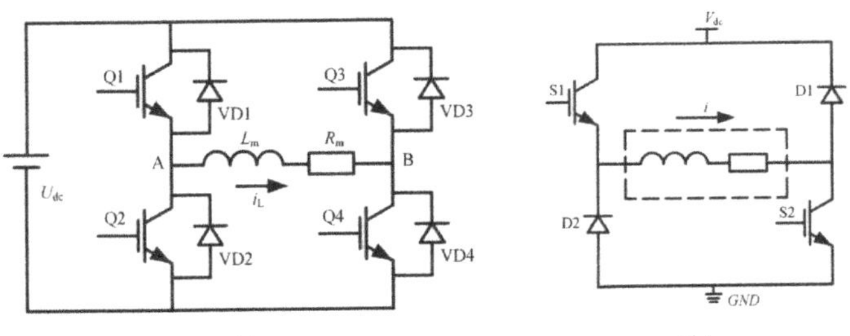

Figure 13. H-bridge drive topology. (**a**) Full bridge [148]; (**b**) Half bridge [149].

However, using an H-bridge topology for each coil requires a large number of electronic devices. Hence, a three-phase full-bridge topology is proposed to control a single degree of freedom (DOF), as shown in Figure 14a [151]. This three-phase topology allows the two coils to share the middle bridge, simplifying the design and reducing the number of switching devices by approximately one-quarter compared to the full-bridge topology.

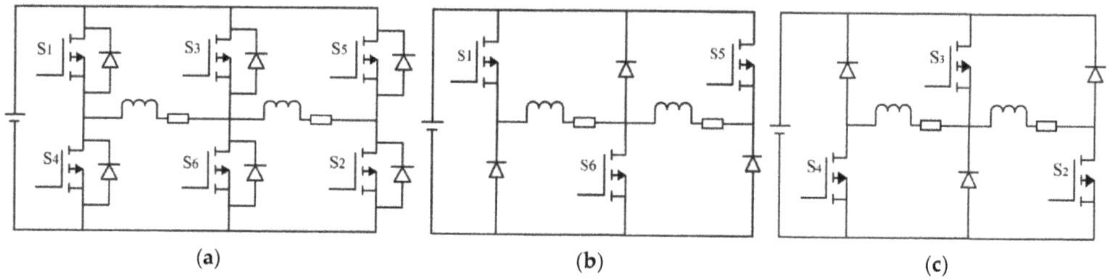

Figure 14. Three-phase drive topology. (**a**) Three-phase full-bridge [151]; (**b**) Three-phase half-bridge t-b-t [152]; (**c**) Three-phase half-bridge b-t-b.

Despite this improvement, the switching devices in the three-phase full-bridge topology are still underutilized, and redundancy remains in the system. To optimize this, a three-phase half-bridge topology is introduced, where each three-phase half-bridge controls two coils on a single DOF [152]. Depending on the configuration of the switching devices, this topology is divided into the b-t-b structure and the t-b-t structure. The b-t-b structure, shown in Figure 14b,c, offers the advantage of reducing the number of common grounds and gate drive power supplies compared to the t-b-t structure. However, since the output voltage on the shared bridge is fixed at Vdc/2, the DC voltage utilization is halved in comparison to the H-bridge structure.

Building on the shared bridge concept in Figure 15, the number of devices can be further reduced by sharing multiple bridges [153]. As illustrated in Figure 16, the switching devices for the shared bridge are positioned in the upper bridge, which connects to the DC bus, while the switching devices of non-shared bridges are located in the lower bridge connected to the negative pole. These bridges share a drive isolation power supply. In the case of an N-axis magnetic bearing system, only $2N + 1$ switching devices and diodes are required. This approach reduces the total number of devices; however, the current on the shared bridge is the sum of the currents from the other bridges, resulting in higher current stress. Consequently, the shared bridge needs to be designed with greater consideration, increasing the risk of failure. To address this, the reverse shared bridge structure is introduced, which modifies the system to reverse half of the coil currents. This configuration causes the currents on the shared bridge to cancel each other out, significantly reducing the current stress. For an N-axis system, this topology requires $2N + 2$ switches and diodes.

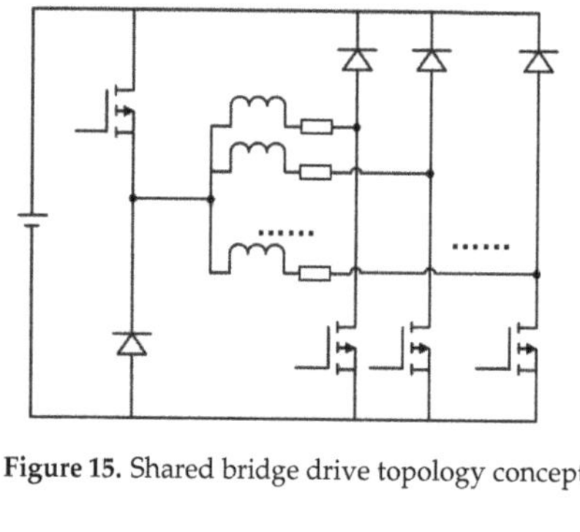

Figure 15. Shared bridge drive topology concept [153].

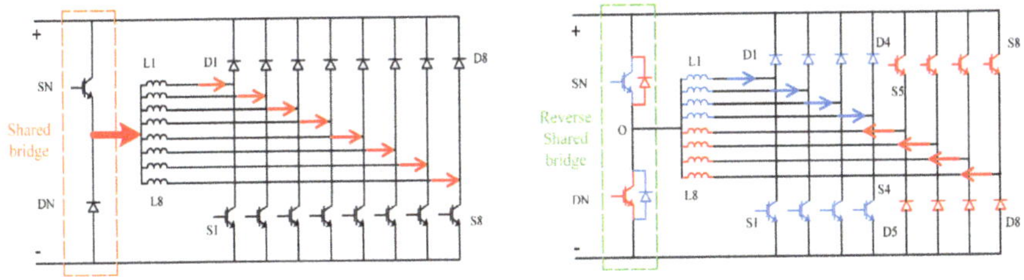

(a) (b)

Figure 16. Shared bridge drive topology and reverse shared bridge drive topology [154]. (**a**) Shared bridge drive. (**b**) reverse shared bridge drive.

In the previously discussed shared bridge topology, a single bridge serves as the common node for all coils. To maintain system stability, the average voltage at the midpoint of this common bridge must remain constant, which inherently limits the DC voltage utilization across all H-bridge topologies. This constraint can reduce the control system's bandwidth. To overcome this limitation, a series coil topology with multiple shared bridges is proposed, allowing the midpoint voltages at both ends of each coil to vary dynamically [154,155]. This approach preserves high DC voltage utilization without increasing the number of switching devices. As shown in Figure 17a, a four-phase, four-bridge topology eliminates the need for a common arm, further reducing the controller's physical footprint while maintaining performance.

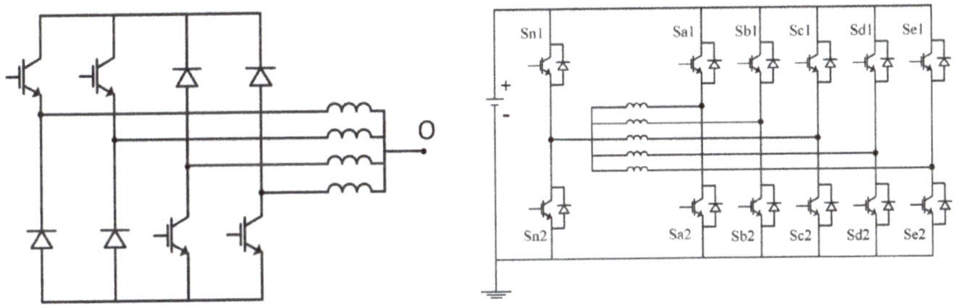

(a) (b)

Figure 17. (**a**) Four-phase four-bridge topology. (**b**) Five-phase six-bridge topology [156].

The above-mentioned three-phase topology, shared bridge topology and series coil topology belong to the unipolar switching power amplifier topology, which is generally applied to the AMB system of current superposition type. For HMB of magnetic field superposition type, it is necessary to use the bipolar switching power amplifier to realize the control of the rotor through the bidirectional flow of current. Therefore, to realize the suspension control of the 5-DOF HMB, the topology of the five-phase six-bridge is proposed in Figure 17b [156].

Table 9 compares the number of devices and DC voltage utilization across various topologies for an *N*-axis system. Compared to the H-bridge structure, all improved topologies demonstrate a reduction of switching devices. Notably, the series coil topology achieves optimal device efficiency without compromising DC voltage utilization, offering a balanced solution in terms of performance and hardware cost.

Table 9. Comparison of the topologies of the *N*-axis system.

Topology	Switches	Diodes	Drive Power	Voltage Utilization
Full bridge [148]	$8N$	$8N$	$4N+1$	1
Half bridge [149]	$4N$	$4N$	$2N+1$	1
Three-phase full bridge [151]	$6N$	$6N$	$3N+1$	0.5
Three-phase half bridge [152]	$3N$	$3N$	$N+1$	0.5
Shared bridge [153]	$2N+1$	$2N+1$	2	0.5
Reverse shared bridge [154]	$2N+2$	$2N+2$	3	0.5
Series coils [155]	$2N+1$	$2N+1$	$N+1$	1

3.3.2. Fault-Tolerant Control and Fault Detection

Magnetic bearing systems are primarily used in applications requiring high-speed and high-precision control, where a rotor drop caused by system failure can lead to severe damage. During a suspension failure, the rotor instability manifests as a mechanical process, whereas the associated current failure in the coils is an electrical process. Since electrical dynamics are significantly faster than mechanical responses, the rotor position undergoes minimal change in the immediate aftermath of a fault. Additionally, given that the coil resistance is typically much lower than its inductive reactance, the power amplifier introduces a natural delay. This delay, combined with the response characteristics of power electronic converters, creates an opportunity for implementing fault-tolerant control strategies within the system.

The electromagnetic force in magnetic bearings is a reluctance force, which depends on the magnitude of the current rather than its direction. Therefore, topologies such as the full-bridge and three-phase full-bridge, despite incorporating redundant switching devices, offer a natural advantage for fault-tolerant operation. These redundant components can be repurposed to serve as a backup for an entire drive channel. In the event of a failure, the current path can be rapidly redirected in the opposite direction without affecting the generated electromagnetic force, enabling continuous system operation. For example, in a three-phase full-bridge topology, if a fault occurs in one switch or current path, the system can quickly switch to the redundant half of the bridge. By reversing the current direction through the alternate path while maintaining the same current magnitude, the magnetic force remains unaffected, thereby achieving fault-tolerant control without compromising suspension performance [153]. Figure 18 shows the fault-tolerant operation for three-phase full-bridge topology.

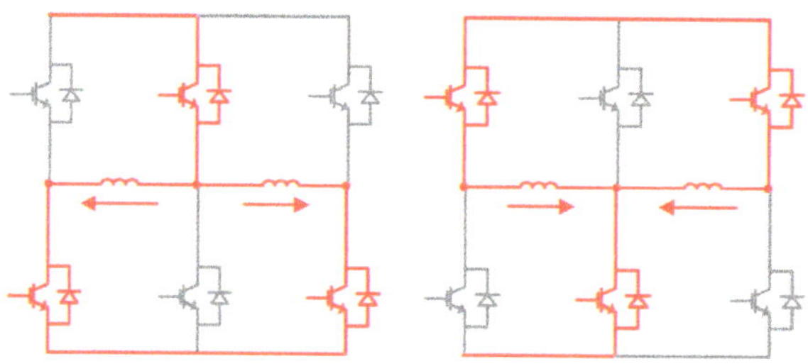

(a) (b)

Figure 18. Fault-tolerant operation for three-phase full-bridge topology [153]. (**a**) normal mode. (**b**) Redundant mode.

The three-phase full-bridge converter shown in Figure 14a can be conceptualized as comprising two sets of three-phase half-bridge modules: (S2, S3, S4) and (S1, S5, S6).

During normal operation, as illustrated in Figure 18a, the first set of switches (highlighted in red) is active, while the remaining switches remain in standby mode without receiving control signals. In this configuration, current flows outward from the shared bridge to the coils. In the event of a failure, the system can switch to the redundant drive mode, as shown in Figure 18b. Here, the second set of switches becomes active (highlighted in red), and the previously used switches are turned off. In this mode, current flows into the shared bridge. Since the electromagnetic force in magnetic bearings is a reluctance force and is independent of the current direction, both drive configurations can generate equivalent suspension forces. This inherent bidirectionality enables fault-tolerant operation, ensuring continuous rotor suspension even under drive system failure.

Common faults in switching devices primarily include open-circuit and short-circuit failures, both of which can cause sudden and abnormal variations in coil currents. As a result, fault detection in magnetic bearing drive systems can be effectively achieved by monitoring and analyzing the total coil current. Significant deviations from expected current profiles may indicate the presence of a switching fault, enabling timely fault diagnosis and the activation of fault-tolerant control mechanisms. To enable rapid fault diagnosis and response, two threshold currents are defined: I_{open} and I_{short}. If the total measured coil current falls below I_{open}, it indicates an open-circuit fault in the switching device. In this case, control of the magnetic bearing rotor can be restored by promptly switching from the faulty drive module to a redundant backup module. Conversely, if the total current exceeds I_{short}, it signifies a short-circuit fault within the module. Under such conditions, all control signals and drive modules should be immediately shut down to prevent further damage to the system and ensure operational safety. This threshold-based strategy enables fast and reliable protection for high-speed magnetic bearing systems.

3.3.3. Controller Hardware Platform

The hardware platform of the magnetic bearing controller is the key carrier to realize high precision and high real-time control. With the complexity of the control algorithm and the increase of the demand for multi-DOF cooperation, the controller hardware has gradually developed from the early analog circuit to a digital architecture with an embedded microcontroller (such as STM32), digital Signal processor (DSP) and Field Programmable Gate Array (FPGA) as the core. The following elaborates on hardware types, performance characteristics, and typical applications

1. Embedded Microcontroller
   The STM32 series core microcontrollers are widely used in medium and low-complexity maglev systems due to their advantages of low power consumption, high integration and cost. For example, the STM32F4 series (main frequency of 180 MHz) captures the Hall sensor signal through a timer and combines the PID algorithm to achieve single-DOF suspension control with a displacement resolution of ± 10 μm [157]. The CAN and Ethernet interfaces of STM32 can realize multi-axis cooperative control, but limited by the computing power, STM32 makes it difficult to meet the real-time requirements of high-speed multi-DOF systems.
2. Digital Signal Processor
   Digital Signal Processor (DSP) has become the mainstream scheme of dynamic control of magnetic bearing because of its high-performance floating-point operation ability and special peripheral equipment. The built-in redundant ADC module supports sensor fault detection and switching, which improves the fault tolerance of the magnetic bearing system. TMS320F28335 (main frequency 150 MHz) is adopted in [158,159]

combined with a fault diagnosis algorithm, which can control the fault switching time of magnetic bearing in 5 ms and significantly improve the stability of the system.

3. Field Programmable Gate Array
   With the characteristics of parallel processing and nanosecond delay, a Field Programmable Gate Array (FPGA) plays a key role in high-speed and multi-variable maglev systems. In the speed-holding mode of a high-speed maglev motor, the observation accuracy of steady-speed frequency is better than 5 Hz, and the response time is less than 50 μs [160,161].

The different types of controller hardware of magnetic bearings are summarized in Table 10. It is necessary to take into account both the performance and cost to select the appropriate controller platform.

Table 10. Comparison of controller hardware platform.

Hardware Types	Advantage	Disadvantages	Applicable Scene
STM32 series [157]	Low power consumption, low cost, short development cycle, easy to control	Limited computing power and limited real-time performance	Low complexity scenarios such as medium and low speed, single/two DOF.
DSP [158,159]	High real-time performance, support for complex algorithms	Relying on special development tools, high power consumption	High-speed, multiple DOF and other high-precision scenes
FPGA [160,161]	Strong real-time performance	Long development cycle, high cost	Ultra-high-speed and complex algorithm scenarios.

4. Application Fields and Typical Cases

Building on the aforementioned key technologies, magnetic bearings have found applications across a wide range of fields. Based on application domains and functional requirements, their use can be classified as follows.

4.1. Industrial Field

4.1.1. High-Speed Motor and Compressor

Fully sealed compressors equipped with shielded magnetic bearings and shielded induction motors can be used for various challenging applications, including acid gas handling, subsea compression, wet gas compression, and CO_2 reinjection. The operating conditions for such sealed compressors are examined. Ref. [51] gives an application where the use of canned magnetic bearings for enhanced recovery in a CO_2 environment can eliminate the shaft end seal and thus improve the reliability of the machine. Table 11 shows its parameters. At the same time, safety is improved due to the reduced risk of gas leakage into the environment, which is also important to underwater applications.

Table 11. The parameters of the hermetically sealed compressor.

Machine Type	5.8 MW Centrifugal Compressor with Canned Electric Motor
Speed range	2850~9975 r/min continuously variable
Rotor mass	1622 kg
Rotor length	3.3 m
Rotor orientation	Vertical
AMB configuration	Single shaft, 2 radial and 1 axial.

4.1.2. Precision Machine Tool Spindle

By tracking the position of the spindle tool, an AMB controller based on μ-synthesis is designed to minimize the difference between the reference position and the estimated position of the tool [48]. The experimental verification is carried out on the spindle AMB boring machine. At 15,000 r/min, the step-shaped, conical, and convex contours are successfully tracked with steady-state errors of about 8.7% and static and position fluctuation of 0.851 μm and 1.887 μm, respectively. The system shows strong robustness under static loads of 3.2 kg and dynamic loads with a 10 mm eccentricity, which verifies the effectiveness of the proposed method for balancing tracking accuracy and anti-interference ability in high-speed machining. Parameters are listed in Table 12.

Table 12. The parameters of the high-speed spindle.

Machine Type	10 kW Industrial Grade AMB High-Speed Spindle
Speed range	0~50,000 r/min continuously variable
Static load	3.2 kg
Tracking range	90 μm
Rotor orientation	Horizontal
AMB configuration	Single shaft, 2 radial and 1 axial.

4.2. *Energy and Transportation Field*

4.2.1. Flywheel Energy Storage System

A 5-DOF AMB system, designed for a high-strength steel energy storage flywheel without a shaft and hub, is presented in Table 13 [162]. By integrating radial, axial, and tilt suspension functions into a single unit, this design replaces the traditional multi-unit magnetic bearing system, significantly simplifying the structure and reducing costs. Experimental tests demonstrate the system's capability to support a flywheel weighing 5440 kg and measuring 2 m in diameter, with an air gap of 1.14 mm. These results confirm the bearing's reliability under high load conditions and validate the design approach. This technology enhances the energy density of the flywheel storage system, achieving twice the capacity of conventional steel flywheels while also reducing material costs, offering strong potential for practical engineering applications.

Table 13. The parameters of flywheel energy storage system.

Machine Type	100 kW Energy Storage Flywheels
Speed range	3225 r/min
Rotor mass	5440 kg
Outer diameter	2.13 m
Rotor orientation	Vertical
AMB configuration	Single shaft, combination 5-DOF AMB.

4.2.2. Wind Turbine Spindle

A design for a small vertical wind turbine based on a magnetic suspension axial flux PM generator is proposed [163]. By integrating PMBs, the turbine achieves frictionless suspension, which minimizes mechanical losses and enhances efficiency. The performance details are provided in Table 14. Experimental results show that the application of PMBs significantly improves the turbine's speed and efficiency, particularly under variable wind conditions, outperforming traditional bearing systems. This design simplifies the system by eliminating the need for gear transmission, reducing complexity and costs, and utilizing low-cost materials, making it ideal for home and small-scale applications.

Table 14. The parameters of the prototype.

Machine Type	Small Scale Vertical Axis Wind Turbine
Speed range	200 r/min
Maximum repelling force	124 N
Outer diameter	500 mm
Rotor orientation	Vertical
AMB configuration	Single shaft, PMB.

4.3. Extreme Environments

4.3.1. Satellite Momentum Wheel

In the aerospace field, spacecraft orientation is stabilized in the desired direction through attitude control. The control moment gyroscope (CMG) serves as a critical actuator for achieving agile and reliable attitude control. The integration of magnetic bearings significantly reduces the external forces transmitted to both the bearing stator and the spacecraft. By utilizing robust control and adaptive feedforward compensation, both internal multi-parameter disturbances and external continuous disturbances are effectively suppressed. This greatly enhances system stability during high-speed and dynamic operations, making it well-suited for spacecraft attitude control, where stringent micro-vibration requirements must be met [54]. The prototype and relevant parameters are provided in Table 15.

Table 15. The parameters of moment gyros.

Machine Type	Magnetically Suspended Control Moment Gyros
Stable operating frequency	100 Hz
Rotor mass	16.7 kg
Outer diameter	250 mm
Rotor orientation	Vertical
AMB configuration	Single shaft, 2 radial and 1 axial.

4.3.2. Nuclear Reactor Cooling

In nuclear reactors, where temperatures and radiation levels are high, gas-cooled high-temperature reactors (HTRs) rely on high-purity helium as both a coolant and heat carrier. The lubricants or wear debris from traditional bearings may fail or contaminate the helium medium, limiting the reliability of these HTRs. Magnetic bearings present a viable solution to address this challenge [164]. Due to the extremely high operating temperatures of gas-cooled HTRs, traditional bearing materials are prone to failure from thermal expansion or fatigue. In contrast, magnetic bearings, with no mechanical contact and no need for lubrication, offer excellent high-temperature resistance. This eliminates the risk of material thermal deformation, ensuring clean, stable, and efficient operation of the HTR under extreme conditions. A practical application is presented in Table 16.

Table 16. The parameters of centrifugal helium circulator.

Machine Type	Centrifugal Helium Circulator
Speed range	800~4800 r/min continuously variable
Rotor mass	4000 kg
Rotor length	3.5 m
Rotor orientation	Vertical
AMB configuration	Single shaft, 2 radial and 1 axial.

5. Technical Challenges and Future Trends

5.1. Technical Challenges

5.1.1. Multi-Field Coupled Modeling and Calculation

The modeling and calculation of magnetic bearings are critical for linking theoretical design, performance optimization, and practical application [165,166]. However, challenges related to three-dimensional (3-D) electromagnetic field calculations [167–169], electromechanical coupling dynamics [170–172] and eddy current [173,174] continue to be major bottlenecks in the current development of this technology.

Currently, the equivalent magnetic circuit method remains the most traditional approach for analyzing the magnetic flux of active magnetic bearings (AMB) and hybrid magnetic bearings (HMB) [44,77,82,83]. In this method, the mathematical model of the suspension force is derived by constructing an equivalent magnetic circuit diagram, with parameters determined using typical reluctance formulas and engineering experience [84–88]. However, this approach assumes ideal conditions, neglecting the effects of local material saturation [175–179], eddy current loss, edge magnetic leakage and other factors [180]. Some studies have explored variations of the equivalent magnetic circuit method that account for material nonlinearity and eddy current effects. These advancements significantly reduce the model's calculation error by incorporating the edge effect, thus improving the accuracy of the 3-D electromagnetic field calculation for magnetic bearings. To further enhance the modeling's accuracy and generality, the Maxwell tensor method [49,89–96] can also be employed. Nonetheless, factors such as magnetic circuit saturation and edge effects must still be carefully considered.

The stiffness and damping characteristics [181–184] of magnetic bearings are influenced by electromagnetic forces, control parameters, and rotor dynamics. Transient modeling of magnetic bearing-rotor systems is crucial for optimizing control and mitigating unbalanced vibrations. While the transient model of the rotor on a rigid foundation has been studied, an accurate multi-DOF transient model that accounts for the modal coupling phenomenon of a flexible rotor crossing the critical speed is still required for further refinement.

The modeling and calculation of magnetic bearings have evolved from static analysis of single physical fields to dynamic coupling of multiple fields. Moving forward, overcoming existing bottlenecks will require the integration of high-performance computing, artificial intelligence, and advanced material technologies. By developing high-precision mathematical models, creating intelligent agent models, and fostering interdisciplinary collaboration, the modeling of magnetic bearings will become more efficient and accurate. This will lay the groundwork for their widespread application in fields such as flywheel energy storage, high-speed motors, and aerospace.

5.1.2. Balance Between System Cost and Reliability

The performance of magnetic bearings relies on the precise detection and timely adjustment of rotor displacement through the closed-loop control system. However, the trade-off between cost and reliability of existing sensors has become a major bottleneck, limiting their widespread application. Eddy current sensors, as the predominant displacement detection solution, offer advantages in accuracy and transient response [185,186]. However, their high cost and reliability concerns have driven the industry to explore alternative solutions, such as inductive sensors and self-sensing technologies.

Eddy current sensors operate at high-frequency excitation ranges (typically 500 kHz to 20 MHz) and impose stringent requirements on material uniformity, surface finish, and temperature stability. The manufacturing costs of key components, such as probe coils and signal processing chips, account for approximately 30% to 40% of the total cost of a magnetic bearing system. Traditional magnetic bearing systems closed-loop control needs to obtain real-time displacement signals in 5-DOF, and any failure of the sensors can directly lead to system instability [187]. Additionally, sensor installation can introduce geometric errors, eventually reducing control accuracy.

Inductive sensors measure displacement by detecting variations in inductance and offer a cost advantage—approximately 1/5 that of eddy current sensors [188]. However, they still face challenges, including nonlinear measurement errors, sensitivity to electromagnetic interference, and limited transient response. Sensorless magnetic bearing technologies estimate rotor position indirectly through high-frequency signal injection or state observers, reducing reliance on physical displacement sensors. Despite this advantage, these approaches involve high algorithmic complexity, posing challenges for real-time implementation and system robustness [66].

Achieving a balance between cost and reliability in magnetic bearing systems requires coordinated advancements in both sensor technologies and control algorithms. In the short term, hybrid sensor configurations combined with control correction algorithms offer practical solutions for cost reduction without compromising performance. In the long term, the development of more intelligent sensorless technologies is expected to transform the industry. With these advancements, magnetic bearings are poised for widespread adoption at manageable costs across high-end manufacturing, renewable energy, and other emerging application areas.

5.2. Frontier Development Direction

5.2.1. Intelligent Control System

Although magnetic bearing rotors can be made to rotate around the inertial axis, interference factors such as mass eccentricity and sudden load changes remain unavoidable due to limitations in machining accuracy and material imbalance [189,190]. Overcoming these challenges requires intelligent control systems, which are key to breaking current performance bottlenecks and enabling large-scale industrial applications. Traditional control strategies, such as PID control, often fall short of maintaining robust vibration suppression across the full speed range. For example, while zero-displacement control can effectively eliminate synchronous vibrations, it may lead to instability under varying speed conditions due to phase lag. Disturbances can be estimated using extended state observers (ESO) [191,192], but their effectiveness is constrained by observation bandwidth, which limits their ability to handle high-frequency disturbances. Moreover, the accuracy of control relies heavily on precise system modeling and parameter tuning.

To address these challenges, the second-order generalized integral frequency-locked loop (SOGI-FLL) is employed for real-time speed estimation. By utilizing a variable gain coefficient to adaptively adjust the control loop bandwidth, SOGI-FLL enables effective vibration suppression over a broader frequency range [193]. The model-aided extended state observer (MESO) further enhances system performance by introducing a feedforward compensation link, which mitigates high-frequency disturbances while reducing control voltage and phase delay [194]. Additionally, integrating inductive sensors with self-sensing technology enables the deployment of nonlinear controllers based on deep learning. These controllers can improve system stability by dynamically tuning PID parameters through multi-sensor data fusion. However, the application of reinforcement learning requires extensive simulation data for training, and the synchronization of multi-sensor data—particularly time alignment—remains an active area of research.

The intelligence of magnetic bearing control systems must overcome the challenges of multidisciplinary coupling within the closed-loop framework of perception, decision-making, and execution. This advancement is essential for shifting from passive response to active prediction, enabling more adaptive, robust, and efficient system behavior.

5.2.2. Miniaturized and Integrated Solutions

The miniaturization and integration of magnetic bearings represent critical breakthroughs for their application in emerging fields such as microrobotics, precision medical devices, and aerospace micro-power systems. However, these advancements are still constrained by challenges in electromagnetic topology optimization, thermal management, and multi-physics coupling modeling.

Traditional magnetic bearing systems often rely on the combination of multiple single-degree-of-freedom electromagnets, resulting in a bulky overall structure. For example, a typical five-degree-of-freedom (5-DOF) suspension configuration requires at least two radial bearings and one axial bearing shown in Figure 19, and load structures such as fans, impeller and pump are installed at the end of the rotating shaft, significantly reducing the effective axial length of the motor relative to the total assembled system length. To enhance integration, multi-DOF hybrid magnetic bearing (HMB) topologies are essential. However, in such compact designs, strong coupling among electromagnetic, mechanical vibration, and thermal fields presents significant complexity. Conventional liquid cooling methods are impractical due to strict volume constraints, while passive thermal management depends on high-conductivity materials, increasing overall cost. Additionally, most current control systems rely on discrete power modules such as H-bridges. In a 5-DOF system, this translates to five independent power circuit sets, further complicating integration and increasing spatial and thermal design burdens.

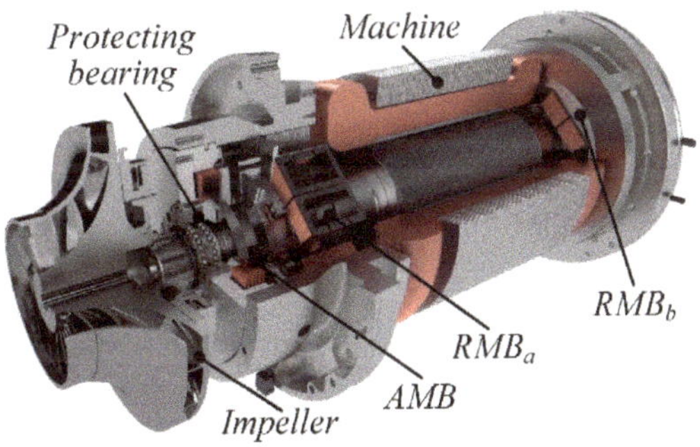

Figure 19. Traditional magnetic suspension machine configuration.

Currently, various multi-DOF integrated magnetic bearing topologies with shared magnetic circuits have been developed, including axial-radial 3-DOF configurations [55,195–199], 4-DOF [200–205] and full 5-DOF [56,61,206–208] magnetic bearing topologies. In addition, bearingless 5-DOF motor architectures have emerged, integrating magnetic bearing functionality directly into the motor to significantly reduce the system's axial length [209–211]. On the control side, system-level packaging techniques offer substantial reductions in the size of power amplifiers, controllers, and sensors, further promoting miniaturization and integration.

Advancing the miniaturization and integration of magnetic bearings necessitates overcoming the electromagnetic-mechanical-thermal limitations inherent in multidisciplinary coupling. Looking forward, permanent magnet–electromagnetic hybrid topologies and system-level packaging are expected to become focal points of innovation in this field. These advancements will drive the evolution of high-end equipment toward non-contact operation, full intelligence, and ultra-compact form factors, enabling broader applications across next-generation technologies.

6. Conclusions and Perspectives

6.1. Technical Summary

With its advantages of non-contact support, zero friction loss, and maintenance-free operation without lubrication, magnetic bearing technology has become a cornerstone for high-speed precision equipment. This paper provides a comprehensive review of its development, covering the comparative analysis of active, passive, and hybrid magnetic bearing topologies, as well as key technological breakthroughs in electromagnetic structure design, advanced intelligent control algorithms, and integrated power drive systems. Together, these aspects highlight the multi-dimensional value and transformative potential of magnetic bearing technology.

In terms of technical principles, the bearing capacity and stability of magnetic bearings have been significantly enhanced through multi-physical field coupling modeling, active-passive hybrid magnetic circuit design, and the use of high-performance permanent magnet (PM) materials. The evolution of intelligent control strategies is particularly noteworthy, progressing from traditional PID control to adaptive sliding mode control and further advancing to parameter self-tuning systems based on deep learning algorithms. These developments have progressively addressed issues related to strong nonlinear vibrations and rotor interference. In power driver technology, the integration of wide bandgap semiconductor devices and high-frequency switching power amplifiers has led to substantial improvements in energy efficiency and volume reduction compared to previous controllers. On the application front, the successful deployment of magnetic bearings in satellite momentum wheels, high-speed motors, and integrated medical devices demonstrates their potential to replace traditional bearings in a variety of critical systems.

However, current technology still faces several key challenges. The accuracy of multi-field coupling modeling remains insufficient, with significant simulation errors in the interactions between magnetic, thermal, and mechanical fields under complex operating conditions. Additionally, the overall system cost remains high, and the reliance on key materials, such as rare-earth permanent magnets (PM), is constrained by supply chain limitations. Furthermore, the adaptability of magnetic bearings to extreme environments is limited, particularly in terms of PM demagnetization at high temperatures, as well as issues related to sealing and corrosion resistance in deep-sea or high-pressure environments.

6.2. Outlook

The future development of magnetic bearing technology may follow several key pathways:

1. Multi-Field Coupled modeling and Accurate Analysis: Magnetic bearings inherently involve the coupling of multiple fields, including electromagnetic, thermodynamic, and mechanical vibration. Future advancements will require the development of more precise multi-field coupling mathematical models to address complex challenges, such as nonlinear vibration and thermal deformation under high-speed and high-load conditions. For example, research could focus on the interaction between magnetic field distribution, thermal effects, and mechanical stress. Additionally, integrating these models with the flexural vibration characteristics of high-speed, flexible rotors

will enable the development of response models capable of predicting instability thresholds at critical speeds.

2. Innovation of Intelligent Control Algorithm and Strategy: To overcome the limitations of traditional PID control, such as insufficient disturbance rejection, it is essential to integrate modern control theory with artificial intelligence technologies to enhance the system's robustness against sudden load changes and parameter variations. Employing deep reinforcement learning for control parameter optimization, along with multi-sensor data fusion (including displacement, current, and temperature data), will enable more precise suspension control and improve overall system stability. This approach promises to significantly enhance the adaptability and performance of magnetic bearing systems in dynamic environments.
3. Efficient Optimization of Power Drive System: As the core execution unit of magnetic bearings, the power electronic converter must achieve breakthroughs in both topology structure and control efficiency. Advancing shared bridge topologies will reduce the number of power devices, thereby minimizing both the volume and cost of the system. The integration of wide-bandgap SiC/GaN devices will enable higher switching frequencies, while advanced PWM modulation strategies will further reduce losses and optimize operational efficiency. These innovations will significantly enhance the performance and cost-effectiveness of magnetic bearing systems.
4. Miniaturization and Integration Design Trend: To meet the demands of medical, aerospace, and precision instrumentation applications, magnetic bearings must be miniaturized and modularized. Integrating the sensor, controller, and power amplifier into a single chip will streamline the system architecture, reduce power consumption, and enhance efficiency. Additionally, optimizing the topology structure is crucial to minimizing weight, which will further support the deployment of magnetic bearings in compact, high-performance systems. These advancements will enable magnetic bearings to be more adaptable and efficient for emerging technologies.

The future development of magnetic bearings will undoubtedly require a multidisciplinary approach. This includes high-precision modeling to guide design, coupled with advanced intelligent control algorithms to enhance transient performance. Additionally, innovation in power electronics will be essential to optimize energy efficiency. Ultimately, the goal is to achieve miniaturization and integration, enabling magnetic bearings to meet the demands of next-generation applications across various industries.

Author Contributions: Motor Data collection, Y.D. and G.Z.; summarization on applications, Y.D. and G.Z.; supervision, G.Z. and W.H.; writing—original draft, Y.D.; writing—review and editing, G.Z. and W.H. All authors have read and agreed to the published version of the manuscript.

Funding: This work is supported in part by the National Nature Science Foundation of China under Grant 52477038.

Conflicts of Interest: The authors declare no conflicts of interest.

References

1. Huang, Z.; Tang, M.; Golovanov, D.; Yang, T.; Herring, S.; Zanchetta, P. Profiling the Eddy Current Losses Variations of High-Speed Permanent Magnet Machines in Plug-In Hybrid Electric Vehicles. *IEEE Trans. Transport. Electrific.* **2022**, *8*, 3451–3463. [CrossRef]
2. Sayed, E.; Abdalmagid, M.; Pietrini, G.; Sa'adeh, N.-M.; Callegaro, A.D.; Goldstein, C. Review of Electric Machines in More-Hybrid-Turbo-Electric Aircraft. *IEEE Trans. Transport. Electrific.* **2021**, *7*, 2976–3005. [CrossRef]
3. Gallicchio, G.; Nardo, M.D.; Palmieri, M.; IlKhani, M.R.; Degano, M.; Gerada, C. Surface Permanent Magnet Synchronous Machines: High Speed Design and Limits. *IEEE Trans. Energy Convers.* **2023**, *38*, 1311–1324. [CrossRef]

4. Nishanth, F.; Van de Ven, J.; Severson, E.L. Evaluation of Torque-Dense Electric Machine Technology for Off-Highway Vehicle Electrification. *IEEE Trans. Ind. Appl.* **2024**, *60*, 3062–3074.
5. Zhang, J.; Peng, F.; Huang, Y.; Yao, Y.; Zhu, Z. Online Inductance Identification Using PWM Current Ripple for Position Sensorless Drive of High-Speed Surface-Mounted Permanent Magnet Synchronous Machines. *IEEE Trans. Ind. Electron.* **2022**, *69*, 12426–12436. [CrossRef]
6. Zang, B.; Chen, Y. Multiobjective Optimization and Multiphysics Design of a 5 MW High-Speed IPMSM Used in FESS Based on NSGA-II. *IEEE Trans. Energy Convers.* **2023**, *38*, 813–824. [CrossRef]
7. Zheng, S.; Wang, C. Rotor Balancing for Magnetically Levitated TMPs Integrated with Vibration Self-Sensing of Magnetic Bearings. *IEEE/ASME Trans. Mechatron.* **2021**, *26*, 3031–3039. [CrossRef]
8. Xiao, Y.; Zhu, Z.Q.; Wang, S.S.; Jewell, G.W.; Chen, J.T.; Wu, D. A Novel Asymmetric Interior Permanent Magnet Machine for Electric Vehicles. *IEEE Trans. Energy Convers.* **2021**, *36*, 2404–2415. [CrossRef]
9. Wang, K.; Ma, X.; Liu, Q.; Chen, S.; Liu, X. Multiphysics Global Design and Experiment of the Electric Machine with a Flexible Rotor Supported by Active Magnetic Bearing. *IEEE/ASME Trans. Mechatron.* **2019**, *24*, 820–831.
10. Dong, T.; Zhu, C.; Zhang, X.; Zhou, F. Hybrid-Strand Winding Topology with Improved Power Density in Automotive Electric Machines. *IEEE Trans. Ind. Electron.* **2022**, *69*, 6036–6045. [CrossRef]
11. Zhang, W.; Yang, H.; Cheng, L.; Zhu, H. Modeling Based on Exact Segmentation of Magnetic Field for a Centripetal Force Type-Magnetic Bearing. *IEEE Trans. Ind. Electron.* **2020**, *67*, 7691–7701. [CrossRef]
12. Earnshaw, S. On the Nature of the Molecular Forces which Regulate the Constitution of the Lumiferous Ether. *Trans. Camb. Philharm. Soc.* **1842**, *3*, 97–112.
13. Kemper, H. Schwebebahn Mit Raederlosen Fahrzeugen, Die an Eisernen Fahrschienen Mittels Magnetischer Felder Schwebend Entlang Gefuehrt Werden. Germany Patent No. 6443316, 5 April 1937.
14. Kemper, H. Suspension by Electromagnetic Forces: A Possibility for A Radically New Method of Transportation. *Elektrotechnische* **1938**, *59*, 391–395.
15. Beams, J.W. Magnetic Suspension for Small Rotors. *Rev. Sci. Instrum.* **1950**, *21*, 182–184. [CrossRef]
16. Sindlinger, R. Magnetic Bearing Momentum Wheels with Vernier Gimballing Capability for 3-Axis Active Attitude Control and Energy Storage. *IFAC Proc. Vol.* **1976**, *9*, 849–860. [CrossRef]
17. Schweitzer, G. Ein Aktives Magnetisches Rotorlager-Auslegung und Anwendung. *at-Automatisierungstechnik* **1978**, *26*, 10–15. [CrossRef]
18. Knospe, C.; Collins, E.G. Introduction to the Special Issue on Magnetic Bearing Control. *IEEE Trans. Control Syst. Technol.* **1996**, *4*, 481–485. [CrossRef]
19. Chu, Y.L.; Wang, X.P.; Lei, Y.F.; Xia, C.Y.; Tang, H.; Wang, L.; Qian, J. Intelligent Method of Information Storage and Interface in Magnetic Bearing System. *Bearing* **2007**, *9*, 14.
20. Sun, J.J.; Fang, J.C. Design on New Permanent Magnet Biased Radial Magnetic Bearing in Magnetic Suspending Flywheel. *Bearing* **2008**, *3*, 8–13.
21. Gong, L.; Zhu, C.S. The Influence of PID Controller Parameters on Polarity Switching Control for Unbalance Compensation of Active Magnetic Bearings Rotor Systems. *IEEE Trans. Ind. Electron.* **2024**, *71*, 8324–8338.
22. Available online: https://www.zionmarketresearch.com/report/magnetic-bearings-market (accessed on 13 April 2025).
23. Available online: https://www.transparencymarketresearch.com/magnetic-bearings-market.html (accessed on 13 April 2025).
24. Available online: https://www.stratviewresearch.com/3279/magnet-bearings-market.html (accessed on 13 April 2025).
25. Available online: https://www.verifiedmarketreports.com/product/magnetic-bearings-market (accessed on 13 April 2025).
26. Prasad, K.N.V.; Narayanan, G. Electro-Magnetic Bearings with Power Electronic Control for High-Speed Rotating Machines: A Review. In Proceedings of the 2019 National Power Electronics Conference (NPEC), Tiruchirappalli, India, 13–15 December 2019; pp. 1–6.
27. Gerada, D.; Mebarki, A.; Brown, N.L.; Gerada, C. High-Speed Electrical Machines: Technologies, Trends, and Developments. *IEEE Trans. Ind. Electr.* **2014**, *61*, 2946–2959. [CrossRef]
28. Dutta, D.; Biswas, P.K.; Debnath, S.; Ahmad, F. Advancements and Challenges in Active Magnetic Bearings: A Comprehensive Review of Performance, Control, and Future Prospects. *IEEE Access* **2025**, *13*, 3051–3071. [CrossRef]
29. Zhao, X.S.; Deng, Z.Q.; Wang, X.L. Research status and development of permanent magnet biased magnetic bearings. *Trans. China Electr. Soc.* **2009**, *24*, 9–20.
30. Samanta, P.; Hirani, H. An Overview of Passive Magnetic Bearings. In Proceedings of the STLE/ASME 2008 International Joint Tribology Conference, Miami, FL, USA, 20–22 October 2008; pp. 465–467.
31. Soni, T.; Dutt, J.K.; Das, A.S. Parametric Stability Analysis of Active Magnetic Bearing Supported Rotor System with a Novel Control Law Subject to Periodic Base Motion. *IEEE Trans. Ind. Electron.* **2020**, *67*, 1160–1170. [CrossRef]

32. Fonseca, C.A.; Santos, I.; Weber, H.I. An experimental and theoretical approach of a pinned and a conventional ball bearing for active magnetic bearings. *Mech. Syst. Signal Proc.* **2020**, *138*, 106541. [CrossRef]
33. Sun, M.; Zheng, S.; Wang, K.; Le, Y. Filter Cross-Feedback Control for Nutation Mode of Asymmetric Rotors with Gyroscopic Effects. *IEEE/ASME Trans. Mechatron.* **2020**, *25*, 248–258.
34. Hsieh, H.C.; Chen, C.W.; Chen, M.C.; Lai, J.S.J.; Lin, J.M. Current Spike Suppression Techniques for Magnetic Bearing Amplifier. In Proceedings of the 2019 IEEE Applied Power Electronics Conference and Exposition (APEC), Anaheim, CA, USA, 17–21 March 2019; pp. 2002–2007.
35. Huang, T.; Zheng, M.; Zhang, G. A Review of Active Magnetic Bearing Control Technology. In Proceedings of the 2019 Chinese Control And Decision Conference (CCDC), Nanchang, China, 3–5 June 2019; pp. 2888–2893.
36. Ravaud, R.; Lemarquand, G.; Lemarquand, V. Force and stifness of passive magnetic bearings using permanent magnets. Part 2: Radial magnetization. *IEEE Trans. Magn.* **2009**, *45*, 3334–3342. [CrossRef]
37. Ravaud, R.; Lemarquand, G.; Lemarquand, V. Force and stifness of passive magnetic bearings using permanent magnets. Part 1: Axial magnetization. *IEEE Trans. Magn.* **2009**, *45*, 2996–3002. [CrossRef]
38. Bachovchin, K.D.; Hoburg, J.F.; Post, R.F. Stable Levitation of a Passive Magnetic Bearing. *IEEE Trans. Magn.* **2013**, *49*, 609–617. [CrossRef]
39. Bekinal, S.I.; Deshwal, D.; Srinivas, V.G. Optimized multi-layer radial permanent magnet bearing with an eddy current damping systems. *J Braz. Soc. Mech. Sci. Eng.* **2022**, *44*, 542. [CrossRef]
40. Hu, J.C.; Wu, H.C.; Fang, K.P.; Li, Q. Overview on Structure of Permanent Magnetic Bearings. *Bearing* **2023**, *7*, 1–7.
41. Paden, B.; Groom, N.; Antaki, J.F. Design Formulas for Permanent-Magnet Bearings. *J. Mech. Design.* **2003**, *125*, 734–738. [CrossRef]
42. Pérez-Loya, J.J.; Santiago, J.; Lundin, U. Construction of A Permanent Magnet Thrust Bearing for A Hydropower Generator Test Setup. In Proceedings of the 1st Brazilian Workshop Magnetic Bearings (BWMB), Rio de Janeiro, Brazil, 18–19 November 2013; pp. 1–13.
43. Beneden, M.V.; Kluyskens, V.; Dehez, B. Optimal Sizing and Comparison of Permanent Magnet Thrust Bearings. *IEEE Trans. Magn.* **2017**, *53*, 1–10. [CrossRef]
44. Han, B.C.; He, Z.; Wen, T.; Zheng, S.Q. Design, optimization and test of RHMB considering dynamic characteristics. *Int. J. Mech. Sci.* **2023**, *238*, 107847. [CrossRef]
45. Li, Q.; Wang, W.M.; Chu, F.L. Damping ratio identification in rotordynamic system with inverse directional Fourier transform, multiple-tests-multiple-outputs and clustering techniques. *Int. J. Mech. Sci.* **2019**, *151*, 797–807. [CrossRef]
46. Amrhein, W.; Gruber, W.; Bauer, W.; Reisinger, M. Magnetic Levitation Systems for Cost-Sensitive Applications—Some Design Aspects. *IEEE Trans. Ind. Appl.* **2016**, *52*, 3739–3752. [CrossRef]
47. Ji, L.; Xu, L.; Jin, C. Research on a Low Power Consumption Six-Pole Heteropolar Hybrid Magnetic Bearing. *IEEE Trans. Magn.* **2013**, *49*, 4918–4926. [CrossRef]
48. Pesch, A.H.; Smirnov, A.; Pyrhönen, O.; Sawicki, J.T. Magnetic Bearing Spindle Tool Tracking Through μ -Synthesis Robust Control. *IEEE/ASME Trans. Mechatron.* **2015**, *20*, 1448–1457. [CrossRef]
49. Zhang, W.Y.; Wang, Z. Accurate modeling and control system design for a spherical radial AC HMB for a flywheel battery system. *Energies* **2023**, *16*, 1973. [CrossRef]
50. Han, X.; Liu, G.; Chen, B.D.; Zheng, S.Q. Surge disturbance suppression of AMB-rotor systems in magnetically suspension centrifugal compressors. *IEEE Trans. Control Syst. Technol.* **2022**, *30*, 1550–1560. [CrossRef]
51. Buse, M.; Marcus, V.A.; Eyad, A.K.; Bader, A.J. *Adapting Hermetically Sealed Compressor Technology to Deal with Sour and Corrosive Gases*; Turbomachinery Laboratories, Texas A&M Engineering Experiment Station: College Station, TX, USA, 2016.
52. He, H.T.; Liu, Y.B.; Ba, L.M. A nonlinear dynamic model of flywheel energy storage systems based on alternative concept of back propagation neural networks. *J. Comput. Nonlinear Dyn.* **2022**, *17*, 091006. [CrossRef]
53. Sun, Y.; Xu, J.; Qiang, H.; Lin, G. Adaptive neural-fuzzy robust position control scheme for Maglev train systems with experimental verification. *IEEE Trans. Ind. Electron.* **2019**, *66*, 8589–8599. [CrossRef]
54. Liu, X.; Ma, X.; Feng, R.; Chen, Y.; Shi, Y.; Zheng, S. Model Reference Adaptive Compensation and Robust Controller for Magnetic Bearing Systems with Strong Persistent Disturbances. *IEEE Trans. Ind. Electron.* **2023**, *70*, 10902–10911. [CrossRef]
55. Ren, X.J.; Le, Y.; Sun, J.J.; Han, B.C.; Wang, K. Magnetic flux leakage modelling and optimisation of a CRAHMB for DC motor. *IET Electr. Power Appl.* **2017**, *11*, 212–221. [CrossRef]
56. Masuzawa, T.; Kojima, J.; Onuma, H.; Okada, Y.; Nishida, M.; Yamane, T. Micro Magnetic Bearing for An Axial Flow Artificial Heart. In Proceedings of the 9th International Symposium on Magnetic Bearings, Lexington, KY, USA, 3–6 August 2004; University of Kentucky: Lexington, KY, USA, 2004; pp. 89–94.

57. Jin, Z.; Sun, X.; Chen, L.; Yang, Z. Robust Multi-Objective Optimization of a 3-Pole Active Magnetic Bearing Based on Combined Curves with Climbing Algorithm. *IEEE Trans. Ind. Electron.* **2022**, *69*, 5491–5501. [CrossRef]
58. Grbesa, B. Low loss and low cost active radial homopolar magnetic bearing. In *Sixth International Symposium on Magnetic Bearings (ISMB)*; Institute of Electrical Drives and Power Electronics: Linz, Austria, 1998; pp. 286–295.
59. Chen, S.L.; Hsu, C.T. Optimal design of a three-pole active magnetic bearing. *IEEE Trans. Magn.* **2002**, *38*, 3458–3466. [CrossRef]
60. Chen, S.L.; Chen, S.H.; Yan, S.T. Experimental validation of a current-controlled three-pole magnetic rotor-bearing system. *IEEE Trans. Magn.* **2005**, *41*, 99–112. [CrossRef]
61. Zhang, W.Y.; Zhu, H.Q. Control System Design for a Five-Degree-of-Freedom Electrospindle Supported With AC Hybrid Magnetic Bearings. *IEEE/ASME Trans. Mechatron.* **2015**, *20*, 2525–2537. [CrossRef]
62. Zhang, W.Y.; Zhu, H.Q.; Yang, Z.B.; Sun, X.D.; Yuan, Y. Nonlinear Model Analysis and "Switching Model" of AC–DC Three-Degree-of-Freedom Hybrid Magnetic Bearing. *IEEE/ASME Trans. Mechatron.* **2016**, *21*, 1102–1115. [CrossRef]
63. Darbandi, S.M.; Behzad, M.; Salarieh, H.; Mehdigholi, H. Linear Output Feedback Control of a Three-Pole Magnetic Bearing. *IEEE/ASME Trans. Mechatron.* **2014**, *19*, 1323–1330.
64. Ju, J.; Zhu, H. Radial Force-Current Characteristics Analysis of Three-Pole Radial-Axial Hybrid Magnetic Bearings and Their Structure Improvement. *Energies* **2016**, *9*, 706. [CrossRef]
65. Zhu, H.; Ding, S.; Jv, J. Modeling for Three-Pole Radial Hybrid Magnetic Bearing Considering Edge Effect. *Energies* **2016**, *9*, 345. [CrossRef]
66. Garcia, P.; Guerrero, J.M.; Mahmoud, I.E.S.; Briz, F.; Reigosa, D.D. Impact of Saturation, Current Command Selection, and Leakage Flux on the Performance of Sensorless-Controlled Three-Pole Active Magnetic Bearings. *IEEE Trans. Ind. Appl.* **2011**, *47*, 1732–1740. [CrossRef]
67. Hemenway, N.R.; Severson, E.L. Exact Force Vector Regulation of the Three-Pole Magnetic Bearing. *IEEE Trans. Ind. Appl.* **2021**, *57*, 7024–7034. [CrossRef]
68. Piłat, A.A. Synergistic Dynamic Model of an Active Magnetic Bearing with three Electromagnets. In Proceedings of the 2013 International Symposium on Electrodynamic and Mechatronic Systems (SELM), Opole-Zawiercie, Poland, 15–18 May 2013; pp. 23–24.
69. Morrison, C.R.; Siebert, M.W.; Ho, E.J. Electromagnetic Forces in a Hybrid Magnetic-Bearing Switched-Reluctance Motor. *IEEE Trans. Magn.* **2008**, *44*, 4626–4638. [CrossRef]
70. Noh, M.; Gruber, W.; Trumper, D.L. Hysteresis Bearingless Slice Motors With Homopolar Flux-Biasing. *IEEE/ASME Trans. Mechatron.* **2017**, *22*, 2308–2318. [CrossRef]
71. Wajnert, D. Analysis of the cross-coupling effect and magnetic force nonlinearity in the 6-pole radial hybrid magnetic bearing. *Int. J. Appl. Electrom.* **2019**, *61*, 43–57.
72. Zhu, H.Q.; Liu, T.T. Rotor Displacement Self-Sensing Modeling of Six-Pole Radial Hybrid Magnetic Bearing Using Improved Particle Swarm Optimization Support Vector Machine. *IEEE Trans. Power Electron.* **2020**, *35*, 12296–12306. [CrossRef]
73. Hu, Y.; Chen, B.; Zhang, X.; Zhang, F.; Gong, G.; Li, X. Analysis of Electromagnetic Characteristics of 16 Pole Radial Active Magnetic Bearings. In Proceedings of the 2019 22nd International Conference on Electrical Machines and Systems (ICEMS), Harbin, China, 11–14 August 2019; pp. 1–6.
74. Hu, J.; Li, W.; Su, Z.; Cheng, S.; Liu, Q.; Wu, C. Radial Displacement Measurement Method for Magnetic Bearing Based on FPC Coils. *IEEE Sens. J.* **2024**, *24*, 29687–29694. [CrossRef]
75. Active Bearings. Available online: https://www.magneticbearings.org/active-bearings (accessed on 27 February 2025).
76. Cao, G.Z.; Li, H.-L.; Hu, H.-J.; Huang, S.-D.; Wang, H.-T.; Liu, K. A Novel Electromagnetic Force Calculation Method for Homopolar Hybrid Magnetic Bearing. In Proceedings of the 2024 IEEE International Magnetic Conference-Short Papers (INTERMAG Short Papers), Rio de Janeiro, Brazil, 5–10 May 2024; pp. 1–2.
77. Yu, C.M.; Deng, Z.Q.; Chen, S.S.; Mei, L.; Peng, C.; Cao, X. A novel subdomain and magnetic circuit modeling method for hybrid homopolar radial magnetic bearings. *Mech. Syst. Signal Process.* **2022**, *170*, 108823. [CrossRef]
78. Yu, C.M.; Deng, Z.Q.; Mei, L.; Chen, S.S.; Peng, C.; Ding, Q. Accurate modelling based on a subdomain method for heteropolar radial magnetic bearing with a solid rotor. *IET Electr. Power Appl.* **2021**, *15*, 1232–1244. [CrossRef]
79. Amrhein, W.; Gruber, W.; Bauer, W.; Reisinger, M. Magnetic levitation systems for cost-sensitive applications. In Proceedings of the 2015 IEEE Workshop on Electrical Machines Design, Control and Diagnosis (WEMDCD), Turin, Italy, 26–27 March 2015; pp. 104–111.
80. Wajnert, D.; Sykulski, J.K.; Tomczuk, B. A Dynamic Simulation Model of a Hybrid Magnetic Bearing. In Proceedings of the 2019 19th International Symposium on Electromagnetic Fields in Mechatronics, Electrical and Electronic Engineering (ISEF), Nancy, France, 29–31 August 2019; pp. 1–2.

81. Zhu, Y.S.; Yang, D.S.; Wang, X.; Hu, B.; Zhang, P.J. Research on fractional-order modeling of nonlaminated electromagnetic bearings considering eddy current effects. *IEEE Trans. Magn.* **2022**, *58*, 8300208. [CrossRef]
82. Le, Y.; Wang, K. Design and optimization method of magnetic bearing for high-speed motor considering Eddy current effects. *IEEE-ASME Trans. Mechatron.* **2016**, *21*, 2061–2072. [CrossRef]
83. Wu, C.; Li, S.S. Modeling, design and suspension force analysis of a novel AC six-pole heteropolar hybrid magnetic bearing. *Appl. Sci.* **2023**, *13*, 1643. [CrossRef]
84. Zhang, H.; Zhu, H.Q.; Wu, M.Y. Multi-objective parameter optimization-based design of six-pole radial hybrid magnetic bearing. *IEEE J. Emerg. Sel. Top. Power Electron.* **2022**, *10*, 4526–4535. [CrossRef]
85. Zhou, J.Y.; Zhu, H.Q. Development and research summary of magnetic bearings. *Micromotors* **2022**, *55*, 93–98.
86. Fu, J.; Wang, X.G.; Hou, L.L.; Pan, A. Structure study and performance analysis on the new laminated core type axial redundant magnetic bearing. *Int. J. Appl. Electromagn. Mech.* **2018**, *58*, 101–120.
87. Seifert, R.; Robenack, K.; Hofmann, W. Rational approximation of the analytical model of nonlaminated cylindrical magnetic actuators for flux estimation and control. *IEEE Trans. Magn.* **2019**, *55*, 8301016. [CrossRef]
88. Romanenko, A.; Smirnov, A.; Jastrzebski, R.P.; Pyrhönen, O. Losses estimation and modelling in active magnetic bearings. In Proceedings of the 2014 16th European Conference on Power Electronics and Applications, Lappeenranta, Finland, 26–28 August 2014; pp. 1–8.
89. Amrhein, M.; Krein, P.T. Force Calculation in 3-D Magnetic Equivalent Circuit Networks with A Maxwell Stress Tensor. *IEEE Trans. Energy Convers.* **2009**, *24*, 587–593. [CrossRef]
90. Zhang, W.Y.; Zhu, H.Q. Radial magnetic bearings: An overview. *Results Phys.* **2017**, *7*, 3756–3766. [CrossRef]
91. Zhang, T.; Liu, X.F.; Mo, L.H.; Ye, X.T.; Ni, W.; Ding, W.H.; Huang, J.; Wang, X. Modeling and analysis of hybrid permanent magnet type bearingless motor. *IEEE Trans. Magn.* **2018**, *54*, 8102804. [CrossRef]
92. Zhang, W.Y.; Zhu, H.Q. Precision modeling method specifically for AC magnetic bearings. *IEEE Trans. Magn.* **2013**, *49*, 5543–5553. [CrossRef]
93. Ma, Z.H.; Liu, G.; Liu, Y.C.; Yang, Z.C.; Zhu, H.Q. Research of a six-pole active magnetic bearing system based on a fuzzy active controller. *Electronics* **2022**, *11*, 1723. [CrossRef]
94. Zhu, H.Q.; Fan, S. Soft-sensing modeling for rotor displacements of six-pole radial active magnetic bearing using improved continuous hidden Markov model. *Proc. CSEE* **2021**, *41*, 3933–3943.
95. Liu, G.; Zhu, H.Q. Calculation model for suspension force of six-pole outer-rotor hybrid magnetic bearing based on Maxwell tensor method. *Bearing* **2022**, *12*, 23–27.
96. Liu, G.; Zhu, H.Q. Displacement estimation of six-pole hybrid magnetic bearing using modified particle swarm optimization support vector machine. *Energies* **2022**, *15*, 1610. [CrossRef]
97. Yonnet, J.P. Analytical Calculation of Magnetic Bearings. In Proceedings of the 5th International Workshop on Rare Earth-Cobalt Permanent Magnets and Their Applications, Roanoke, VA, USA, 7–10 June 1981; Volume 3, pp. 199–216.
98. Li, J.G.; Tan, Q.C.; Zhang, Y.Q.; Zhang, K. Study on the Calculation of Magnetic Force Based on the Equivalent Magnetic Charge Method. *Phys. Procedia* **2012**, *24*, 190–197. [CrossRef]
99. Wang, N.X.; Wang, D.X.; Chen, K.S.; Wu, H.C. Research on Analytical Model and Design Formulas of Permanent Magnetic Bearings Based on Halbach Array with Arbitrary Segmented Magnetized Angle. *J. Magn. Magn. Mater.* **2016**, *410*, 257–264. [CrossRef]
100. Chen, Y.; Zhang, W.L.; Bird, J.Z.; Paul, S.; Zhang, K.L. A 3-D Analytic-Based Model of a Null-Flux Halbach Array Electrodynamic Suspension Device. *IEEE Trans. Magn.* **2015**, *51*, 1–5. [CrossRef]
101. Tian, L.L.; Ai, X.P.; Tian, Y.Q. Analytical Model of Magnetic Force for Axial Stack Permanent Magnet Bearings. *IEEE Trans. Magn.* **2012**, *48*, 2592–2699. [CrossRef]
102. Shakibapour, F.; Rahideh, A.; Mardaneh, M. 2D Analytical Model for Heteropolar Active Magnetic Bearings Considering Eccentricity. *IET Electr. Power Appl.* **2018**, *12*, 614–626. [CrossRef]
103. Du, T.; Geng, H.; Zhang, Y.; Lin, H.; Li, Y.; Yu, L. Exact Analytical Method for Active Magnetic Bearings with Rotor Eccentricity. *IEEE Trans. Magn.* **2019**, *55*, 1–12. [CrossRef]
104. Wang, K.; Wang, D.; Shen, Y.; Zhang, X.; Chen, J.; Zhang, Y. Subdomain Method for Permanent Magnet Biased Homopolar Radial Magnetic Bearing. *IEEE Trans. Magn.* **2016**, *52*, 1–5.
105. Yu, W.; Li, X. A magnetic levitation system for advanced control education. *IFAC Proc. Vol.* **2014**, *47*, 9032–9037. [CrossRef]
106. Alves, P. Magnetic Bearings-A Primer. In Proceedings of the 27th Turbomachinery Symposium, College Station, TX, USA, 15–17 September 1998.
107. Wei, C. Survey of magnetic bearing control. *Electr. Eng.* **2022**, *20*, 33–35.

108. Srinivas, R.S.; Tiwari, R.; Kannababu, C. Application of active magnetic bearings in flexible rotordynamic systems—A state-of-the-art review. *Mech. Syst. Signal Proc.* **2018**, *106*, 537–572. [CrossRef]
109. Chen, M.; Knospe, C.R. Control approaches to the suppression of machining chatter using active magnetic bearings. *IEEE Trans. Control Syst. Technol.* **2007**, *15*, 220–232. [CrossRef]
110. Basaran, S.; Sivrioglu, S. Novel repulsive magnetic bearing flywheel system with composite adaptive control. *IET Electr. Power Appl.* **2019**, *13*, 676–685. [CrossRef]
111. Barthod, C.; Lamarquand, G. Degrees of Freedom Control of a Magnetically Levitated Rotor. *IEEE Trans. Magn.* **1995**, *31*, 4202–4204. [CrossRef]
112. Lee, A.C.; Fan, Y.H. Decentralized PID Control of Magnetic Bearings in a Rotor System. Decentralized Control of A Rotor System Supported by Magnetic Bearings. *Int. J. Mach. Tools Manuf.* **1995**, *35*, 445–458.
113. Hutterer, M.; Wimmer, D.; Schrodl, M. Stabilization of a magnetically levitated rotor in the case of a defective radial actuator. *IEEE/ASME Trans. Mechatron.* **2020**, *25*, 2599–2609. [CrossRef]
114. Chen, L.L.; Zhu, C.S.; Zhong, Z.X.; Wang, C.K.; Li, Z.N. Radial position control for magnetically suspended high-speed flywheel energy storage system with inverse system method and extended 2-DOF PID controller. *IET Electr. Power Appl.* **2020**, *14*, 71–81. [CrossRef]
115. Wu, H.C.; Yu, M.Y.; Song, C.S.; Wang, N.X. Unbalance vibration suppression of maglev high-speed motor based on the least-mean-square. *Actuators* **2022**, *11*, 348. [CrossRef]
116. Laldingliana, J.; Biswas, P.K. Artificial intelligence based fractional order PID control strategy for active magnetic bearing. *J. Electr. Eng. Technol.* **2022**, *17*, 3389–3398. [CrossRef]
117. Gupta, S.; Biswas, P.K.; Debnath, S.; Ghosh, A.; Babu, T.S.; Zawbaa, H.M.; Kamel, S. Metaheuristic optimization techniques used in controlling of an active magnetic bearing system for high-speed machining application. *IEEE Access* **2023**, *11*, 12100–12118. [CrossRef]
118. Gupta, S.; Debnath, S.; Biswas, P.K. Control of an active magnetic bearing system using swarm intelligence-based opti-mization techniques. *Electr. Eng.* **2023**, *105*, 935–952. [CrossRef]
119. Wardi, M.L.; Amairi, M.; Abdelkrim, M.N. Fractional PID controller design for nonlinear systems based on singular perturbation technique. *Int. J. Digit. Signals Smart. Syst.* **2018**, *2*, 95–120.
120. Tang, J.Q.; Zhang, M.; Cui, X.; Sun, J.J.; Zhou, X.X. Neural network sliding model control of radial translation for magneti-cally suspended rotor (MSR) in control moment gyro. *Actuators* **2023**, *12*, 217. [CrossRef]
121. Chen, S.Y.; Lin, F.J. Decentralized PID neural network control for five degree-of-freedom active magnetic bearing. *Eng. Appl. Artif. Intel.* **2013**, *26*, 962–973. [CrossRef]
122. Noshadi, A.; Shi, J.; Lee, W.S.; Shi, P.; Kalam, A. Optimal PID-type fuzzy logic controller for a multi-input multi-output active magnetic bearing system. *Neural Comput. Appl.* **2016**, *27*, 2031–2046. [CrossRef]
123. Dhyani, A.; Panda, M.K.; Jha, B. Moth-flame optimization-based fuzzy-PID controller for optimal control of active magnetic bearing system. *IJST-T Electr. Eng.* **2018**, *42*, 451–463. [CrossRef]
124. Ye, Y.Y.; Hu, J.D. Design of magnetic suspended rotor controller based on fuzzy-PID double mode control. *Electr. Eng.* **2020**, *12*, 15–16.
125. Fang, J.; Ren, Y. High-Precision Control for a Single-Gimbal Magnetically Suspended Control Moment Gyro Based on Inverse System Method. *IEEE Trans. Ind. Electron.* **2011**, *58*, 4331–4342. [CrossRef]
126. Xie, J.J.; Liu, G.; Liu, H. High-Precision Control for Magnetically Suspended Rotor of A DGMSCMG Based on Motion Separation. *J. Eng. Gas Turbines Power.* **2015**, *137*, 41–50. [CrossRef]
127. Ahrens, M.; Kucera, L.; Larsonneur, R. Performance of A Magnetically Suspended Flywheel Energy Storage Device. *IEEE Trans. Control Syst. Technol.* **1996**, *4*, 494–502. [CrossRef]
128. Chen, X.C.; Ren, Y. Modal Decoupling Control for A Double Gimbal Magnetically Suspended Control Moment Gyroscope Based on Modal Controller and Feedback Linearization Method. *Proc. Inst. Mech. Eng. Part C J. Mech. Eng. Sci.* **2014**, *228*, 2303–2313. [CrossRef]
129. Pesch, A.H.; Sawicki, J.T. Active Magnetic Bearing Online Levitation Recovery Through μ-Synthesis Robust Control. *Actuators* **2017**, *6*, 2. [CrossRef]
130. Kuseyri, İ.S. Robust control and unbalance compensation of rotor/active magnetic bearing systems. *J. Vib. Control* **2012**, *18*, 817–832. [CrossRef]
131. Lusty, C.; Keogh, P. Active vibration control of a flexible rotor by flexibly mounted internal-stator magnetic actuators. *IEEE-ASME Trans. Mechatron.* **2018**, *23*, 2870–2880. [CrossRef]
132. Zhao, J.; Zhang, H.T.; Fan, M.C.; Wu, Y.; Zhao, H. Control of a Constrained Flexible Rotor On Active Magnetic Bearings. *IFAC PapersOnLine* **2015**, *48*, 156–161. [CrossRef]

133. Qi, W.; Liu, J.; Chen, X. Supervisory predictive control of standalone wind/solar energy generation systems. *IEEE Trans. Control Syst. Technol.* **2011**, *19*, 199–207. [CrossRef]
134. Morsi, A.; Abbas, H.S.; Ahmed, S.M.; Mohamed, A.M. Imbalance compensation of active magnetic bearing systems using model predictive control based on linear parameter-varying models. *J. Vib. Control* **2022**, *29*, 3516–3527. [CrossRef]
135. Wang, Z.; Zhu, C. Active Control of Active Magnetic Bearings for Maglev Flywheel Rotor System Based on Sliding Mode Control. In Proceedings of the 2016 IEEE Vehicle Power and Propulsion Conference (VPPC), Hangzhou, China, 17–20 October 2016; pp. 1–6.
136. Dominguez, J.R.; Mora-Soto, C.; Ortega-Cisneros, S.; Panduro, J.J.R.; Loukianov, A.G. Copper and core loss minimization for induction motors using high-order sliding-mode control. *IEEE Trans. Ind. Electron.* **2012**, *59*, 2877–2889. [CrossRef]
137. Wang, X.Y.; Zhang, Y.P.; Gao, P. Design and analysis of second-order sliding mode controller for active magnetic bearing. *Energies* **2020**, *13*, 5965. [CrossRef]
138. Kandil, M.S.; Dubois, M.R.; Bakay, L.S.; Trovao, J.P.F. Application of second-order sliding-mode concepts to active magnetic bearings. *IEEE Trans. Ind. Electron.* **2018**, *65*, 855–864. [CrossRef]
139. Chen, S.Y.; Song, M.H. Energy-Saving Dynamic Bias Current Control of Active Magnetic Bearing Positioning System Using Adaptive Differential Evolution. *IEEE Trans. Syst. Man Cybern. Syst.* **2017**, *99*, 942–953. [CrossRef]
140. Tang, J.Q.; Zhao, X.F.; Wang, Y.; Cui, X. Adaptive neural network control for rotor's stable suspension of Vernier-gim-balling magnetically suspended flywheel. *Proc. Inst. Mech. Eng. Part I-J. Syst. Control. Eng.* **2019**, *233*, 1017–1029.
141. Yu, Y.J.; Yang, Z.H.; Han, C.; Liu, H. Fuzzy adaptive back-stepping sliding mode controller for high-precision deflection control of the magnetically suspended momentum wheel. *IEEE Trans. Ind. Electron.* **2018**, *65*, 3530–3538. [CrossRef]
142. Xu, Y.H.; Wang, X.Y.; Liu, M.X.; Li, N.; Yu, T. Adaptive robust control of active magnetic bearings rigid rotor systems. *J. Power Electron.* **2023**, *23*, 1004–1015. [CrossRef]
143. Sun, Z.R.; Zhang, Y.B.; Wang, H.F. Variable universe fuzzy PID control based on genetic optimization for magnetic bearings. *Mod. Electron. Tech.* **2020**, *43*, 94–97.
144. Yao, D.; Wang, J.W.; Liu, Y. Enhancing working performance of active magnetic bearings using improved fuzzy control and Kalman-LMS filter. *J. Intell. Fuzzy Syst.* **2015**, *29*, 1343–1353. [CrossRef]
145. Chen, K.Y.; Tung, P.C.; Tsai, M.T.; Fan, Y.H. A self-tuning fuzzy PID-type controller design for unbalance compensation in an active magnetic bearing. *Expert Syst. Appl.* **2009**, *36*, 8560–8570. [CrossRef]
146. Han, J.Q. From PID to active disturbance rejection control. *IEEE Trans. Ind. Electron.* **2009**, *56*, 900–906. [CrossRef]
147. Jin, C.W.; Guo, K.X.; Xu, Y.P.; Cui, H.B.; Xu, L.X. Design of magnetic bearing control system based on active disturbance rejection theory. *J. Vib. Acoust.* **2019**, *141*, 011009. [CrossRef]
148. Ren, Y.; Fang, J. Current-Sensing Resistor Design to Include Current Derivative in PWM H-Bridge Unipolar Switching Power Amplifiers for Magnetic Bearings. *IEEE Trans. Ind. Electron.* **2012**, *59*, 4590–4600. [CrossRef]
149. Liu, Z.; Cao, X.; Cai, J. A Magnetic Bearing Switched Reluctance Motor with Simultaneous Excitation by a Modified Half-Bridge Converter. *IEEE Transa. Magn.* **2019**, *55*, 1–5. [CrossRef]
150. Zhang, Y.; Zhang, W.; Han, Y.; Yu, J. A Quasi-Three-Level PWM Modulation Method with Suppressed Coil Terminal Overvoltage for Active Magnetic Bearing. In Proceedings of the 2022 25th International Conference on Electrical Machines and Systems (ICEMS), Chiang Mai, Thailand, 29 November–2 December 2022; pp. 1–6.
151. Jiang, D.; Kshirsagar, P. Analysis and Control of a Novel Power Electronics Converter for Active Magnetic Bearing Drive. *IEEE Trans. Ind. Appl.* **2017**, *53*, 2222–2232. [CrossRef]
152. Zhou, M.; Jiang, D.; Shuai, Y.; Liu, Z. Double Five-Phase-Four-Leg Topology Converter for Five-Axis Active Magnetic Bearing with Optimized Current Control. *IEEE Trans. Ind. Electron.* **2024**, *71*, 3953–3963. [CrossRef]
153. Jiang, D.; Li, T.; Hu, Z.; Sun, H. Novel Topologies of Power Electronics Converter as Active Magnetic Bearing Drive. *IEEE Trans. Ind. Electron.* **2020**, *67*, 950–959. [CrossRef]
154. Yang, J.; Jiang, D.; Sun, H.; Li, A.; Liu, Z. Series-Winding Topology Converter for Active Magnetic Bearing Drive. *IEEE Trans. Ind. Electron.* **2021**, *68*, 11772–11782. [CrossRef]
155. Yang, J.; Jiang, D.; Sun, H.; Ding, J.; Li, A.; Liu, Z. A Series-Winding Topology Converter with Capability of Fault-Tolerant Operation for Active Magnetic Bearing Drive. *IEEE Trans. Ind. Electron.* **2022**, *69*, 6678–6687. [CrossRef]
156. Liu, C.; Zhan, J.; Wang, J.; Yang, Y.; Liu, Z. An Improved One-Cycle Control Algorithm for a Five-Phase Six-Leg Switching Power Amplifier in Active Magnetic Bearings. *IEEE Trans. Ind. Electron.* **2022**, *69*, 12564–12574. [CrossRef]
157. Aboki, N.H. IoT Integration of a Voltage Controlled Active Magnetic Bearing for Real-Time Remote Monitoring and Control. Master's Thesis, Youngstown State University, Youngstown, OH, USA, 2023.
158. Cheng, X.; Deng, S.; Cheng, S.B.; Hu, Y.F.; Wu, H.C.; Zhou, R.G. Design and Implementation of a Fault-Tolerant Magnetic Bearing Control System Combined With a Novel Fault-Diagnosis of Actuators. *IEEE Access* **2021**, *9*, 2454–2465. [CrossRef]

159. Liu, P. Study on Optimal Controller for Nonlinear Active Magnetic Bearing Systems Based on DSP Active Queue Management. In Proceedings of the 2023 11th International Conference on Intelligent Computing and Wireless Optical Communications (ICWOC), Chongqing, China, 16–18 June 2023; pp. 50–54.
160. Abrahamsson, J.; Hedlund, M.; Kamf, T.; Bernhoff, H. High-Speed Kinetic Energy Buffer: Optimization of Composite Shell and Magnetic Bearings. *IEEE Trans. Ind. Electron.* **2014**, *61*, 3012–3021. [CrossRef]
161. Wang, Y.; Zhang, K.; Dong, J.P. FPGA-Based Expanded Circuit Design for DSP Signal Processing. In Proceedings of the 2014 Fifth International Conference on Intelligent Systems Design and Engineering Applications, Hunan, China, 15–16 June 2014; pp. 511–516.
162. Li, X.; Palazzolo, A.; Wang, Z. A Combination 5-DOF Active Magnetic Bearing for Energy Storage Flywheels. *IEEE Trans. Transport. Electrific.* **2021**, *7*, 2344–2355. [CrossRef]
163. Ahmad, G.; Amin, U. Design, Construction and Study of Small Scale Vertical Axis Wind Turbine Based on a Magnetically Levitated Axial Flux Permanent Magnet Generator. *Renew. Energy* **2017**, *101*, 286–292. [CrossRef]
164. Jayawant, R. From Natural gas to Nuclear: The Diverse Application of Active Magnetic Bearing Systems. *Chin. J. Turbomach.* **2018**, *60*, 52–61.
165. Zhang, W.Y.; Zhu, H.Q.; Yuan, Y. Study on key technologies and applications of magnetic bearings. *Trans. China Electrotech. Soc.* **2015**, *30*, 12–20.
166. Zhang, W.Y.; Zhu, H.Q.; Ju, J.T.; Chen, T. Research status and development of magnetic bearings. *Bearing* **2016**, *12*, 56–63.
167. Jiang, H.; Su, Z.; Wang, D. Analytical Calculation of Active Magnetic Bearing Based on Distributed Magnetic Circuit Method. *IEEE Trans. Energy Convers.* **2021**, *36*, 1841–1851. [CrossRef]
168. Smirnov, A.; Uzhegov, N.; Sillanpää, T.; Pyrhönen, J.; Pyrhönen, O. High-Speed Electrical Machine with Active Magnetic Bearing System Optimization. *IEEE Trans. Ind. Electron.* **2017**, *64*, 9876–9885. [CrossRef]
169. Abdollha, M.B.; Tokhi, M.O. Nonlinear full-order observer-based controller design for active magnetic levitation via LQR-feedback linearisation. *Int. J. Model Ident. Control* **2016**, *26*, 59–67.
170. Espenhahn, T.; Sparing, M.; Niklas, C.; Nielsch, K.; Hühne, R. Dynamic Characteristics of a Superconducting Magnetic Bearing Under μm Displacements. *IEEE Trans. Appl. Supercond.* **2021**, *31*, 1–5. [CrossRef]
171. Fang, J.; Zheng, S.; Han, B. Attitude Sensing and Dynamic Decoupling Based on Active Magnetic Bearing of MSDGCMG. *IEEE Trans. Instrum. Meas.* **2012**, *61*, 338–348. [CrossRef]
172. Ji, J.C.; Hansen, C.H.; Zander, A.C. Nonlinear dynamics of magnetic bearing systems. *J. Intel Mat. Syst. Str.* **2008**, *19*, 1471–1491. [CrossRef]
173. Zhou, L.; Li, L. Modeling and Identification of a Solid-Core Active Magnetic Bearing Including Eddy Currents. *IEEE/ASME Trans. Mechatron.* **2016**, *21*, 2784–2792. [CrossRef]
174. Zhu, Y.; Yang, D.; Gao, X.; Ma, Z. Fractional Modeling and Identification of Active Magnetic Bearings with Laminated and Nonlaminated Structures Including Eddy Currents. *IEEE Trans. Transport. Electrific.* **2024**, *10*, 2423–2433. [CrossRef]
175. Ye, P.Z.; Li, H.W.; Tian, J.; Yu, W.T. Influence of rotor vibration on magnetic field distribution of radial active magnetic bearings. In Proceedings of the 2019 14th IEEE Conference on Industrial Electronics and Applications (ICIEA), Xi'an, China, 19–21 June 2019; pp. 2399–2404.
176. Le, Y.; Sun, J.J.; Han, B.C. Modeling and design of 3-DOF magnetic bearing for high-speed motor including eddy-current effects and leakage effects. *IEEE Trans. Ind. Electron.* **2016**, *63*, 3656–3665. [CrossRef]
177. Zhang, P.F.; Wang, Z.H.; Xi, G. A multi-objective optimal design methodology for solid core no-thrust-disk/thrust hybrid magnetic bearings considering eddy-current effects and leakage. *Int. J. Appl. Electromagn. Mech.* **2019**, *61*, 13–42. [CrossRef]
178. Wang, D.X.; Wang, N.X.; Ye, C.; Chen, K.S. Research on analytical bearing capacity model of active magnetic bearings based on magnetic saturation. *IET Electr. Power Appl.* **2017**, *11*, 1548–1557. [CrossRef]
179. Cao, Z.; Huang, Y.; Guo, B.; Peng, F.; Dong, J.; Hemeida, A. A Novel Hybrid Analytical Model of Active Magnetic Bearing Considering Rotor Eccentricity and Local Saturation Effect. *IEEE Trans. Ind. Electron.* **2022**, *69*, 7151–7160. [CrossRef]
180. Zhong, Y.; Wu, L.; Huang, X.; Fang, Y.; Zhang, J. An Improved Magnetic Circuit Model of a 3-DOF Magnetic Bearing Considering Leakage and Cross-Coupling Effects. *IEEE Trans. Magn.* **2017**, *53*, 1–6. [CrossRef]
181. Wu, M.; Zhu, H.; Zhang, W. Parameter Design of Six-Pole Hybrid Magnetic Bearing Considering Variable Stiffness. *IEEE Trans. Appl. Supercond.* **2020**, *30*, 1–5. [CrossRef]
182. Liu, C.; Liu, G. Equivalent Damping Control of Radial Twist Motion for Permanent Magnetic Bearings Based on Radial Position Variation. *IEEE Trans. Ind. Electron.* **2015**, *62*, 6417–6427. [CrossRef]
183. Asami, K.; Chiba, A.; Rahman, M.A.; Hoshino, T.; Nakajima, A. Stiffness analysis of a magnetically suspended bearingless motor with permanent magnet passive positioning. *IEEE Trans. Magn.* **2005**, *41*, 3820–3822. [CrossRef]

184. Chiba, A.; Kiryu, K.; Rahman, M.A.; Fukao, T. Radial force and speed detection for improved magnetic suspension in bearingless motors. *IEEE Trans. Ind. Appl.* **2006**, *42*, 415–422. [CrossRef]
185. Xie, Y.; Jiang, D.; Ding, J.; Liu, Z. Modeling, Analyzing, and Suppression of EMI Effects in Eddy Current Sensors of Active Magnetic Bearings. *IEEE Trans. Power Electron.* **2025**, *40*, 8093–8108. [CrossRef]
186. Wang, K.; Zhang, L.; Zheng, S.; Zhou, J.; Liu, X. Analysis and Experiment of Self-Differential Eddy-Current Sensor for High-Speed Magnetic Suspension Electric Machine. *IEEE Trans. Ind. Appl.* **2019**, *55*, 2538–2547. [CrossRef]
187. Hutterer, M.; Schroedl, M. Stabilization of Active Magnetic Bearing Systems in the Case of Defective Sensors. *IEEE/ASME Trans. Mechatron.* **2022**, *27*, 3672–3682. [CrossRef]
188. Wang, K.; Zhang, L.; Le, Y.; Zheng, S.; Han, B.; Jiang, Y. Optimized Differential Self-Inductance Displacement Sensor for Magnetic Bearings: Design, Analysis and Experiment. *IEEE Sens. J.* **2017**, *17*, 4378–4387. [CrossRef]
189. Zhang, P.; Zhu, C. Vibration Control of Base-Excited Rotors Supported by Active Magnetic Bearing Using a Model-Based Compensation Method. *IEEE Trans. Ind. Electron.* **2024**, *71*, 261–270. [CrossRef]
190. Yang, R.; Deng, Z.; Peng, C.; Li, K. Frequency-Varying Suppression of Vibration for Active Magnetic Bearing Using Improved Resonant Controller. *IEEE Trans. Ind. Electron.* **2022**, *69*, 13494–13503. [CrossRef]
191. Liu, C.; Liu, G.; Fang, J. Feedback Linearization and Extended State Observer-Based Control for Rotor-AMBs System With Mismatched Uncertainties. *IEEE Trans. Ind. Electron.* **2017**, *64*, 1313–1322. [CrossRef]
192. Wang, S.S.; Zhu, H.Q.; Wu, M.Y.; Zhang, W.Y. Active disturbance rejection decoupling control for three-degree-of- freedom six-pole active magnetic bearing based on BP neural network. *IEEE Trans. Appl. Supercon.* **2020**, *30*, 1–5. [CrossRef]
193. Cui, P.; Li, Y.; Li, J.; Du, L.; Wu, Y. Synchronous Vibration Force Suppression of Magnetically Suspended CMG Based on Modified Double SOGI-FLL. *IEEE Trans. Ind. Electron.* **2023**, *70*, 11566–11575. [CrossRef]
194. Xu, B.; Zhou, J.; Xu, L. A Backstepping Controller Based on a Model-Assisted Extended State Observer for a Slice Rotor Supported by Active Magnetic Bearings. *Actuators* **2022**, *11*, 266. [CrossRef]
195. Fang, J.C.; Sun, J.J.; Liu, H.; Tang, J.Q. A Novel 3-DOF Axial Hybrid Magnetic Bearing. *IEEE Trans. Magn.* **2010**, *46*, 4034–4045.
196. Zhong, L.; Wu, Y.; Fang, X.; Huang, X. Investigation of Cross-Coupling Effect of A 3-DOF Magnetic Bearing Using Magnetic Circuit Method. In Proceedings of the 2017 20th International Conference on Electrical Machines and Systems (ICEMS), Sydney, Australia, 11–14 August 2017; pp. 1–6.
197. Shi, J.; Zhu, H. Control Study for Compensating Rotor Vibration of Four-DOF Six-Pole Hybrid Magnetic Bearings Based on Variable Step Size LMS Algorithm. *IEEE J. Emerg. Sel. Top. Power Electron.* **2023**, *11*, 1616–1626. [CrossRef]
198. Xiao, L.; He, X.W.; Cheng, W.J.; Li, M. Structural optimization and dynamic characteristics of the new type 3-degrees of freedom axial and radial hybrid magnetic bearing. *Proc. Inst. Mech. Eng. Part C J. Mech. Eng. Sci.* **2022**, *236*, 5097–5110.
199. Zhu, H.; Wang, S. Decoupling control based on linear/nonlinear active disturbance rejection switching for three-degree-of-freedom six-pole active magnetic bearing. *IET Electr. Power Appl.* **2020**, *14*, 1818–1827. [CrossRef]
200. Hemenway, N.R.; Gjemdal, H.; Severson, E.L. New Three-Pole Combined Radial–Axial Magnetic Bearing for Industrial Bearingless Motor Systems. *IEEE Trans. Ind. Appl.* **2021**, *57*, 6754–6764. [CrossRef]
201. Ye, X.; Le, Q.; Zhou, Z. A Novel Homopolar Four Degrees of Freedom Hybrid Magnetic Bearing. *IEEE Trans. Magn.* **2020**, *56*, 1–4. [CrossRef]
202. Xu, S.L.; Sun, J.J.; Ren, H.L. An Active Magnetic Bearing with Controllable Permanent-Magnet Bias Field. *IEEE/ASME Trans. Mech.* **2022**, *27*, 3474–3481. [CrossRef]
203. Krishnan, S.; Harikrishnan, A.; Sajeev Kumar, A.; Ravichandran, M.H. Design of a Novel Four DOF Active Magnetic Bearing for Space Applications. In Proceedings of the 2022 IEEE International Conference on Power Electronics, Drives and Energy Systems (PEDES), Jaipur, India, 14–17 December 2022; pp. 1–5.
204. Sun, J.J.; Ju, Z.Y.; Han, W.T.; Liu, G. A novel integrated 4-DOF radial hybrid magnetic bearing for MSCMG. *J. Magn. Magn. Mater.* **2016**, *421*, 86–97.
205. Sun, J.J.; Ju, Z.Y.; Peng, C.; Le, Y.; Ren, H.L. A Novel 4-DOF Hybrid Magnetic Bearing for DGMSCMG. *IEEE Trans. Ind. Electron.* **2017**, *64*, 2196–2204. [CrossRef]
206. Blumenstock, K.; Brown, G. Novel Integrated Radial and Axial Magnetic Bearing. In Proceedings of the 7th International Symposium on Magnetic Bearings, Zurich, Switzerland, 23–25 August 2000; ETH Zurich: Zurich, Switzerland, 2000; pp. 467–471.
207. Zhang, W.Y.; Wang, J.W.; Li, A.; Xiang, Q. Multiphysics Fields Analysis and Optimization Design of a Novel Saucer-Shaped Magnetic Suspension Flywheel Battery. *IEEE Trans. Transp. Electrif.* **2024**, *10*, 5473–5483. [CrossRef]
208. Cheng, X.; Cheng, B.X.; Deng, S.; Zhou, R.G.; Lu, M.Q.; Wang, B. State-feedback decoupling control of 5-DOF magnetic bearings based on α-order inverse system. *Mechatronics* **2020**, *68*, 102358. [CrossRef]
209. Chen, J.; Zhu, J.; Severson, E.L. Severson, Review of Bearingless Motor Technology for Significant Power Applications. *IEEE Trans. Ind. Appl.* **2020**, *56*, 1377–1388. [CrossRef]

210. Sun, X.D.; Chen, L.; Yang, Z.B. Overview of bearingless permanent-magnet synchronous motors. *IEEE Trans. Ind. Electron.* **2013**, *60*, 5528–5538. [CrossRef]
211. Kato, H.; Komori, M.; Asami, K.; Sakai, N. Development of one-axis active controlled bearingless motor working at extremely low temperature. *J. Adv. Mech. Des. Syst.* **2020**, *14*, JAM DSM0034. [CrossRef]

energies

MDPI

Article

Optimal Split Ratio in Double-Stator Permanent-Magnet Motors Considering Loss Limitations for Robot Joint Applications

Tianran He [1], Yang Shen [2,*], Wei Li [1] and Dawei Liang [3]

1 School of Electronic and Information Engineering, Tongji University, Shanghai 200092, China; hetianran@tongji.edu.cn (T.H.); liweimail@tongji.edu.cn (W.L.)
2 Strategic Assessment & Consultation Institute of AMS, Beijing 100071, China
3 Electrical Machines and Drives Group, The University of Sheffield, Sheffield S10 2TN, UK; dliang4@sheffield.ac.uk
* Correspondence: shenyang20240317@163.com

Abstract

Recently, humanoid robots with personification behavior and high working efficiency have received significant attention. Meanwhile, high-torque-density motors, which serve as the core power source for robot joints, have also been widely researched. In this paper, a high-torque-density double-stator permanent-magnet (DSPM) motor is designed for robot joint applications, and its outer stator (OS) split ratio (the ratio between the inner and outer diameters of the OS) and inner stator (IS) split ratio (the ratio between the inner and outer diameters of the IS) are analyzed and optimized. Since the DSPM motor has different heat dissipation capabilities for the OS and IS, their different loss limitations should be considered to avoid the risk of local overheating, especially for the IS. This paper shows that the loss limitations affect the optimal OS and IS split ratios, as well as the maximum average torque. The IS loss limitation increases the optimal OS split ratio and decreases the optimal IS split ratio; however, the OS loss limitation has the opposite effect. Additionally, an investigation into the electromagnetic characteristics of the optimized DSPM motor was conducted using the finite element method. Finally, a prototype was manufactured, and the results of the temperature rise experiments verified the feasibility of the proposed DSPM motor and the effectiveness of the optimal method.

Keywords: double-stator permanent-magnet motor; electromagnetic performance; high torque density; split ratio

1. Introduction

Dual-stator machines have been widely used in various industrial applications due to their high torque and power density, such as wind power generation [1–3], robotics [4], electric vehicles [5–7], electric motorcycles [8], and the servo industry [9–11]. To achieve high dynamic characteristics in humanoid robots, dual-stator permanent-magnet (DSPM) motors are attractive for applications in robot joint actuators [12,13].

In traditional single-stator PM motors, one of the most crucial parameters is the split ratio, which is defined as the ratio of the stator's inner diameter to its outer diameter. By balancing the magnetic load and electrical load of the motor, the split ratio can be optimized to maximize the output torque, power density, and efficiency. Reference [14] achieved the minimum copper loss by analyzing the optimal split ratio, and the authors of [15] investigated the influence of the optimal split ratio on the motor magnetic circuit and electromagnetic performance by fixing the copper loss. Reference [16] studied the factors affecting the optimal split ratio, including the drive mode, winding configuration, end-winding, and stator tooth tips. Based on surface PM motors, the authors of [17,18] presented analytical models for computing the optimal split ratio in motors with inner and external rotor configurations. Based on interior PM motors, the optimal split ratio was investigated in [19]. Compared with single-stator motors, dual-stator motors not only need to balance the magnetic load and electrical load but also consider the power ratio of the inner and outer stators. Therefore, the optimal design of DSPM motors can also adopt the optimal split ratio to achieve the maximum torque.

To maximize torque per volume, the authors of [20,21] introduced two split ratios: the outer stator split ratio (outer stator inner diameter/outer stator outer diameter) and the inner stator split ratio (inner stator outer diameter/outer stator outer diameter). In addition, the literature shows that this optimization can be divided into two steps: the first step is the optimization of the outer stator split ratio, and the second step is the optimization of the inner stator split ratio, within the demanded value range. However, this optimal method considers the outer and inner stators separately but neglects the relationship between the inner and outer stators. Based on the same winding connection, packing factor, and wire diameter of the inner and outer stators, the turn ratio (the ratio of the inner and outer stator winding turns) was used as a substitute for the split ratio in [22], and the turn ratio was determined as 1:1, i.e., the same power of the inner and outer air gaps. In [23], only one split ratio was optimized to achieve the maximum torque density, i.e., the ratio of the outer stator inner diameter to the inner stator outer diameter. This split ratio considers the inner and outer stators simultaneously by fixing the sum of the inner and outer stators' surface current densities and the maximum flux densities of the inner and outer stators. Although the relationship between the inner and outer stator is considered in this method, the proportion of the inner and outer stator power is neglected. Reference [24] introduced the power split ratio, i.e., the ratio of the power of the inner and outer air gaps, to achieve the maximum output power under different magnetic loads. It showed that the optimal power split ratio decreases with an increase in the magnetic load, indicating that the power provided by the inner stator decreases. Compared to the outer stator winding, the inner stator winding exhibits significantly poorer heat dissipation characteristics, and the authors of [25] determined that the power split ratio, i.e., the ratio of the power on the outer stator winding to the rated power of the DSPM motor, should be larger than 0.5. However, ">0.5" is only a qualitative result, not a quantitative result. Therefore, to quantitatively describe the power split ratio while considering the different heat dissipation of the IS and OS, the IS and OS loss limitations are considered during the optimization of DSPM motors.

This paper is organized as follows: The DSPM motor topology is described, and the main parameters of the proposed motor are given in Section 2. Section 3 investigates the optimal split ratio of the DSPM motor considering the IS and OS loss limitations. In addition, the influence of the current density on the optimal split ratio is shown. The electromagnetic performance of the optimized DSPM motor is analyzed in Section 4. Section 5 shows the prototype and its experimental results. Finally, Section 6 provides the conclusion.

2. Motor Topology

A 36-slot/42-pole (36s/42p) DSPM motor is shown in Figure 1a, and it includes an outer stator (OS), an inner stator (IS), and a cup-rotor with outer and inner PMs (OPM and IPM). In addition, a house is designed to support the double stator and rotor, as shown in Figure 1b. With the size limitation of the robot joint, the outer diameter of the house is 114 mm, and its thickness is designed as 4 mm due to the interference fit between the house and OS. Therefore, the outer diameter of the OS is 106 mm. Considering the standardized bearing, the inner diameter of the IS is 54 mm. It is worth noting that the circular fins are designed on the surface of the house to enhance the heat dissipation capacity of the OS by increasing the heat dissipation area. The material of the stator is 20JNEH1200 with 0.2 mm lamination, and that of the PM is N42SH with 1.3 remanence (B_r = 1.3). To have a lightweight design, the material of the rotor disk and frame is aluminum alloy (6061-T6). However, a magnetically permeable material, i.e., 10# steel, is used for the rotor yoke. The key design parameters of the proposed DSPM motor are summarized in Table 1.

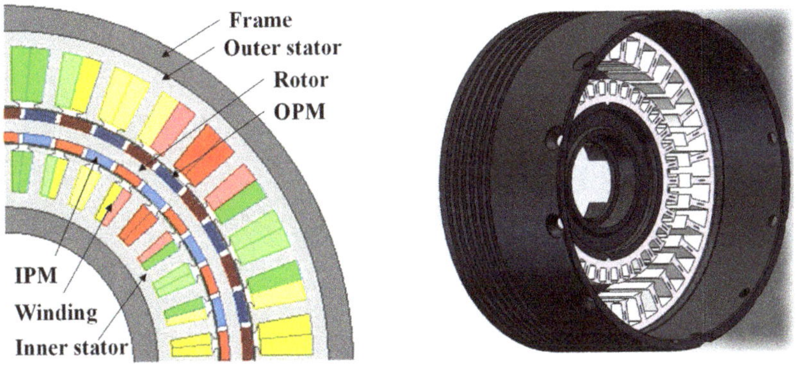

(a) (b)

Figure 1. Motor topology of a 36s/42p DSPM motor. (**a**) Two-dimensional diagram of the proposed motor; (**b**) three-dimensional diagram of the house and stators. The red, yellow, and green colors represent the three phase windings.

Table 1. Main parameters of the proposed DSPM motor.

Parameter	Data	Parameter	Data
Outer diameter of OS, mm	106	Rated speed, r/min	1200
Inner diameter of IS, mm	54	Stator active length, mm	20
Air-gap length, mm	0.4	Pole arc coefficient	0.84

3. Optimal Split Ratio

3.1. Loss Limitation

Due to two stators and two air gaps, the DSPM motor has an OS split ratio (outer stator inner diameter/outer stator outer diameter), i.e., $\lambda_{OS} = D_{in\_OS}/D_{out\_OS}$, and an IS split ratio (inner stator inner diameter/inner stator outer diameter), i.e., $\lambda_{IS} = D_{in\_IS}/D_{out\_IS}$. Both split ratios are optimized in this paper. For robot joint applications, the maximum winding temperature rise is limited, which depends on the heat dissipation capability, copper loss, and current density. Compared with the conventional single-stator PM motor, the inner and outer stator losses should be considered respectively due to different heat dissipation capabilities, especially for the inner stator. The drive system and end cover obstruct the inner stator's limited cooling surface, dramatically impairing heat rejection. Therefore, the loss limitation should be considered in the optimization.

Under a uniform winding temperature and exclusive convection heat dissipation, the maximum allowed loss (P_{limit}) is derived from the motor's thermal limits [26]:

$$P_{limit} = hV_m \pi D_o l_s, \quad (1)$$

where h represents the stator effective heat transfer coefficient, a cooling-system-dependent parameter. Its empirical range spans 25–100 W/(m^2·K) [27], for natural and forced-air cooling, respectively. V_m is the maximum winding temperature rise, which is fixed by the insulation class (F class, 100K). D_o and L_s are the stator outer diameter and stator active length. Therefore, '$\pi D_o L_s$' indicates the heat dissipation area. Table 2 shows the loss limitations of the OS and IS in the DSPM motor. It is worth noting that the heat transfer coefficients of the OS and IS are different, considering the house design.

Table 2. Loss limitations of the OS and IS in the DSPM motor.

Parameter	OS	IS
Heat transfer coefficient, W/(m^2·K)	50	25
Maximum winding temperature rise, K	100	100
Stator outer/inner diameter, mm	106	54
Stator active length, mm	20	20
Loss limitation, W	33.28	8.48
K_{loss} = Loss_OS/Loss_IS	3.51/1	

3.2. Torque Calculation

The RMS phase currents (I_{rms}) in windings of the OS and IS are the same due to the series connection, and they can be calculated by the fixed copper loss limitation (P_{cu}).

$$P_{cu} = 3{I_{rms}}^2 R_{ph} = 3{I_{rms}}^2 \rho_{cu} \frac{2N_a(L_s + L_{ew})}{S_{wire} N_{pb}}, \quad (2)$$

$$S_{wire} = \frac{k_{fill}(S_{slot} N_{slot}/3)}{N_a}, \quad (3)$$

$$P_{cu} = \frac{18(N_a I_{rms})^2 \rho_{cu}(L_s + L_{ew})}{k_{fill} S_{slot} N_{slot} N_{pb}}, \quad (4)$$

where R_{ph} is the phase resistance, ρ_{cu} is the copper resistivity, N_a is the number of series turns per phase, L_s is the stator active length, L_{ew} is the one side end-winding length, S_{wire} is the copper wire area, N_{pb} is the number of parallel branches, k_{fill} is the slot fill factor, and S_{slot} and N_{slot} are the stator slot area and slot number.

Therefore, the electromagnetic torque can be rewritten as

$$T_e = 3k_w N_a D_{air} L_s B_g I_{rms}, \quad (5)$$

where K_w is the winding factor, D_{air} is the air-gap diameter, D_o is the outer diameter of the stator, λ is the split ratio, $D_0\lambda$ indicates the air-gap diameter, and B_g is the amplitude of the air-gap flux density.

In the DSPM motor, the total torque can be divided into an OS torque and an IS torque, and they can be calculated by the analytical method. It is assumed that the motor has a rectangular open-circuit airgap flux density distribution under FOC control. The OS, IS, and total electromagnetic torque (T_{OS}, T_{IS}, and T_{total}) are given by

$$\begin{cases} T_{OS} = 3k_w N_{a\_os} D_{air\_outer} L_s B_{g\_outer} I_{rms} \\ T_{IS} = 3k_w N_{a\_is} D_{air\_inner} L_s B_{g\_inner} I_{rms} \\ T_{total} = T_{OS} + T_{IS} \end{cases} \quad (6)$$

where D_{air_outer} and D_{air_inner} are the outer and inner air-gap diameters, B_{g_outer} and B_{g_inner} are the outer and inner air-gap flux densities, and N_{a_os} and N_{a_is} are the number of series turns per phase in the OS and IS.

In this paper, the air gap length is fixed at 0.4 mm. It is well-known that a smaller thickness of the rotor yoke leads to a larger active slot area. It is designed as 2.2 mm, considering the mechanical strength and manufacturing. In addition, for simplicity, it is assumed that the inner air-gap diameter is equal to the outer diameter of the IS, and the outer air-gap diameter is equal to the inner diameter of the OS.

3.3. Optimal Results

Considering the different stator loss limitations, the optimal results are given and analyzed. The ranges of the IS and OS split ratios can be determined by the ranges of the inner and outer diameters of the OS and IS, respectively. Therefore, in this optimization, the range of the inner diameter of the OS is from 70 mm to 90 mm, and that of the OS split ratio is from 0.66 to 0.85; the range of the outer diameter of the IS is from 60 mm to 80 mm, and that of the IS split ratio is from 0.675 to 0.9.

3.3.1. Torque Considering Outer Stator Loss Limitation Only

When only considering the outer stator loss limitation, Figure 2a shows that the average torque versus OS split ratio follows a parabolic trend (initial increase followed by decrease) for all tested IS split ratios. Therefore, an optimal OS split ratio exists to maximize torque for each IS split ratio. As shown in Figure 3b, the maximum torque diminishes as the IS split ratio decreases.

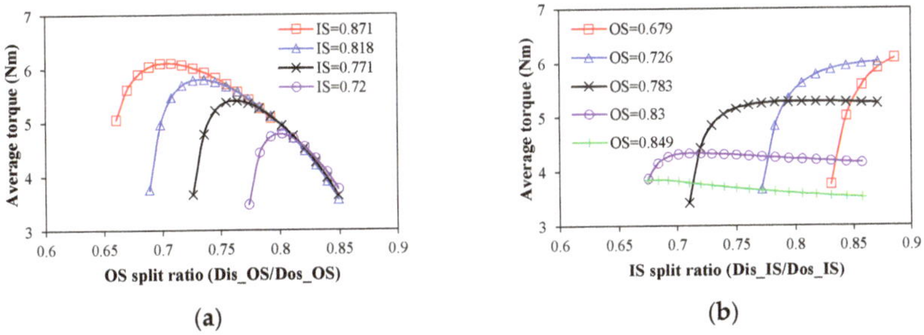

Figure 2. Variation in average torque with OS and IS split ratios considering outer stator loss limitation only. (**a**) OS split ratio; (**b**) IS split ratio.

However, the trend of the average torque with the IS split ratio depends on the OS split ratio. Figure 2b shows that with a small OS split ratio, such as 0.679 and 0.726, the average torque increases sharply with the increase in the IS split ratio due to the rise of PM thickness. With the relatively large OS split ratio, such as 0.783 and 0.830, the average torque increases sharply at first and then tends to stabilize or even decreases slightly. The reason is that the rise of the PM thickness leads to the rise of the magnetic load and torque at first, while, as the PM thickness increases further, the effect on the magnetic load is weakened, and the electric load decreases. With a large OS split ratio, such as 0.849, the torque decreases with the increase in the IS split ratio. Consequently, for each fixed OS split ratio, an optimal IS split ratio maximizes torque. As shown in Figure 3a, this maximum torque initially rises then diminishes with increasing OS split ratios.

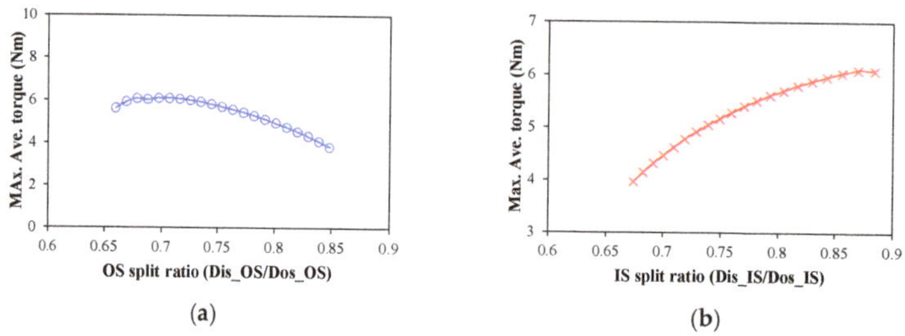

Figure 3. Maximum average torque versus OS/IS split ratios considering outer stator loss limitation only. (**a**) OS split ratio; (**b**) IS split ratio.

According to Figures 2 and 3, an optimal combination of IS and OS split ratios (λ_{IS} = 54/62 = 0.871; λ_{OS} = 75/106 = 0.707) exists to maximize the torque, T_{total_max} = 6.104 Nm. However, under this scenario, the outer stator loss is 136.09 W, which significantly exceeds the loss limitation of the IS and results in a high temperature.

3.3.2. Torque Considering Inner Stator Loss Limitation Only

When considering the inner stator loss limitation only, mathematically, the variation in the torque with different IS and OS split ratios is given in Figure 4. It shows that the maximum average torque decreases linearly with the increase in the IS split ratio, and it also increases linearly with the increase in the OS split ratio. In other words, when neglecting the outer stator loss limitation, the torque increases with the increase in the diameter of the air gap. The optimal combination of IS and OS split ratios is λ_{IS} = 54/80 = 0.675 and λ_{OS} = 90/106 = 0.849, as shown in Figure 4. However, under this scenario, the outer stator loss is 175.49 W, which will lead to a high temperature of the OS.

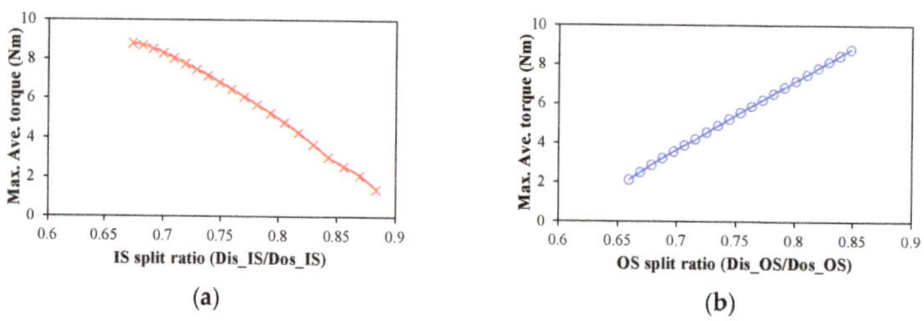

Figure 4. Maximum average torque versus OS/IS split ratios considering inner stator loss limitation only. (**a**) OS split ratio; (**b**) IS split ratio.

Therefore, both the inner and outer stator losses should be considered in the optimization of DSPM motors.

3.3.3. Torque Considering Outer and Inner Stator Loss Limitations

Previous analyses show that the torque variation with IS/OS split ratios exhibits distinctly different behaviors under different limiting conditions. To achieve the maximum torque while maintaining a reasonable temperature rise of the DSPM motor, both the inner and outer stator loss limitations should be considered simultaneously.

The variations in the maximum average torque with IS and OS split ratios considering both inner and outer stator loss limitations are shown in Figure 5. It shows that, for the IS and OS split ratios, the available designs can be divided into two parts: one is dominated by the OS loss, and the other one is dominated by the IS loss. It is worth noting that the available designs in Figure 5 are obtained through a mathematical method. However, the real optimal process is fixing the outer stator loss limitation first and then comparing the inner stator loss limitation. In general, there is a rare available design that has the same inner stator loss as the limitation, as shown in Figure 6. Therefore, there is an optimal combination of IS and OS split ratios, i.e., $\lambda_{IS} = 54/70 = 0.771$ and $\lambda_{OS} = 81/106 = 0.764$, to achieve the optimal design, as shown in Figure 6.

Figure 7 shows three optimized motor designs considering the inner stator loss limitation only, the outer stator loss limitation only, and both the outer and inner stator loss limitations, respectively. It indicates that considering either IS or OS loss alone will result in an uneven spatial distribution between the IS and OS, whereas only simultaneous consideration of both IS and OS losses can achieve an optimal motor design.

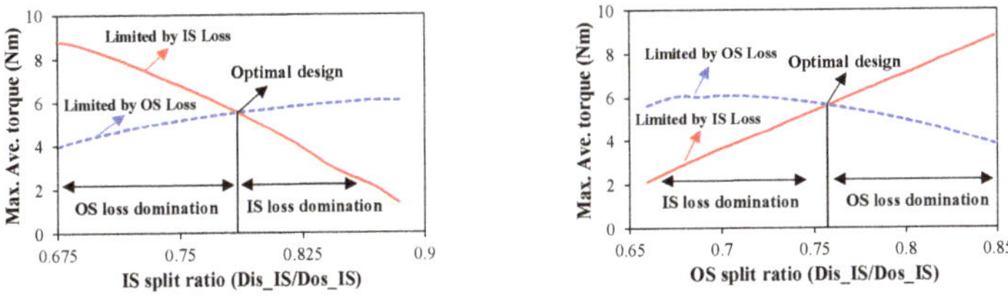

(a) (b)

Figure 5. Maximum average torque versus OS/IS split ratios considering both outer and inner stator loss limitations. (**a**) IS split ratio; (**b**) OS split ratio.

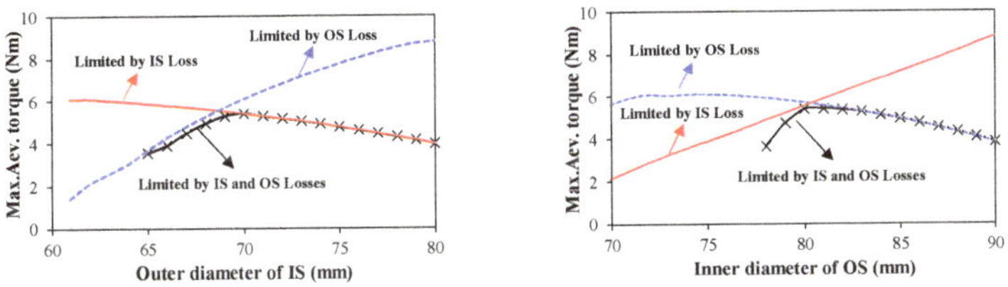

(a) (b)

Figure 6. Maximum average torque versus outer diameter of IS and inner diameter of OS considering outer stator loss limitation firstly and then inner stator loss limitation. (**a**) Outer diameter of IS; (**b**) Inner diameter of OS.

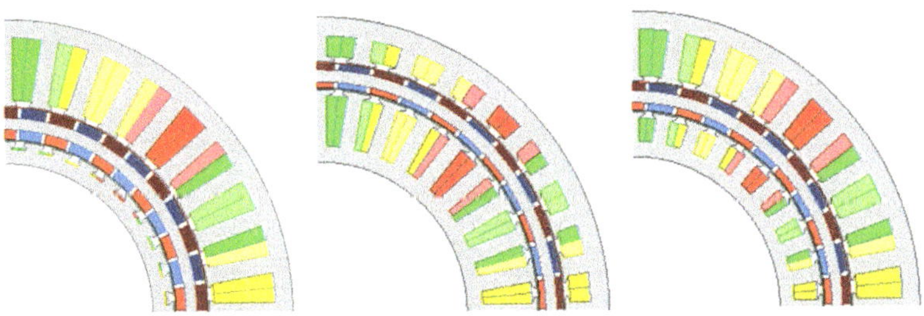

(a) (b) (c)

Figure 7. Motor topologies of the optimized 36s/42p DSPM motors considering different loss limitations. (**a**) Considering outer stator loss only; (**b**) considering inner stator loss only; (**c**) considering inner and outer stator loss. In the OS and IS, the red, yellow, and green colors represent the three phase windings. In the rotor, the red (brown) and blue (light/dark) represent the N-and S-pole.

3.3.4. Torque Considering Current Density

If the copper loss is below the limitation, but the current density is large, the local temperature rise may lead to the insulation breakdown in windings and subsequent short-circuiting. Therefore, after considering the IS and OS losses, the current density also should be limited. Figure 8 shows that the limitation of current density will exclude several designs. When the current density is limited to less than 12 A/mm^2, the optimal split ratio remains unchanged. When the current density is limited to less than 10 A/mm^2, the optimal split ratio is changed, and the optimal combination of the outer diameter of IS and inner diameter of OS changes from (70 mm, 81 mm) to (71 mm, 80 mm). In high-torque-density DSPM motors, the system operates with a current density not exceeding 12 A/mm^2 by engineering experience. The parameters of the optimized DSPM motor are shown in Table 3.

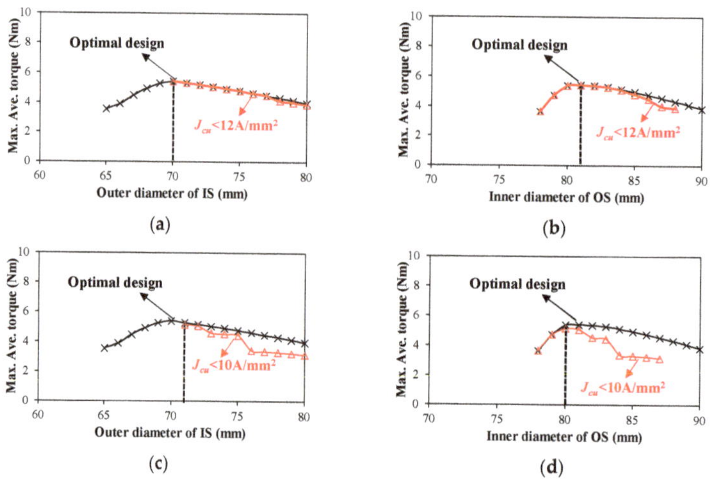

Figure 8. Maximum average torque versus outer diameter of IS and inner diameter of OS considering different current density limitations. (**a**) Outer diameter of IS, J_{cu} < 12 A/mm^2; (**b**) inner diameter of OS, J_{cu} < 12 A/mm^2; (**c**) outer diameter of IS, J_{cu} < 10 A/mm^2; (**d**) inner diameter of OS, J_{cu} < 10 A/mm^2.

Table 3. Parameters of the optimized DSPM motor.

Parameter	Data	Parameter	Data
Outer diameter of OS (mm)	106	RMS phase current (A)	5.93
Inner diameter of OS (mm)	81	Outer/inner packing factor	0.43/0.39
Outer diameter of IS (mm)	70	Max. stator flux density (T)	1.8
Inner diameter of IS (mm)	54	Copper losses of OS (W)	33.28
Rotor yoke thickness (mm)	2.2	Copper losses of IS (W)	8.05
PM thickness (mm)	1.25	Current density of OS/IS (A/mm^2)	10.5/10.3
Tooth width of OS (mm)	3.21	Torque_total (Nm)	5.41
Yoke thickness of OS (mm)	3.21	Torque_IS (IS/Total) (Nm)	1.01 (18.67%)
Tooth width of IS (mm)	2.77	Torque_OS (OS/Total) (Nm)	4.40 (81.33%)
Yoke thickness of IS (mm)	2.77	KT = Torque_OS/Torque_IS	4.36/1

4. Electromagnetic Performance

This section examines the electromagnetic characteristics of the optimized DSPM motor, including open-circuit air-gap flux density, back electromotive force (back-EMF), cogging torque, and electromagnetic torque.

The flux distribution pattern in Figure 9 demonstrates balanced magnetic loading, with theinner stator teeth carrying slightly higher flux density (1.36 T) than the outer stator teeth (1.29 T), while both stator yokes maintain similar sub-0.8 T densities, as shown in Figure 10. The maximum tooth-tip flux densities in the IS and OS are almost 2.0 T, which lead to local magnetic saturation.

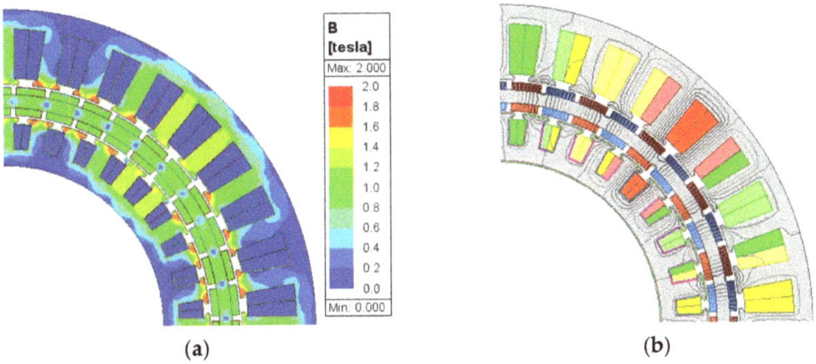

Figure 9. Equal potential and flux distribution of the optimized DSPM motor. (**a**) Equal potential. (**b**) Flux distribution.

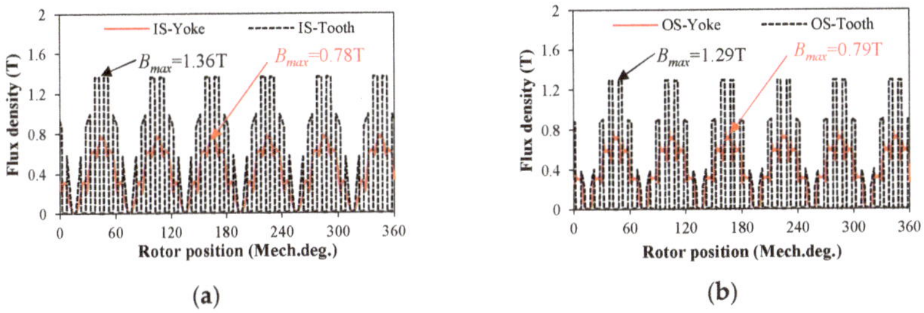

Figure 10. Flux densities of the inner and outer stator yoke and tooth of the optimized 36s/42p DSPM motor. (**a**) Inner stator; (**b**) outer stator.

Figure 11 shows the air-gap flux density and harmonics of the optimized DSPM motor. The air-gap flux density waveforms are trapezoidal, and their harmonics mainly are third order due to the slot effect and local saturation, Figure 11b. It is worth noting that the fundamental component of the inner air-gap flux density exceeds that of the outer air-gap flux density (1.16 T vs. 1.09 T).

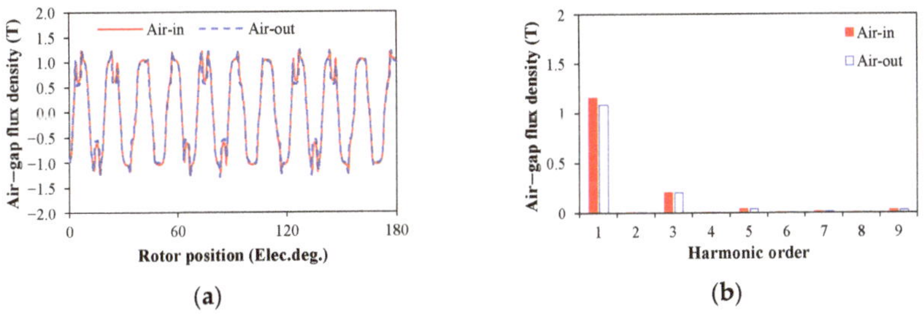

Figure 11. Air-gap flux densities of the inner and outer air-gaps. (**a**) Air-gap flux density waveforms; (**b**) harmonics.

Under the rated operation speed, i.e., n = 1200 r/min, the three-phase back-EMF waveforms of the IS and OS and harmonics of the winding A are shown in Figure 12. The maximum back EMFs of the IS and OS are 8.55 V and 36.1 V, and their back-EMF coefficients are 7.12 V/krpm and 30.08 V/krpm, respectively. The total three-phase back-EMF waveforms of the DSPM motor and harmonics of the winding A are shown in Figure 13. Since the windings of the IS and OS are connected in series, the total three-phase back-EMFs of the DSPM motor are the sum of the three-phase back-EMFs of the IS and OS. Therefore, the maximum phase back EMF of the DSPM motor is 44.65 V.

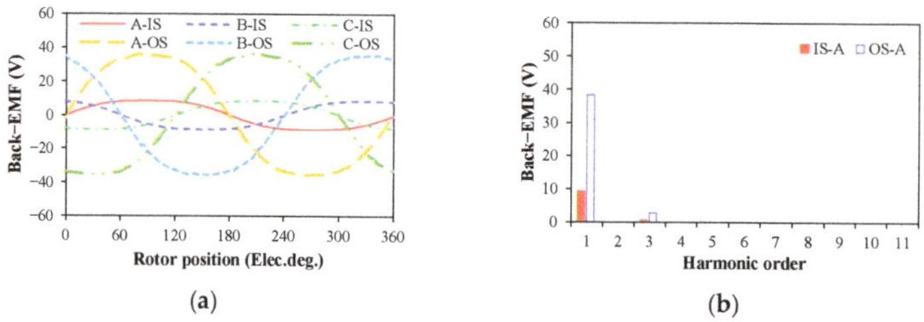

Figure 12. Three-phase back-EMFs of IS and OS windings. (**a**) Three-phase back-EMF waveforms; (**b**) harmonics.

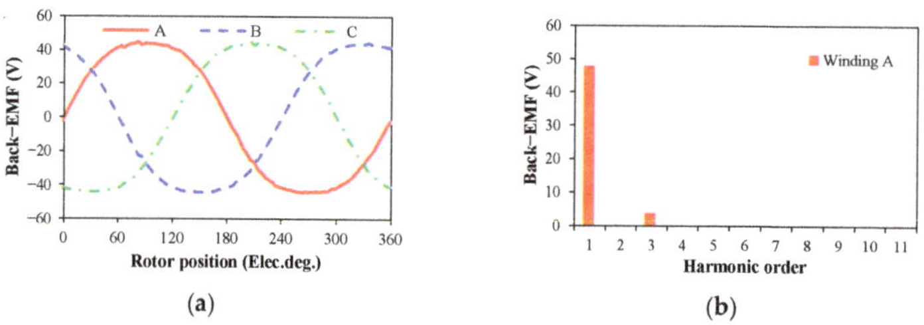

Figure 13. Three-phase back-EMFs of the optimized DSPM motor. (**a**) Three-phase back-EMF waveforms; (**b**) harmonics.

The cogging torque of the optimized motor is shown in Figure 14, and its maximum value is 50 mNm. Figure 15a shows the electromagnetic torque waveform of the optimized motor. The average torque is 4.75 Nm under the rated current (I_{rms} = 5.93 A), and the torque ripple is 3.34%. Figure 15b shows the variation in the average torque with the rms phase current. With the increase in the phase current, the average torque increases, but the slope of the rise decreases due to the electromagnetic saturation.

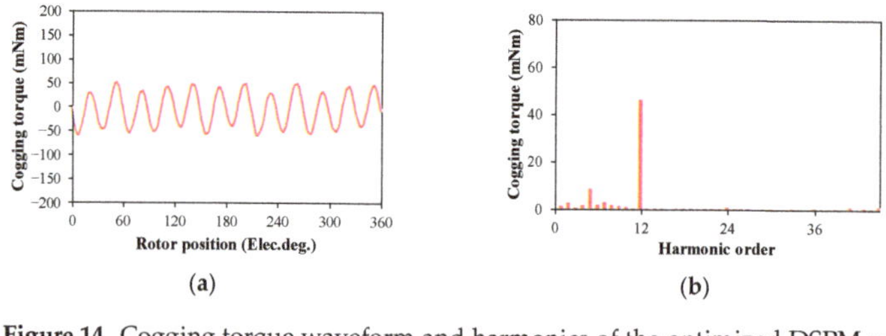

Figure 14. Cogging torque waveform and harmonics of the optimized DSPM motor. (**a**) Cogging torque waveform; (**b**) harmonics.

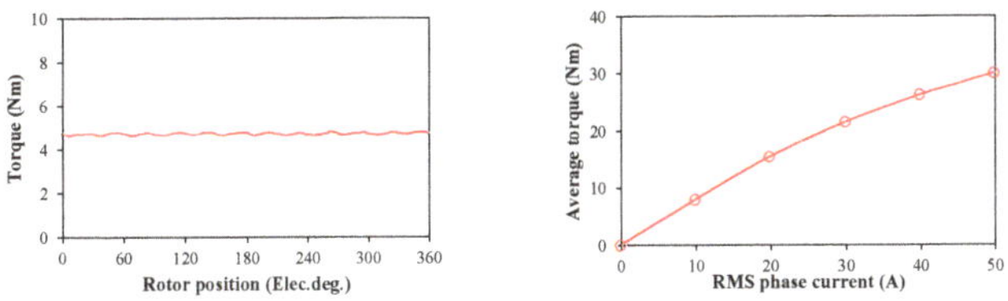

(**a**) (**b**)

Figure 15. Electromagnetic torque and torque-current curve of the optimized DSPM motor. (**a**) Electromagnetic torque waveform; (**b**) torque–current curve.

5. Experimental Validation

The optimized DSPM motor was manufactured, and its double-stator and rotor structures are shown in Figure 16. Figure 16d shows the 18:1 planetary gearbox. The experiments were carried out, and the measured results validated the analytical and FE predictions. Table 4 presents the measured and FE predicted parameters of the DSPM motor, including the phase resistances and inductances of the IS, OS, and the entire motor. The measured and predicted results have a good agreement.

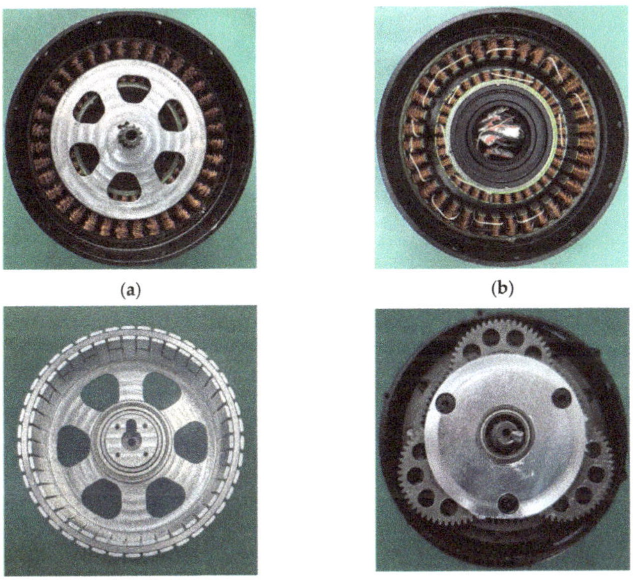

Figure 16. Prototype of the optimized DSPM motor. (**a**) Stator and rotor; (**b**) double-stator structure; (**c**) rotor structure; (**d**) 18:1 dual-stage planetary gearbox.

Table 4. Measured and FE predicted parameters of the DSPM motor.

Parameter	Measured	Predicted
Phase resistance of OS (Ω)	0.273	0.315
Phase resistance of IS (Ω)	0.074	0.076
Phase resistance (Ω)	0.347	0.396
Phase inductance of OS (μH)	364.0	456.19
Phase inductance of IS (μH)	27.8	32.22
Phase inductance (μH)	391.8	488.41

The close correspondence between the measured and simulated back-EMF waveforms in Figure 17a confirms the accuracy of the FE model's electromagnetic formulation and parameter assumptions. The FE predicted results of torque and *T-I* curves of the actuator were validated by the torque test platform. It includes a 200 Nm magnetic particle brake, a torque-speed sensor, and a temperature monitor, as shown in Figure 18. The measured and FE-predicted *T-I* curves of the DSPM motor are shown in Figure 17b. It shows that the measured torque is smaller than the FE predicted torque of the joint actuator, and the difference mainly is caused by the gearbox losses, the transmission losses of the test bench, the machinal losses of the DSPM motor, and the end effect. Since the efficiency of the 18:1 dual-stage planetary gearbox is almost 85% (The data are provided by the supplier), the measured *T-I* curve is less than the FE-predicted result, but a satisfactory agreement is achieved, especially for the 85% FE prediction.

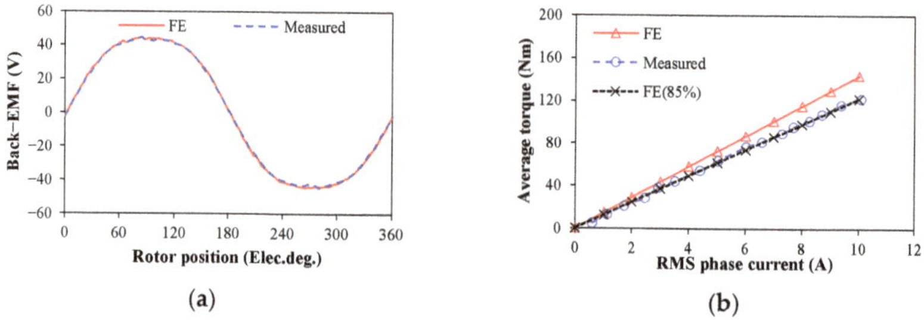

Figure 17. Measured and FE-predicted back-EMF waveforms and torque–current curves of the DSPM motor. (**a**) Measured and FE-predicted back-EMF waveforms; (**b**) measured and FE-predicted torque–current curves.

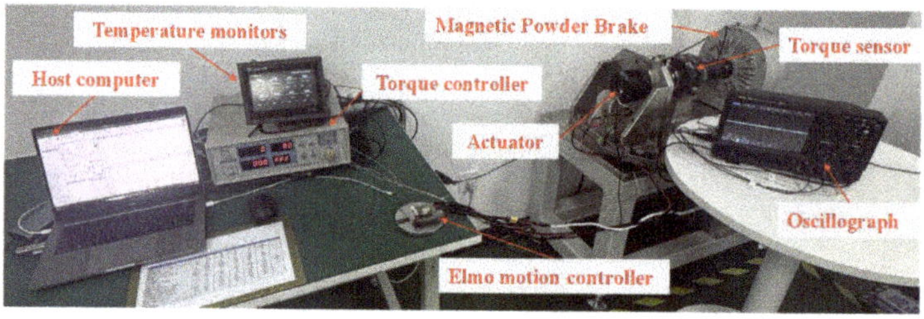

Figure 18. 200 Nm torque test platform.

Additionally, the temperature rise of the various components of the DSPM motor under the rated condition was tested. Figure 19 shows the temperature test points, including the three-phase windings in the IS and OS and the outer and inner walls of the house. The temperature rise curves of various components are shown in Figure 20. After 1500 s (25 min), those temperatures remain almost unchanged, and the winding temperatures of the IS (almost 80.4 °C) are lower than that of the OS (almost 84 °C). In addition, the inner and outer walls of the house remain at 74.3 °C and 60.7 °C, respectively. The temperature results indicate the optimized DSPM motor has good thermal management under the rated condition.

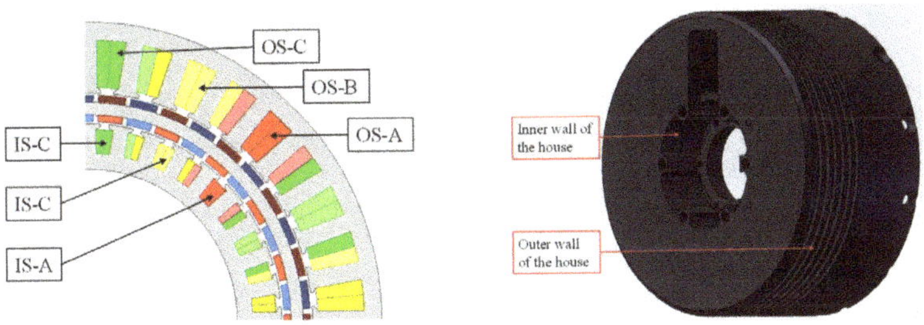

(a) (b)

Figure 19. Temperature test points of the DSPM motor. (**a**) Test points of the three-phase windings in IS and OS. (**b**) Test points of the inner and outer walls of the house.

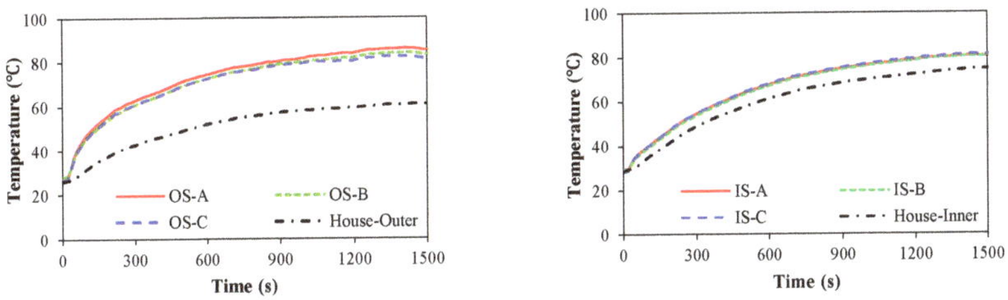

(a) (b)

Figure 20. Temperature test of the DSPM motor. (**a**) Outer stator; (**b**) inner stator.

6. Conclusions

This paper proposes a new analytical optimal method for two split ratios of the DSPM motor for humanoid robot joints. Since the heat dissipation capabilities of the inner and outer stators are different, the allowed maximum losses are limited in this optimization, respectively. It shows that both OS and IS loss limitations have a significant influence on the electromagnetic and thermal performance and the optimal combination of OS and IS split ratios. Only limiting the OS loss, the optimal OS split ratio is small, the PM thickness is large, and the IS windings have a high temperature. Only limiting the IS loss, the OS optimal split ratio is large, the PM thickness is small, and the OS windings have a high temperature. Considering both IS and OS losses, the IS and OS windings have the allowed temperature, and the motor has a maximum torque. Finally, the experimental measurements confirm excellent agreement with both FE simulations and analytical predictions, validating the proposed models.

Author Contributions: Conceptualization, T.H.; methodology, T.H.; software, T.H.; validation, Y.S. and W.L.; formal analysis, D.L.; writing—original draft preparation, T.H.; writing—review and editing, Y.S. and D.L. All authors have read and agreed to the published version of the manuscript.

Funding: This research was funded by the National Natural Science Foundation of China (Grant No. 52407062), the Aeronautical Science Foundation of China (Grant No. 20240007038001), and the Shanghai Pujiang Program (Grant No. 22PJD080).

Data Availability Statement: No new data were created or analyzed in this study.

Conflicts of Interest: The authors declare no conflicts of interest.

References

1. Levy, D. Analysis of a double-stator induction machine used for a variable-speed/constant-frequency small-scale hydro/wind electric power generator. *Electr. Power Syst. Res.* **1986**, *11*, 205–223. [CrossRef]
2. Niu, S.; Chau, K.T.; Jiang, J.Z.; Liu, C. Design and control of a new double-stator cup-rotor permanent-magnet machine for wind power generation. *IEEE Trans. Magn.* **2007**, *43*, 2501–2503. [CrossRef]
3. Liu, X.P.; Lin, H.Y.; Yang, C.F. 3-D FEA and experiment study of novel dual-stator hybrid excited wind generator. *Proc. CSEE* **2008**, *20*, 142–146.
4. Welburn, R. Ultra high torque motor system for direct drive robotics. In Applications for Today, Proceedings of the Robots 8: Conference Proceedings, Detroit, MI, USA, 4–7 June 1984; Robotics International of SME: Dearborn, MI, USA, 1984.
5. Chai, F.; Xia, J.; Guo, B.; Cheng, S.K.; Zhang, J.G. Double-Stator Permanent Magnet Synchronous in-Wheel Motor for Hybrid Electric Drive System. *IEEE Trans. Magn.* **2009**, *45*, 278–281. [CrossRef]
6. Zhao, Y.; Huang, W.; Jiang, W.; Lin, X.; Dong, D. Optimal design of double stator permanent magnet motors used in electric vehicle. In Proceedings of the 22nd International Conference on Electrical Machines and Systems (ICEMS), Harbin, China, 11–14 August 2019.
7. Fan, D.; Quan, L.; Zhu, X.; Xiang, Z.; Que, H. Airgap-harmonic-based multilevel design and optimization of a double-stator flux-modulated permanent-magnet motor. *IEEE Trans. Ind. Electron.* **2021**, *68*, 10534–10545. [CrossRef]
8. Wang, H.; Yan, X.; Huang, S.; Sui, T. Design and torque analysis of dual-stator permanent magnet motor for electric motorcycle. *Micromotors* **2011**, *4*, 7–10.
9. Sugiura, K.; Noguchi, T. Ultra-High-Power-Rate PM motor with double stator structure. In Proceedings of the IEEE Transportation Electrification Conference and Expo, Asia-Pacific (ITEC Asia-Pacific), Chiang Mai, Thailand, 28 November–1 December 2023.
10. Zheng, B.; Yan, D.; Song, P.; Zhang, Z.; Yan, Y. Optimal design of dual-stator permanent magnet synchronous motors under multiple working conditions. *Electric Drive* **2025**, 1–7. [CrossRef]
11. Kandil, A.; Hou, L.; Sharaf, M.; Arafa, A.A. Configuration angle effect on the control process of an oscillatory rotor in 8-pole active magnetic bearings. *AIMS Math.* **2024**, *9*, 12928–12963. [CrossRef]
12. Li, R.; Xu, J.; Sun, X.; Chen, J.; Zhang, C. Design of Double Stator Permanent Magnet Synchronous Motor for Robot Joint. *Small Spec. Electr. Mach.* **2021**, *49*, 14–18.
13. He, T.; Liang, D.; Tian, J.; Li, W.; Zhu, Z.Q. Comparative study of permanent magnet synchronous motors with single and double stators for robot joint applications. In Proceedings of the 3th International Conference on Sustainable Mobility Applications, Renewables and Technology (SMART), Dubai, United Arab Emirates, 22–24 November 2024.
14. Hesmondhalgh, D.E.; Tipping, D.; Amrani, M. Design and construction of a high-speed high-performance direct drive handpiece. *IEE Proc.* **1987**, *134*, 286–296. [CrossRef]
15. Chaaban, F.B. Determination of the optimum rotor/stator diameter ratio of permanent magnet machines. *Electr. Mach. Power Syst.* **1993**, *1*, 521–531. [CrossRef]
16. Pang, Y.; Zhu, Z.; Howe, D. Howe. Analytical determination of optimal split ratio for permanent magnet brushless motors. *IEE Proc. Electr. Power Appl.* **2006**, *153*, 7–13. [CrossRef]
17. Shen, Y.; Zhu, Z.Q. Analytical prediction of optimal split ratio for fractional-slot external rotor PM brushless machines. *IEEE Trans. Magn.* **2011**, *47*, 4187–4190. [CrossRef]
18. Chu, W.; Zhu, Z. Optimal split ratio and torque comparison of surface-mounted permanent magnet machines having inner or outer rotor. In Proceedings of the 6th IET International Conference on Power Electronics, Machines and Drives, Bristol, UK, 27–29 March 2012.
19. Wu, L.J.; Zhu, Z.Q.; Chen, J.T.; Xia, Z.P.; Jewell, G.W. Optimal split ratio in fractional-slot interior permanent-magnet machines with non-overlapping windings. *IEEE Trans. Magn.* **2009**, *46*, 1235–1242. [CrossRef]
20. Chai, F. *Fundamental Research on Double-Stator Integrated Starter Generator for Hybrid Electric Vehicle*; Harbin Institute of Technology: Harbin, China, 2003.
21. Chai, F.; Xia, J.; Gong, H.L.; Guo, B.; Cheng, S.K. Torque analysis of double-stator permanent magnet synchronous for hybrid electric vehicle. In Proceedings of IEEE Vehicle Power and Propulsion Conference, Harbin, China, 3–5 September 2008.
22. Zhang, D. Design and Analysis of a Novel Double-Stator Permanent Magnet Machine. Ph.D. Thesis, Shanghai University, Shanghai, China, 2007.
23. Wang, Y.B.; Cheng, M.; Fan, Y.; Chau, K.T. Design and analysis of double-stator permanent magnet brushless motor for hybrid electric vehicles. In Proceedings of International Conference on Electrical Machines and Systems, Wuhan, China, 17–20 October 2008.
24. Wang, Y.B.; Cheng, M.; Hua, W. Analysis and optimization of split ratio for double-stator permanent magnet brushless machine. *Proc. CSEE* **2010**, *30*, 62–67.

25. Yang, S.; Zhang, F.; Zhang, Z.; Liu, H. Design of double stator permanent magnet synchronous motor with low speed large torque. In Proceedings of IEEE Student Conference on Electric Machines and Systems, Huzhou, China, 14–16 December 2018.
26. Bianchi, N.; Bolognani, S.; Luise, F. Potentials and limits of high-speed PM motors. *IEEE Trans. Ind. Appl.* **2004**, *40*, 1570–1578. [CrossRef]
27. Levi, E. *Polyphase Motors*; Wiley: New York, NY, USA, 1984.

energies

Article

Performance Comparison of Coreless PCB AFPM Topologies for Duct Fan

Seung-Hoon Ko [1], Min-Ki Hong [2], Na-Rim Jo [1], Ye-Seo Lee [1] and Won-Ho Kim [3,*]

1 Department of Next Generation Smart Energy System Convergence, Gachon University, Seongnam 13120, Republic of Korea; tmdgns437@naver.com (S.-H.K.); rimna9922@gmail.com (N.-R.J.); yeseo9909@gmail.com (Y.-S.L.)
2 Department of Electrical Engineering, Hanyang University, Seoul 04763, Republic of Korea; asas231322@hanyang.ac.kr
3 Department of Electrical Engineering, Gachon University, Seongnam 13120, Republic of Korea
* Correspondence: wh15@gachon.ac.kr

Abstract

Duct fan motors must provide high torque within limited space to maintain airflow while requiring low vibration characteristics to minimize fluid resistance caused by fan oscillation. Axial Flux Permanent Magnet Motor (AFPM) offers higher torque performance than Radial Flux Permanent Magnet Motor (RFPM) due to their large radial and short axial dimensions. In particular, the coreless AFPM structure enables superior low-vibration performance. Conventional AFPM typically employs a core-type stator, which presents manufacturing difficulties. In core-type AFPM, applying a multi-stator configuration linearly increases winding takt time in proportion to the number of stators. Conversely, a Printed Circuit Board (PCB) stator AFPM significantly reduces stator production time, making it favorable for implementing multi-stator topologies. The use of multi-stator structures enables various topological configurations depending on (1) stator placement, (2) magnetization pattern of permanent magnets, and (3) rotor arrangement—each offering specific advantages. This study evaluates and analyzes the performance of different topologies based on efficient arrangements of magnets and stators, aiming to identify the optimal structure for duct fan applications. The validity of the proposed approach and design was verified through three-dimensional finite element analysis (FEA).

Keywords: permanent magnet motor; axial flux permanent magnet synchronous motor; printed circuit board; coreless motor

1. Introduction

To effectively drive airflow within the duct, the duct fan must maintain a constant static pressure during operation. Accordingly, the motor must be capable of delivering sufficient rated torque. Moreover, due to installation space constraints, the duct fan motor requires a compact form factor in the axial direction [1]. Vibrations generated during fan rotation are transmitted through the motor and increase the fluid resistance experienced by the fan blades. Therefore, low-vibration characteristics that minimize such effects are essential. The AFPM, with its large rotor diameter and short axial length, is well-suited to deliver high torque, making it an appropriate choice for duct fan applications [2,3].

AFPM can be classified into core type and coreless type. In core-type AFPM, design strategies such as laminated structures are required to minimize eddy current paths generated in the stator core. However, the stator core typically features complex shapes such as teeth and shoes, making it difficult to manufacture using rolled electrical steel sheets or laminations, as commonly done in RFPM. To address these challenges, research is being conducted using materials such as amorphous alloy sheet and soft magnetic composite (SMC). Specifically, amorphous alloy sheet can be formed into rolls, then cut or punched and re-rolled to create an effective laminated structure for reducing eddy current losses. However, when applied to an AFPM, the use of amorphous alloy sheet becomes disadvantageous for mass production due to interlayer alignment issues that worsen with increasing outer diameter and high mechanical forming complexity. SMC is fabricated by press-molding insulated magnetic powder, offering ease of manufacturing and excellent high-frequency core loss characteristics. Nevertheless, for motors operating in relatively low-speed regions, SMC exhibits inferior magnetic properties and mechanical strength compared to electrical steel. Additionally, the insulation coating on the powder particles can be damaged during the pressing process [4]. Furthermore, core-type AFPM requires sufficient axial length to accommodate windings on the stator core, limiting axial compactness. The use of complex geometries such as teeth, shoes, and slot openings also causes magnetic reluctance variation, leading to vibration issues due to cogging torque.

The coreless type AFPM, which does not include a stator core, offers the advantage of significantly lower cogging torque, resulting in superior vibration characteristics. However, it presents limitations in winding implementation, and active research is ongoing to address this challenge [5,6]. A coreless type AFPM utilizing a PCB stator eliminates the need for separate coil-winding machinery, as the winding process is integrated into standard PCB fabrication steps. This drastically reduces the production takt time relative to core-type AFPM, making the design well suited to high-volume manufacturing; moreover, it is well established that PCB-stator fabrication is significantly faster than conventional coil-winding processes [7]. Additionally, while the thickness of a PCB substrate varies depending on its oz rating, a 6-layer board with a 2 oz specification typically achieves a thickness of approximately 1 mm, thereby minimizing the magnetic air gap. Figure 1 illustrates the stator structures and cross-sectional views of a core-type AFPM stator and a PCB stator for a coreless-type AFPM.

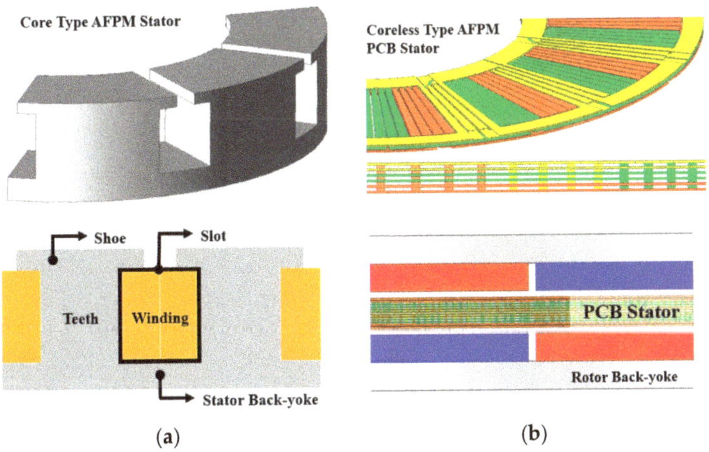

Figure 1. Stator Structures of Core Type AFPM and Coreless Type AFPM (**a**) 3D and 2D views of a core-type AFPM stator, illustrating the stator structure including shoe, slot, and teeth. (**b**) 3D and 2D views of a coreless-type AFPM PCB stator.

Active research is underway on PCB stator motors, focusing on various aspects inherent to the coreless motor structure. These include studies on AC loss caused by the time-varying flux linkage directly interacting with the windings [8,9], designs utilizing Halbach magnets to reduce leakage flux and enhance air-gap flux density [10,11], coil pattern optimization with slit structures for AC loss reduction [12], thermal design investigations considering PCB vias and pads [13], and comparisons between PCB-stator and laminated-core-stator axial flux permanent magnet motors [14]. When the rotor of a coreless motor rotates, eddy currents are induced in the conductive windings due to the time-varying magnetic flux generated by the permanent magnets. This is one of the major sources of loss in a coreless motor. In a conventional AFPM with a stator core, most of the magnetic flux follows the high-permeability core path, resulting in minimal flux passing directly through the windings. In contrast, a coreless motor has a stator with permeability similar to that of air, allowing more flux from the magnets to flow directly through the windings. This leads to additional AC losses caused by the skin effect and proximity effect. The aforementioned AC loss refers to the loss occurring in the stator winding. Although AC losses in the permanent magnets should ideally be considered, they are neglected in this study due to the use of ferrite magnets, which have relatively low electrical conductivity and therefore exhibit negligible eddy current losses. In addition, since the study focuses on a small-scale AFPM, rotor core iron losses are also considered negligible.

Therefore, it is essential to identify methods for reducing AC loss in PCB stator AFPM. Since AC loss primarily arises from the skin and proximity effects, it can be mitigated through modeling based on finite element analysis and trend analysis of these effects induced by external magnetic fields. Figure 2 illustrates the magnetic flux linkage between the magnets and the stator in both core-type and coreless-type AFPM.

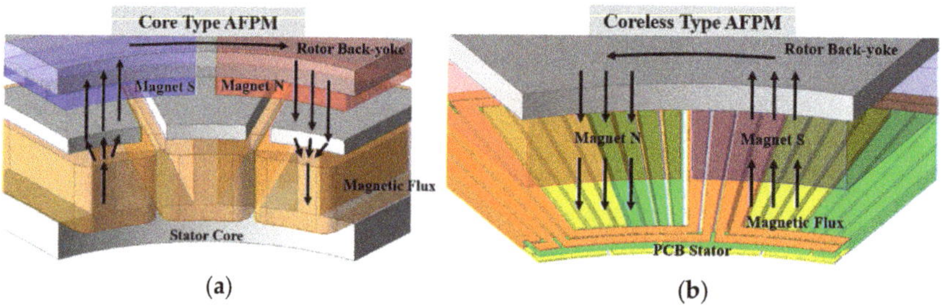

Figure 2. Flux Path in AFPM (**a**) Schematic of magnetic flux linkage in a core-type AFPM; (**b**) Schematic of magnetic flux linkage in a coreless-type AFPM.

In addition, various studies have been conducted on AFPM configurations that utilize multi-stator structures. For example, research has been conducted on high-speed axial flux induction motors employing dual-stator structures [15], comparative analyses between core-type and coreless type configurations in dual-stator single-rotor AFPM [16], and comparative study on novel dual stator radial flux and axial flux permanent magnet motors [17]. However, when a multi-stator structure is applied to a conventional core type AFPM, additional winding processes are required, resulting in increased takt time. Moreover, sufficient axial length for teeth is necessary to accommodate the windings. In contrast, a PCB stator can be reused with identical boards, which significantly reduces production time when implementing various topologies such as multi-stator configurations compared to the core type AFPM. In addition, the PCB stator offers high design flexibility, allowing various topologies to be implemented by simply changing its placement depending on the desired configuration. A PCB stator motor can be categorized into different topologies based on the magnetization pattern of the rotor and the arrangement of the stator and rotor. As examples, the stator can be arranged as a single module composed of four 6-layer PCB stator units connected in series, or as two separate modules, each consisting of two 6-layer PCB stator units connected in series. The most representative examples are the single stator–single rotor type and the single stator–double rotor type [7,18,19]. In the single stator–single rotor type, the PCB stator is attached to the stator yoke, and the rotor is placed on one side, forming a single-sided air gap. This structure minimizes the air gap length, which most significantly affects electromagnetic performance, and enables heat dissipation through contact between the stator back yoke and the housing. However, since only one magnetic source is present, phase back-EMF imbalance may occur. The single stator–double rotor structure is the most commonly utilized configuration. It does not require a stator back yoke, and the presence of magnetic sources on both sides results in minimal phase back-EMF imbalance. In PCB stator AFPM, relatively simple methods such as increasing the magnet thickness or connecting multiple PCB stators in series to increase the total number of series turns can be used to improve performance within a limited outer diameter. While these approaches are effective for initial performance enhancement, stacking multiple PCB layers increases the magnetic air gap, which may degrade electromagnetic performance. Additionally, when the magnet thickness exceeds a certain threshold, performance improvement no longer scales linearly due to magnetic flux saturation. Accordingly, this study investigates a topology design approach for enhancing performance by adjusting the arrangement of the rotor and PCB substrate. The most widely used single stator–double rotor structure was selected as the reference model, and the electromagnetic performance of different topologies was analyzed based on the same number of PCB stators and the same amount of magnets. The structure of this paper is as follows: (1) analysis of performance variation according to pole-slot combinations in the single stator–double rotor type PCB stator AFPM based on conventional motor design specifications; (2) selection of pattern end-turn parameters and design of a linear conductor structure for AC loss reduction based on the selected pole-slot combination; and (3) design of several PCB stator AFPM topologies using the same PCB stator as the reference model, with no-load back-EMF and load torque, and loss analysis to determine the optimal model. The proposed models and design process were validated using finite element analysis (FEA).

2. Selection of the Baseline Model

Table 1 presents the design specifications for a conventional duct fan motor system. The target rotational speed is 3350 rpm, and a rated torque of 0.48 Nm is required. Under constant speed conditions, torque is the primary factor determining motor output. Equation (1) expresses the torque of an AFPM in terms of motor size and magnetic flux density. Equation (2) expresses the torque of an AFPM in terms of motor size and magnetic flux density. Commonly in both equations, $k_{\omega 1}$ denotes the winding factor, B_{avg} represents the average air-gap flux density, and ac refers to the specific electric loading, defined as the total electric loading divided by the air-gap circumference. Here, the total electric loading is expressed as the product of the equivalent series turns and the armature current. In Equation (1), D_g represents the air-gap diameter, while in Equation (2), D_{out} indicates the rotor outer diameter. Furthermore, K_d is a constant that denotes the ratio of the rotor inner diameter to the rotor outer diameter. Since the torque of an AFPM is proportional to the cube of the rotor outer diameter, the motor was designed using the maximum allowable outer diameter while minimizing the lamination stack height. This approach was used to select an appropriate motor size.

$$T_{RFPM} = \left(\frac{\pi}{4} ack_{\omega 1} B_{avg}\right) D_g^2 L_{stk} \quad (1)$$

$$T_{AFPM} = (\frac{1}{8} ac\pi k_{\omega 1} B_{avg})(1 - K_d^2) D_{out}^3 \quad (2)$$

Table 1. Motor specifications for 168 W Target Duct fan.

Dimensions	Conventional Motor	Unit
Motor Outer Diameter/stack length Permanent Magnet	107/43 Ferrite Magnet	mm/mm -
Parameters	**Conventional Motor**	**Unit**
Rotating Speed	3350	rpm
Rated Torque	0.48	Nm
Efficiency	82	%

In the PCB stator proposed in this study, the winding is distributed, and electrical connection between the active conductor and end-turn sections is essential. Since the end-turn does not contribute to torque generation, the permanent magnets on the rotor are applied only up to the outer region, excluding the end-turn area. Accordingly, the PCB stator AFPM is designed such that the stator outer diameter is larger than the rotor outer diameter, and the previously selected maximum allowable outer diameter is used as the stator outer diameter. The maximum lamination height was determined based on the lamination size of a conventional motor, as summarized in Table 1 along with the overall motor specifications and dimensional constraints. The thickness of the PCB copper pattern was selected as 2 oz, which provides optimal electrical resistance and dimensional characteristics (approximately 1 mm in axial direction for 6 layers). For the base model analysis, a 12-layer PCB stator was configured by serially connecting two sets of 6-layer boards with 6 parallel windings. Figure 3 shows the series connection schematic of two 6-layer PCB, along with copper thickness and the structure of a 12-layer PCB.

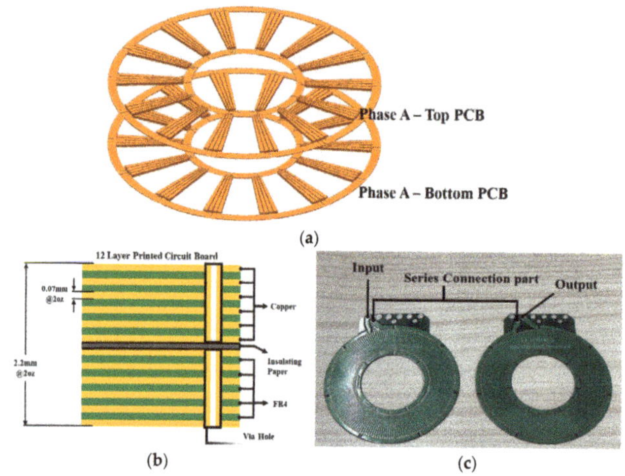

Figure 3. (**a**) 3D view of two 6-layer PCB substrates; (**b**) Cross-sectional view of the 12-layer PCB structure, indicating copper thickness, total substrate thickness, and via holes; (**c**) Schematic diagram of the series connection between two 6-layer PCB.

The winding layout of the PCB stator proposed in this study adopts a distributed winding configuration with a pole-to-slot ratio of 1:3, which enables the winding factor to be set as 1. Within the selected design dimensions, performance analysis was conducted based on combinations of optimal pole and slot numbers and the number of conductors per slot. Using the 8-pole configuration of the conventional model as a reference, analysis was performed on a PCB stator AFPM with 8 poles and 24 slots, with 6 conductors per slot, as well as other combinations that provide similar or higher total series turns. In a PCB stator AFPM, via holes are required to establish electrical connections between conductors in the axial direction. However, implementation of via holes requires a minimum spacing between them, which imposes a constraint on the minimum inner diameter. Therefore, the analysis was carried out using only combinations of pole number, slot number, and conductors per slot that satisfy the via hole spacing requirements. Figure 4 shows the variation in the number of conductors per slot under the same pole-slot combination. Table 2 presents the performance of comparison models based on the analysis of pole-slot combinations and conductors per slot in the PCB stator AFPM. Under the 1:3 pole-to-slot configuration, an increase in the number of poles leads to an increase in the number of slots, which in turn increases the total number of series turns on the board and results in a higher no-load back-EMF. However, increasing the number of conductors per slot extends the current path on the PCB, leading to higher resistance. Conversely, when the number of poles decreases, the total number of series turns is reduced, which lowers the resistance. However, the effective conductor area on the PCB increases, resulting in greater AC loss in the winding. Additionally, the reduction in series turns also leads to a decrease in no-load back-EMF. Based on this analysis, the model with 16 poles, 48 slots, and 4 conductors per slot was selected as optimal, as it delivers the highest output within the allowable current density limit. Although the selected current density exceeds the values commonly applied in conventional motors considering standard cooling approaches, it was validated through prototype fabrication and performance evaluation, confirming that the design operates without thermal or functional issues [16].

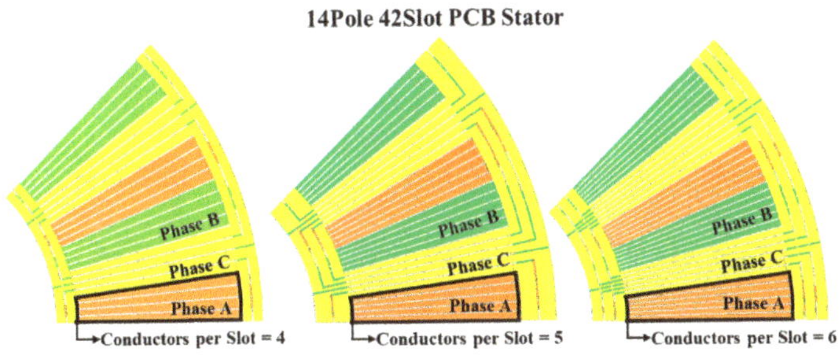

Figure 4. Variation in the number of conductors per slot in a 14 pole, 42 slot PCB stator.

Table 2. Performance table according to pole-slot combinations and conductor per slot.

Parameters	Single Stator-Doble Rotor									Unit
Conductor per Slot	3	4			5			6		-
Pole/Slot	16/48	16/48	14/42	12/36	14/42	12/36	10/30	10/30	8/24	-/-
Output power	55.6	77.6	60.8	42.4	60.2	44.2	27.7	39.1	32.5	W
No Load Back EMF	7.6	10.2	9.0	7.6	11.0	9.4	7.3	6.7	5.1	Vrms
Current	3.8	3.0	2.9	2.8	2.1	2.0	1.9	2.3	2.7	Arms
Power factor	0.99									-

Table 2. *Cont.*

Parameters	Single Stator-Doble Rotor									Unit
Current Density	21									Arms/mm^2
Phase Resistance	0.36	0.61	0.58	0.23	0.96	0.88	0.85	0.78	0.60	Ohm
Copper Loss	15.8	9.8	9.2	8.6	4.8	4.4	3.9	5.8	8.0	W
AC Loss	31.6	14.3	17.8	21.9	9.5	12.0	13.7	7.3	9.2	W
Efficiency	54.0	76.3	69.2	58.2	80.8	73.0	61.1	75.0	65.4	%

3. Design of the Geometry for a Single Stator–Double Rotor

3.1. Output Improvement Through Analysis of PCB End-Turn Parameters

Following the selection of the pole-slot and conductors-per-slot combination, the basic PCB stator design was carried out through the end-turn pattern layout. The end-turn regions were analyzed separately for the outer and inner diameter sides, which are referred to as the outer end-turn and inner end-turn, respectively. Initially, an analysis of the outer end-turn thickness was conducted. As the outer end-turn thickness increases, the conductor area in the current path also increases, resulting in reduced resistance of the PCB stator. However, since the outer diameter of the PCB stator was constrained by a predefined limit, any increase in the outer end-turn thickness was implemented in the inward direction, toward the inner diameter. This design concept is illustrated in Figure 5. As the outer end-turn thickness increases, the resistance decreases, allowing a higher maximum current to be applied within the current density limit, thereby increasing the output. Conversely, the increase in end-turn thickness reduces the effective conductor length, which results in a smaller rotor outer diameter and a slight decrease in the no-load back-EMF. However, the effect of the increased current due to reduced resistance is more significant than the decrease in back-EMF. Figure 6 illustrates the output according to the variation in the outer end-turn thickness. Based on the analysis of the trade-off between these two parameters, the optimal point for maximizing output was identified, and the outer end-turn thickness was set to 2.75 mm.

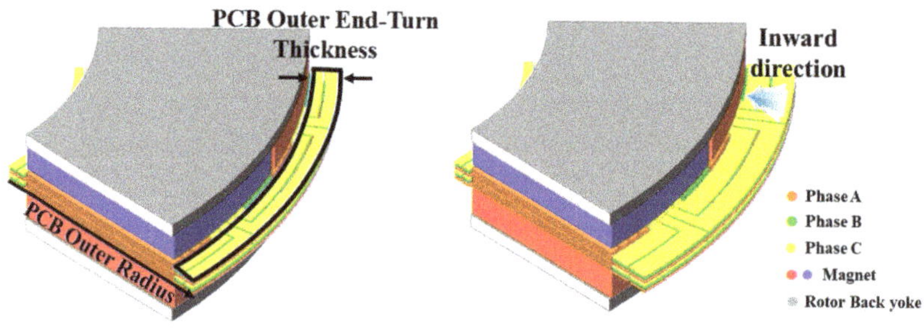

(a) (b)

Figure 5. (**a**) Thickness of the PCB stator outer end-turn and the outer radius of the PCB; (**b**) Conceptual diagram of increasing the outer end-turn thickness in the inward direction.

Figure 7 illustrates the method of modifying the inner end-turn structure. The selection of the inner end-turn thickness was conducted using two approaches. Similar to the analysis of the outer end-turn, performance was compared for cases in which the inner end-turn thickness was increased either toward the outer diameter or the inner diameter, as shown in Figure 7. In the former case, a trend similar to that observed in the outer end turn analysis was identified, where the output increases initially and then begins to decrease beyond a certain point. However, as long as the inner diameter of the PCB stator is larger than the shaft outer diameter, there is no design constraint. Therefore, the inner end-turn thickness can be increased in the inward direction, constrained only by the shaft diameter. This allows the inner end-turn thickness to be increased while maintaining the effective conductor length, which helps preserve the no-load back-EMF and reduce the resistance of the PCB stator. As the end-turn area increases, the allowable current within the limited current density also increases. A trade-off occurs where copper loss increases due to higher current, but the increase in copper loss becomes saturated due to the decreasing resistance. Figure 8 illustrates the output according to the variation in the inner end-turn configuration. Consequently, a maximum output point was observed, and the inner end-turn thickness was finally set to 2.75 mm to achieve the highest output.

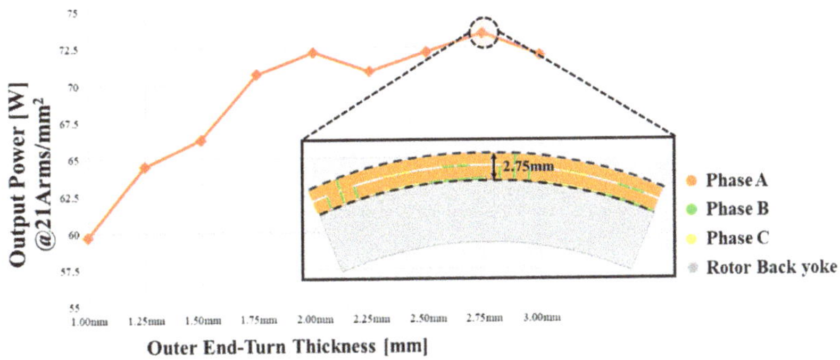

Figure 6. This figure shows the motor output as a function of the PCB outer end-turn thickness.

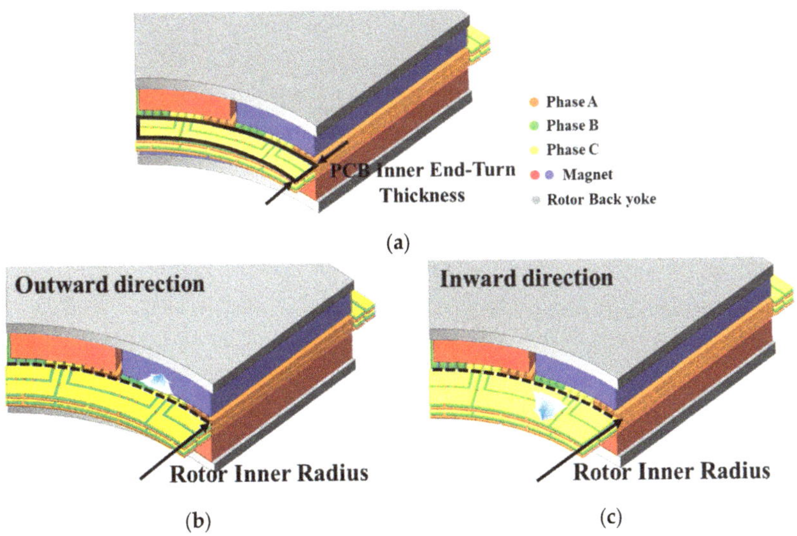

Figure 7. (**a**) Thickness of the PCB stator inner end-turn and the inner radius of the PCB; (**b**) Conceptual diagram of increasing the inner end-turn thickness in the outward direction; (**c**) Conceptual diagram of increasing the inner end-turn thickness in the inward direction.

Although the optimal PCB stator specifications were determined through conductor pattern design, a higher voltage utilization was required to meet the target output within the allowable current density. The current configuration consists of two 6-layer PCB boards connected in series, each composed of six parallel windings through vias, resulting in a total of 64 series turns. As shown in Table 1, the stator outer diameter is limited to 107 mm, which restricts further output enhancement through diameter expansion. Therefore, to increase the voltage utilization, additional axial length was employed by connecting four 6-layer PCB boards in series. As a result, when a total of 24 PCB layers were used, the base model with a single stator–double rotor type satisfied the target output of 168 W under the current density constraint. Figure 9 shows the schematic diagrams of the 12-layer and 24-layer PCB AFPM, and the performance comparison of these two models is summarized in Table 3.

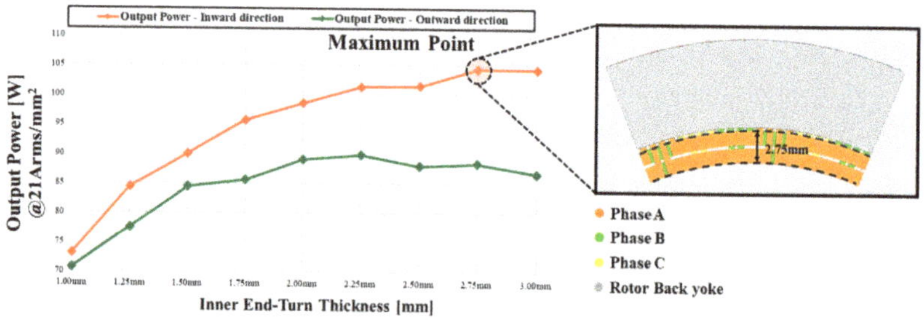

Figure 8. Motor output for different methods of varying the PCB inner end-turn thickness.

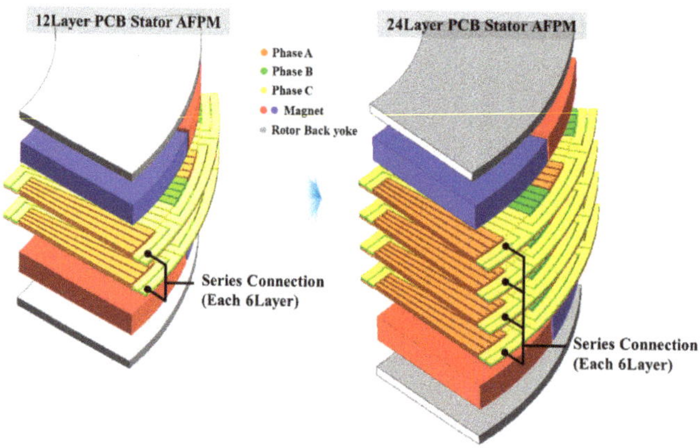

Figure 9. Structure and series connection between substrates for 12-layer and 24-layer PCB stator AFPM.

Table 3. Performance comparison between 12-layer PCB AFPM and 24-layer PCB AFPM.

Dimensions	12 Layer PCB AFPM	24 Layer PCB AFPM	
Output	104	168	W
Rated speed	3350	3350	rpm
No Load Back-emf	9.45	15.13	Vrms
Pole/slot	16/48	16/48	-/-
Input Current (@21 Arms/mm^2)	4.1	4.1	Arms
Current Density	21	21	Arms/mm^2

3.2. Enhancing Efficiency by Reducing AC Losses

AC loss is one of the major losses in the PCB stator AFPM, which is a type of coreless structure. One effective method to reduce this loss is the application of a linear conductor pattern, which removes portions of the coil area where AC loss is concentrated [16]. Figure 10 shows the effective conductor shape when applying the linear conductor pattern, the eddy current paths induced in the PCB stator by the time-varying magnetic flux, and a comparison of the eddy current paths with and without the linear pattern. The linear conductor method involves removing portions of the traditional circular sector-shaped coil pattern that are significantly affected by the external magnetic field and redesigning the winding into a linear form. This approach can be easily implemented using standard PCB manufacturing processes.

While the application of the linear conductor pattern reduces the cross-sectional area of the conductor and leads to an increase in winding resistance, thereby increasing copper loss, it also results in a reduction in AC loss. As such, a trade-off point exists between the two types of losses. When the 24-layer PCB board, as previously selected, is used, the increase in AC loss becomes more significant than the increase in copper loss compared to the 12-layer configuration. This occurs because the increase in no-load back-EMF due to the higher number of series turns reduces the input current, thereby limiting the increase in copper loss. To identify the trade-off point, the linear conductor pattern was applied to the original PCB stator conductor shape, and performance trends were analyzed based on changes in conductor width, focusing on the increase in copper loss and the decrease in AC loss. Figure 11a shows the trends of copper loss and AC loss as a function of linear thickness, while Figure 11b presents the efficiency variation. When the conductor width is 0.7 mm, the optimal balance between copper loss and AC loss is achieved, enabling maximum motor efficiency. Therefore, the conductor thickness was set to 0.7 mm.

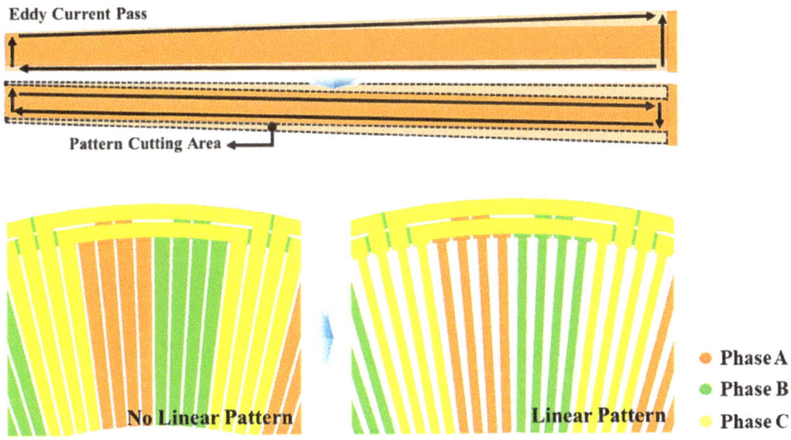

Figure 10. Eddy current paths and shapes depending on PCB conductor pattern.

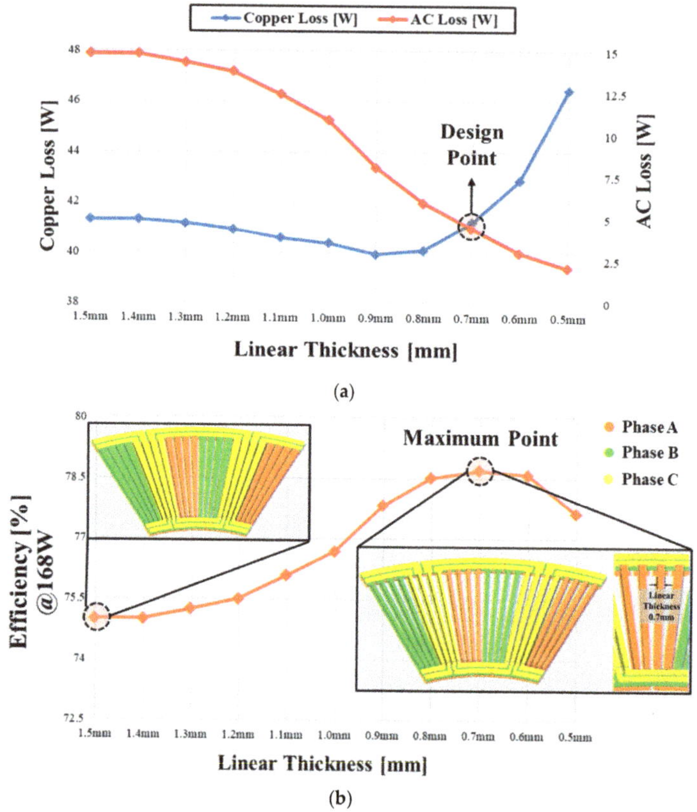

Figure 11. (**a**) the trend between copper loss and AC loss according to linear thickness, along with the design point; (**b**) the variation in motor efficiency with respect to changes in linear thickness.

4. Analysis of the APFM Topology in PCB Stators

4.1. Selection of Topology and Comparison of Model-Specific Characteristics

Based on the previously selected baseline model, the Single Stator–Double Rotor type, a comparative analysis of PCB stator AFPM topologies was conducted while maintaining the same number of PCB layers and the same amount of permanent magnets. The analysis focused on improving the efficiency of magnet and PCB stator placement and implementing additional structural modifications for new topology configurations. Model A is structured by separating the 24-layer PCB board used in the conventional single stator–double rotor type into two 12-layer boards, with a stator back-yoke inserted between them. When the 24-layer PCB boards are stacked continuously, the total magnetic air gap of the PCB stator becomes 8.6 mm, which constitutes a relatively large air gap from the motor's perspective. By inserting a stator back-yoke with high magnetic reluctance between the boards, this configuration compensates for the excessive magnetic air gap, offering advantages in terms of magnetic flux path control and electromagnetic performance.

Figure 12 shows the structure and magnetic flux diagram of Model A. A key distinction between the two types of Model A lies in the magnetization direction of the magnets, allowing classification into NN-type and NS-type configurations. However, this topology requires the presence of a stator core. To reduce core loss in the stator back-yoke, SMC or amorphous steel sheets are typically considered. In this study, SMC was selected as the stator core material due to its suitability for the simple geometry of the stator back-yoke. Unlike the complex shape of conventional core-type stators, the back-yoke in Model A features a form that is much easier to fabricate.

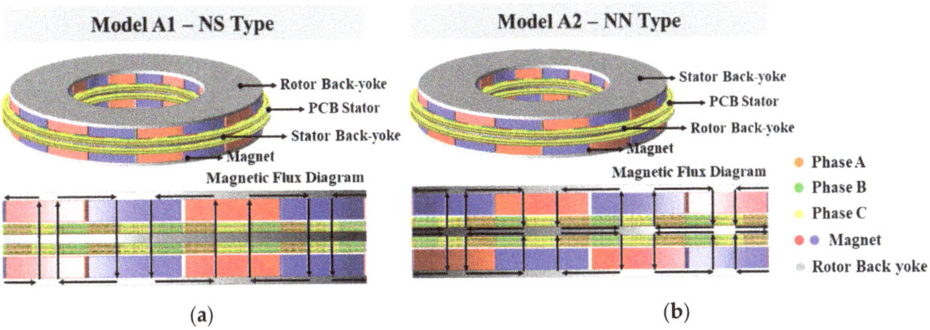

Figure 12. (**a**) NS-type configuration of Model A; (**b**) NN-type configuration of Model A.

Figure 13 illustrates the structure and magnetic flux flow of Model B. Model B is a modified configuration based on the placement of the 24-layer PCB board and the rearrangement of the rotor and stator from the conventional single stator–double rotor structure. In this configuration, the magnetic air gap from the magnetic source to the stator back-yoke is reduced by half compared to the original structure. One distinguishing feature of Model B is its central rotor configuration, which allows structural variations. Specifically, it can be categorized into two types: one in which the magnets are attached to both sides of the rotor core, and another in which the rotor core is removed to reduce motor weight, with additional structures used to support the magnets. For the type using a rotor core, the volume of the core was determined to maintain the same magnetic saturation level as in the original single stator–double rotor structure.

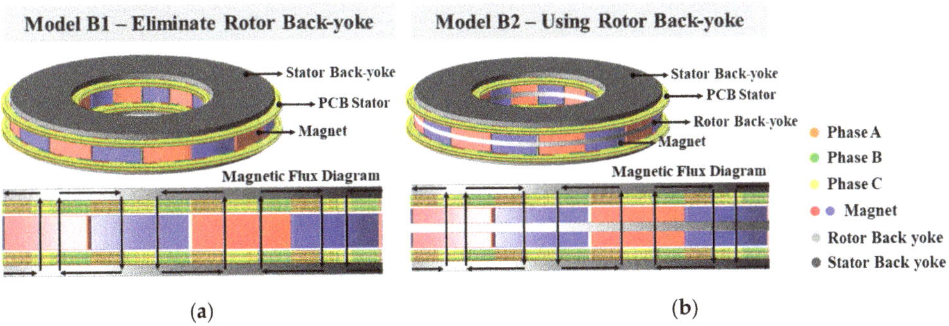

Figure 13. (**a**) Model B without a rotor back-yoke; (**b**) Model B with a rotor back-yoke.

Figure 14 illustrates the structure and magnetic flux flow of Model C. Model C is configured by dividing the PCB board into two sets of 12-layer boards and placing an additional rotor at the center of the motor by splitting the magnets while maintaining the same total amount of magnet usage. This configuration is generally referred to as a multi-stator multi-rotor structure. Similar to Model B, if the central rotor core in the motor is removed and a support structure is employed to secure the magnets, the axial length can be reduced and the motor can be made lighter. Based on the selected models, performance analysis was conducted using finite element analysis (FEA).

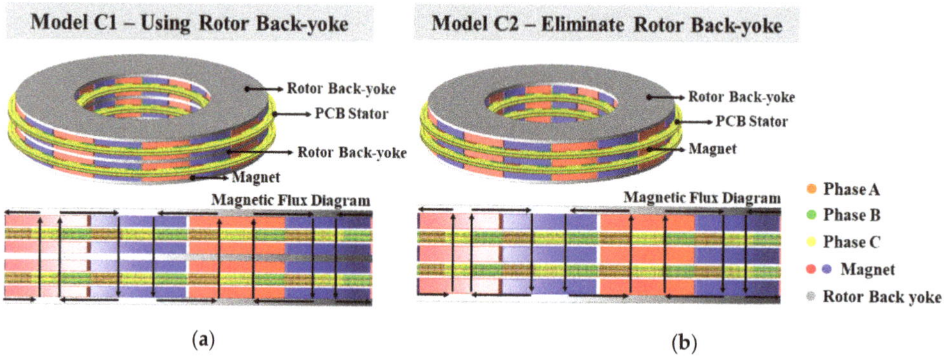

Figure 14. (**a**) Model C with a central rotor back-yoke; (**b**) Model C without a central rotor back-yoke.

4.2. Comparative Performance Analysis Based on Topology

Figure 15 presents the graph of no-load back-EMF according to each topology. All topology comparisons were conducted based on the same number of PCB stator layers and the same amount of magnet usage. For the rotor core, performance analysis was carried out using the rotor back-yoke thickness that yields a magnetic saturation level equivalent to that of the rotor back-yoke in the conventional single stator–double rotor structure. Model A is designed to compensate for the magnetic flux path between the PCB layers using a stator back-yoke with high magnetic reluctance. However, despite the intended improvement in the magnetic flux path, the presence of magnetic leakage within the high-reluctance stator back-yoke resulted in a decrease in performance. Specifically, the no-load back EMF RMS value was approximately 3% lower than that of the original type. In the case of Model B, both types exhibited a decrease in no-load back EMF compared to the original model. This is attributed to the less efficient arrangement of the rotor and stator compared to the single stator–double rotor configuration. When comparing the two types—one with a rotor back-yoke and one without—the type with the rotor back-yoke showed a slightly higher no-load back EMF RMS value by approximately 0.2%. Although the target output was not achieved, the use of a rotor back-yoke with high magnetic reluctance was found to be advantageous in terms of performance under the given amount of magnet usage. In the case of Model C, the no-load back EMF RMS value improved by up to 5.3% compared to the original type, due to the more efficient arrangement of magnets. Both types of Model C exhibited higher no-load back EMF than the Single Stator–Double Rotor configuration. Specifically, the structure without a rotor back-yoke showed an additional 1.8% increase in no-load back EMF compared to the one with a back-yoke. However, beyond a certain magnet thickness, the use of a rotor back-yoke for the central rotor magnets was found to be more effective in terms of efficiency.

Since all configurations share the same PCB stator geometry, the maximum allowable current based on the current density limit of 21 Arms/mm^2 remains constant. As a result, both topologies of Model C, which showed the greatest increase in no-load back emf, enable higher output compared to the original type. Among them, Model C-1 exhibits the highest output, with a 7.5% increase compared to the conventional model. Additionally, Table 4 summarizes the specifications of the final analyzed models, indicating that the stacking length can be reduced by approximately 50% relative to the stacking limit of the conventional model. Accordingly, this study selects Model C-1 as the optimal topology for the PCB stator AFPM. Figure 16 presents the THD analysis of Model C-1, further highlighting its performance advantages.

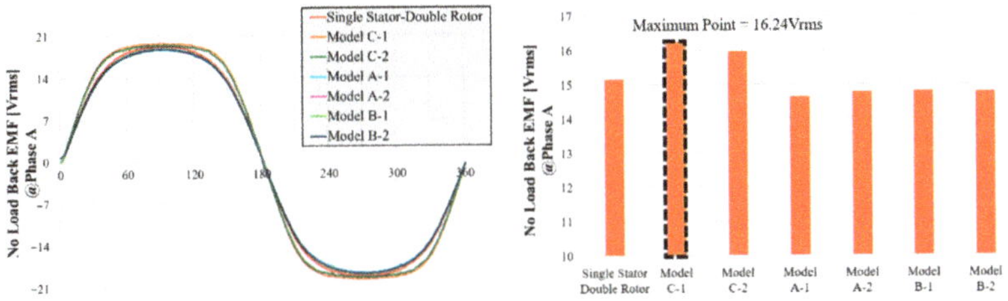

(**a**) (**b**)

Figure 15. (**a**) Waveform differences of phase-A no-load back-EMF across topologies; (**b**) Magnitude differences of phase-A no-load back-EMF across topologies.

Table 4. Performance comparison of PCB stator AFPM with different topologies.

Dimensions	SSDR	Model A-1	Model A-2	Model B-1	Model B-2	Model C-1	Model C-2	Unit
Output	168	163.3	165	164.8	164.6	180.6	177.7	W
Rated speed				3350				rpm
Torque	0.48	0.46	0.47	0.47	0.47	0.51	0.5	Nm
No Load Back EMF	15.15	14.63	14.78	14.79	14.76	16.24	15.95	Vrms
Current				3.81				Arms
Power factor				0.99				-
Current Density				21				Arms/mm^2
Copper Loss				41.2				W
AC Loss	4.47	3.93	3.89	4.28	4.21	4.99	4.54	W
Core Loss (SMC 700 3P)	-	3.41	2.41	4.31	4.54	-	-	W
Efficiency	79	77	77	76	76	80	80	%

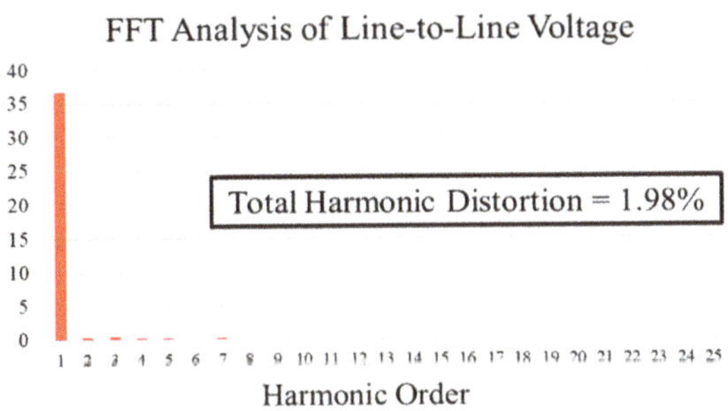

Figure 16. Total harmonic distortion (THD) of the line-to-line voltage for Model C-1.

5. Conclusions

This study presents the design of a PCB stator-based AFPM optimized for duct fan applications under size constraints and analyzes the performance of various topologies based on different rotor and stator arrangements. To address the manufacturing limitations and structural constraints of conventional core-type AFPM, a coreless structure was adopted. The stator was designed using PCB fabrication processes, enabling automated production and facilitating the implementation of diverse topologies.

In the early design stage, combinations of pole-slot numbers and the number of conductors per slot were evaluated to identify the configuration that delivers the highest output. Subsequently, outer and inner end-turn patterns were optimized to increase the allowable current, thereby enhancing output. Additionally, a linear conductor pattern was applied to effectively reduce AC loss. Based on this, a baseline single stator–double rotor topology was applied, and comparisons among topologies with different rotor and stator arrangements were conducted while maintaining equal magnet usage and an identical number of PCB stators. As a result, Model C-1 exhibited approximately 5.3% higher no-load back EMF and 7.5% higher output compared to the baseline, demonstrating its superior efficiency. While previous studies have primarily focused on topology comparisons in core-type AFPM, this study contributes by proposing an efficient rotor–stator arrangement specifically designed for PCB stator-based AFPM. This result demonstrates that performance improvement is achievable under identical resource constraints and provides valuable insights for future motor designs utilizing PCB stators in similarly constrained environments.

Author Contributions: Conceptualization, W.-H.K.; methodology, S.-H.K.; software, N.-R.J. and Y.-S.L.; validation, S.-H.K.; formal analysis, M.-K.H.; investigation, M.-K.H.; resources, S.-H.K.; data curation, N.-R.J.; writing—original draft preparation, S.-H.K.; writing—review and editing, M.-K.H. and N.-R.J. visualization, N.-R.J. and Y.-S.L.; supervision, W.-H.K. All authors have read and agreed to the published version of the manuscript.

Funding: This work was supported by the Korea Institute of Energy Technology Evaluation and Planning (KETEP) grant funded by the Korea government (MOTIE) (20214000000060), Department of Next Generation Energy System Convergence based-on Techno-Economics—STEP and in part by National Research Foundation of Korea (NRF) grant funded by the Ministry of Education (No. NRF-2022R1I1A3068863).

Data Availability Statement: The original contributions presented in the study are included in the article, further inquiries can be directed to the corresponding author.

Conflicts of Interest: The authors declare no conflicts of interest.

References

1. Fawzal, A.S.; Cirstea, R.M.; Gyftakis, K.N.; Woolmer, T.J.; Dickison, M.; Blundell, M. Fan Performance Analysis for Rotor Cooling of Axial Flux Permanent Magnet Machines. *IEEE Trans. Ind. Appl.* **2017**, *53*, 3295–3304. [CrossRef]
2. Gao, B.; Cheng, Y.; Wang, Y.; Zhao, T.; Ding, L.; Cui, S.; Liu, X.; Shi, Y. Optimal Design of PCB Coreless Axial Flux Permanent Magnet Synchronous Motor with Arc Windings. *IEEE Trans. Energy Convers.* **2023**, *39*, 567–577. [CrossRef]
3. Zhao, J.; Wang, Y.; He, W.; Dou, Y.; Ding, Y.; Hu, N. A 3D Semi-Analytical Calculation Method for a Slotless PMSM with End-Wrapped FPCB Concentrated Winding. *IEEE Trans. Energy Convers.* **2024**, *39*, 2324–2336. [CrossRef]
4. Pyo, H.-J.; Jeong, J.W.; Yu, J.; Lee, S.G.; Kim, W.-H. Design of 3D-Printed Hybrid Axial-Flux Motor Using 3D-Printed SMC Core. *IEEE Trans. Appl. Supercond.* **2020**, *30*, 5202004. [CrossRef]
5. Hong, H.; Lee, J.-K.; Jung, D.-H.; Lee, J.; Kim, H. Comparative Study on Electromagnetic Performance of Slotted and Slotless PM Synchronous Motors According to Aspect Ratios. *IEEE Trans. Magn.* **2025**, *61*, 8201604. [CrossRef]
6. Choi, D.-H.; Kim, D.-H.; Han, H.-S.; Jung, D.-H.; Kim, W.-H. Design of a Slotless Structure for Minimizing Cogging Torque and Torque Ripple in a Column Type EPS Motor for Vehicles. *IEEE Trans. Magn.* **2024**, *60*, 8700305. [CrossRef]
7. Choi, D.-H.; Han, H.-S.; Hong, M.-K.; Jung, D.-H.; Kim, W.-H. Design for Loss Reduction in a Compact AFPM Electric Water Pump with a PCB Motor. *Energies* **2025**, *18*, 2538. [CrossRef]
8. Chulaee, Y.; Lewis, D.; Mohammadi, A.; Heins, G.; Patterson, D.; Ionel, D.M. Circulating and Eddy Current Losses in Coreless Axial Flux PM Machine Stators with PCB Windings. *IEEE Trans. Ind. Appl.* **2023**, *59*, 4010–4020. [CrossRef]

9. Zhao, J.; Wang, Y.; Ma, T.; Liu, X.; Li, J. Losses and Thermal Analysis of an Integrated PCB Coreless Axial Flux PMSM with the Drive System. *IEEE Trans. Ind. Electron.* **2023**, *70*, 11022–11032. [CrossRef]
10. Liu, N.-W.; Hung, K.-Y.; Lyu, B.-T.; Yang, S.-C.; Lin, Y.-P.; Chan, C.-Y. Reaction Wheel Design for Precise Attitude Control of CubeSats Using Flexible PCB Windings and Halbach Magnet Array. *IEEE Trans. Aerosp. Electron. Syst.* **2024**, *60*, 3496–3508. [CrossRef]
11. Wang, C.; Han, J.; Zhang, Z.; Hua, Y.; Gao, H. Design and Optimization Analysis of Coreless Stator Axial-Flux Permanent Magnet In-Wheel Motor for Unmanned Ground Vehicle. *IEEE Trans. Transp. Electrif.* **2021**, *8*, 1053–1062. [CrossRef]
12. Francois, G.; Dehez, B. Impact of Slit Configuration on Eddy Current and Supply Current Losses in PCB Winding of Slotless PM Machines. *IEEE Trans. Ind. Appl.* **2022**, *58*, 6035–6044. [CrossRef]
13. Shen, Y.; Wang, H.; Blaabjerg, F.; Zhao, H.; Long, T. Thermal Modeling and Design Optimization of PCB Vias and Pads. *IEEE Trans. Power Electron.* **2019**, *35*, 882–900. [CrossRef]
14. S., N.; Nikam, S.P.; Pal, S.; Wankhede, A.K.; Fernandes, B.G. Performance Comparison Between PCB-Stator and Laminated-Core-Stator-Based Designs of Axial Flux Permanent Magnet Motors for High-Speed Low-Power Applications. *IEEE Trans. Ind. Electron.* **2019**, *67*, 5269–5277. [CrossRef]
15. Hong, C.; Huang, W.; Hu, Z. Performance Calculation of a Dual Stator Solid Rotor Axial Flux Induction Motor Using the Multi-Slice and Multi-Layer Method. *IEEE Trans. Magn.* **2018**, *55*, 8100709. [CrossRef]
16. Bandarkar, A.W.; Bin Azam, K.M.; Sozer, Y. Comparative Performance Analysis of Slotted and Slotless Dual-Stator and Single-Rotor Axial-Flux Permanent Magnet Motor for Integrated Motor-Compressor System. *IEEE Trans. Ind. Appl.* **2025**, *61*, 4234–4245. [CrossRef]
17. Zhao, W.; Lipo, T.A.; Kwon, B.-I. Comparative Study on Novel Dual Stator Radial Flux and Axial Flux Permanent Magnet Motors with Ferrite Magnets for Traction Application. *IEEE Trans. Magn.* **2014**, *50*, 8104404. [CrossRef]
18. Tokgoz, F.; Gulsuna, O.; Karakaya, F.; Cakal, G.; Keysan, O. Mechanical and Thermal Design of an Optimized PCB Motor for an Integrated Motor Drive System with GaNFETs. *IEEE Trans. Energy Convers.* **2022**, *38*, 653–661. [CrossRef]
19. Sun, H.; Li, Y.; Zhang, L.; Xue, Z.; Hu, W.; Li, G.; Guo, Y. Multi-Parameter Optimization of Stator Coreless Disc Motor Based on Orthogonal Response Surface Method. *Electronics* **2023**, *12*, 3020. [CrossRef]

energies

Review

Real-Time Digital Twins for Intelligent Fault Diagnosis and Condition-Based Monitoring of Electrical Machines †

Shahin Hedayati Kia [1,*,‡], Larisa Dunai [2,‡], José Alfonso Antonino-Daviu [3,‡] and Hubert Razik [4,‡]

1 MIS Laboratory UR4290, University of Picardie "Jules Verne", 33 rue St Leu, 80039 Amiens, France
2 Departamento de Ingeniería Gráfica, Universitat Politecnica de Valencia, Camino de Vera s/n, 46022 Valencia, Spain; ladu@upv.es
3 Instituto de Tecnología Eléctrica, Universitat Politecnica de Valencia, Camino de Vera s/n, 46022 Valencia, Spain; joanda@die.upv.es
4 Laboratory Ampère UMR5005, University of Lyon 1, 69622 Villeurbanne, France; hubert.razik@univ-lyon1.fr
* Correspondence: shdkia@u-picardie.fr
† This paper is an extended version of our paper published in 2024 IECON—50th Annual Conference of the IEEE Industrial Electronics Society, Chicago, IL, USA, 3–6 November 2024; pp. 1–6.
‡ These authors contributed equally to this work.

Abstract

This article presents an overview of selected research focusing on digital real-time simulation (DRTS) in the context of digital twin (DT) realization with the primary aim of enabling the intelligent fault diagnosis (FD) and condition-based monitoring (CBM) of electrical machines. The concept of standalone DTs in conventional multiphysics digital offline simulations (DoSs) is widely utilized during the conceptualization and development phases of electrical machine manufacturing and processing, particularly for virtual testing under both standard and extreme operating conditions, as well as for aging assessments and lifecycle analysis. Recent advancements in data communication and information technologies, including virtual reality, cloud computing, parallel processing, machine learning, big data, and the Internet of Things (IoT), have facilitated the creation of real-time DTs based on physics-based (PHYB), circuit-oriented lumped-parameter (COLP), and data-driven approaches, as well as physics-informed machine learning (PIML), which is a combination of these models. These models are distinguished by their ability to enable real-time bidirectional data exchange with physical electrical machines. This article proposes a predictive-level framework with a particular emphasis on real-time multiphysics modeling to enhance the efficiency of the FD and CBM of electrical machines, which play a crucial role in various industrial applications.

Keywords: condition-based monitoring; digital simulation; digital twins; electric machines; hardware-in-the-loop simulation; fault diagnosis; machine learning; predictive maintenance; real-time systems

1. Introduction

Today, the design and implementation of industrial products are almost inconceivable without the initial phase of virtual development supported by digital simulation (DS). DS enables the creation of a reference model that provides a precise description of a physical entity. This is commonly known as a digital twin (DT), which serves as a virtual representation of the product prior to physical prototyping [1–4]. This approach provides access to a broad range of variables that may not be directly measurable or may require sophisticated instrumentation. Using this method, industrial products can be developed more efficiently and cost-effectively through a comprehensive analysis of their entire lifecycle. An important feature of the DS world is that every test and decision can be implemented, and undesired conditions can be anticipated. This can facilitate tasks such as configuration, FD, and CBM for both preventive and predictive maintenance. Compared with real-world testing, DS offers significant time and cost savings while minimizing waste and reducing potential risks. Conducting experiments in a physical lab requires substantial time owing to the preparation and construction of the experimental test benches. Additionally, laboratory testing is often constrained by facility capabilities, capacity limitations, and lack of flexibility when modifications are required. In a design process that involves multiple iterations, in which each stage depends on the outcomes of the previous test, the resulting time delay can be substantial. Beyond saving time and cost, DS enables the exploration of a much wider range of variants. This gives product developers greater flexibility to innovate, test numerous combinations, and experiment with unconventional approaches, ultimately leading to a more refined and optimized solution. To accurately represent the physical world, the DS must incorporate interactions across multiple physical domains, including electromagnetics, fluid dynamics, mechanics, thermodynamics, and materials science [5]. If DS results fail to accurately represent real-world conditions, they lose their value. Confidence in DS increases when its outcomes align with those obtained from physical tests and field experiences. The core idea of DS relies on solving particularly partial differential equations (PDEs), that is, differential equations in multi-variables of time and space which often represent coupled equations derived from multiphysics modeling. Apart from certain simplified cases in which these PDEs can be solved analytically, their complexity in most real-world scenarios necessitates the utilization of discretization approaches for numerical computation. Over time, various methods have been developed, among which the most widely applied are the finite difference method (FDM), the finite element method (FEM), and the finite volume method (FVM) [6,7]. The finite difference equation for differential operators can be expressed using forward, backward, or central differences, whereas the Galerkin method is commonly employed to formulate finite element equations. The finite volume method (FVM) can be derived from either the FDM or FEM [6]. The FEM has become the standard approach for numerically solving PDEs that describe physical fields, including multiphysics couplings. Consequently, static, transient, and time-harmonic analyses involving both material and geometrical nonlinearities can be efficiently performed [8]. These formulations are widely implemented in many open-source and commercial multiphysics simulation packages (OpenFOAM, FEniCS, Elmer, FreeFEM, ANSYS, COMSOL, STAR-CCM+, etc.) [9].

Electrical machines are widely used in various industrial applications, including electric traction systems, hybrid and electric vehicles, wind and marine energy conversion, power generation, energy storage, shipboard systems, and aerospace electrification. To enhance their design and performance, digital offline simulation (DoS) is frequently used to optimize structural configurations, develop controllers, and analyze the dynamic behavior of innovative electrical machines [10]. The modeling of electrical machines can be broadly categorized into two main approaches: physics-based (PHYB) and data-driven approaches. PHYB models primarily rely on the observation of physical phenomena and aim to represent them using mathematical formulations, which are then solved, either analytically or numerically. By contrast, data-driven models require a large number of data during the training phase, which can be generated from partial PHYB models or experimental measurements [9]. The multiphysics modeling of electrical machines can capture various fields of physics, including electromagnetism, mechanics, vibro-acoustics, and thermal phenomena, and can be composed of several interacting PHYB or data-driven models [11–14]. Focusing on the electrical and magnetic aspects, the literature commonly reports both simplified PHYB models using circuit-oriented lumped-parameter (COLP) approaches and detailed PHYB (DPHYB) models for electrical machine modeling [10]. The former rely on an electrically and magnetically coupled circuit and are typically used for analyzing the dynamic behavior of electrical machines and electromagnetic transient studies. This type of model assumes perfect symmetry in both the field and construction, a constant air gap that is small relative to the rotor radius, a linear magnetic system, and the omission of hysteresis effects. In contrast, the second method considers complex geometry, magnetic saturation, eddy current, and hysteresis effects in detail and is employed for design purposes [10,15]. Clearly, DPHYB models offer greater accuracy than COLP models, but this comes at the cost of increased computation time. COLP models include the dynamic 'abc' and 'dq0' reference frame models and the dynamic voltage-behind-reactance model, in which the variables of the stator are described in the 'abc' reference frame and the variables of the rotor are given in the 'dq0' reference frame [15,16]. The 'dq0' reference frame model can be derived from the 'abc' reference model by using the Park transform, which eliminates time-varying inductances in the 'abc' frame and simplifies the voltage equations in AC electrical machines. The self- and mutual inductances of the 'abc' reference model are commonly determined based on the winding function approach (WFA) or modified WFA (MWFA) when some data regarding the geometry of the machine are known [17]. Another approach is to perform basic experimental tests through which it is possible to obtain these parameters for a simplified internal equivalent circuit model [18]. DS of COLP models has been widely conducted using Electromagnetic Transient Program-type (EMTP-type) software, initially developed by H.W. Dommel in the late 1960s. Well-established examples include EMTP-ATP, EMTP-RV, and EMTDC/PSCAD, which are circuit-theory-based simulation tools for studying power system transients and dynamics. The universal COLP machine model was later incorporated into EMTP by H.K. Lauw and W.S. Meyer, enabling the analysis of various types of electric machines within a single modeling framework [19]. The most common formulation methods in EMTP-type tools fall into two main categories: nodal analysis and state-space methods for computing the response of power systems in the time domain. The discretization of ordinary differential equations (ODEs) using integration techniques yields difference equations. The combination of the method of characteristics and trapezoidal rule in a generalized algorithm enabled the accurate simulation of transient networks involving both distributed and lumped parameters in Dommel's EMTP. However, the use of a constant time-step length presents some problems when modeling switching elements and can produce spikes and numerical oscillations (or chatter). In this regard, critical damping adjustment, in which the integration method is changed to backward Euler for two steps after discontinuity, has been proposed [20]. In this respect, two-stage diagonally implicit Runge–Kutta, a mathematically oscillation-free method, was utilized in Expandable Transient Analysis Program (XTAP) [19]. A guideline for choosing an appreciate integration method for modeling COLP models is presented in [21].

DPHYB models encompass both FEM-based and analytical models. In FEM-based models of electrical machines, Maxwell's equations must be solved numerically for each mesh element, whether in 2D triangular or 3D tetrahedral configurations. Conversely, analytical models rely on explicit solutions of Maxwell's equations and are widely employed to aid in design and analysis, particularly when an accurate calculation of the magnetic field distribution across various machine regions is essential. The accurate modeling of electric machines often relies on the FEM. However, finite element analysis (FEA) can be highly time-consuming, particularly during the initial design stages when conducting parametric studies. Several tools have been proposed in the domain of electrical machine design (MotorAnalysis, Flux-2D, Code Carmel, JMAG, Altair Flux/FluxMotor, etc.). Two commonly used analytical models for design purposes are magnetic equivalent circuits (MECs), which rely on a permanence element network, and models derived from the formal resolution of Maxwell's equations in regions with constant permeability [22,23]. In an MEC-based model, the mesh network must be carefully designed to achieve an accuracy comparable to that of a FEM-based model. When considering the effect of iron magnetic saturation, the differential equation governing the MEC-based model transforms into a nonlinear representation of magnetic scalar potentials. To solve this problem, the Newton–Raphson method is typically employed, often in combination with an adaptive simulation time step [24].

Data-driven modeling has gained significant attention owing to the development of advanced open-source machine learning (ML) and artificial intelligence (AI) tools, user-friendly and affordable computational resources, and extensive training materials. Unlike PHYB models, this approach operates under the premise that the data encapsulate both explicit and implicit physical behaviors. Therefore, when trained on sufficiently large datasets, data-driven models can independently uncover underlying physical relationships. In particular, deep learning has enabled models to achieve near-human or even superhuman performance in tasks once considered out of reach for computers. These models can be typically categorized into six types: supervised and unsupervised linear models, supervised and unsupervised nonlinear models, and unsupervised deep learning approaches [25]. Although the performance of data-driven models depends on the widespread availability of large-scale/big data, the proposed methodology seeks to address the limitations of relying solely on either PHYB or data-driven modeling techniques. This integrated approach, known as hybrid modeling, combines the clarity, theoretical grounding, and insight of PHYB models with the precision, computational efficiency, and pattern recognition capabilities of advanced ML and AI methods. Hybrid modeling can be conceptualized as existing at the intersection of large-scale/big data, physics-informed modeling, and data-driven techniques, as illustrated in Figure 1 [9,26]. For example, instead of relying on computationally intensive FEA, a deep neural network (DNN) is employed as a surrogate model. The DNN was trained in a supervised manner using a large dataset generated from precomputed FEA results. During inference, the outputs of the network intermediate are utilized as inputs for the PHYB post-processing step, which computes characteristic maps and key performance indicators. This hybrid strategy demonstrates a significant reduction in computational time while preserving the flexibility within the simulation workflow of electrical machines [27].

The reliability and availability of electrical machines are critical in many industrial sectors, where they operate under specific stresses and loads under real-world conditions, often subject to random variations. CBM plays a pivotal role in preventing failures and minimizing downtime, making it an essential component of most maintenance strategies [28,29]. FD involves the detection, isolation, and identification of faults. The main challenge is to perform the incipient FD online under transient operating conditions as early as possible. FD approaches can be broadly classified into three categories based on the algorithms used: signal-based, model-based, and data-driven. Machine learning techniques have been widely applied to automate data-driven approaches in three main steps: sensor data acquisition, statistical feature extraction, and fault classification [30–34]. Statistical feature extraction involves the analysis of collected data across various domains of time, frequency, and time–frequency. After dimensionality reduction, the extracted features are typically used in small-scale machine learning models for unsupervised clustering, such as K-means clustering, K-nearest neighbors, self-organizing map, supervised classification using random forest, support vector machine, backpropagation neural networks, and decision trees [30,32,35]. However, the selection of meaningful features remains a challenging issue because it is labor-intensive and must be adapted to different fault classification tasks and varying machine operating conditions. Deep learning (DL) has been significant in overcoming the shortcomings of traditional ML methods owing to its ability to automatically extract representative features from raw data and accurately establish nonlinear mappings of different health conditions. Popular DL models and methods utilized in the field of FD include the deep belief network, the stacked autoencoder, and the convolutional neural network [31].

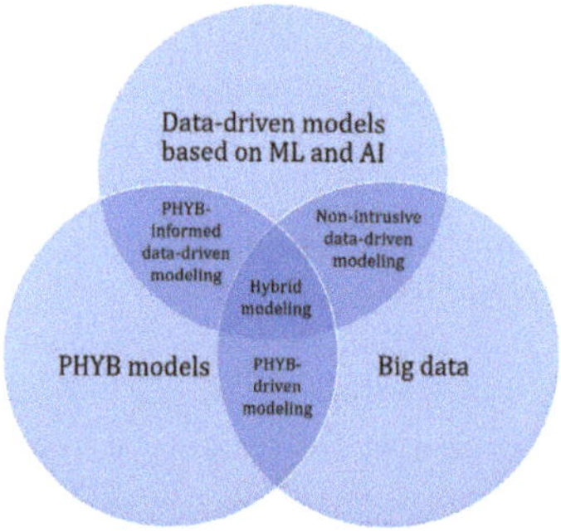

Figure 1. Scheme for hybrid modeling at the intersection of PHYB models, big data, and data-driven models [9].

Although learning can be supervised, semi-supervised, or reinforcement learning [36], training a DL model from scratch for classical ML tasks requires a large number of data that cover all possible machine fault conditions. This is a major challenge in practical applications because the measured data for only a few fault conditions can be insufficient. To solve the issue of limited training data, transfer learning (TL) has been introduced, which has shown remarkable capability in obtaining a satisfactory deep architecture by fine-tuning a DL model that has been pre-trained in other tasks [37]. Data-driven methods mostly focus on the design of models, with the same aim as signal-based methods, to extract fault features with high quality and avoid noise interference. Mamba was proposed in the field of DL [38], the core idea of which is the state-space model (SSM), where the coefficient matrices of the SSM are learnable parameters of the model. It achieves the selective processing of input information by parameterizing the input of the SSM [32]. The CBM framework incorporates not only condition monitoring data but also event data. Event data record occurrences such as installations, breakdowns, and overhauls, along with their causes, as well as actions taken on the asset, such as minor repairs, preventive maintenance, or oil changes. Condition monitoring data, on the other hand, consist of measurements that reflect the health status or operational state of an electrical machine. Analyzing event data alone typically falls under the domain of reliability analysis, which involves fitting the data to the probability distributions of time between events and using these models for further assessment. In CBM, however, condition monitoring data are available in addition to event data. Combining these datasets enables a more comprehensive analysis, as it leverages both the occurrence of events and underlying asset conditions. This integrated analysis is achieved by developing mathematical models that accurately represent fault or failure mechanisms. Models built from both data types form the foundation for maintenance decision making, including diagnostics and prognostics [28].

The DoS concept has been widely applied in the development of FD and CBM strategies for electrical machines. This approach primarily focuses on creating models that accurately incorporate various fault types, including bearing defects; static, dynamic, or mixed eccentricity; open or unbalanced circuits; short circuits in stator or rotor windings; insulation degradation; damage to rotor bars and end rings in squirrel-cage induction machines (SCIMs); and demagnetization in permanent magnet synchronous machines (PMSMs) [39–42]. DoS enhances the clarity of fault signatures in physical variables that are often influenced by noise and the inherent imperfections of the sensors used in experimental test setups. This makes it possible to effectively analyze early-stage faults. A digital real-time simulation (DRTS) corresponds to a DoS that must operate within real-time processing constraints. In other words, the discretized model must be sufficiently swiftly computed to synchronize with real-world timing [43]. Therefore, the selected solver must be optimized to resolve the system equation within a specified time step. DRTS is more advantageous than DoS because it can be linked to external equipment for design, test control, and robustness analysis in a hardware-in-the-loop (H-i-L) framework. H-i-L systems have been essential in the aerospace industry and flight simulations for several decades [44,45]. Their application has expanded significantly in various fields, including powertrain modeling for electric vehicles, energy management in microgrids, power and energy systems, and the dynamic performance analysis of power electronic devices. Recently, this concept was proposed to develop advanced fault diagnosis methods for electrical machines [18,46–48]. DRTS contributes to the FD and CBM of electrical machines by incorporating technologies such as virtual reality, DT, data analytics, large-scale/big data, Internet of Things (IoT), and ML, which have been widely adopted in the Industry 4.0 revolution. DT technology, which defines the adaptable mapping of a physical system to a virtual replication [49], is a powerful and emerging technology that enables the representation of the state of health of complex systems and facilitates CBM [50,51]. This article provides an overview of selected papers on the topic of multiphysics DRTS in the context of DTs for the intelligent FD and CBM of electrical machines. Section 2 reviews the definition of DT, its emerging concepts, and the various levels and scales at which it is applied. Section 3 discusses the main challenges of DRTS as an enabling tool within an electrical machine DT platform. Section 4 explores whether a multiphysics RTDS can contribute to the intelligent FD and CBM of electrical machines. Finally, the conclusion summarizes the key points and offers perspectives for future research in this domain.

2. Electrical Machine Digital Twins

2.1. Digital Twin Definition

The notion of a DT can be traced back to NASA in the 1960s, when it emerged as a dynamic representation of the Apollo missions. Following the explosion of an oxygen tank and damage to the main engine during Apollo 13, NASA employed several simulators to investigate the root causes of the malfunction. They enhanced their physical model of the spacecraft by incorporating digital elements, creating an early version of what is now known as a DT. This system allows for continuous data input, supporting both forensic analysis of the incident and the planning of subsequent actions [52]. Since its initial development, the concept of DT has gained significant attention from both academia and the industry. Figure 2 illustrates the timeline of the DT evolution, starting from its inception [53,54]. Numerous reviews and state-of-the-art articles have focused on enabling technologies, existing challenges, and methodologies for leveraging DTs in product lifecycle management and innovation [53,55–57]. Scholars and institutions have proposed various definitions and interpretations of DT; comprehensive discussions can be found in [25,53,56–58]. For instance, Vrabic et al. [59] characterized DT as a digital counterpart of a physical asset or system enhanced with integrated simulations and service-related data. The digital model aggregates data from multiple sources throughout the product lifecycle, which is regularly updated and represented in different formats to forecast both current and future states in design and operational contexts, thereby supporting improved decision making [56]. According to Tao et al. [55], a DT can be characterized as a high-fidelity simulation that integrates multiphysics and multi-scale modeling, incorporating probabilistic elements. It dynamically mirrors the state of its physical counterpart using a combination of historical information, real-time sensor inputs, and physical modeling.

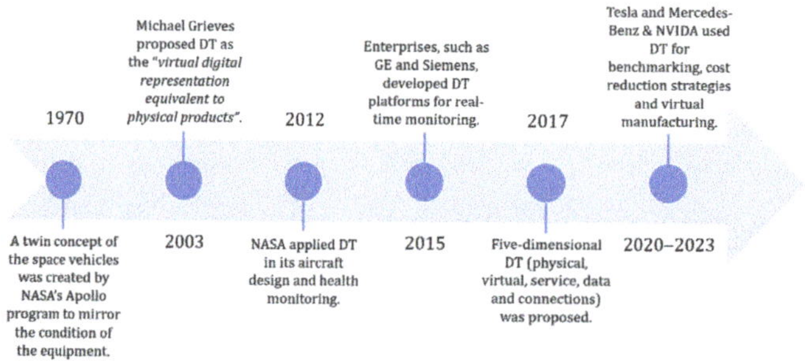

Figure 2. Timeline of DT evolution [53,54].

Similarly, Rasheed et al. [9,25] defined DT as a virtual model of a physical system that utilizes real-time data and digital simulations to enable monitoring, control, optimization, prediction, and informed decision making. Unlike traditional digital models or digital shadows, DTs are distinguished by a fully bidirectional data flow between their physical and digital counterparts that allows any change in the physical system to be immediately reflected in its digital representation [56]. More recently, Rasheed et al. [25] presented a comprehensive survey of the emerging applications of DT technology in the wind energy industry. Their study presents widely recognized definitions of DT from leading organizations and researchers, including Gartner, NVIDIA, IBM, DNV, GE Digital, Siemens, Oracle, Microsoft, the Digital Twin Consortium, Trauer et al., Grieves and Vickers, and the Industrial Digital Twin Association. Table 1 summarizes these recent definitions based on key aspects: 'Things,' which indicates the scope of application covered by each definition; 'Representation,' which refers to the realization space of the digital model; 'Data,' which outlines how information is utilized; and 'Purposes,' which identifies the primary objectives of DT development.

Table 1. Summary of most popular DT definitions [25].

	Things	Representation	Data	Purposes
Gartner	Process, physical object, organization, person, or any abstraction	Encapsulated software	Information from several DTs can be collected to provide a unified perspective of real-world objects	Simulate an entity in real time
NVIDA	Real-world physical things, people, and systems	Virtual	Information collected from connected sensors, processed through edge computing, enables the replication of physical equipment behavior	Enable the autonomy of systems through machine learning

Table 1. *Cont.*

	Things	Representation	Data	Purposes
IBM	Objects and systems	Virtual	Two-way flow of information	Decision making based on simulation, machine learning, and reasoning
DNV	Assets and systems	Virtual	Provide system information through a unified modeling and data solution	Offer guidance for decision making throughout the asset lifecycle
GE Digital	Physical assets, systems, and processes	Software	Real-time analytics	Enhance business outcomes through proactive detection, prevention, prediction, and optimization
Siemens	Physical products and processes	Virtual	Data is used throughout the product lifecycle to simulate, predict, and optimize products before any prototyping	Understand and predict the physical counterpart's performance characteristics
Oracle	Physical assets and devices	Digital	Updated with operational data and can be combined with physics-based models	Virtual sensors to detect anomalous behavior and prevent anomalies
Microsoft	Objects	Digital exact replicas	Data from monitoring devices for real-time view of assets	Improve the real-life version
Digital twin consortium	Real-world entities and processes	Virtual counterpart that is synchronized at a specified frequency and fidelity	Use real-time and historical data to represent the past and present	Transform business and simulate predicted futures
Trauer et al.	Physical systems	Virtual dynamic	Bidirectional information exchange and connection along the entire lifecycle	Improve product development by refining requirements, easing troubleshooting, or after-sales support
Grieves and Vickers	Physical manufactured products	Virtual equivalent from the micro-atomic level to the macro-geometrical level	Link between physical system and its replica	Understand system behavior
Industrial digital twin association	Assets	Digital	Updating throughout the lifecycle based on real-time data	Emulation, simulation, integration, testing, monitoring, and maintenance

The definition of DT can fall within its capabilities and levels, which are classified as standalone, descriptive, diagnostic, prescriptive, and autonomous [25,60]. The standalone DT represents the foundational level and is primarily utilized during the design phase. It relies on a DoS approach to enable a cost-effective system evaluation prior to physical construction. At the descriptive level, a CAD model combined with real-time sensor data is used to reflect physical assets. This setup relies on a precise PHYB that supports data interpolation in the targeted areas of interest. At the diagnostic level, powerful ML tools can be applied to the data to support FD and CBM. Using insights from diagnostic DTs, human experts can intervene early to make the necessary adjustments and prevent minor issues from escalating into major problems. Unlike standalone, descriptive, and diagnostic DTs, which do not offer foresight, the predictive DT continuously delivers updated predictions by leveraging real-time data streams from the physical entity. The prescriptive DT is valuable for optimizing asset control, as it generates recommendations based on what-if scenarios, risk assessments, and uncertainty analysis. This capability is especially beneficial for decision support systems, offering guidance to experts, who then determine the appropriate course of action. In the final stage, the DT and its physical counterpart engage in continuous two-way communication. The physical asset updates the DT in real time, whereas the twin, in turn, issues control commands to steer the physical system toward optimal performance. This enables rapid decision making without requiring human intervention. Referred to as the autonomous DT, this fifth level requires a high degree of technological maturity, particularly for deployment in safety-critical applications [25]. Industrial platforms such as General Electric's 'Predix,' Siemens' 'MindSphere,' and 'ThingWorx' support the development of DTs by leveraging Internet of Things (IoT) technologies. The IoT extends digital connectivity into the physical environment through the widespread use of radio-frequency identification (RFID) chips embedded in real-world devices [61]. Although the IoT offers significant benefits, it also presents major challenges related to IT infrastructure, connectivity, data privacy, security, trust, and data management [56]. In this context, data analytics plays a critical role in collecting, cleaning, and processing data for further analysis. The data cleaning stage, particularly through imputation techniques, is essential to addressing errors or missing values and ensuring the quality and reliability of the data before analysis.

2.2. Benefits and Constraints

The technical and economic benefits of DTs can be highlighted across five phases of the product lifecycle: product design, manufacturing, operation, maintenance, and recycling [53]. This means that if the DT can replace the physical entity, it can be used during the design, testing, and operation phases. This capability enables a better understanding of how an asset performs under both normal operating conditions and undesired stress, helps to determine material weaknesses, and assesses its suitability for its intended purpose. This has at least two beneficial effects: decreasing the cost of development and testing while increasing the number of situations that can be considered, assessed, and evaluated and extend/augmenting physical objects with software capabilities that will substantially enrich them. The greater the accuracy and fidelity of the DT, the more effectively it can replicate and predict the behavior of the physical asset in all scenarios. Furthermore, the DT approach facilitates the systematic collection and organization of data related to physical assets. This is particularly valuable, as historical records of an asset's behavior under various conditions can offer insights for optimizing its performance and guiding future improvements. Analyzing and leveraging this information can significantly support the refinement and evolution of products over time [62]. In modern manufacturing, DTs are increasingly integrated into smart factories to model, analyze, and manage diverse processes with greater efficiency. Their adoption offers several advantages, such as real-time machine condition monitoring, enhanced system availability and reliability, advanced CBM capabilities, and the provision of tailored services to stakeholders [9,56]. These advantages are made possible through continuous data acquisition and transmission through IoT platforms. The collected data serve as a foundation for essential functions, including production scheduling, equipment maintenance, logistics, and strategic decision making. Within the framework of Industry 4.0, the overarching objective is to enable intelligent, data-driven decisions that strengthen both profitability and sustainability in smart manufacturing environments [63].

Although the DT approach offers several advantages for enterprises and communities, it also carries significant risks. The core purpose of the DT is to replicate the behavior of a physical entity. Consequently, one major challenge lies in determining the aspects of reality that are critical to accurately representing a digitalized model. Furthermore, DT development and structuring within an enterprise may require considerable time and effort. Secure data collection, communication, manipulation, and representation also require infrastructure that requires competent teams in information technology, which may not be available in a company. In addition, from a security and privacy perspective, a DT gathers and represents a significant amount of information that holds considerable value to the owner of the corresponding physical object. It not only reflects how the object is built and operated but also captures patterns of its usage and behavior. Such data are valuable for both competitive reasons and personal insights. For manufacturers and vendors, understanding how a product is used provides critical intelligence, whereas for users, this usage data can reveal detailed profiles of their preferences and habits. Therefore, clearly defining the ownership of a DT and regulating the collection of behavioral data are essential to ensuring proper security and protection for both users and owners. Another critical concern is software programmability and control, as unauthorized access to an object poses significant risks that can lead to disruptions or even serious accidents that impact both individuals and operations. One can easily imagine scenarios in which DTs might be exploited to interfere with people's daily lives or manipulate specific environments. Therefore, expanding the use of DTs beyond tightly controlled settings requires the implementation of a strong and reliable security framework capable of detecting malicious activities or harmful actions, whether triggered intentionally by hackers or inadvertently by owners [62].

2.3. Electrical Machine DT Realization

Drawing from the diverse range of existing DT definitions, the authors propose a predictive-level framework with the following vision for the DT design of an electrical machine in the service phase [51,55,58]:

- The DT of an electrical machine is a synchronized, ultra-high-fidelity replica of it, incorporating multiphysics, multi-scale, and probabilistic modeling.
- An automated, bidirectional, real-time flow of data occurs between the DT and the electrical machine through appropriate instrumentation and an IoT platform.
- The twin encompasses data from the service stage of the electrical machine's lifecycle and remains connected to this phase through to the retirement stage.

Figure 3 represents the scheme of electrical machine DT realization based on IoT and cloud computing technologies for FD and CBM. The physical electrical machine in this configuration is fully instrumented using RFID smart sensors to collect data on voltage, current, vibration, acoustics, temperature, speed, and mechanical torque. Quantities such as the active, reactive, and apparent power can be computed based on the measured stator voltages and currents. The implemented sensors describe the conditions of the electrical machine supplied by an electrical drive. This device adjusts the operating point of the electrical machine based on its health state. The RFID smart sensors and electrical drive are connected via an IoT platform to a cloud computing platform, as shown in Figure 3. The cyber layer of this structure enables secure bidirectional data transfer among smart sensors, electrical drives, and cloud computing platforms. In the next section, the DRTS challenges are discussed, as they are considered a fundamental stage for the predictive-level framework for the DT design of electrical machines in the illustrated configuration (Figure 3).

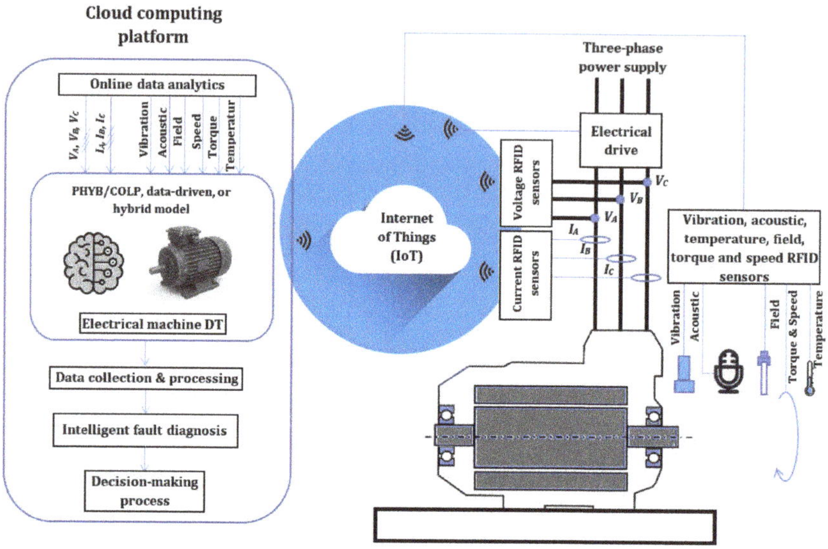

Figure 3. Scheme of electrical machine DT for intelligent fault diagnosis [4].

3. Electrical Machine DRTS Challenges

3.1. Electrical Machine Models

In addition to DT, other terms have been introduced to clarify its workings in the real world, including physical entities, physical objects, physical things, physical assets, physical processes, physical products, real-world entities, physical systems, and physically manufactured products, as well as its workings in the digitally generated virtual world, including virtual entities, virtual dynamics, virtual representations, digital representations, and digital exact replicas (Table 1). DPHYB, COLP, or data-driven models or a combination of these models can be used to build the DTs of electrical machines. As mentioned in the Introduction, data-driven models, despite exhibiting stable behavior after the training process, require extensive historical data (big data), which can be obtained from the collected information. They are simpler to configure, typically do not require detailed physical entity parameters, and operate faster than real-time models. However, they operate as black boxes, and the presence of biased data can manifest in the model, resulting in unpredictable errors and uncertainties. PHYB and COLP models have significant advantages because they operate based on a solid physics foundation, resulting in estimated errors and uncertainties, which are important criteria for fault management systems.

The PHYB and COLP modeling approaches can be categorized into experimental and numerical models. Experimental modeling involves conducting reduced- or full-scale experiments to determine the main parameters of the electrical machine model [18,64]. To enhance the physical authenticity of a DT, it is necessary to solve the governing equations obtained from the physical modeling. In some cases, analytical solutions can be derived using simplified equations. However, owing to their complexity, numerical solutions are often obtained using computers [9,65,66]. The main aim is to analyze quantities that are commonly difficult or costly to measure directly. These models can be subsequently employed, as shown in Figure 3, as the DTs of electrical machines. The traditional approach involves creating a COLP dynamic 'dq0' reference model for AC electrical machines [67,68]. The estimation of model parameters can be used for fault diagnosis (Table 2). Although several techniques have been proposed for the identification of AC electrical machine parameters, they are commonly performed offline, limiting their application to DT realization. Another approach relies on residual computations using parity equations, state observers, or state estimators. This technique has been used extensively in sensor fault detection and diagnosis systems. Such dynamic models are sufficiently simple to be used in DRTS and DT realizations, as presented in [15]. However, they omitted the magnetic saturation effect and assumed a sinusoidal winding distribution. Moreover, the machine is assumed to be perfectly symmetrical. The 'dq0' reference frame models are commonly preferred for simulating electromagnetic transients because the 'abc' reference frame models have time-varying inductances. This means that to solve the governing equations, the inductance matrices must be inverted at each time step during the digital simulation, in addition to the computational burden of each digital simulation step [15]. Despite the aforementioned inconvenience, the 'abc' reference frame has the advantage of being easily applicable for modeling multi-phase electrical machines, regardless of asymmetry in both stator and rotor circuits, and can account for all space harmonics in the electrical machine [69–71]. In this regard, it is possible to include certain inherent asymmetries in the physical electrical machines in the model. Induction machines (IMs), which are broadly utilized in the industry, are good examples [72].

Table 2. COLP modeling of various fault types in 'dq0' reference frame.

Fault Types	References
Broken rotor bar and end ring	[73–78]
Stator/rotor windings unbalance	[79]
Stator/rotor windings short circuit	[17,74,75,80–83]
Static, dynamic, or mixed eccentricity	[74,84]
Ball bearing and race	[85]
Magnetization-related	[86]

The equations representing the dynamic of a p-pole, three-phase, wye-connected IM in the 'abc' reference frame is given by [87]

$$\begin{bmatrix} \mathbf{v}_{abcs} \\ \mathbf{v}_{abcr} \end{bmatrix} = \begin{bmatrix} \mathbf{r}_s & 0 \\ 0 & \mathbf{r}_r \end{bmatrix} \cdot \begin{bmatrix} \mathbf{i}_{abcs} \\ \mathbf{i}_{abcr} \end{bmatrix} + \frac{d}{dt} \begin{bmatrix} \boldsymbol{\lambda}_{abcs} \\ \boldsymbol{\lambda}_{abcr} \end{bmatrix} \tag{1}$$

with

$$\begin{bmatrix} \boldsymbol{\lambda}_{abcs} \\ \boldsymbol{\lambda}_{abcr} \end{bmatrix} = \begin{bmatrix} \mathbf{L}_s & \mathbf{L}_{sr}(\theta_r) \\ \mathbf{L}_{sr}(\theta_r)^T & \mathbf{L}_r \end{bmatrix} \cdot \begin{bmatrix} \mathbf{i}_{abcs} \\ \mathbf{i}_{abcr} \end{bmatrix} \tag{2}$$

$$T_e = \left(\frac{p}{2}\right) \mathbf{i}_{abcs}^T \frac{\partial}{\partial \theta_r} \left[\mathbf{L}'_{sr}(\theta_r)\right] \mathbf{i}'_{abcr} \tag{3}$$

where $\mathbf{v} = [\mathbf{v}_{abcs}\ \mathbf{v}_{abcr}]^T$, $\mathbf{i} = [\mathbf{i}_{abcs}\ \mathbf{i}_{abcr}]^T$, and $\boldsymbol{\lambda} = [\boldsymbol{\lambda}_{abcs}\ \boldsymbol{\lambda}_{abcr}]^T$ denote the voltage, current, and flux linkages, respectively. The subscripts $abcs$ and $abcr$ denote variables related to the stator and rotor circuits, respectively. $\mathbf{r}_s$ and $\mathbf{r}_r$ represent the stator and rotor resistance matrices, respectively, in diagonal form. $\mathbf{L}_s$ and $\mathbf{L}_r$ denote stator and rotor winding inductances, respectively. θ_r is the angular separation of the stator as and rotor's ar axes. $\mathbf{L}_{sr}(\theta_r)$ represents the mutual inductance between the stator and the rotor windings. The superscript $'$ represents the variables that refer to stator windings. The relationship between the torque and rotor speed is defined as

$$T_e - T_l = J \cdot \frac{d}{dt}\omega_m + B \cdot \omega_m \tag{4}$$

where J denotes the moment of inertia (kgm^2), B denotes the viscous friction (N.m.s), and $\omega_m = \frac{2}{p} d\theta_r/dt$. The governing differential algebraic equations represented by (1) and (2) can be reformulated into a first-order ODE (5) for the DS of the IM model in the 'abc' reference frame [16]:

$$\frac{d}{dt}\mathbf{i} = \mathbf{L}^{-1}(\theta_r)\left\{\mathbf{v} - \mathbf{r}\cdot\mathbf{i} - \omega_r \cdot \frac{d}{d\theta_r}\mathbf{L}(\theta_r)\cdot\mathbf{i}\right\} \tag{5}$$

with

$$\mathbf{L}(\theta_r) = \begin{bmatrix} \mathbf{L}_s & \mathbf{L}_{sr}(\theta_r) \\ \mathbf{L}_{sr}(\theta_r)^T & \mathbf{L}_r \end{bmatrix}, \ \mathbf{r} = \begin{bmatrix} \mathbf{r}_{abcs} & 0 \\ 0 & \mathbf{r}_{abcr} \end{bmatrix} \tag{6}$$

Relations (1) and (2) can be considered a baseline for modeling various types of faults such as broken rotor bars, end rings, and stator winding ITSCs in IMs (Table 3). For instance, the COLP model of a three-phase IM can be developed to include an ITSC in one of its stator windings [80] (Figure 4). The main parameters of the three-phase IM are modified in (1) and (2) owing to the stator-related matrices of the stator winding ITSC: $\mathbf{r}_s$, $\mathbf{L}_s$, and $\mathbf{L}_{sr}$. In these matrices, the parameters corresponding to the faulty segment are separated from those of the remaining windings. The proposed model can be extended to study the simultaneous occurrence of ITSC faults in the stator windings of IMs [17]. To investigate broken rotor bars or end-ring defects in squirrel-cage induction machines (SCIMs) using the COLP modeling approach, the rotor structure can be represented by the configuration shown in Figure 5. In this model, $\mathbf{r}_r$ accounts for the resistance of the rotor bars and end-ring segments, whereas $\mathbf{L}_r$ includes self-, leakage, and mutual inductances. Additionally, $\mathbf{L}_{sr}$ denotes the mutual inductances between each rotor loop or end-ring segment and each phase of the stator winding [88].

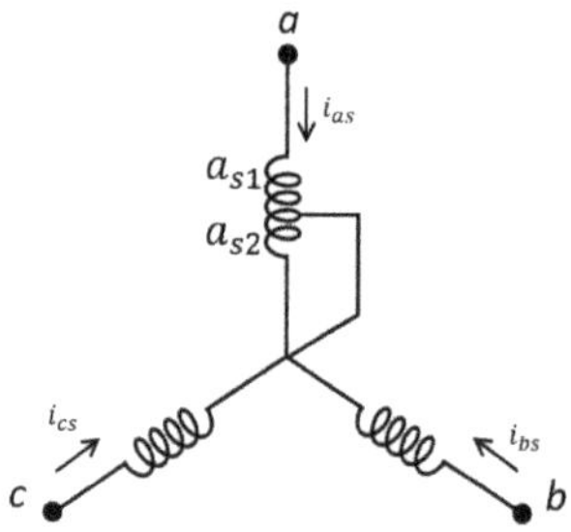

Figure 4. Stator windings of an IM with an ITSC fault in phase 'a' [80].

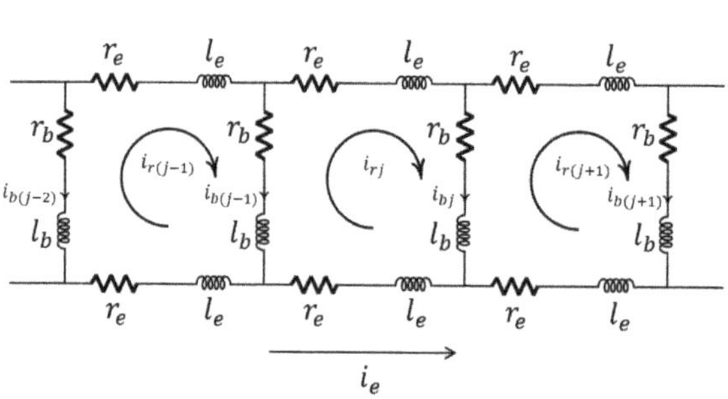

Figure 5. Equivalent circuit of a squirrel-cage rotor showing rotor loop currents [88].

Table 3. COLP modeling of various fault types in 'abc' reference frame.

Fault Types	References
Broken rotor bar and end ring	[89–96]
Stator/rotor windings unbalance	[18,48]
Stator/rotor windings short circuit	[17,89,97,98]
Static, dynamic, or mixed eccentricity	[99–102]
Ball bearing and race	[103–105]
Magnetization-related	[106]

The COLP modeling approach was initially developed for multi-phase winding structures of IMs, designed to operate even in scenarios where one or more of the stator phases are open-circuit, as discussed in [70]. The winding function approach (WFA) for computing self- and mutual inductances, which considers all space harmonics in a multi-phase IM, is presented. The proposed model includes asymmetry resulting from inter-turn faults in the stator phase windings, as well as any issue and number of failures in the rotor bars and end rings [71]. This type of modeling is crucial to fault diagnosis in AC electrical machines, where in many cases, space harmonics are required for efficient fault identification [107]. A universal mathematical model for a five-phase IM encompassing the influence of higher space and time harmonics in the air-gap field is presented in a similar manner. Given the significant role of space harmonics in AC electrical machines with more than three phases, mathematical relations for computing self- and mutual inductances are derived [108]. To simplify the DRTS of the 'abc' reference frame model of the IM based on (5), both $\mathbf{L}^{-1}(\theta_r)$ and $\frac{d}{d\theta_r}\mathbf{L}(\theta_r)$ are described through look-up tables as functions of θ_r [109]. This modeling approach can be extended to other electrical machine classes, such as permanent magnet synchronous machines (PMSMs) [48] (Table 3). In this regard, the classical 'abc' reference frame models of the IM and PMSM can be utilized for the DRTS of stator winding asymmetry faults. This illustrates the efficacy of such simple representations in detecting winding imbalance faults [17,18,110,111].

The use of PHYB models, especially FEM-based models that incorporate spatial and nonlinear aspects, as well as non-sinusoidal winding distribution phenomena inside the machine, can enhance the fault detection performance. These models provide comprehensive information regarding the health states of machines by using residual signals. However, the computational intensity of these models and their size and time-step limitations render them unsuitable for DRTS application. MEC-based models are generally considered to be a middle ground between FEM-based and COLP models in terms of computational performance [10]. The concept of MEC-based modeling relies on the analogy between electric and magnetic circuits. A deeper analysis of the electric and magnetic fields revealed that magnetic circuits typically function in a saturated nonlinear mode, whereas the majority of the elements in electric circuits exhibit linearity. In this regard, the machine's 2D or 3D structure must be divided into small elements that describe the MEC. Each element of the MEC model includes voltage and current sources and reluctances, which can be expressed as

$$\Re_m = \int_0^l \frac{dx}{\mu(x)A(x)} \tag{7}$$

where l represents the length of the flux tube and A denotes its cross-sectional area. μ denotes the permeability of the magnetic material. Figure 6 represents the 3D MEC reluctance-based network and the circuit model of its elements. The voltage source, F_{ij}, represents the magnetomotive force (MMF), and the current source, ϕ_{PM-ij}, represents the flux of the permanent magnet. This network must be solved to obtain magnetic scalar potential u at each node. Based on Figure 6b, it is possible to derive the following [10]:

$$\Re_{ij} \cdot (\phi_{ij} - \phi_{PM\text{-}ij}) = (u_i - u_j - F_{ij}) \tag{8}$$

Once the permeance and MMF are computed, the MEC relations can be determined based on classical circuit nodal analysis.

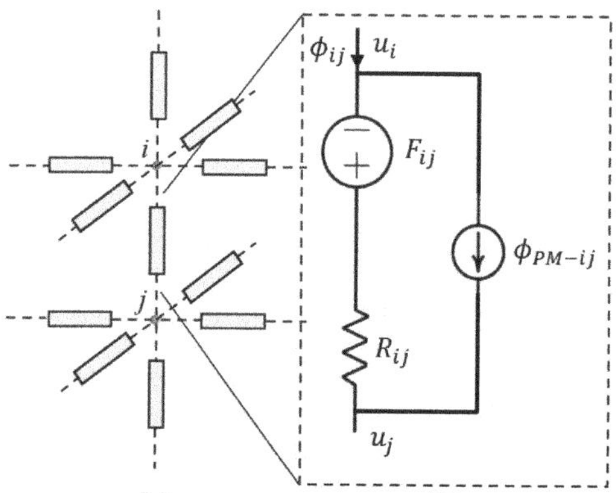

(a) (b)

Figure 6. Scheme of MEC model: (**a**) 3D reluctance network and (**b**) equivalent circuit of each element of the network [4].

Flux tubes serve as the foundation for MEC modeling techniques. A flux tube constitutes a geometrical region where all flux lines are perpendicular to their bases and no flux lines cut their sides. This requires knowledge of the machine geometry, including the effects of the stator and rotor slots, skewing, winding connections, and magnetic nonlinearity of the electrical machine cores. Consequently, it produces more accurate results than the COLP modeling technique under both healthy and faulty conditions (Table 4). For instance, permeance network and nonlinear MEC-based models have been introduced for real-time simulation of IMs [112,113]. A DRTS model was developed for PMSMs with shaped poles by utilizing the analytical solution of field equations incorporating space harmonics in the air-gap flux density distribution. The model was constructed under the assumption of linear superposition [114].

Hybrid modeling, as shown in Figure 1, is an emerging research topic that has received significant attention in recent years, especially since traditional multiphysics modeling approaches account for only a fraction of the relevant physical phenomena, which often explains the mismatch between their predictions and experimental observations [26]. With the availability of large number of data, open-source cutting-edge ML and data analytics libraries, and computation resources, the utilization of a hybrid approach to modeling allows for predictions of all aspects owing to its ability to combine the strengths of physics-based models and data-driven approaches, enabling more accurate, robust, and generalizable solutions across various engineering and scientific domains. Some active areas of research in this context include cost function modification to accommodate the Jacobian model, grow-when-required networks, physics-informed machine learning (PIML), embedding hard physical constraints in a neural network, leveraging uncertainty information, developing visual tools for network analysis, and nonparametric modeling approaches for bridging data science and dynamical systems [26]. In particular, PIML, initially introduced by Raissi et al. [115], is a significant advancement at the intersection of ML and PHYB modeling [116]. PIML distribution terms refer to physics-informed, physics-based, physics-guided, physics-infused, and physics-aware machine learning [117]. The architecture of physics-based neural networks (PINNs) is similar to that of conventional ML perceptrons (MLPs), as they consist of an input layer, multiple hidden layers, and an output layer. The main distinction between PINNs and standard MLPs is the inclusion of a differentiation layer (Figure 7). Each layer plays a specific and essential role in the PINN training process. Spatiotemporal data, represented by coordinates x and time t, are provided to the network through the input layer and then processed by the hidden layers. Then, the network predicts the latent solution y of the PDE through the output layer. These predicted values y are subsequently passed to the differentiation layer, where automatic differentiation is applied with respect to both x and t to generate the differential terms of the PDE. This process leads to the formulation of a loss function used to train the network. The total loss function of a PINN is expressed as a combination of a term that penalizes the residuals of the PDE and the term y loss function, which ensures agreement between the model predictions and given data points [118]. This concept has been widely employed in various applications, such as electrical impedance tomography [119], power system dynamic modeling and analysis [120,121], the modeling and prediction of temperature fluctuations in proton exchange membrane electrolysis [122], state estimation in unbalanced distribution power systems [123], high-speed transportation [124], and hydrogen-powered electric aircraft, among others [125]. A new partitioned PINN framework was proposed to enhance both the prediction accuracy and computational efficiency of electromagnetic field analysis in PMSMs. In this regard, separate neural network models were developed for the stator, air-gap, and rotor regions, with strict boundary conditions applied to maintain magnetic field continuity between neighboring subdomains [126]. A PINN architecture was designed to tackle the challenges of developing DTs for PMSMs in intelligent condition-based monitoring [127]. A physics-informed generative adversarial network-based approach was proposed for the optimal design of permanent magnet linear synchronous machines, achieving speeds up to 40 times faster than the FEM method while maintaining comparable accuracy [128].

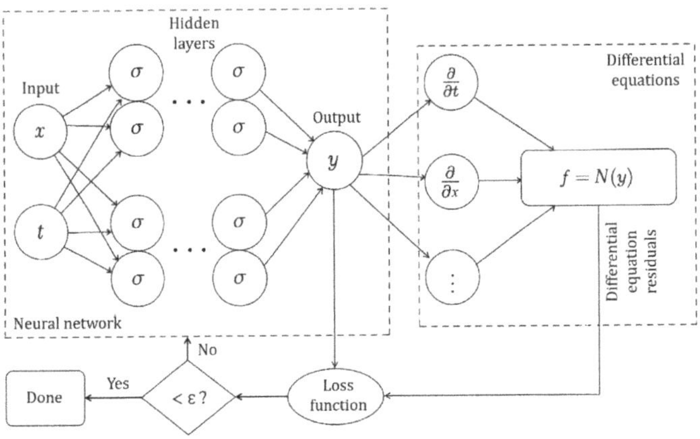

Figure 7. Schematic representation of a PINN framework [116].

Table 4. MEC modeling of various fault types.

Fault Types	References
Broken rotor bar and end ring	[129–131]
Stator/rotor windings ITSC	[132–134]
Static, dynamic, or mixed eccentricity	[135]
Ball bearing and race	[136,137]
Magnetization-related	[138–144]

3.2. RTDS Hardware Platforms

A real-time simulation is defined as a process in which computational tasks are completed within the same time interval as the physical processes that are represented. This section explores the main aspects of the practical implementation of RTDS across various hardware platforms, emphasizing the constraints, trade-offs, and architectural capabilities. RTDS assumes that the state of a system changes only at fixed discrete time steps, synchronized with a real-time clock. The execution of RTDS involves three critical stages: receiving/sending signals through input/output (I/O) interfaces; executing computations, particularly solving the ODEs that define machine behavior; and transferring data between processing units and memory. To preserve real-time behavior, all these processes must be completed within the simulation time step. In this regard, among the various implicit and explicit numerical integration methods, the Runge–Kutta (RK), backward Euler, forward Euler, and trapezoidal rule are more widely used in power system simulations, including electrical machines. When selecting an integration method, the key factors include numerical precision, computational time, and resource usage. Higher-order methods, such as RK4, offer better precision but require more computational resources. In contrast, the FE method is faster and simpler but less accurate, especially for larger step sizes [10,145,146].

There are two types of RTDS, H-i-L and S-i-L, in the context of electrical machine DT design. In the H-i-L configuration, a physical component interacts with simulated components on a hardware platform, whereas in the S-i-L configuration, all components are simulated digitally on the same platform, ensuring signal integrity and computational flexibility. H-i-L is advantageous for testing in high-risk or inaccessible environments, whereas S-i-L allows for pure software validation [13,147]. Simulation fidelity is governed by the time-step size. A small time step ensures higher accuracy but increases the risk of overrun if computations cannot be completed in time. This is particularly challenging for high-frequency simulations such as those involving PWM drives, which may require nanosecond-level precision.

Hardware platforms can be categorized into chip single-core processor units (CSPUs), chip multiprocessor units (CMPUs), computer clusters, field-programmable gate arrays (FPGAs), and graphics processing units (GPUs). Each platform presents a unique set of advantages and trade-offs in terms of execution speed, resource usage, scalability, latency, and programming complexity. CMPUs or multicore CPUs include several processing cores within a single chip. Each core executes the instructions in parallel using multithreading. The CMPUs support shared memory and offer high clock speeds, making them suitable for medium-complexity simulations. Their advantages include user-friendly programming with tools like MATLAB/Simulink, cost-effectiveness and wide availability, and effectiveness for models like 'dq0' and MECs [113,148–152]. Their limitations include high I/O latency owing to PCI bus delays, difficulty in handling time steps smaller than one microsecond, and suboptimal performance for high-frequency simulations [150,153]. Computer clusters consist of multiple interconnected computers (nodes), each with multiple CPUs. They are scalable and suitable for simulating large-scale electric systems such as wind farms. Their architecture is based on a master node that manages the simulation, whereas the slave nodes execute parallel computations. The nodes communicate via high-speed Ethernet or Infiniband links. Challenges include high communication latency, complex programming models involving message-passing interfaces (MPIs), and node synchronization, which can become a bottleneck. Computer clusters are widely employed in the RTDS of electrical machines [148,154–157].

FPGAs consist of configurable logic blocks, interconnection resources, and I/O interfaces. These platforms allow for deterministic execution and very fine time-step control, making them suitable for the real-time simulation of electric machines and drives. Key features include low-latency I/O interfaces without PCIs, parallel or pipelined architectures for high performance, and support for both fixed-point and floating-point arithmetic operations. The development approach is based on either textual programming languages (e.g., VHDL and Verilog) or schematic/block-based tools such as the Xilinx System Generator. Limitations include limited hardware resources, the need for expertise in digital hardware design, and high costs when scaling with multiple FPGAs. They are widely used in the DRTS of electrical machines and drives [87,112,114,158–166]. Multiple interconnected FPGAs or pipelining schemes can be considered for high-order RTDS models [10,68]. Graphics processing units (GPUs) are well suited for handling large-scale numerical simulations owing to their highly parallel architecture and strong floating-point processing capabilities. They consist of numerous cores arranged in blocks and grids, with threads executing instructions in a Single-Instruction, Multiple-Data (SIMD) manner and working alongside a central host CPU. These features make GPUs particularly effective in speeding up FEM simulations and managing models with thousands of ODEs. Additionally, they offer high computational throughput and can be programmed using languages such as Compute Unified Device Architecture (CUDA) or Open Computing Language (OpenCL). However, GPUs have drawbacks, such as significant initialization and data transfer overhead, and tend to be less efficient when applied to small-scale or rapidly changing models. Owing to their inherent parallel processing capabilities, GPUs have been adopted in various numerical analyses related to electrical engineering. Their applications include the numerical field analysis and simulation of electric machines [167–171]. In comparison, CMPUs provide adaptable but moderately performing solutions with higher latency; computer clusters offer scalability but require complex coordination and programming; FPGAs deliver fast, low-latency performance but are constrained by limited resources; and GPUs excel in processing large models, although they are less optimal for high-frequency operations.

4. Intelligent FD and CBM of Electrical Machines

The multiphysics modeling of electrical machines integrates various physical domains, including electromagnetics, thermal dynamics, mechanics, and acoustics. For instance, lumped models (LMs) are typically used to represent magnetic, electrical, electronic, and thermal components, whereas analytical models are applied to describe the vibro-acoustic and mechanical behaviors of a PMSM. Although such coupled modeling approaches are commonly used during the design phase of electrical machines, their applications in FD and CBM remain limited [11,172]. In particular, the resistances in the COLP model presented in [172] were defined as functions of the supply frequency and winding temperature, thereby allowing for improved accuracy. The frequency effect accounts for both skin and proximity effects. A multiphysics model of an induction machine is shown in Figure 8 [173].

This model comprises five interdependent sub-models. All of these sub-models are either analytical or semi-analytical to achieve a good compromise among computational speed, accuracy, and flexibility. The core of this approach is the electromagnetic model, which provides key insights, such as induction in the air gap, radial forces acting on the stator, and current distributions. The outputs of this model can be utilized in two ways. First, they serve as inputs for mechanical vibration and acoustic models, which are designed to predict the vibration behavior of the machine and the resulting electromagnetic noise emissions. Second, the electromagnetic model outputs are fed into the loss model, which quantifies the various losses occurring within the motor. The calculated losses are subsequently used in a fluid flow and thermal coupled (FFTC) model to simulate the evolution of internal motor temperature over time. A dynamic interaction exists between the electromagnetic and FFTC models: the temperature predictions from the FFTC model influence certain electromagnetic parameters, which in turn affect the loss estimations, that is, the same losses that act as heat sources in FFTC analysis.

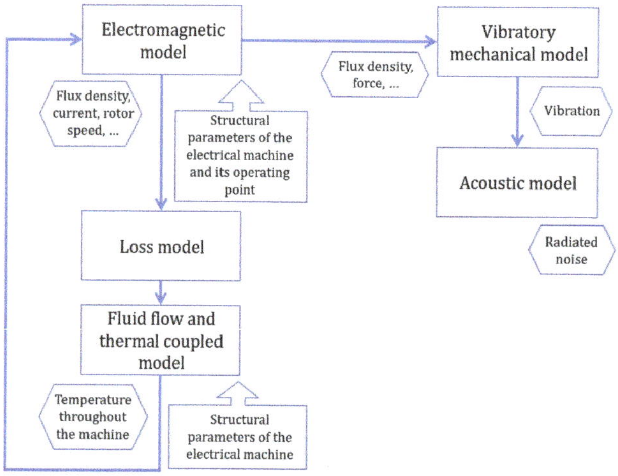

Figure 8. A schematic of a multiphysics model of an induction machine [173].

In this regard, the DT design procedure for the FD and CBM of complex systems is well defined in [174]. In the first stage, it is essential to establish a high-fidelity reference model (a DT) for the electrical machine. Table 5 summarizes the advantages and disadvantages of PHYB/COLP techniques compared with data-driven models for DT development [9,66]. A multiphysics model that considers all the important aspects of the electrical, mechanical, and thermal behaviors of a physical entity enables the reference model to provide a wide spectrum of signals. Although data-driven models have weaknesses in representing all aspects of complex systems, they can be combined with PHYB/COLP techniques to optimize and solve the ODE governing the model [175,176]. In the second stage, an IoT infrastructure must be built to facilitate the evolution of the DT through available data from sensors installed on a physical electrical machine. In the final stage, the DT data are fused with well-known data-driven models for further processing, FD, and CBM [31]. The main features of the DT of an electrical machine that contribute to intelligent FD and CBM are highlighted as follows:

- DT parameters can be updated in real time based on voltage, current, vibration, acoustic, field, speed, and temperature measurements.

- The DT can be supplied by the measured phase (V_A, V_B, and V_C) or line (V_{AB}, V_{BC}, and V_{CA}) voltages.
- The DT provides a wide range of inaccessible signals that commonly require sophisticated instrumentation.
- More clear fault signatures can be detected in the physical variables of the DT.
- Intelligent FD and CBM become possible by processing the DT data outcomes.
- Remote monitoring and control become feasible via the IoT infrastructure.

To achieve a high level of DT performance, it is crucial to adopt a strategy based on multiphysics modeling and integration of the main fault types (Figure 9a). This approach involves coupling all physical phenomena that contribute to lifecycle analysis, including thermal, mechanical, electrical, and chemical aspects. The parameters of the electrical machine DT can be updated using measured signals. Furthermore, the DRTS data can be used for DNN pre-training. The parameters of the DNN, namely, the weights, biases, and hyper-parameters, can then be transferred to a new DNN for fine-tuning using the measured data for FD, as shown in Figure 9b. The DRTS data, measured data, and event data which represent the FD results can be leveraged to perform the intelligent FD and CBM of electrical machines, but they have rarely been investigated in the literature [31,53].

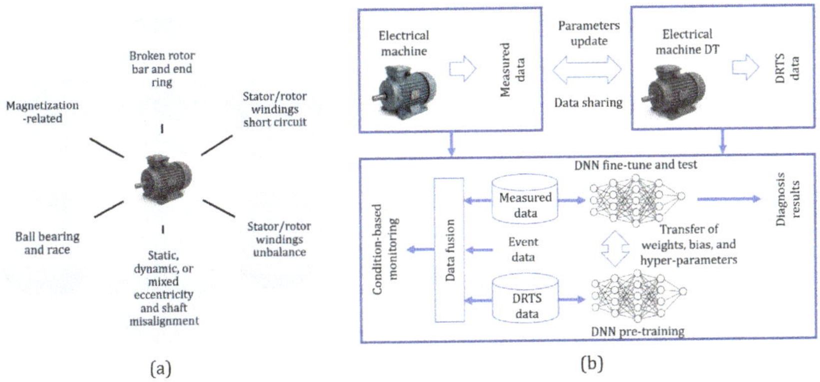

Figure 9. (**a**) Integration of the main fault types into the DT of an AC electrical machine. (**b**) DT-assisted simplified framework for the intelligent FD and CBM of AC electrical machines [31,53].

Table 5. Comparison of PHYB/COLP modeling techniques with data-driven approaches for electrical machines DT design [4,9,66].

PHYB/COLP Techniques	Data-Driven Approaches
+ Solid foundation in physics	– Black-box concept
– Need for partial or entire geometric data of the electrical machine	+ No need for any knowledge about the electrical machine
+ No need for data for training	– A lot of data need to be provided for machine learning
– Need for optimization algorithms for continuous updates of model parameters	+ Neural network update
– Numerical instability of the model	+ Stable for a trained model
+ Less prone to bias	– Bias in the data can be reflected in the model
– Difficulty in assimilating extensive historical data	+ They integrate easily extensive historical data
+ Developed model can be used for similar electrical machines	– New model needs to be trained for each electrical machine
+ Several variables are available from the developed model	– Only the trained variables are available

5. Conclusions and Perspectives

This paper presents a summary of articles focusing on DRTS in the context of emerging DT technologies for intelligent FD and CBM in electrical machines. A predictive-level framework with the following vision is proposed for DT design during the service phase:

- The DT of an electrical machine is a synchronized, ultra-high-fidelity replica of it, incorporating multiphysics, multi-scale, and probabilistic modeling.
- An automated, bidirectional, real-time flow of data occurs between the DT and the electrical machine through appropriate instrumentation and an IoT platform.
- The twin encompasses data from the service stage of the electrical machine's lifecycle and remains connected to this phase through to the retirement stage.

Based on the above definition, achieving an ultra-high-fidelity replica requires a multiphysics modeling approach with online parameter updating, whether this is a PHYB, data-driven, or hybrid approach. Furthermore, implementing intelligent FD and CBM requires simulation of key fault scenarios using the DT of an electrical machine, which offers a wide range of variables. The model must also operate in real time to ensure proper synchronization and bidirectional data flow with the physical system through the IoT infrastructure. Notably, the PIML approach is an increasingly attractive research direction, aiming to bridge machine learning and PHYB modeling to achieve a precise, real-time DT. In this context, hardware platforms such as GPUs, FPGAs, PC clusters, and CMPUs, which are commonly used in RTDS for electromagnetic transient studies of electrical machines, are well suited to support this requirement.

Author Contributions: These authors contributed equally to this work. All authors have read and agreed to the published version of the manuscript.

Funding: This research study was funded by the Spanish 'Ministerio de Ciencia e Innovación', Agencia Estatal de Investigación and FEDER program in the framework of 'Proyectos de Generación de Conocimiento 2021' of 'Programa Estatal para Impulsar la Investigación Científico-Técnica y su Transferencia', belonging to 'Plan Estatal de Investigación Científica, Técnica y de Innovación 2021-2023' (ref: PID2021-122343OB-I00).

Conflicts of Interest: The authors declare no conflicts of interest.

Abbreviations

The following abbreviations are used in this manuscript:

AC	Alternating Current
COLP	circuit-oriented lumped-parameter
CBM	condition-based monitoring
CMPU	chip multiprocessor unit
CSPU	chip single-core processor unit
CUDA	Compute Unified Device Architecture
DNN	deep neural network
DoS	digital offline simulation
DRTS	digital real-time simulation
DS	digital simulation
DT	digital twin
EMTP	Electromagnetic Transient Program
FD	fault diagnosis
FDM	finite difference method

FEM	finite element method
FFTC	fluid flow and thermal coupled
FPGA	field-programmable gate array
FVM	finite volume method
GPU	Graphics Processor Unit
H-i-L	hardware-in-the-loop
IM	induction machine
IoT	Internet of Things
MEC	magnetic equivalent circuit
ML	machine learning
MMF	magnetomotive force
MPI	message-passing interface
MWFA	modified winding function approach
ODE	ordinary differential equation
PDE	partial differential equation
PHYB	physics-based
PIML	physics-informed machine learning
PINN	physics-informed neural network
PMSM	permanent magnet synchronous machine
RFID	radio-frequency identification
RK	Runge–Kutta
RTDS	Real-Time Digital Simulator
PMSG	permanent magnet synchronous generator
SCIM	squirrel-cage induction machine
SIMD	Single-Instruction, Multiple-Data
SSM	state-space model
TL	transfer learning
TSR	Tip–Speed Ratio
MPPT	Maximum Power Point Tracking
P-H-i-L	power-hardware-in-the-loop
P-i-L	processor-in-the-loop
S-i-L	software-in-the-loop
WECS	Wind Energy Conversion System
WFA	winding function approach
WT	Wind Turbine
XATP	Expandable Transient Analysis Program

References

1. Semenkov, K.; Promyslov, V.; Poletykin, A.; Mengazetdinov, N. Validation of Complex Control Systems with Heterogeneous Digital Models in Industry 4.0 Framework. *Machines* **2021**, *9*, 62. [CrossRef]
2. Markwirth, T.; Jancke, R.; Sohrmann, C. Dynamic Fault Injection into Digital Twins of Safety-Critical Systems. In Proceedings of the 2021 Design, Automation and Test in Europe Conference and Exhibition (DATE), Virtual, 1–5 February 2021; pp. 446–450. [CrossRef]
3. Guo, K.; Wan, X.; Liu, L.; Gao, Z.; Yang, M. Fault Diagnosis of Intelligent Production Line Based on Digital Twin and Improved Random Forest. *Appl. Sci.* **2021**, *11*, 7733. [CrossRef]
4. Kia, S.H.; Razik, H.; Dunai, L. Electrical Machines: Real-time Simulation for Intelligent Fault Diagnosis, Prognostics and Health Management. In Proceedings of the IECON 2024—50th Annual Conference of the IEEE Industrial Electronics Society, Chicago, IL, USA, 3–6 November 2024; pp. 1–6. [CrossRef]
5. Florkowski, M. Digital Twins and Simulations: World of Simulation. In *ABB Report*; ABB: Zurich, Switzerland, 2019; pp. 8–13.
6. Chung, T. Computational Fluid Dynamics. *Computational Fluid Dynamics*, 2nd ed.; McGraw-hill: New York, NY, USA, 2010. [CrossRef]
7. Allaire, G. *Numerical Analysis and Optimization: An Introduction to Mathematical Modelling and Numerical Simulation*; Oxford University Press: Oxford, UK, 2007. [CrossRef]

8. Kaltenbacher, M. *Numerical Simulation of Mechatronic Sensors and Actuators: Finite Elements for Computational Multiphysics*; Springer: Berlin/Heidelberg, Germany, 2015. [CrossRef]
9. Rasheed, A.; San, O.; Kvamsdal, T. Digital Twin: Values, Challenges and Enablers from a Modeling Perspective. *IEEE Access* **2020**, *8*, 21980–22012. [CrossRef]
10. Mojlish, S.; Erdogan, N.; Levine, D.; Davoudi, A. Review of Hardware Platforms for Real-time Simulation of Electric Machines. *IEEE Trans. Transp. Electrif.* **2017**, *3*, 130–146. [CrossRef]
11. Le Besnerais, J.; Fasquelle, A.; Hecquet, M.; Pelle, J.; Lanfranchi, V.; Harmand, S.; Brochet, P.; Randria, A. Multiphysics Modeling: Electro-Vibro-Acoustics and Heat Transfer of PWM-Fed Induction Machines. *IEEE Trans. Ind. Electron.* **2010**, *57*, 1279–1287. [CrossRef]
12. Le Besnerais, J. Vibroacoustic Analysis of Radial and Tangential Air-Gap Magnetic Forces in Permanent Magnet Synchronous Machines. *IEEE Trans. Magn.* **2015**, *51*, 8105609. [CrossRef]
13. Omar Faruque, M.D.; Strasser, T.; Lauss, G.; Jalili-Marandi, V.; Forsyth, P.; Dufour, C.; Dinavahi, V.; Monti, A.; Kotsampopoulos, P.; Martinez, J.A.; et al. Real-time Simulation Technologies for Power Systems Design, Testing, and Analysis. *IEEE Power Energy Technol. Syst. J.* **2015**, *2*, 63–73. [CrossRef]
14. Boglietti, A.; Cavagnino, A.; Lazzari, M.; Pastorelli, M. A Simplified Thermal Model for Variable-speed Self-cooled Industrial Induction Motor. *IEEE Trans. Ind. Appl.* **2003**, *39*, 945–952. [CrossRef]
15. Matar, M.; Iravani, R. Massively Parallel Implementation of AC Machine Models for FPGA-Based Real-time Simulation of Electromagnetic Transients. *IEEE Trans. Power Deliv.* **2011**, *26*, 830–840. [CrossRef]
16. Wang, L.; Jatskevich, J.; Dinavahi, V.; Dommel, H.W.; Martinez, J.A.; Strunz, K.; Rioual, M.; Chang, G.W.; Iravani, R. Methods of Interfacing Rotating Machine Models in Transient Simulation Programs. *IEEE Trans. Power Deliv.* **2010**, *25*, 891–903. [CrossRef]
17. Agah, G.R.; Rahideh, A.; Faradonbeh, V.Z.; Kia, S.H. Stator Winding Interturn Short-Circuit Fault Modeling and Detection of Squirrel-Cage Induction Motors. *IEEE Trans. Transp. Electrif.* **2024**, *10*, 5725–5734. [CrossRef]
18. Hedayati Kia, S. Detection of Stator and Rotor Asymmetries Faults in Wound Rotor Induction Machines: Modeling, Test and Real-time Implementation. In *Emerging Electric Machines*; Zobaa, A.F., Aleem, S.H.A., Eds.; IntechOpen: Rijeka, Croatia, 2021; Chapter 3. [CrossRef]
19. Ametani, A. *Numerical Analysis of Power System Transients and Dynamics*; The Institution of Engineering and Technology: Stevenage, UK, 2015. [CrossRef]
20. Watson, N.; Arrillaga, J. *Power Systems Electromagnetic Transients Simulation*; The Institution of Engineering and Technology: Stevenage, UK, 2003. [CrossRef]
21. Hardware Emulation Building Blocks for Power System Components. In *Real-Time Electromagnetic Transient Simulation of AC–DC Networks*; John Wiley & Sons, Ltd.: Hoboken, NJ, USA, 2021; Chapter 2, pp. 17–77. [CrossRef]
22. Tiegna, H.; Amara, Y.; Barakat, G. Overview of Analytical Models of Permanent Magnet Electrical Machines for Analysis and Design Purposes. *Math. Comput. Simul.* **2013**, *90*, 162–177. [CrossRef]
23. Devillers, E.; Le Besnerais, J.; Lubin, T.; Hecquet, M.; Lecointe, J.P. A Review of Subdomain Modeling Techniques in Electrical Machines: Performances and Applications. In Proceedings of the 2016 XXII International Conference on Electrical Machines (ICEM), Lausanne, Switzerland, 4–7 September 2016; pp. 86–92. [CrossRef]
24. Rotating Machines. In *Real-Time Electromagnetic Transient Simulation of AC–DC Networks*; John Wiley & Sons, Ltd.: Hoboken, NJ, USA, 2021; Chapter 4, pp. 143–191. [CrossRef]
25. Stadtmann, F.; Rasheed, A.; Kvamsdal, T.; Johannessen, K.A.; San, O.; Kölle, K.; Tande, J.O.; Barstad, I.; Benhamou, A.; Brathaug, T.; et al. Digital Twins in Wind Energy: Emerging Technologies and Industry-Informed Future Directions. *IEEE Access* **2023**, *11*, 110762–110795. [CrossRef]
26. San, O.; Rasheed, A.; Kvamsdal, T. Hybrid Analysis and Modeling, Eclecticism, and Multifidelity Computing Toward Digital Twin Revolution. *GAMM-Mitteilungen* **2021**, *44*, e202100007. [CrossRef]
27. Parekh, V.; Flore, D.; Schöps, S. Performance Analysis of Electrical Machines Using a Hybrid Data- and Physics-Driven Model. *IEEE Trans. Energy Convers.* **2023**, *38*, 530–539. [CrossRef]
28. Jardine, A.K.; Lin, D.; Banjevic, D. A review on machinery diagnostics and prognostics implementing condition-based maintenance. *Mech. Syst. Signal Process.* **2006**, *20*, 1483–1510. [CrossRef]
29. Frosini, L. Novel Diagnostic Techniques for Rotating Electrical Machines—A Review. *Energies* **2020**, *13*, 5066. [CrossRef]
30. Kumar, R.R.; Andriollo, M.; Cirrincione, G.; Cirrincione, M.; Tortella, A. A Comprehensive Review of Conventional and Intelligence-Based Approaches for the Fault Diagnosis and Condition Monitoring of Induction Motors. *Energies* **2022**, *15*, 8938. [CrossRef]
31. Xia, M.; Shao, H.; Williams, D.; Lu, S.; Shu, L.; de Silva, C.W. Intelligent Fault Diagnosis of Machinery Using Digital Twin-Assisted Deep Transfer Learning. *Reliab. Eng. Syst. Saf.* **2021**, *215*, 107938. [CrossRef]

32. Wang, S.; Xu, Q.; Zhang, K.; Liu, Y.; Liu, H. Selection refines diagnosis: Mamba for acoustic weak fault diagnosis combining feature mode decomposition and selection. *Adv. Eng. Inform.* **2025**, *66*, 103421. [CrossRef]
33. Li, D.; Kakosimos, P.; Peretti, L. Machine-Learning-Based Condition Monitoring of Power Electronics Modules in Modern Electric Drives. *IEEE Power Electron. Mag.* **2023**, *10*, 58–66. [CrossRef]
34. Nasiri, A.; Rahideh, A.; Agah, G.R.; Kia, S.H. Ball-Bearing Fault Detection of Squirrel-Cage Induction Motors Based on Single-Phase Stator Current Using Wavelet Packet Decomposition and Statistical Features. *IEEE Trans. Energy Convers.* **2025**, *40*, 1529–1537. [CrossRef]
35. Kumar, R.R.; Waisale, L.O.; Tamata, J.L.; Tortella, A.; H. Kia, S.; Andriollo, M. Advanced Fault Detection and Severity Analysis of Broken Rotor Bars in Induction Motors: Comparative Classification and Feature Study Using Dimensionality Reduction Techniques. *Machines* **2024**, *12*, 890. [CrossRef]
36. Ouali, Y.; Hudelot, C.; Tami, M. An Overview of Deep Semi-Supervised Learning. *arXiv* **2020**, arXiv:2006.05278. [CrossRef]
37. Xu, Y.; Sun, Y.; Liu, X.; Zheng, Y. A Digital-Twin-Assisted Fault Diagnosis Using Deep Transfer Learning. *IEEE Access* **2019**, *7*, 19990–19999. [CrossRef]
38. Gu, A.; Dao, T. Mamba: Linear-Time Sequence Modeling with Selective State Spaces. *arXiv* **2024**, arXiv:2312.00752.
39. Terron-Santiago, C.; Martinez-Roman, J.; Puche-Panadero, R.; Sapena-Bano, A. A Review of Techniques Used for Induction Machine Fault Modelling. *Sensors* **2021**, *21*, 4855. [CrossRef]
40. Singh, A.; Grant, B.; DeFour, R.; Sharma, C.; Bahadoorsingh, S. A Review of Induction Motor Fault Modeling. *Electr. Power Syst. Res.* **2016**, *133*, 191–197. [CrossRef]
41. Liu, Z.; Zhang, P.; He, S.; Huang, J. A Review of Modeling and Diagnostic Techniques for Eccentricity Fault in Electric Machines. *Energies* **2021**, *14*, 4296. [CrossRef]
42. Usman, A.; Joshi, B.M.; Rajpurohit, B.S. Review of Fault Modeling Methods for Permanent Magnet Synchronous Motors and Their Comparison. In Proceedings of the 2017 IEEE 11th International Symposium on Diagnostics for Electrical Machines, Power Electronics and Drives (SDEMPED), Tinos, Greece, 29 August–1 September 2017; pp. 141–146. [CrossRef]
43. Bélanger, J.; Venne, P.; Paquin, J.N. The What, Where, and Why of Real-time Simulation. *Planet RT* **2010**, *1*, 37–49.
44. Mihalič, F.; Truntič, M.; Hren, A. Hardware-in-the-Loop Simulations: A Historical Overview of Engineering Challenges. *Electronics* **2022**, *11*, 2462. [CrossRef]
45. Horri, N.; Pietraszko, M. A Tutorial and Review on Flight Control Co-Simulation Using Matlab/Simulink and Flight Simulators. *Automation* **2022**, *3*, 486–510. [CrossRef]
46. Terron-Santiago, C.; Martinez-Roman, J.; Puche-Panadero, R.; Sapena-Bano, A. Low-Computational-Cost Hybrid FEM-Analytical Induction Machine Model for the Diagnosis of Rotor Eccentricity, Based on Sparse Identification Techniques and Trigonometric Interpolation. *Sensors* **2021**, *21*, 6963. [CrossRef]
47. Sapena-Bano, A.; Chinesta, F.; Pineda-Sanchez, M.; Aguado, J.; Borzacchiello, D.; Puche-Panadero, R. Induction Machine Model with Finite Element Accuracy for Condition Monitoring Running in Real Time Using Hardware in the Loop System. *Int. J. Electr. Power Energy Syst.* **2019**, *111*, 315–324. [CrossRef]
48. Yousefzadeh, M.; Hedayati Kia, S.; Hoseintabar Marzebali, M.; Arab Khaburi, D.; Razik, H. Power-Hardware-in-the-Loop for Stator Windings Asymmetry Fault Analysis in Direct-Drive PMSG-Based Wind Turbines. *Energies* **2022**, *15*, 6896. [CrossRef]
49. Liu, L.; Guo, Y.; Lei, G.; Yin, W.; Ba, X.; Zhu, J. Construction Method and Application Prospect of Electrical Machine Digital Twin. In Proceedings of the 2022 25th International Conference on Electrical Machines and Systems (ICEMS), Chiang Mai, Thailand, 29 November–2 December 2022; pp. 1–6. [CrossRef]
50. Hosamo, H.H.; Svennevig, P.R.; Svidt, K.; Han, D.; Nielsen, H.K. A Digital Twin Predictive Maintenance Framework of Air Handling Units Based on Automatic Fault Detection and Diagnostics. *Energy Build.* **2022**, *261*, 111988. [CrossRef]
51. Falekas, G.; Karlis, A. Digital Twin in Electrical Machine Control and Predictive Maintenance: State-of-the-Art and Future Prospects. *Energies* **2021**, *14*, 5933. [CrossRef]
52. Allen, B.D. Digital Twins and Living Models at NASA. In Proceedings of the Digital Twin Summit, Virtual, 3–4 November 2021.
53. Hu, W.; Zhang, T.; Deng, X.; Liu, Z.; Tan, J. Digital Twin: A State-of-the-Art Review of Its Enabling Technologies, Applications and Challenges. *J. Intell. Manuf. Spec. Equip.* **2021**, *2*, 1–34. [CrossRef]
54. Chen, Z.; Choudhury, M.D.; Blincoe, K.; Dhupia, J.S. A Digital Twin Based Framework for Real-Time Machine Condition Monitoring. In Proceedings of the 2023 IEEE 19th International Conference on Automation Science and Engineering (CASE), Auckland, New Zealand, 26–30 August 2023; pp. 1–6. [CrossRef]
55. Tao, F.; Zhang, H.; Liu, A.; Nee, A.Y.C. Digital Twin in Industry: State-of-the-Art. *IEEE Trans. Ind. Inform.* **2019**, *15*, 2405–2415. [CrossRef]
56. Fuller, A.; Fan, Z.; Day, C.; Barlow, C. Digital Twin: Enabling Technologies, Challenges and Open Research. *IEEE Access* **2020**, *8*, 108952–108971. [CrossRef]

57. Liu, M.; Fang, S.; Dong, H.; Xu, C. Review of Digital Twin About Concepts, Technologies, and Industrial Applications. *J. Manuf. Syst.* **2021**, *58*, 346–361. [CrossRef]
58. Trauer, J.; Schweigert-Recksiek, S.; Engel, C.; Spreitzer, K.; Zimmermann, M. What is a Digital Twin?—Definitions and Insights From an Industrial Case Study in Technical Product Developement. *Proc. Des. Soc. Des. Conf.* **2020**, *1*, 757–766. [CrossRef]
59. Vrabič, R.; Erkoyuncu, J.A.; Butala, P.; Roy, R. Digital Twins: Understanding the Added Value of Integrated Models for Through-life Engineering Services. *Procedia Manuf.* **2018**, *16*, 139–146. [CrossRef]
60. Shao, G.; Kibira, D.; Frechette, S., Digital Twins for Advanced Manufacturing: The Standardized Approach. In *Digital Twins, Simulation, and the Metaverse: Driving Efficiency and Effectiveness in the Physical World through Simulation in the Virtual Worlds*; Grieves, M., Hua, E.Y., Eds.; Springer: Cham, Switzerland, 2024; pp. 145–169. [CrossRef]
61. Yang, C.; Shen, W.; Wang, X. The Internet of Things in Manufacturing: Key Issues and Potential Applications. *IEEE Syst. Man, Cybern. Mag.* **2018**, *4*, 6–15. [CrossRef]
62. Crespi, N.; Drobot, A.T.; Minerva, R. The Digital Twin: What and Why? In *The Digital Twin*; Crespi, N., Drobot, A.T., Minerva, R., Eds.; Springer International Publishing: Cham, Switzerland, 2023; pp. 3–20. [CrossRef]
63. Friederich, J.; Francis, D.P.; Lazarova-Molnar, S.; Mohamed, N. A framework for data-driven digital twins of smart manufacturing systems. *Comput. Ind.* **2022**, *136*, 103586. [CrossRef]
64. Alvarez-Gonzalez, F.; Griffo, A.; Sen, B.; Wang, J. Real-time Hardware-in-the-Loop Simulation of Permanent-Magnet Synchronous Motor Drives Under Stator Faults. *IEEE Trans. Ind. Electron.* **2017**, *64*, 6960–6969. [CrossRef]
65. Chen, H.; Zhang, Z.; Karamanakos, P.; Rodriguez, J. Digital Twin Techniques for Power Electronics-Based Energy Conversion Systems: A Survey of Concepts, Application Scenarios, Future Challenges, and Trends. *IEEE Ind. Electron. Mag.* **2023**, *17*, 20–36. [CrossRef]
66. Mendes, G.; Ferreira, A. Extending the Multiphysics Modelling of Electric Machines in a Digital Twin Concept. In Proceedings of the 2021 11th IEEE International Conference on Intelligent Data Acquisition and Advanced Computing Systems: Technology and Applications (IDAACS), Cracow, Poland, 22–25 September 2021; Volume 2, pp. 689–693. [CrossRef]
67. Iranian, M.E.; Mohseni, M.; Aghili, S.; Parizad, A.; Baghaee, H.R.; Guerrero, J.M. Real-time FPGA-Based HIL Emulator of Power Electronics Controllers Using NI PXI for DFIG Studies. *IEEE J. Emerg. Sel. Top. Power Electron.* **2022**, *10*, 2005–2019. [CrossRef]
68. Chen, Y.; Dinavahi, V. Hardware Emulation Building Blocks for Real-time Simulation of Large-Scale Power Grids. *IEEE Trans. Ind. Inform.* **2014**, *10*, 373–381. [CrossRef]
69. Levi, E. Multiphase Electric Machines for Variable-Speed Applications. *IEEE Trans. Ind. Electron.* **2008**, *55*, 1893–1909. [CrossRef]
70. Zhao, Y.; Lipo, T. Modeling and Control of a Multi-phase Induction Machine with Structural Unbalance. *IEEE Trans. Energy Convers.* **1996**, *11*, 570–577. [CrossRef]
71. Toliyat, H.; Lipo, T. Transient Analysis of Cage Induction Machines Under Stator, Rotor Bar and End Ring faults. *IEEE Trans. Energy Convers.* **1995**, *10*, 241–247. [CrossRef]
72. Bouzid, S.; Viarouge, P.; Cros, J. Real-Time Digital Twin of a Wound Rotor Induction Machine Based on Finite Element Method. *Energies* **2020**, *13*, 5413. [CrossRef]
73. Ambrožič, V.; Fišer, R.; Nemec, M.; Drobnič, K. Dynamic Model of Induction Machine with Faulty Rotor in Field Reference Frame. In Proceedings of the 2013 9th IEEE International Symposium on Diagnostics for Electric Machines, Power Electronics and Drives (SDEMPED), Valencia, Spain, 27–30 August 2013; pp. 142–149. [CrossRef]
74. Baccarini, L.M.R.; de Menezes, B.R.; Caminhas, W.M. Fault Induction Dynamic Model, Suitable for Computer Simulation: Simulation Results and Experimental Validation. *Mech. Syst. Signal Process.* **2010**, *24*, 300–311. [CrossRef]
75. Bachir, S.; Tnani, S.; Trigeassou, J.C.; Champenois, G. Diagnosis by Parameter Estimation of Stator and Rotor Faults Occurring in Induction Machines. *IEEE Trans. Ind. Electron.* **2006**, *53*, 963–973. [CrossRef]
76. Bossio, G.; De Angelo, C.; Garcia, G.; Solsona, J.; Valla, M. Effects of Rotor Bar and End-Ring Faults Over the Signals of a Position Estimation Strategy for Induction Motors. *IEEE Trans. Ind. Appl.* **2005**, *41*, 1005–1012. [CrossRef]
77. Nemec, M.; Drobnič, K.; Fišer, R.; Ambrožič, V. Simplified model of induction machine with broken rotor bars. In Proceedings of the 2016 IEEE International Power Electronics and Motion Control Conference (PEMC), Varna, Bulgaria, 25–28 September 2016; pp. 1085–1090. [CrossRef]
78. Magagula, G.S.; Nnachi, A.F.; Akumu, A.O. Broken Rotor Bar Fault Simulation And Analysis In D-q Reference Frame. In Proceedings of the 2020 IEEE PES/IAS PowerAfrica, Virtual, 25–28 August 2020; pp. 1–4. [CrossRef]
79. Jannati, M.; Idris, N.; Salam, Z. A new method for modeling and vector control of unbalanced induction motors. In Proceedings of the 2012 IEEE Energy Conversion Congress and Exposition (ECCE), Raleigh, NC, USA, 15–20 September 2012; pp. 3625–3632. [CrossRef]
80. Tallam, R.; Habetler, T.; Harley, R. Transient Model for Induction Machines with Stator Winding Turn Faults. *IEEE Trans. Ind. Appl.* **2002**, *38*, 632–637. [CrossRef]

81. Yassa, N.; Rachek, M. Modeling and detecting the stator winding inter turn fault of permanent magnet synchronous motors using stator current signature analysis. *Math. Comput. Simul.* **2020**, *167*, 325–339. [CrossRef]
82. Choudhary, A.; Meena, D.C.; Patra, A.K. Asynchronous Motor Modeling in Simulink for Stator and Rotor Fault Analysis. In Proceedings of the 2019 International Conference on Green and Human Information Technology (ICGHIT), Kuala Lumpur, Malaysia, 16–18 January 2019; pp. 82–85. [CrossRef]
83. Guezmil, A.; Berriri, H.; Pusca, R.; Sakly, A.; Romary, R.; Mimouni, M. Detecting Inter-Turn Short-Circuit Fault in Induction Machine Using High-Order Sliding Mode Observer: Simulation and Experimental Verification. *J. Control. Autom. Electr. Syst.* **2017**, *28*, 532–540. [CrossRef]
84. Bindu, S.; Thomas, V.V. Detection of Static Air-Gap Eccentricity in Three-Phase Squirrel Cage Induction Motor Through Stator Current and Vibration Analysis. In *Advances in Power Systems and Energy Management: ETAEERE-2016*; Springer: Singapore, 2018; pp. 511–518. [CrossRef]
85. Zhang, S.; Wang, B.; Kanemaru, M.; Lin, C.; Liu, D.; Miyoshi, M.; Teo, K.H.; Habetler, T.G. Model-Based Analysis and Quantification of Bearing Faults in Induction Machines. *IEEE Trans. Ind. Appl.* **2020**, *56*, 2158–2170. [CrossRef]
86. Ishikawa, T.; Seki, Y.; Kurita, N. Analysis for Fault Detection of Vector-Controlled Permanent Magnet Synchronous Motor With Permanent Magnet Defect. *IEEE Trans. Magn.* **2013**, *49*, 2331–2334. [CrossRef]
87. Roshandel Tavana, N.; Dinavahi, V. A General Framework for FPGA-Based Real-time Emulation of Electrical Machines for HIL Applications. *IEEE Trans. Ind. Electron.* **2015**, *62*, 2041–2053. [CrossRef]
88. Didier, G.; Razik, H.; Rezzoug, A. An Induction Motor Model Including the First Space Harmonics for Broken Rotor Bar Diagnosis. *Eur. Trans. Electr. Power* **2005**, *15*, 229–243. [CrossRef]
89. Tang, J.; Chen, J.; Dong, K.; Yang, Y.; Lv, H.; Liu, Z. Modeling and Evaluation of Stator and Rotor Faults for Induction Motors. *Energies* **2020**, *13*, 133. [CrossRef]
90. Zouzou, S.E.; Ghoggal, A.; Aboubou, A.; Sahraoui, M.; Razik, H. Modeling of Induction Machines with Skewed Rotor Slots Dedicated to Rotor Faults. In Proceedings of the 2005 5th IEEE International Symposium on Diagnostics for Electric Machines, Power Electronics and Drives, Vienna, Austria, 7–9 September 2005; pp. 1–6. [CrossRef]
91. Kaikaa, M.Y.; Hadjami, M.; Khezzar, A. Effects of the Simultaneous Presence of Static Eccentricity and Broken Rotor Bars on the Stator Current of Induction Machine. *IEEE Trans. Ind. Electron.* **2014**, *61*, 2452–2463. [CrossRef]
92. Jung, J.H.; Kwon, B.H. Corrosion Model of a Rotor-Bar-Under-Fault Progress in Induction Motors. *IEEE Trans. Ind. Electron.* **2006**, *53*, 1829–1841. [CrossRef]
93. Bossio, G.R.; De Angelo, C.H.; Pezzani, C.M.; Bossio, J.M.; Garcia, G.O. Evaluation of Harmonic Current Sidebands for Broken Bar Diagnosis in Induction Motors. In Proceedings of the 2009 IEEE International Symposium on Diagnostics for Electric Machines, Power Electronics and Drives, Cargèse, France, 31 August–3 September 2009; pp. 1–6. [CrossRef]
94. Houdouin, G.; Barakat, G.; Dakyo, B.; Destobbeleer, E. A Winding Function Theory Based Global Method for the Simulation of Faulty Induction Machines. In Proceedings of the IEEE International Electric Machines and Drives Conference, IEMDC'03, Chicago, IL, USA, 12–15 May 2003; Volume 1, pp. 297–303. [CrossRef]
95. Ojaghi, M.; Sabouri, M.; Faiz, J. Performance Analysis of Squirrel-Cage Induction Motors Under Broken Rotor Bar and Stator Inter-Turn Fault Conditions Using Analytical Modeling. *IEEE Trans. Magn.* **2018**, *54*, 8203705. [CrossRef]
96. Benninger, M.; Liebschner, M.; Kreischer, C. Automated Parameter Identification for Multiple Coupled Circuit Modeling of Induction Machines. In Proceedings of the 2022 International Conference on Electrical Machines (ICEM), Valencia, Spain, 5–8 September 2022; pp. 1307–1313. [CrossRef]
97. Devanneaux, V.; Dagues, B.; Faucher, J.; Barakat, G. An Accurate Model of Squirrel Cage Induction Machines Under Stator Faults. *Math. Comput. Simul.* **2003**, *63*, 377–391. [CrossRef]
98. Vaseghi, B.; Takorabet, N.; Meibody-Tabar, F. Analytical Circuit-Based Model of PMSM Under Stator Inter-Turn Short-Circuit Fault Validated by Time-Stepping Finite Element Analysis. In Proceedings of the The XIX International Conference on Electrical Machines—ICEM 2010, Rome, Italy, 6–8 September 2010; pp. 1–6. [CrossRef]
99. Ilamparithi, T.; Nandi, S. Comparison of Results for Eccentric Cage Induction Motor Using Finite Element Method and Modified Winding Function Approach. In Proceedings of the 2010 Joint International Conference on Power Electronics, Drives and Energy Systems & 2010 Power India, Delhi, India, 20–23 December 2010; pp. 1–7. [CrossRef]
100. Faiz, J.; Ojaghi, M. Unified Winding Function Approach for Dynamic Simulation of Different Kinds of Eccentricity Faults in Cage Induction Machines. *IET Electr. Power Appl.* **2009**, *3*, 461–470. [CrossRef]
101. Joksimovic, G.; Durovic, M.; Penman, J.; Arthur, N. Dynamic Simulation of Dynamic Eccentricity in Induction Machines-Winding Function Approach. *IEEE Trans. Energy Convers.* **2000**, *15*, 143–148. [CrossRef]
102. Pal, R.S.C.; Mohanty, A.R. A Simplified Dynamical Model of Mixed Eccentricity Fault in a Three-Phase Induction Motor. *IEEE Trans. Ind. Electron.* **2021**, *68*, 4341–4350. [CrossRef]

103. Wang, C.; Wang, M.; Yang, B.; Song, K. A Model-Based Method for Bearing Fault Detection Using Motor Current. *J. Physics: Conf. Ser.* **2020**, *1650*, 032130. [CrossRef]
104. Ojaghi, M.; Yazdandoost, N. Winding Function Approach to Simulate Induction Motors Under Sleeve Bearing Fault. In Proceedings of the 2014 IEEE International Conference on Industrial Technology (ICIT), Busan, Republic of Korea, 26 February–1 March 2014; pp. 158–163. [CrossRef]
105. Ojaghi, M.; Sabouri, M.; Faiz, J. Analytic Model for Induction Motors Under Localized Bearing Faults. *IEEE Trans. Energy Convers.* **2018**, *33*, 617–626. [CrossRef]
106. Ajily, E.; Ardebili, M.; Abbaszadeh, K. Magnet Defect and Rotor Eccentricity Modeling in Axial-Flux Permanent-Magnet Machines via 3-D Field Reconstruction Method. *IEEE Trans. Energy Convers.* **2016**, *31*, 486–495. [CrossRef]
107. Strangas, E.; Clerc, G.; Razik, H.; Soualhi, A. *Fault Diagnosis, Prognosis, and Reliability for Electrical Machines and Drives*; IEEE Press Series; John Wiley & Sons: Hoboken, NJ, USA, 2021.
108. Pereira, L.A.; Scharlau, C.C.; Pereira, L.F.A.; Haffner, J.F. General Model of a Five-Phase Induction Machine Allowing for Harmonics in the Air Gap Field. *IEEE Trans. Energy Convers.* **2006**, *21*, 891–899. [CrossRef]
109. Sapena-Bano, A.; Martinez-Roman, J.; Puche-Panadero, R.; Pineda-Sanchez, M.; Perez-Cruz, J.; Riera-Guasp, M. Induction Machine Model with Space Harmonics for Fault Diagnosis Based on the Convolution Theorem. *Int. J. Electr. Power Energy Syst.* **2018**, *100*, 463–481. [CrossRef]
110. Hu, R.; Wang, J.; Mills, A.R.; Chong, E.; Sun, Z. Current-Residual-Based Stator Interturn Fault Detection in Permanent Magnet Machines. *IEEE Trans. Ind. Electron.* **2021**, *68*, 59–69. [CrossRef]
111. Jeong, H.; Moon, S.; Kim, S.W. An Early Stage Interturn Fault Diagnosis of PMSMs by Using Negative-Sequence Components. *IEEE Trans. Ind. Electron.* **2017**, *64*, 5701–5708. [CrossRef]
112. Tavana, N.R.; Dinavahi, V. Real-time Nonlinear Magnetic Equivalent Circuit Model of Induction Machine on FPGA for Hardware-in-the-Loop Simulation. *IEEE Trans. Energy Convers.* **2016**, *31*, 520–530. [CrossRef]
113. Asghari, B.; Dinavahi, V. Experimental Validation of a Geometrical Nonlinear Permeance Network Based Real-time Induction Machine Model. *IEEE Trans. Ind. Electron.* **2012**, *59*, 4049–4062. [CrossRef]
114. Tavana, N.R.; Dinavahi, V. Real-time FPGA-Based Analytical Space Harmonic Model of Permanent Magnet Machines for Hardware-in-the-Loop Simulation. *IEEE Trans. Magn.* **2015**, *51*, 8106609. [CrossRef]
115. Raissi, M.; Perdikaris, P.; Karniadakis, G.E. Physics Informed Deep Learning (Part I): Data-driven Solutions of Nonlinear Partial Differential Equations. *arXiv* **2017**, arXiv:1711.10561. [CrossRef]
116. Farea, A.; Yli-Harja, O.; Emmert-Streib, F. Understanding Physics-Informed Neural Networks: Techniques, Applications, Trends, and Challenges. *AI* **2024**, *5*, 1534–1557. [CrossRef]
117. DENG, W.; NGUYEN, K.T.; MEDJAHER, K.; GOGU, C.; MORIO, J. Physics-informed machine learning in prognostics and health management: State of the art and challenges. *Appl. Math. Model.* **2023**, *124*, 325–352. [CrossRef]
118. Son, S.; Jeong, J.; Jeong, D.; ho Sun, K.; Oh, K.Y. Physics-Informed Neural Network: Principles and Applications. In *Recent Advances in Neuromorphic Computing*; Bai, K.J., Yi, Y., Eds.; IntechOpen: Rijeka, Croatia, 2024; Chapter 4. [CrossRef]
119. Smyl, D.; Tallman, T.N.; Homa, L.; Flournoy, C.; Hamilton, S.J.; Wertz, J. Physics Informed Neural Networks for Electrical Impedance Tomography. *Neural Netw.* **2025**, *188*, 107410. [CrossRef]
120. Misyris, G.S.; Venzke, A.; Chatzivasileiadis, S. Physics-Informed Neural Networks for Power Systems. *arXiv* **2020**, arXiv:1911.03737. [CrossRef]
121. Ventura Nadal, I.; Stiasny, J.; Chatzivasileiadis, S. Physics-Informed Neural Networks: A plug and play integration into power system dynamic simulations. *Electr. Power Syst. Res.* **2025**, *248*, 111885. [CrossRef]
122. Zerrougui, I.; Li, Z.; Hissel, D. Physics-Informed Neural Network for modeling and predicting temperature fluctuations in proton exchange membrane electrolysis. *Energy AI* **2025**, *20*, 100474. [CrossRef]
123. Iliadis, P.; Petridis, S.; Skembris, A.; Rakopoulos, D.; Kosmatopoulos, E. Physics-Informed Neural Networks for Enhanced State Estimation in Unbalanced Distribution Power Systems. *Appl. Sci.* **2025**, *15*, 7507. [CrossRef]
124. Zhang, Y.; Zhang, S.; Dinavahi, V. Physics-Informed Machine Learning Modeling and Inferencer-in-The-Loop Based Real-Time Digital-Twin Emulation for a High-Speed Maglev Transportation System. 2025. Available online: https://ssrn.com/abstract=5389685 (accessed on 1 August 2025).
125. Zhang, S.; Dinavahi, V.; Liang, T. Towards hydrogen-powered electric aircraft: Physics-informed machine learning based multi-domain modeling and real-time digital twin emulation on FPGA. *Energy* **2025**, *322*, 135451. [CrossRef]
126. Qiao, Z.; Wang, D.; Ni, Y.; Song, K.; Li, Y.; Wang, S. A partitioned modeling approach using a physics-informed neural network for PMSM. *Eng. Anal. Bound. Elem.* **2025**, *179*, 106379. [CrossRef]
127. Son, S.; Lee, H.; Jeong, D.; Oh, K.Y.; Ho Sun, K. A novel physics-informed neural network for modeling electromagnetism of a permanent magnet synchronous motor. *Adv. Eng. Inform.* **2023**, *57*, 102035. [CrossRef]

128. Wu, H.; Niu, S.; Zhang, Y.; Fu, W. Physics-Informed Generative Adversarial Network-Based Modeling and Simulation of Linear Electric Machines. *Appl. Sci.* **2022**, *12*, 10426. [CrossRef]
129. Sizov, G.Y.; Yeh, C.C.; Demerdash, N.A.O. Magnetic Equivalent Circuit Modeling of Induction Machines Under Stator and Rotor Fault Conditions. In Proceedings of the 2009 IEEE International Electric Machines and Drives Conference, Miami, FL, USA, 3–6 May 2009; pp. 119–124. [CrossRef]
130. Naderi, P. Modified Magnetic-Equivalent-Circuit Approach for Various Faults Studying in Saturable Double-Cage-Induction Machines. *IET Electr. Power Appl.* **2017**, *11*, 1224–1234. [CrossRef]
131. Hemeida, A.; Billah, M.M.; Kudelina, K.; Asad, B.; Naseer, M.U.; Guo, B.; Martin, F.; Rasilo, P.; Belahcen, A. Magnetic Equivalent Circuit and Lagrange Interpolation Function Modeling of Induction Machines Under Broken Bar Faults. *IEEE Trans. Magn.* **2024**, *60*, 8200704. [CrossRef]
132. Naderi, P.; Shiri, A. Rotor/Stator Inter-Turn Short Circuit Fault Detection for Saturable Wound-Rotor Induction Machine by Modified Magnetic Equivalent Circuit Approach. *IEEE Trans. Magn.* **2017**, *53*, 8107013. [CrossRef]
133. Faiz, J.; Moosavi, S.M.; Abadi, M.B.; Cruz, S.M. Magnetic Equivalent Circuit Modelling of Doubly-Fed Induction Generator with Assessment of Rotor Inter-Turn Short-Circuit Fault Indices. *IET Renew. Power Gener.* **2016**, *10*, 1431–1440. [CrossRef]
134. Roshanfekr, R.; Jalilian, A. An Approach to Discriminate Between Types of Rotor and Stator Winding Faults in Wound Rotor Induction Machines. In Proceedings of the Electrical Engineering (ICEE), Iranian Conference on, Mashhad, Iran, 8–10 May 2018; pp. 1067–1070. [CrossRef]
135. Faiz, J.; Ghasemi-Bijan, M.; Ebrahimi, B.M. Modeling and Diagnosing Eccentricity Fault Using Three-dimensional Magnetic Equivalent Circuit Model of Three-phase Squirrel-cage Induction Motor. *Electr. Power Components Syst.* **2015**, *43*, 1246–1256. [CrossRef]
136. Han, Q.; Ding, Z.; Xu, X.; Wang, T.; Chu, F. Stator Current Model for Detecting Rolling Bearing Faults in Induction Motors Using Magnetic Equivalent Circuits. *Mech. Syst. Signal Process.* **2019**, *131*, 554–575. [CrossRef]
137. Gong, Z.; Desenfans, P.; Pissoort, D.; Hallez, H.; Vanoost, D. Multiphysics Coupling Model to Characterise the Behaviour of Induction Motors With Eccentricity and Bearing Faults. *IEEE Trans. Energy Convers.* **2024**, *39*, 146–159. [CrossRef]
138. Faiz, J.; Mazaheri-Tehrani, E. Demagnetization Modeling and Fault Diagnosing Techniques in Permanent Magnet Machines Under Stationary and Nonstationary Conditions: An Overview. *IEEE Trans. Ind. Appl.* **2017**, *53*, 2772–2785. [CrossRef]
139. Raminosoa, T.; Farooq, J.; Djerdir, A.; Miraoui, A. Reluctance Network Modelling of Surface Permanent Magnet Motor Considering Iron Nonlinearities. *Energy Convers. Manag.* **2009**, *50*, 1356–1361. [CrossRef]
140. Abbaszadeh, K.; Saied, S.; Hemmati, S.; Tenconi, A. Inverse Transform Method for Magnet Defect Diagnosis in Permanent Magnet Machines. *IET Electr. Power Appl.* **2014**, *8*, 98–107. [CrossRef]
141. Farooq, J.; Srairi, S.; Djerdir, A.; Miraoui, A. Use of Permeance Network Mmethod in the Demagnetization Phenomenon Modeling in a Permanent Magnet Motor. *IEEE Trans. Magn.* **2006**, *42*, 1295–1298. [CrossRef]
142. Momma, D.; Yoshida, Y.; Tajima, K. Demagnetization Analysis of Ferrite Magnet Motor Based on Reluctance Network Analysis. In Proceedings of the 2016 19th International Conference on Electrical Machines and Systems (ICEMS), Chiba, Japan, 13–16 November 2016; pp. 1–4.
143. Guo, L.; Xia, C.; Wang, H.; Wang, Z.; Shi, T. Improved Equivalent Magnetic Network Modeling for Analyzing Working Points of PMs in Interior Permanent Magnet Machine. *J. Magn. Magn. Mater.* **2018**, *454*, 39–50. [CrossRef]
144. Mahmouditabar, F.; Vahedi, A.; Ojaghlu, P.; Takorabet, N. Irreversible Demagnetization Analysis of RWAFPM Motor Using Modified MEC Algorithm. *COMPEL Int. J. Comput. Math. Electr. Electron. Eng.* **2020**, *39*, 1227–1239. [CrossRef]
145. Paja, C.A.R.; Romero, A.; Giral, R. Evaluation of Fixed-Step Differential Equations Solution Methods for Fuel Cell Real-Time Simulation. In Proceedings of the 2007 International Conference on Clean Electrical Power, Capri, Italy, 21–23 May 2007; pp. 480–487. [CrossRef]
146. Grégoire, L.A.; Blanchette, H.F.; Bélanger, J.; Al-Haddad, K. Real-Time Simulation-Based Multisolver Decoupling Technique for Complex Power-Electronics Circuits. *IEEE Trans. Power Deliv.* **2016**, *31*, 2313–2321. [CrossRef]
147. Bouscayrol, A. Different Types of Hardware-In-the-Loop simulation for Electric Drives. In Proceedings of the 2008 IEEE International Symposium on Industrial Electronics, Cambridge, UK, 30 June–2 July 2008; pp. 2146–2151. [CrossRef]
148. Pak, L.F.; Dinavahi, V. Real-Time Simulation of a Wind Energy System Based on the Doubly-Fed Induction Generator. *IEEE Trans. Power Syst.* **2009**, *24*, 1301–1309. [CrossRef]
149. Champagne, R.; Dessaint, L.A.; Fortin-Blanchette, H.; Sybille, G. Analysis and Validation of a Real-Time AC Drive Simulator. *IEEE Trans. Power Electron.* **2004**, *19*, 336–345. [CrossRef]
150. Dufour, C. A Real-Time Simulator for Doubly Fed Induction Generator based Wind Turbine Applications. In Proceedings of the 2004 IEEE 35th Annual Power Electronics Specialists Conference (IEEE Cat. No.04CH37551), Aachen, Germany, 20–25 June 2004; Technical Report.

151. Abourida, S.; Dufour, C.; Belanger, J.; Yamada, T.; Arasawa, T. Hardware-In-the-Loop Simulation of Finite-Element Based Motor Drives with RT-LAB and JMAG. In Proceedings of the 2006 IEEE International Symposium on Industrial Electronics, Montreal, QC, Canada, 9–13 July 2006; Volume 3, pp. 2462–2466. [CrossRef]
152. Dufour, C.; Abourida, S.; Belanger, J. Hardware-In-the-Loop Simulation of Power Drives with RT-LAB. In Proceedings of the 2005 International Conference on Power Electronics and Drives Systems, Kuala Lumpur, Malaysia, 28 November–1 December 2005; Volume 2, pp. 1646–1651. [CrossRef]
153. Dufour, C.; Cense, S.; Jalili-Marandi, V.; Bélanger, J. Review of State-of-the-Art Solver Solutions for HIL Simulation of Power Systems, Power Electronic and Motor Drives. In Proceedings of the 2013 15th European Conference on Power Electronics and Applications (EPE), Lille, France, 2–6 September 2013; pp. 1–12. [CrossRef]
154. Paquin, J.N.; Li, W.; Belanger, J.; Schoen, L.; Peres, I.; Olariu, C.; Kohmann, H. A Modern and Open Real-Time Digital Simulator of All-Electric Ships with a Multi-Platform Co-Simulation Approach. In Proceedings of the 2009 IEEE Electric Ship Technologies Symposium, Baltimore, MD, USA, 20–22 April 2009; pp. 28–35. [CrossRef]
155. Faruque, M.O.; Dinavahi, V. An Advanced PC-Cluster Based Real-Time Simulator for Power Electronics and Drives. In Proceedings of the 2006 IEEE International Symposium on Industrial Electronics, Montreal, QC, Canada, 9–13 July 2006; Volume 3, pp. 2579–2584. [CrossRef]
156. Pak, L.F.; Faruque, M.O.; Nie, X.; Dinavahi, V. A Versatile Cluster-Based Real-Time Digital Simulator for Power Engineering Research. *IEEE Trans. Power Syst.* **2006**, *21*, 455–465. [CrossRef]
157. Larose, C.; Guerette, S.; Guay, F.; Nolet, A.; Yamamoto, T.; Enomoto, H.; Kono, Y.; Hasegawa, Y.; Taoka, H. A fully digital real-time power system simulator based on PC-cluster. *Math. Comput. Simul.* **2003**, *63*, 151–159. [CrossRef]
158. Chen, H.; Sun, S.; Aliprantis, D.C.; Zambreno, J. Dynamic Simulation of Electric Machines on FPGA Boards. In Proceedings of the 2009 IEEE International Electric Machines and Drives Conference, Miami, FL, USA, 3–6 May 2009; pp. 1523–1528. [CrossRef]
159. Ponce, P.; Ibarra, L.; Molina, A.; MacCleery, B. Real Time Simulation for DC and AC Motors Based on Lab VIEW FPGAs. In Proceedings of the IFAC Proceedings Volumes (IFAC-PapersOnline). IFAC Secretariat, Bucharest, Romania, 23–25 May 2012; Volume 14, pp. 1777–1784. [CrossRef]
160. Jandaghi, B.; Dinavahi, V. Hardware-in-the-Loop Emulation of Linear Induction Motor Drive for MagLev Application. *IEEE Trans. Plasma Sci.* **2016**, *44*, 679–686. [CrossRef]
161. Schmitt, A.; Richter, J.; Jurkewitz, U.; Braun, M. FPGA-based Real-Time Simulation of Nonlinear Permanent Magnet Synchronous Machines for Power Hardware-in-the-Loop Emulation Systems. In Proceedings of the IECON 2014—40th Annual Conference of the IEEE Industrial Electronics Society, Dallas, TX, USA, 29 October–1 November 2014; pp. 3763–3769. [CrossRef]
162. Dufour, C.; Belanger, J.; Abourida, S.; Lapointe, V. FPGA-Based Real-Time Simulation of Finite-Element Analysis Permanent Magnet Synchronous Machine Drives. In Proceedings of the 2007 IEEE Power Electronics Specialists Conference, Orlando, FL, USA, 17–21 June 2007; pp. 909–915. [CrossRef]
163. Parma, G.G.; Dinavahi, V. Real-Time Digital Hardware Simulation of Power Electronics and Drives. *IEEE Trans. Power Deliv.* **2007**, *22*, 1235–1246. [CrossRef]
164. Kredo, K.; Zenor, J.; Bednar, R.; Crosbie, R. FPGA-Accelerated Simulink Simulations of Electrical Machines. In Proceedings of the 2015 IEEE Electric Ship Technologies Symposium (ESTS), Old Town Alexandria, VA, USA, 21–24 June 2015; pp. 74–79. [CrossRef]
165. Rajne, P.A.; Ramanarayanan, V. Programming an FPGA to Emulate the Dynamics of DC Machines. In Proceedings of the 2006 India International Conference on Power Electronics, Chennai, India, 19–21 December 2006; pp. 120–124. [CrossRef]
166. Herrera, L.; Li, C.; Yao, X.; Wang, J. FPGA-Based Detailed Real-Time Simulation of Power Converters and Electric Machines for EV HIL Applications. *IEEE Trans. Ind. Appl.* **2015**, *51*, 1702–1712. [CrossRef]
167. O'Connell, T.C.; Krein, P.T.; Yilmaz, M.; Friedl, A. On the Feasibility of Using Large-Scale Numerical Electric Machine Field Analysis Software in Complex Electric Drive System Design Tools. In Proceedings of the 2008 11th Workshop on Control and Modeling for Power Electronics, Zurich, Switzerland, 17–20 August 2008; pp. 1–8. [CrossRef]
168. Adzima, A.J.; Krein, P.T.; O'Connell, T.C. Investigation of Accelerating Numerical-Field Analysis Methods for Electric Machines with the Incorporation of Graphic-Processor Based Parallel Processing Techniques. In Proceedings of the 2009 IEEE Electric Ship Technologies Symposium, Baltimore, MD, USA, 20–22 April 2009; pp. 59–64. [CrossRef]
169. Rodrigues, A.W.O.; Guyomarc'h, F.; Dekeyser, J.L.; Le Menach, Y. Automatic Multi-GPU Code Generation Applied to Simulation of Electrical Machines. *IEEE Trans. Magn.* **2012**, *48*, 831–834. [CrossRef]
170. Rodrigues, A.W.D.O.; Chevallier, L.; Menach, Y.L.; Guyomarch, F. Test harness on a preconditioned conjugate gradient solver on GPUs: An efficiency analysis. *IEEE Trans. Magn.* **2013**, *49*, 1729–1732. [CrossRef]
171. Liu, Z.H.; Li, X.H.; Wu, L.H.; Zhou, S.W.; Liu, K. GPU-Accelerated Parallel Coevolutionary Algorithm for Parameters Identification and Temperature Monitoring in Permanent Magnet Synchronous Machines. *IEEE Trans. Ind. Inform.* **2015**, *11*, 1220–1230. [CrossRef]

172. Bracikowski, N.; Hecquet, M.; Brochet, P.; Shirinskii, S.V. Multiphysics Modeling of a Permanent Magnet Synchronous Machine by Using Lumped Models. *IEEE Trans. Ind. Electron.* **2012**, *59*, 2426–2437. [CrossRef]
173. Fasquelle, A. Coupled Electromagnetic, Acoustic and Thermal-Flow Modelling of an Induction Motor of Railway Traction. Ph.D. Thesis, Ecole Centrale de Lille, Villeneuve-d'Ascq, France, 2007.
174. Tao, F.; Zhang, M.; Liu, Y.; Nee, A. Digital Twin Driven Prognostics and Health Management for Complex Equipment. *CIRP Ann.* **2018**, *67*, 169–172. [CrossRef]
175. Lagaris, I.; Likas, A.; Fotiadis, D. Artificial Neural Networks for Solving Ordinary and Partial Differential Equations. *IEEE Trans. Neural Netw.* **1998**, *9*, 987–1000. [CrossRef]
176. Wu, D.; Chamoin, L.; Lisser, A. Solving Large-scale Variational Inequalities with Dynamically Adjusting Initial Condition in Physics-informed Neural Networks. *Comput. Methods Appl. Mech. Eng.* **2024**, *429*, 117156. [CrossRef]

energies

MDPI

Article

Design and Analysis of an IE6 Hyper-Efficiency Permanent Magnet Synchronous Motor for Electric Vehicle Applications

Hayatullah Nory [1,2,*], Ahmet Yildiz [3], Serhat Aksun [1] and Cansu Aksoy [1]

1 Volt Weg Group, Izmir 35735, Turkey; serhat.aksun@volt.weg.net (S.A.); cansu.aksoy@volt.weg.net (C.A.)
2 Department of Electrical and Electronics Engineering, Faculty of Engineering, Firat University, Elazig 23119, Turkey
3 Department of Mechatronics Engineering, Faculty of Engineering, Firat University, Elazig 23119, Turkey; ayildiz@firat.edu.tr
* Correspondence: hayatullah.nory@volt.weg.net

Abstract

In this study, a high-efficiency permanent magnet synchronous motor (PMSM) was designed for a geared electric vehicle. The motor was developed for use in an L-category electric vehicle with four wheels and a two-passenger capacity. During the design process, application-specific dimensional constraints, electromagnetic requirements, and material limitations were taken into consideration. A spoke-type rotor structure was adopted to achieve both mechanical robustness and high efficiency with minimized leakage flux. In addition, the combination of a 12-stator slot and a 10-rotor pole was selected to suppress low-order harmonic components and improve torque smoothness. The motor model was analyzed using Siemens Simcenter SPEED software (Product Version 2020.3.1), and an efficiency above 94% was achieved, meeting the IE6 efficiency class. Magnetic flux analysis results showed that the selected core material operated within the magnetic saturation limits. The findings demonstrate that a compact and high-efficiency PMSM design is feasible for electric vehicle applications.

Keywords: electric vehicle; hyper efficiency; IE6 efficiency class; permanent magnet synchronous motor; spoke rotor structure

1. Introduction

The rapid development of electric vehicle (EV) technologies has been driven by the urgent need to reduce greenhouse gas emissions and improve energy efficiency in the transportation sector. This transition is further accelerated by regulations that are increasingly geared towards environmental factors, such as the European Union's CO_2 emission targets and the global adoption of higher efficiency standards for electric drive systems [1–4]. As the electrification of transportation continues to expand, the traction motor has emerged as a critical component influencing vehicle performance, energy consumption, and operational range. Consequently, the design and optimization of traction motors have become a central research focus in the industry.

Electric traction motors for EV applications must satisfy multiple and often conflicting design requirements, including high efficiency across a wide operating range, a compact size to fit within constrained chassis geometries, and sufficient thermal robustness to ensure reliability under continuous load [5]. Permanent magnet synchronous motors

(PMSMs) have become the preferred choice for many EV manufacturers due to their high efficiency, high torque density, and wide constant power speed range compared to induction machines or switched reluctance motors [6–9]. Within this category, interior permanent magnet synchronous motors (IPMSMs) offer further advantages through rotor configurations such as spoke-type or V-shaped topologies, which enhance flux focusing, improve torque output, and extend flux-weakening capability [10].

Over the past decade, numerous studies have investigated PMSM and IPMSM designs for traction applications, focusing on aspects such as rotor topology optimization [11–13], fractional-slot concentrated windings [14–16], and advanced cooling strategies [17–19]. Spoke-type PMSM configurations, in particular, have been shown to reduce torque ripple and suppress harmonics [20], while fractional-slot concentrated windings can decrease copper loss through shorter end-turns and improve the slot fill factor [21]. Moreover, the integration of finite element analysis (FEA) with analytical modeling has enabled multi-objective optimization simultaneously targeting efficiency, torque density, and manufacturability. Recent advances in computational tools have also enabled more accurate prediction of iron losses, magnet demagnetization risk, and thermal behavior, further extending the applicability of simulation-based design approaches. Despite these advancements, several technical gaps remain unaddressed.

High-efficiency PMSMs have been successfully developed for industrial applications, but these machines typically employ larger frame sizes or active cooling systems that are impractical for compact EV applications. PMSM designs for EV applications often aim to achieve high efficiency while also meeting strict constraints on size, weight, and integration within limited space. Various PMSM designs targeting EV applications place primary emphasis on specific technical features, such as harmonic mitigation or flux-weakening capability [22], while offering limited benchmarking of overall efficiency relative to other high-performance machines. However, adapting such designs to the compact packaging requirements of EVs often demands a re-evaluation of rotor geometry, cooling methods, and material selection, which introduces additional multidisciplinary challenges. This underscores the importance of broader performance comparisons to assess the practical applicability of such designs in real-world EV scenarios.

In addition to electromagnetic performance, mechanical robustness and manufacturability are also key considerations in the design of PMSMs for EV applications. Spatial constraints in specific vehicles, such as L-category urban EVs, impose further design challenges. Motors for these applications must deliver high torque density within a limited envelope while ensuring manufacturability and cost-effectiveness. Reports in the literature describing PMSM designs that meet such spatial limitations while simultaneously achieving high efficiency remain limited. Existing industrial high-efficiency machines often cannot be directly adapted due to their larger frame sizes, active cooling requirements, or incompatible mounting configurations. This combination of requirements (compactness, high efficiency, and the potential for passive cooling) thus defines a clear gap in the current state of the art. In particular, there is a lack of studies that explicitly address the simultaneous optimization of slot/pole number selection, spoke-type rotor topology, and winding design under the dimensional and operational constraints of L-category EVs. While achieving high efficiency and torque density is a general requirement across traction motor designs, the present study addresses this gap by combining these requirements with the strict packaging limitations and manufacturability constraints of L-category EVs.

The present study focuses on the design and analysis of a 12-slot/10-pole spoke-type PMSM optimized for electric vehicle applications with geared transmission systems. The primary objectives are to achieve high efficiency consistent with IE6 efficiency classification targets, maintain a compact geometry suitable for limited installation spaces, and ensure smooth torque output through appropriate rotor topology selection. The design process combines analytical pre-design calculations with detailed finite element method (FEM) simulations to refine the electromagnetic parameters, evaluate torque and efficiency characteristics, and estimate the associated loss components. The proposed approach aims to provide a balanced solution that addresses both performance and manufacturability requirements, offering practical insights for the development of high-efficiency traction motors in next-generation electric vehicles.

In summary, the key contributions of this work are

(1) The design of a high-efficiency spoke-type PMSM optimized for the dimensional constraints and operational requirements of L-category electric vehicles;
(2) Demonstrating high torque density and low torque ripple within restricted installation spaces;
(3) Providing a simulation-based design framework for evaluating electromagnetic performance, torque production, and loss distribution in compact traction motor applications.

The remainder of this paper is organized as follows. Section 2 presents the design requirements and methodology. Section 3 discusses the simulation results and performance evaluation, while Section 4 summarizes the main conclusions and outlines directions for future work.

2. Materials and Methods

In this study, a high-efficiency permanent magnet synchronous motor was designed for an electric vehicle with a geared transmission system. Within the design process, both electromagnetic and mechanical criteria were considered to size the motor, select appropriate materials, and analyze its conformity to performance targets and operating conditions. The design criteria and characteristics of the proposed motor are presented in this section.

2.1. Design Requirements

The PMSM is designed specifically for the traction requirements of L-category electric vehicles, which are compact urban mobility solutions typically intended for two passengers. These vehicles are characterized by their lightweight structure, limited installation space, and modest speed range, making them well-suited for short-range city commuting. In this context, the motor must deliver high torque at low speeds to ensure sufficient acceleration in urban driving conditions, while simultaneously maintaining high efficiency to maximize vehicle range within the limited battery capacity.

The proposed machine is designed to operate with a geared transmission system, enabling an effective balance between torque generation and vehicle speed. Through the integration with a single-speed gearbox, the motor's shaft power is efficiently transferred to the wheels, ensuring the high starting torque required in stop-and-go traffic, while also providing smooth operation at the rated top speed of the vehicle. This configuration enhances driving comfort, reduces energy losses, and ensures compatibility with the overall dynamics of compact EV drivetrains.

Moreover, the design requirements place particular emphasis on compactness, manufacturability, and cost-effectiveness. Given the limited chassis space of L-category EVs, the motor geometry must be carefully optimized to achieve high torque density within a restricted envelope. At the same time, thermal robustness is essential, since the compact design reduces the possibility of employing large active cooling systems. Therefore, the proposed design aims to ensure reliable operation under natural or simplified forced-air cooling, consistent with the practical constraints of lightweight urban vehicles.

Table 1 summarizes the key technical specifications of the selected reference vehicle, which guided the definition of motor power, torque, and operational parameters in the design process.

Table 1. Technical specifications of the electric vehicle.

Parameter	Value	Unit
Motor power	2.5	kW
Nominal motor torque	3.5	Nm
Battery capacity	5.5	kWh
Battery technology	Lithium-Ion	
Charging time (0–100%)	4	h
Range	75	km
Top speed	45	km/h
Gearbox type	Single-speed automatic transmission	

2.2. Design of the PMSM

In the motor design process, the outer diameter of the stator was determined to be 144 mm, considering the available mounting space within the vehicle. A 12-stator-slot and 10-rotor-pole configuration was selected to achieve high efficiency, compact structure, and low harmonics. The selection of a 12-slot/10-pole spoke-type PMSM topology was motivated by its favorable electromagnetic performance characteristics, including high torque density, wide constant-power speed range, and the inherent potential for low torque ripple. While this combination has been reported in previous studies, its direct adaptation to the spatial and operational constraints of L-category electric vehicles presents unique engineering challenges. The compact chassis configuration of such vehicles imposes strict limitations on the motor's axial length and overall envelope, while also restricting the integration of active cooling systems. To address these constraints, the rotor geometry, stator slot design, and winding configuration were tailored to achieve the required torque output and efficiency targets within the restricted installation space. The spoke-type rotor layout was specifically optimized to suppress low-order harmonics, ensuring smooth torque delivery without the need for additional structural complexity. Analytical pre-design calculations provided the initial dimensions and electromagnetic loading estimates, which were then refined through high-fidelity finite element analysis to evaluate the effects of magnet placement, flux concentration, and tooth geometry on torque ripple, efficiency, and manufacturability. Special attention was given to the winding end-turn length to improve the slot fill factor and minimize copper losses, which is critical in compact designs where space efficiency directly impacts performance. This engineering adaptation demonstrates that the 12-slot/10-pole spoke-type topology, when systematically optimized for compact EV constraints, can simultaneously satisfy high efficiency, favorable torque per volume ratio, and manufacturability requirements without active cooling. The approach bridges the gap between theoretically optimal electromagnetic designs and practical implementation in space-limited traction applications.

The electromagnetic and geometric considerations, the material selection for both the stator/rotor cores and the permanent magnets is a decisive factor in achieving the targeted performance objectives. The materials employed in the designed PMSM are summarized in Table 2. The magnetic cores of both the stator and rotor are composed of M250-50A non-oriented electrical steel laminations. This grade was selected due to its favorable compromise between magnetic permeability and specific core losses under the design operating frequency. Its relatively low hysteresis and eddy current losses contributed to the efficiency requirements of the machine while maintaining manufacturability and cost-effectiveness. Furthermore, the use of M250-50A allowed sufficient magnetic flux conduction capacity without pushing the laminations into premature saturation, which is a crucial concern for compact traction motors where high flux densities are concentrated in a confined volume. For the rotor permanent magnets, NdFeB grade N30SH was selected. NdFeB alloys are widely recognized for their superior remanence and coercivity compared to ferrite or SmCo alternatives, making them particularly suitable for high-torque-density traction motors. The N30SH grade, with a remanence flux density (B_r) on the order of 1.05–1.10 T, combines strong magnetic properties with enhanced thermal stability. The "SH" classification corresponds to a maximum operating temperature of approximately 150 °C, which significantly reduces the risk of irreversible demagnetization during continuous traction operation in compact electric vehicle environments. This high-temperature capability was considered essential for the intended geared EV application, as forced cooling options are limited in the restricted chassis layout. The spoke-type rotor was designed with rectangular NdFeB permanent magnets having dimensions of 27 mm (axial length) × 16 mm (width) × 3.8 mm (height/thickness) for each insert. This corresponds to a per magnet volume of approximately 1641.6 mm^3 (≈1.64 cm^3). The relatively small thickness of 3.8 mm was selected to support flux focusing and reduce leakage across the rotor bridges, while the 16 mm width ensured adequate radial flux interaction with the stator teeth. The length of 27 mm was matched to the active stack height of the motor, thereby maximizing the effective utilization of the magnet volume in the electromagnetic design. The selected geometry was finalized after iterative evaluation of flux distribution, torque density, and manufacturability, ensuring that both electromagnetic and mechanical requirements were satisfied within the restricted motor envelope. This magnet configuration was determined through a parametric design process, in which variations in insertion depth and tangential span were evaluated to balance torque density, efficiency, and manufacturability under the geometric restrictions of L-category electric vehicles. Particular attention was given to avoiding local saturation in the stator teeth and rotor bridges, as well as to mitigating mechanical stress concentrations around the magnet cavities. The adopted geometry reflects a compromise between electromagnetic performance and structural robustness, ensuring reliable operation in continuous duty.

Table 2. Materials of the designed PMSM.

Component	Material	Grade
Stator core material	Slicon steel	M250-50A
Stator coil material	Copper	Class 180 (H)
Rotor core material	Slicon steel	M250-50A
Rotor magnet material	NdFeB	N30SH

The motor designed with this configuration is targeted to achieve an efficiency level exceeding 94%. This objective aligns with the IE6 efficiency class discussed in the draft version of IEC 60034-30-3 [23], which, although not yet officially standardized, has been proposed in the literature. The IE6 class aims to reduce motor losses by approximately 20% compared to the currently highest standardized efficiency class, IE5 [24]. Thanks to the reduced losses in the motor, the resulting thermal load within the system will be significantly decreased, eliminating the need for an active cooling system. Since the motor is planned to be mounted at the front of the electric vehicle, it is intended to utilize natural airflow for passive cooling during operation. Accordingly, the motor has been designed with a structure featuring only two end covers, allowing direct contact between the internal components and the air. As there is no specific requirement for a protection class in this application, the motor was designed with an IP0 protection level, and ventilation holes were added to the front and rear covers to promote air circulation. This configuration enables effective heat dissipation and eliminates the need for additional active cooling systems, ensuring that the thermal assumptions used in the analyses are consistent with the real installation environment.

2.3. Design Constraints

The design process of an electric motor is shaped by constraints arising from both the physical installation space and the selected materials. Dimensional parameters, such as the stator outer diameter and core length, are generally limited by the available mounting space within the application environment. In this study, the overall motor envelope was constrained to 150 × 150 × 54 mm (W × H × L), corresponding to a total volume of 1.215×10^{-3} m^3 (≈1.2 L). This compact envelope was dictated by the installation requirements of the target vehicle. Additionally, constraints related to materials and performance stem from factors such as the motor's thermal endurance and magnetic properties. In this context, design parameters including torque per rotor volume (TRV), electrical loading (Ac), the ratio of inner to outer diameter (D_{si}/D_{so}), the aspect ratio (L_{stk}/D_{si}), supply voltage, and current density (J_{rms}) are taken into consideration. These parameters define the design boundaries while ensuring optimal performance within the physical and material limitations of the motor, as well as compatibility with application-specific requirements.

2.4. Outputs of PMSM

The motor model was developed in accordance with the specific application requirements and the constraints imposed by the selected materials. To maintain the clarity and conciseness of this paper, detailed analytical and numerical calculations involved in the design process have been omitted. Instead, the key design parameters are summarized in Table 3. For a comprehensive explanation of the design methodology and the underlying computational procedures, readers are referred to [25–27].

Table 3 presents the results obtained from analytical calculations related to the motor's geometrical, electrical, and electromagnetic characteristics. These results represent the design outputs based on the input parameters outlined in Table 1 and serve as foundational data for subsequent motor performance analyses. In this context, the PMSM was modeled using the determined design parameters, and motor analyses were conducted using Siemens Simcenter SPEED software (Product Version 2020.3.1). The designed motor model is shown in Figure 1. In this design, a spoke-type rotor structure was preferred. This configuration offers significant advantages in terms of mechanical strength, compact motor construction, and achieving high efficiency. By utilizing the spoke-type rotor structure, leakage fluxes are minimized, enabling magnetic flux interaction between the rotor and stator to occur with minimal leakage. Furthermore, a 12-slot stator and 10-pole rotor combination was selected for the motor topology. This configuration prevents the occurrence of low-order odd harmonic components, such as the 5th, 7th, and 9th, thereby reducing torque ripple and enhancing the overall performance of the motor.

To further benchmark the electromagnetic performance, the torque per rotor volume (TRV) was evaluated. The designed PMSM achieves a TRV of 25.1 kNm/m^3, as reported in Table 3. This metric was adopted instead of envelope-based torque density, since it provides a gearbox-independent and machine-intrinsic normalization of torque capability.

Reported TRV values for totally enclosed rare-earth permanent magnet motors generally lie within the range of 14–42 kNm/m^3 [26], confirming that the proposed design demonstrates a competitive performance.

Table 3. Design parameters of the proposed motor.

Parameter	Value	Unit
Number of Stator Slots	12	
Number of Rotor Poles	10	
Winding Topology	Concentrated Winding	
Slots per Pole per Phase	0.4	
Winding Symmetry Number	2	
Winding Factor	0.933	
Number of Turns per Phase	16	
Conductor Cross-Sectional Area	11.4	mm^2
LCM (Least Common Multiple)	60	
Coil Span (Throw)	1	
Torque per Rotor Volume	25.1	kNm/m^3
Electrical Loading	22.67	kA/m
Average Airgap Flux Density	0.62	T
Current density	5.2	A/mm^2
Split Ratio	0.56	
Aspect Ratio	0.34	
Stator Tooth Width	11	mm
Tooth Width/Tooth Pitch	0.4	
Number of Phases	3	
Stack Length	27	mm
Rotor Volume	135,716.8	mm^3

Table 3 summarizes the electrical, geometrical, and electromagnetic outputs of the designed PMSM. The selected winding configuration ensures high fundamental component utilization while maintaining compact coil geometry. The relatively high split ratio and aspect ratio were determined based on vehicle constraints, ensuring a balance between torque capability and axial compactness. The average air-gap flux density of 0.62 T and the electrical loading of 22.67 kA/m indicate a moderate magnetic and electric loading, selected to limit saturation and thermal stress under continuous duty. These outputs, together with the geometric parameters such as tooth width and stack length, form the baseline inputs for the FEM-based performance analysis presented in the subsequent section.

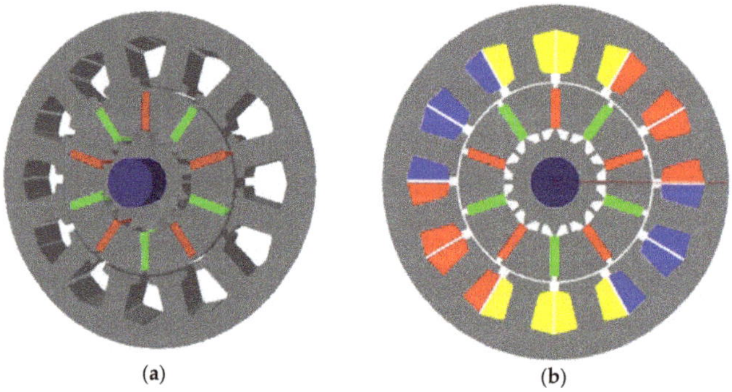

(a) (b)

Figure 1. The designed PMSM: (**a**) 3D structural view; (**b**) winding layout.

3. Results

This section presents the analysis results of the PMSM designed for the electric vehicle. The analyses were carried out using the FEM-based environment of Siemens Simcenter SPEED software. The simulations were performed under the rated operating point in continuous duty, considering a winding temperature of 100 °C. In this process, both the temperature-dependent reduction in the winding resistance and the temperature-dependent decrease in the magnet remanence were taken into account. The copper resistance was calculated by including the end-winding extension height (≈5 mm per side), while the coil configuration of 16 turns per phase with an RMS current density of 5.2 A/mm^2 was applied. The efficiency evaluation followed the total losses approach, as defined in IEC 60034-2-1 [28], where core losses were obtained from FEM using the material B–H and loss characteristics and AC copper losses were verified to remain negligible compared to the DC component. The operating point was determined using a maximum torque per ampere (MTPA) current phasing strategy. The current vector angle was determined to be $\gamma_d = 90.95°$ with respect to the d-axis, corresponding to a small deviation of 0.95° from the q-axis, indicating that $id \approx 0$. The electromagnetic load angle was calculated as $\delta = 33.88°$. These two angles have distinct definitions: γ_d represents the current vector angle relative to the rotor d-axis, whereas δ denotes the load angle between the rotor flux and the stator flux linkage. At rated operating conditions, such values are typical for spoke-type PMSMs and confirm that the current vector remains predominantly aligned along the q-axis, as expected under MTPA control.

Figure 2 shows the results of the electromagnetic analysis. The primary objective of this analysis is to evaluate whether the core material reaches magnetic saturation. Upon examination of the obtained results, it is observed that magnetic saturation does not occur in the motor core. Remaining below the critical saturation threshold of 1.6 T indicates that the core material has been appropriately selected and that the motor does not pose a risk of saturation under nominal operating conditions. Moreover, the general orientation and symmetry of the flux lines suggest that the motor possesses a magnetically balanced structure.

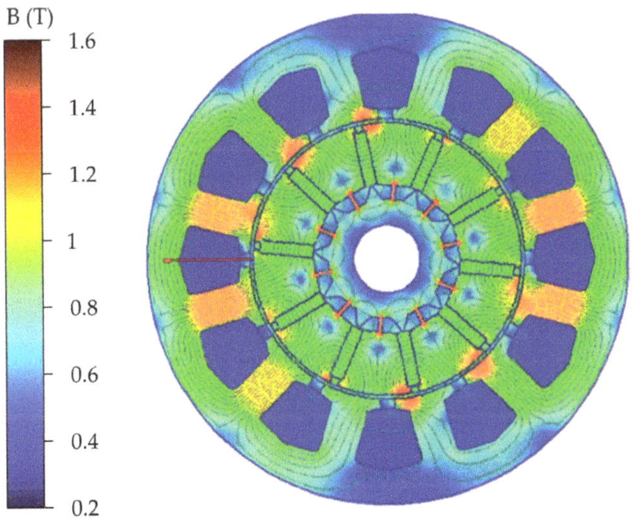

Figure 2. Magnetic flux density distribution.

The current waveforms of the three-phase stator windings as a function of the rotor's electrical position are shown in Figure 3. Each phase of the three-phase system exhibits a sinusoidal waveform with a 120° phase shift between them, indicating that the motor operates under balanced and ideal conditions. The root mean square (RMS) value of the current for each phase is obtained as 59.4 A.

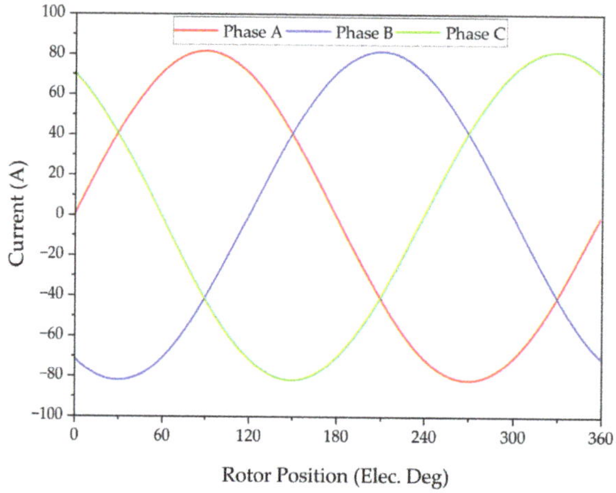

Figure 3. Three phase stator current waveforms of designed PMSM.

The back-electromotive force (back-EMF) waveforms induced in each phase as a function of the rotor's electrical position are shown in Figure 4. Each phase exhibits a sinusoidal waveform with a 120° phase shift, indicating a balanced and ideal electromagnetic design. The RMS value of the back-EMF per phase is determined as 16.2 V.

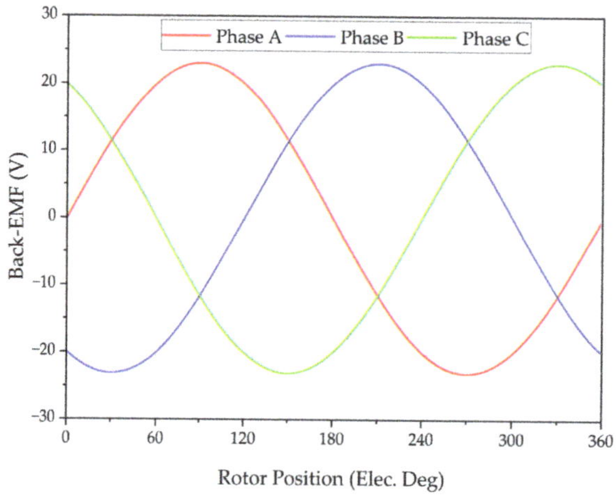

Figure 4. Three phase back-EMF waveforms of PMSM.

The nominal torque of the motor is approximately 3.504 Nm, as shown in Figure 5. The torque ripple is calculated to be around 0.46% using (1). The low amplitude of torque ripple indicates that the motor has a well-balanced magnetic design and offers high torque stability. From the perspective of electric vehicle applications, this feature provides significant advantages by enhancing driving comfort and the reducing vibrations and noise that may affect mechanical components.

$$T_r = (T_{max} - T_{min})/T_{avg} \quad (1)$$

where T_{max}, T_{min}, and T_{avg} are maximum torque, minimum torque, and average torque, respectively.

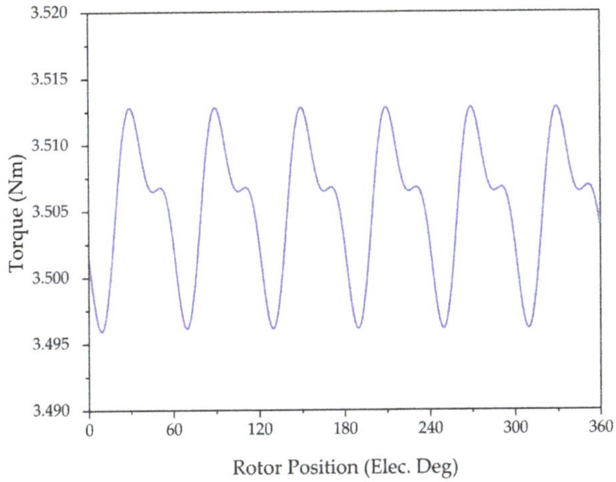

Figure 5. Electromagnetic torque variation versus rotor electrical position.

Figure 6 shows the cogging torque waveform obtained over a mechanical rotation of 0–30°. The peak cogging torque is approximately 0.18 Nm. The periodic and symmetrical nature of the waveform indicates a proper alignment between the rotor magnets and stator teeth. In electric vehicle applications, such a low cogging torque level is particularly beneficial at low speeds, where it contributes to smoother and vibration-free operation.

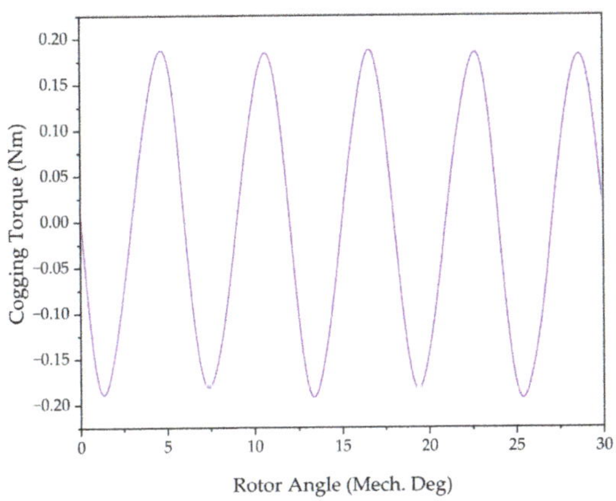

Figure 6. Cogging torque variation versus rotor electrical position.

The torque and power characteristics of the designed PMSM are shown in Figure 7 as a function of motor speed. At low speeds, the torque remains nearly constant around 3.5 Nm, but it begins to decrease beyond a certain speed. In contrast, the power increases with speed and reaches approximately 2900 W. At the nominal operating point of 7000 rpm, the torque is 3.5 Nm and the output power is approximately 2.5 kW. This reflects a transition from the constant torque region to the constant power region, which is a desirable performance profile for electric vehicle applications.

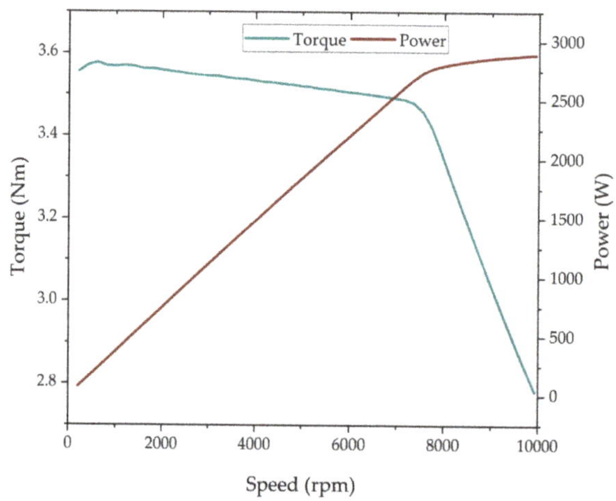

Figure 7. Operating characteristics of designed PMSM.

The efficiency map obtained as a function of shaft torque and motor speed is presented in Figure 8. It is observed that the motor operates with an efficiency above 90% over a wide operating range. In particular, high efficiency is achieved within the torque range of 1.5–3.5 Nm and the speed range of 1500–8000 rpm. Under nominal operating conditions, the motor reaches an efficiency of approximately 95.35%, which exceeds the level defined by the highest officially recognized IE5 efficiency class. In the literature, the proposed IE6 classification is associated with approximately 20% lower losses compared to IE5 [24]. Accordingly, the obtained efficiency value meets the expected performance of the IE6 efficiency class, indicating that the motor offers a highly advantageous structure in terms of energy efficiency and overall performance for electric vehicle applications.

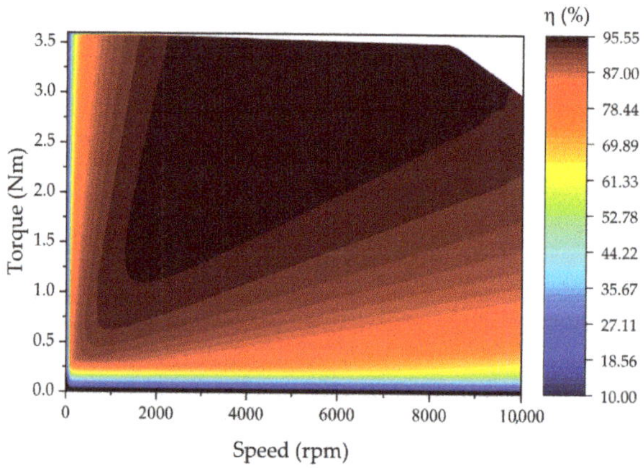

Figure 8. Efficiency map of the designed PMSM.

The summary of the key electromagnetic and performance parameters of the motor is presented in Table 4. These results highlight that the proposed PMSM achieves a nominal torque of 3.5 Nm and an output power of 2.56 kW at 7000 rpm, with an efficiency of 95.35%. The relatively high power factor confirms the effectiveness of the MTPA-oriented current phasing strategy, while the low winding resistance ensures that copper losses remain small compared to iron losses. The inductance values indicate a saliency ratio close to unity, which is consistent with the spoke-type rotor design and beneficial for stable torque production. Furthermore, the back-EMF constant aligns well with the targeted DC bus voltage, ensuring compatibility with the intended EV drive system. Overall, the parameters confirm that the designed motor offers a balanced combination of high efficiency, compact size, and suitability for geared electric vehicle applications.

Table 4. Output parameters of the designed PMSM.

Parameter	Value	Unit
Nominal torque	3.5	Nm
Nominal speed	7000	rpm
Nominal output power	2565	W
Input power	2690	W
Efficiency	95.35	%
Phase winding current	59.4	A_{rms}
Power factor	0.83	
Frequency	583	Hz
Iron loss	84.7	W
Winding loss	39.8	W
Back EMF	16.2	V_{rms}
D-axis inductance	0.0394	mH
Q-axis inductance	0.0469	mH
Phase winding resistance (20 °C)	0.00285	Ω

4. Conclusions

In this study, a permanent magnet synchronous motor with a spoke-type rotor topology was designed for application in an L-category electric vehicle intended for the urban transportation of two passengers. The motor was tailored to meet the specific torque and speed requirements of the reference vehicle, while also ensuring a compact construction suitable for gearbox integration.

Simulation results showed that the proposed PMSM achieves a peak efficiency of 95.35% under continuous operation, thereby confirming its hyper-efficiency level. The adopted 12-slot/10-pole configuration significantly suppresses low-order harmonics, which reduces torque ripple and contributes to stable drive performance. Magnetic flux analysis further demonstrated that the selected core material remains well below saturation limits, ensuring electromagnetic stability during nominal operation.

In addition, the torque-per-rotor-volume (TRV) of 25.1 kNm/m^3 highlights the competitive torque density of the proposed design compared with state-of-the-art permanent magnet machines, while maintaining a lightweight and manufacturable structure.

In conclusion, this study demonstrates the feasibility of achieving both compactness and high efficiency in a PMSM designed for geared electric vehicle applications. In future work, prototype production is planned, and the design will be validated experimentally through efficiency mapping, torque ripple characterization, and thermal performance tests at the electrical laboratory of Volt WEG Group.

Author Contributions: Conceptualization, H.N., S.A., and C.A.; methodology, H.N., S.A., and C.A.; validation, H.N., A.Y., S.A., and C.A.; formal analysis, H.N., S.A., and C.A.; investigation, H.N., A.Y., S.A., and C.A.; writing—original draft preparation, H.N. and A.Y.; writing—review and editing, A.Y.; visualization, A.Y.; supervision, A.Y.; funding acquisition, H.N., S.A. and C.A. All authors have read and agreed to the published version of the manuscript.

Funding: This research was funded by the European Union's Horizon Europe research and innovation programme under the Grant Agreement 101189783.

Data Availability Statement: The original contributions presented in this study are included in the article. Further inquiries can be directed to the corresponding author.

Acknowledgments: The authors would like to thank Volt WEG Group for their valuable support.

Conflicts of Interest: Authors Hayatullah Nory, Serhat Aksun and Cansu Aksoy were employed by Volt Weg Group. The remaining author declares that the research was conducted in the absence of any commercial or financial relationships that could be construed as a potential conflict of interest.

References

1. Fayyazbakhsh, A.; Bell, M.L.; Zhu, X.; Mei, X.; Koutný, M.; Hajinajaf, N.; Zhang, Y. Engine Emissions with Air Pollutants and Greenhouse Gases and Their Control Technologies. *J. Clean. Prod.* **2022**, *376*, 134260. [CrossRef]
2. Rabbi, M.F.; Popp, J.; Máté, D.; Kovács, S. Energy Security and Energy Transition to Achieve Carbon Neutrality. *Energies* **2022**, *15*, 8126. [CrossRef]
3. El, E.; Yildiz, C.; Dandil, B.; Yildiz, A. Effect of Wind Turbine Designed for Electric Vehicles on Aerodynamics and Energy Performance of the Vehicle. *Therm. Sci.* **2022**, *26*, 2907–2917. [CrossRef]
4. Hainsch, K.; Löffler, K.; Burandt, T.; Auer, H.; Crespo del Granado, P.; Pisciella, P.; Zwickl-Bernhard, S. Energy Transition Scenarios: What Policies, Societal Attitudes, and Technology Developments Will Realize the EU Green Deal? *Energy* **2022**, *239*, 122067. [CrossRef]
5. Un-Noor, F.; Padmanaban, S.; Mihet-Popa, L.; Mollah, M.N.; Hossain, E. A Comprehensive Study of Key Electric Vehicle (EV) Components, Technologies, Challenges, Impacts, and Future Direction of Development. *Energies* **2017**, *10*, 1217. [CrossRef]
6. Yang, Y.; He, Q.; Fu, C.; Liao, S.; Tan, P. Efficiency Improvement of Permanent Magnet Synchronous Motor for Electric Vehicles. *Energy* **2020**, *213*, 118859. [CrossRef]
7. Polat, M.; Yildiz, A.; Akinci, R. Performance Analysis and Reduction of Torque Ripple of Axial Flux Permanent Magnet Synchronous Motor Manufactured for Electric Vehicles. *IEEE Trans. Magn.* **2021**, *57*, 8106809. [CrossRef]
8. Agamloh, E.; von Jouanne, A.; Yokochi, A. An Overview of Electric Machine Trends in Modern Electric Vehicles. *Machines* **2020**, *8*, 20. [CrossRef]
9. Monadi, M.; Nabipour, M.; Akbari-Behbahani, F.; Pouresmaeil, E. Speed Control Techniques for Permanent Magnet Synchronous Motors in Electric Vehicle Applications Toward Sustainable Energy Mobility: A Review. *IEEE Access* **2024**, *12*, 119615–119632. [CrossRef]
10. Vlachou, V.I.; Sakkas, G.K.; Xintaropoulos, F.P.; Pechlivanidou, M.S.C.; Kefalas, T.D.; Tsili, M.A.; Kladas, A.G. Overview on Permanent Magnet Motor Trends and Developments. *Energies* **2024**, *17*, 538. [CrossRef]
11. Niu, L.; Zhang, M. Rotor Topology Optimization and Electromagnetic Performance Analysis of Interior Permanent Magnet Synchronous Motors for Electric Vehicle Applications. *J. Eng.* **2023**, *2023*, e212240. [CrossRef]
12. Cheng, Y.; Ding, L.; Zhao, T.; Cui, S. Design and Optimization of Electric Vehicle Traction Motor Considering Rotor Topology and Manufacturing Uncertainty. *IEEE Trans. Ind. Electron.* **2024**, *71*, 5034–5044. [CrossRef]
13. Chu, G.; Dutta, R.; Xiao, D.; Fletcher, J.E.; Rahman, M.F. Development and Optimization of a Mechanically Robust Novel Rotor Topology for Very-High-Speed IPMSMs. *IEEE Trans. Energy Convers.* **2023**, *38*, 1781–1792. [CrossRef]
14. Zhou, C.; Huang, X.; Li, Z.; Cao, W. Design Consideration of Fractional Slot Concentrated Winding Interior Permanent Magnet Synchronous Motor for EV and HEV Applications. *IEEE Access* **2021**, *9*, 64116–64126. [CrossRef]
15. Xi, Z.; Niu, L.; Yan, X.; Xu, L. Research on Performance of Interior Permanent Magnet Synchronous Motor with Fractional Slot Concentrated Winding for Electric Vehicles Applications. *World Electr. Veh. J.* **2024**, *15*, 470. [CrossRef]
16. Fan, X.; Zhang, B.; Qu, R.; Li, D.; Li, J.; Huo, Y. Comparative Thermal Analysis of IPMSMs With Integral-Slot Distributed-Winding (ISDW) and Fractional-Slot Concentrated-Winding (FSCW) for Electric Vehicle Application. *IEEE Trans. Ind. Appl.* **2019**, *55*, 3577–3588. [CrossRef]

17. Tikadar, A.; Johnston, D.; Kumar, N.; Joshi, Y.; Kumar, S. Comparison of Electro-Thermal Performance of Advanced Cooling Techniques for Electric Vehicle Motors. *Appl. Therm. Eng.* **2021**, *183*, 116182. [CrossRef]
18. Shewalkar, A.G.; Dhoble, A.S.; Thawkar, V.P. Review on Cooling Techniques and Analysis Methods of an Electric Vehicle Motor. *J. Therm. Anal. Calorim.* **2024**, *149*, 5919–5947. [CrossRef]
19. Gundabattini, E.; Mystkowski, A.; Idzkowski, A.; Singh, R.R.; Solomon, D.G. Thermal Mapping of a High-Speed Electric Motor Used for Traction Applications and Analysis of Various Cooling Methods—A Review. *Energies* **2021**, *14*, 1472. [CrossRef]
20. Shah, S.H.; Wang, Y.-C.; Shi, D.; Shen, J.-X. Investigation of Torque and Reduction of Torque Ripples through Assisted-Poles in Low-Speed, High-Torque Density Spoke-Type PMSMs. *Machines* **2024**, *12*, 327. [CrossRef]
21. EL-Refaie, A.M.; Shah, M.R.; Alexander, J.P.; Galioto, S.; Huh, K.-K.; Gerstler, W.D. Rotor End Losses in Multiphase Fractional-Slot Concentrated-Winding Permanent Magnet Synchronous Machines. *IEEE Trans. Ind. Appl.* **2011**, *47*, 2066–2074. [CrossRef]
22. Miguel-Espinar, C.; Heredero-Peris, D.; Villafafila-Robles, R.; Montesinos-Miracle, D. Review of Flux-Weakening Algorithms to Extend the Speed Range in Electric Vehicle Applications With Permanent Magnet Synchronous Machines. *IEEE Access* **2023**, *11*, 22961–22981. [CrossRef]
23. *IEC 60034-30-3*; Rotating Electrical Machines—Part 30-3: Efficiency Classes of High Voltage AC Motors (IE-Code). International Electrotechnical Commission (IEC): Geneva, Switzerland, 2024.
24. Wang, J.; Cao, H.; Xie, P. Design of Variable Speed Three-Phase PMSM Series IE6 Efficiency Class. In Proceedings of the 2024 IEEE International Conference on Electrical Energy Conversion Systems and Control (IEECSC), Shanghai, China, 8–10 November 2024; pp. 97–101.
25. Hendershot, J.R.; Miller, T.J.E. *Design Studies in Electric Machines*; Motor Design Books LLC: Venice, FL, USA, 2022; ISBN 9780984068746.
26. Hendershot, J.R.; Miller, T.J.E. *Design of Brushless Permanent-Magnet Machines*, 2nd ed.; Motor Design Books LLC: Venice, FL, USA, 2010; ISBN 978-0-9840687-0-8.
27. Miller, T.J.E. *SPEED's Electric Motors: An Outline of Some of the Theory in the Speed Software for Electric Machine Design: With Problems and Solutions*; Magna Physics Publishing: Lebanon, OH, USA, 2002; ISBN 978-1-881855-10-1.
28. *IEC 60034-2-1*; Rotating Electrical Machines—Part 2-1: Standard Methods for Determining Losses and Efficiency from Tests (Excluding Machines for Traction Vehicles). International Electrotechnical Commission (IEC): Geneva, Switzerland, 2024.

energies

Article

Three-Phase Small-Power Low-Speed Induction Motor with Can-Type Rotor

Krzysztof Sołtys * and Krzysztof Kluszczyński

Faculty of Electrical and Computer Engineering, Cracow University of Technology, Warszawska Str. 24, 31-155 Cracow, Poland; krzysztof.kluszczynski@pk.edu.pl
* Correspondence: krzysztof.soltys@pk.edu.pl

Abstract

To explore possible design solutions for induction motors, we designed and tested a three-phase small-power induction motor with a can-type rotor and a stationary internal ferromagnetic core, a design not previously described in the technical literature. This three-phase motor combines certain features of a reliable solid-rotor motor, a two-rotor layer motor, and a motor in which the rotating thin aluminium layer is separated from the stationary inner ferromagnetic core. The motor prototype was based on a mass-produced, small-power, three-phase squirrel-cage motor. Its operating properties and characteristics were tested, highlighting its potential application as a special-purpose drive or a very interesting case for teaching purposes in laboratories of electrical machines. Measurements confirmed theoretical predictions and enabled the formation of a motor equivalent circuit with shunt and series branch parameters, among which magnetization reactance and rotor resistance varied with rotational speed. The main advantages of the motor are its simple rotor construction, low rotational speed, low-rotor inertia and good dynamics, as well as reliable operation across the entire range of useful torque from no-load to short-circuit conditions, without the risk of overheating.

Keywords: special induction motor; can-type rotor; motor equivalent circuit; eddy currents; laboratory stand; measurement tests

1. Introduction

Asynchronous motors are among the most commonly used types of electric motors. The most popular mass-produced motors include three-phase squirrel-cage (induction) motors and three-phase slip-ring motors. In both cases, the rotors are slotted. In slip-ring motors, the slots contain distributed three-phase winding. In squirrel-cage motors, the rotor contains a cage made of copper or bronze bars short-circuited by end rings or formed by casting aluminium directly into the rotor slots, simultaneously forming the end rings and fan blades on the rotor's side surfaces [1–3]. The rotor of a slip-ring motor contains easily damageable components, such as distributed windings with insulated wires, slot insulation, front connections, etc. In the squirrel-cage motor, motor failures often result from poor slot filling during the aluminium casting process or mechanical cracking at the connections between the bars and the end rings of the cage caused by rotor overheating or by mechanical bar vibrations [4,5]. Regarding slip-ring motors, their important advantage is the ease of speed control (without the use of power electronic devices). This can be achieved by inserting external resistance of increasing value into the rotor circuit using slip rings and brushes. A slip-ring motor with a very high value of external resistance in the rotor can be treated as a low-speed motor.

Three-phase motors with solid rotors (which are also termed massive rotors [6]) are less commonly used. They have worse electromechanical properties (electromagnetic torque is a result of interactions between circular revolving magnetic field developed by a stator and eddy currents induced in the solid rotor), but because of simple rotor construction and lack of easily damageable components, they are regarded as motors with increased reliability. Their rotors may be constructed as either solid ferromagnetic rotors or as two-layer rotors, in which a ferromagnetic core is covered by an aluminium (or copper) layer with higher electrical conductivity (rotor designs with a larger number of layers are also possible) [6–10].

One example of an unconventional electromechanical converter is a two-phase induction motor (with an elliptical magnetic field revolving along the periphery of the air gap) with a can-type rotor and a stationary internal ferromagnetic core located inside the can (which is also named an inner stator). This type of motor is used as a control system component: a proportional or integral element in automatic control systems. The advantage of this design is a low electromechanical time constant due to the low inertia of the can rotor [2–12].

The authors decided to combine the construction features of different types of the above-described asynchronous machines to create a small power, low-speed, and low-rotor inertia motor. The thin-walled can provides high rotor resistance, making the motor a low-speed motor (just like a slip-ring asynchronous motor with a high value of external resistance inserted into the rotor circuit). A rotating thin aluminium layer in the form of a can, separated from the stationary inner ferromagnetic core (inner stator), ensures low rotor mass and low rotor inertia. Such a three-phase motor, combining certain features of a reliable solid-rotor motor, a two-rotor layer (aluminium and ferromagnetic) motor and a motor in which the rotating thin aluminium layer is separated from the stationary inner ferromagnetic core, has not been previously described in the technical literature. The combination of the above-listed design features leads to the creation of a motor that can be described as a low-speed motor, a motor with low rotor inertia and a motor with increased reliability (resulting from the lack of easily damageable rotor components).

Driven by curiosity and research inquisitiveness, the authors decided to construct and study the above-described three-phase induction motor (with a circular magnetic field revolving along the periphery of the air gap) with a can-type rotor and a stationary internal ferromagnetic core. The motor was built using a mass-produced, small-power, three-phase squirrel-cage motor as its base. Its operational properties were investigated, with emphasis on potential use as a special-purpose drive as well as an electromechanical converter for teaching purposes in laboratories of electrical machines. Due to its relatively low developed electromagnetic torque and low efficiency, the motor is not suitable for general use; nevertheless, it offers an interesting case from a research perspective and exhibits characteristics that may be useful in special-purpose drive systems. These characteristic features include simple rotor construction and low rotational speed, achievable through direct grid supply without using a power electronics or gearbox. The motor can also operate safely and continuously across the entire range of useful torque, from no-load to short-circuit conditions, without the risk of overheating. Thanks to such properties the motor can be used in household appliances, for example, in the process of thickening, grinding, or crushing food products.

The interesting operating properties of the motor (for example, a slight decrease in the stator current as the motor load increases resulting from a change in the depth of magnetic field penetration into the aluminium can) mean that the motor can also be used in electrical machine laboratories as an interesting and instructive case for educational purposes.

2. Materials and Methods

Construction of a Three-Phase, Low-Speed Induction Motor with a Can-Type Rotor and a Measurement Stand

The prototype of the three-phase, low-speed induction motor with a can-type rotor was constructed using a commercially available low-power, three-phase induction motor (presented in Figure 1) with the following rated data: power P = 90 W, supply voltage 230/400 V (Δ/Y), rated current 0.68/0.39 A (Δ/Y), frequency f = 50 Hz, rotational speed n = 1400 rpm, and power factor cos φ= 0.59. The number of pole pairs is p = 2. As such, the synchronous speed is n_0 = 1500 rpm.

Figure 1. Three-phase squirrel-cage motor used as the prototype's base, with the side cover removed.

After removing the squirrel-cage rotor, a thin-walled rotor made from an aluminium tube with a wall thickness $\Delta = 1.5 \times 10^{-3}$ m was placed inside of the stator, along with a stationary ferromagnetic core (an internal stator) in the form of a solid cylinder. The can-type rotor was mounted on two bearings located on both sides of the internal stator, secured in place by two Seeger rings and two pressure rings. The procedure for constructing the prototype motor is shown in Figure 2. This configuration will be referred to hereafter as a can-type motor.

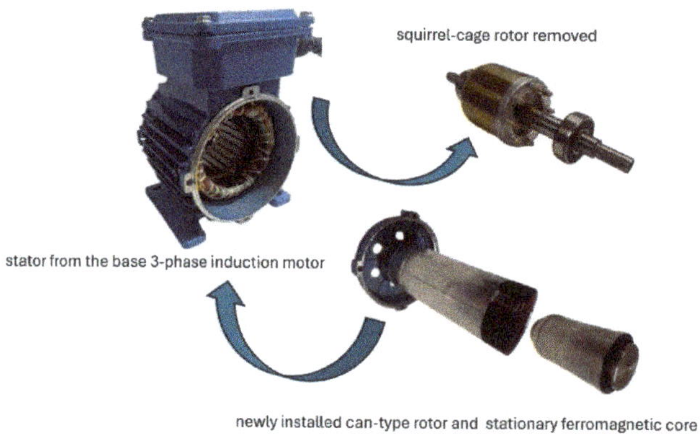

Figure 2. Procedure for transforming three-phase squirrel-cage motor into three-phase induction motor with can-type rotor.

In the base motor, the air gap was 3×10^{-4} m. As seen in Figure 3, in the prototype, the main air gap between the inner surface of the stator and the outer surface of the aluminium tube was $\delta_1 = 2 \times 10^{-3}$ m. The thickness of the can was $\Delta = 1.5 \times 10^{-3}$ m, and the internal air gap between the inner surface of the can and the outer surface of the internal ferromagnetic core was $\delta_2 = 5 \times 10^{-4}$ m. This increase in the main air gap from 3×10^{-4} m (in the base motor) to $\delta_1 = 2 \times 10^{-3}$ m (in the can-type motor) significantly increased the magnetization current and reduced the developed electromagnetic torque. However, the motor also acquired a new property, as its no-load speed shifted significantly into the low-speed range. The stationary internal ferromagnetic core was machined from magnetic steel, as using laminated sheets would be too costly. The total geometric air gap (the sum of the two air gaps and can wall's thickness) was thus $\delta_1 + \Delta + \delta_2 = 4 \times 10^{-3}$ m. In calculations, it is necessary to account for the effective air gap, not just the geometric one. The difference between the geometric and effective air gaps in conventional induction motors is due to the saturation phenomenon in the ferromagnetic circuit (the saturation factor) and the slotting of the inner stator surface (the Carter factor). In the case of the can-type motor, it is also necessary to consider the change in the depth of magnetic field penetration into the can rotor, an effect known as the shielding effect [7–12].

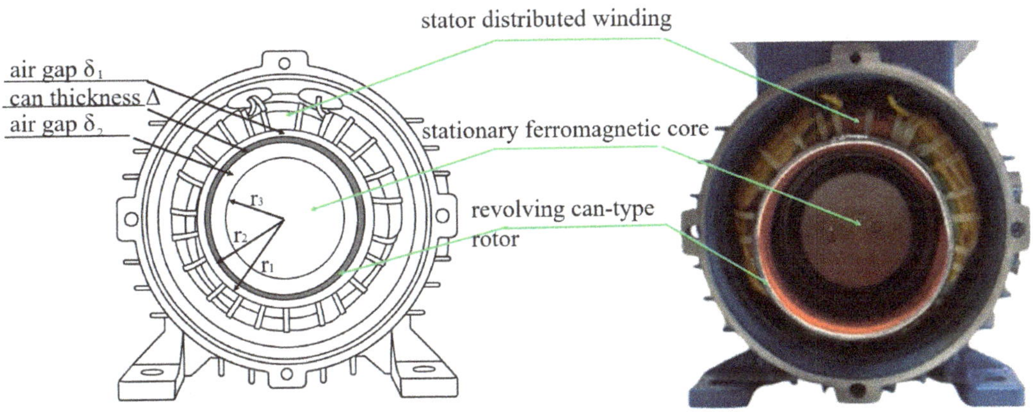

Figure 3. Cross-section of the can-type motor with essential geometric dimensions marked: main air gap $\delta_1 = 2 \times 10^{-3}$ m, can thickness $\Delta = 1.5 \times 10^{-3}$ m, internal air gap $\delta_2 = 5 \times 10^{-4}$ m, inner radius of the stator $r_1 = 2.7 \times 10^{-3}$ m, outer radius of the can rotor $r_2 = 2.5 \times 10^{-3}$ m, and outer radius of the internal stator $r_3 = 2.3 \times 10^{-3}$ m.

A specially designed laboratory stand was used to test the three-phase can-type motor and the base three-phase motor, used as a reference. This made it possible to measure stator voltage and current, input electrical power, rotational speed, useful (load) torque, starting torque, and mechanical friction torque [1,3]. Some components of the laboratory stand were manufactured using 3D printing. Torque measurements were performed using standard methods, including a friction brake and a force gauge (for load torque—Figure 4b) and an electronic measuring scale with a blocking arm (for starting torque—Figure 4a). The laboratory stand was equipped in 4-channel ultra vision oscilloscope and thermal imaging camera.

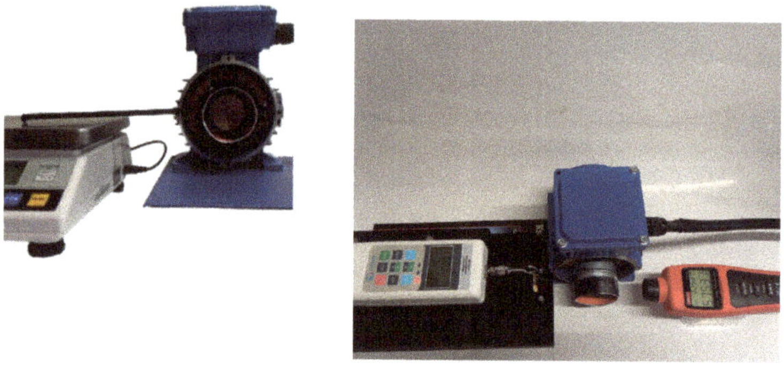

(**a**) (**b**)

Figure 4. Laboratory stand (**a**) part for measuring starting torque (**b**) part for measuring torque and speed.

3. Results

3.1. Operating Characteristics of the Three-Phase Can-Type Motor

Before the development of power electronics, three-phase slip-ring asynchronous motors were widely used because their speed is easily controlled. This was achieved by inserting external resistance of increasing value into the rotor circuit using slip rings and brushes. In this method of speed control, which is still used today, increasing rotor resistance decreases the slope of the torque–speed characteristic relative to the negative part of the horizontal axis, while the breakdown torque remains unchanged (Figure 5a). Increasing rotor circuit resistance is also accompanied by a reduction in stator current (Figure 5a). The downside of achieving lower rotational speed is increased electrical losses in the rotor and reduced motor efficiency.

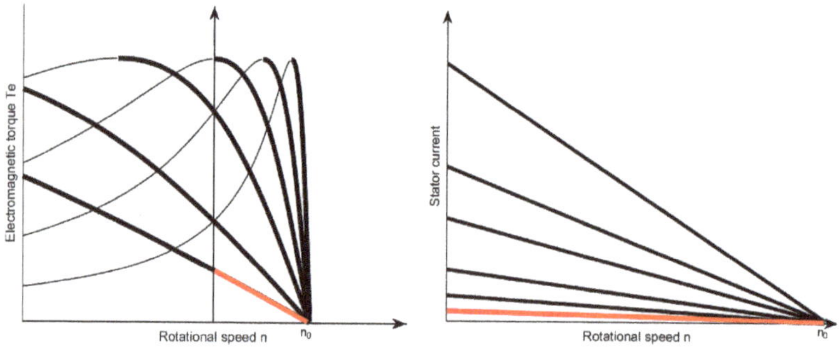

(**a**) (**b**)

Figure 5. Speed control of a slip-ring motor achieved by increasing rotor resistance: (**a**) electromagnetic torque as a function of rotational speed, (**b**) change in stator current with change in rotor resistance.

Replacing the slotted rotor in an induction motor with a can-type rotor is an extreme form of this method achieved through structural changes to the rotor (indicated by the red lines in Figure 5). The only response to the stator's rotating magnetic field is eddy currents induced in the thin-walled can, which has high electrical resistance.

To experimentally determine the operating properties of the three-phase can-type motor, it is crucial to determine the allowable supply voltage, which is directly related to the rated current of the base motor. For a star connection, the rated current is I_n = 0.39 A. All experiments were carried out with the stator winding connected in a star configuration (Y), and all currents, voltages, and parameters were recalculated per phase. Figure 6 shows the rotational speed n and the stator current I as a function of voltage U measured during no-load operation of the can-type motor (at useful torque T_{us} = 0).

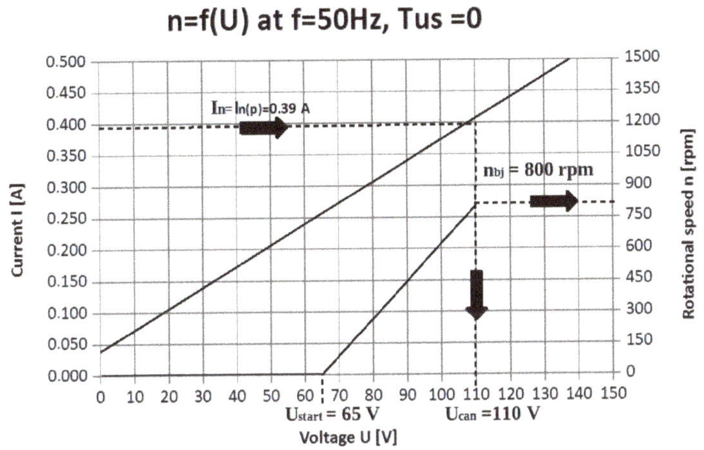

Figure 6. Stator current I and speed n of the can-type motor as a function of supply voltage U at no-load (T_{us} = 0).

The motor starts at a voltage U_{start} = 65 V based on an average of several measurements taken while increasing and decreasing the voltage. At this value, the developed electromagnetic torque exceeds the mechanical loss torque (friction torque). Note that the base three-phase squirrel-cage motor starts at 45 V. The motor current reaches the rated value I_n = 0.39 A at a supply voltage of 110 V. This value is considered the rated voltage of the can rotor motor and is denoted as U_{can} = 110 V. For the rated voltage, the useful torque–speed curve and stator current–speed curve were measured (Figure 7).

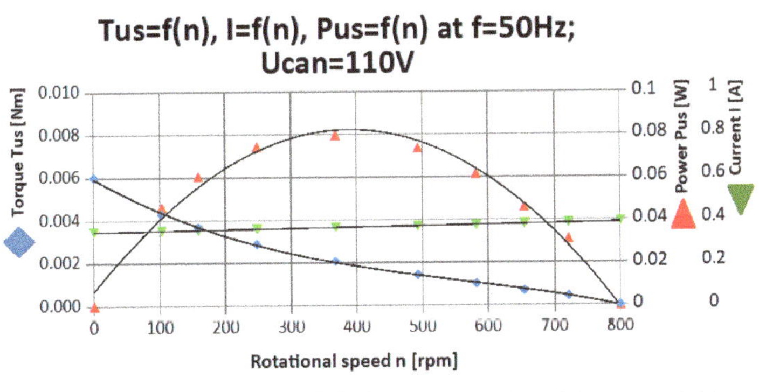

Figure 7. Can-type motor curves: useful torque Tus, current I, and output power Pus as functions of speed n.

As shown in Figure 7, the useful torque–speed curve differs from the linear form and is slightly concave compared to the red curve shown in Figure 5a. The maximum torque equals the starting torque, T_{start} = 0.006 Nm. The no-load speed is 800 rpm. The current–speed curve is particularly interesting. As seen in Figure 7, as speed decreases from no-load (n_0 = 800 rpm) to zero (n = 0), the current noticeably decreases by approximately 2.5%. The reason for this behaviour will be explained in the next section. A key conclusion is that the motor can operate safely and continuously throughout the entire torque range, from standstill to no-load speed (n = 0 to n = 800 rpm).

Additionally, Figure 7 includes the output mechanical power curve P_{us} as a function of speed n. The curve takes the shape of a slightly flattened inverted parabola, with maximum output power occurring at n = 400 rpm and equal to Pus max = 0.8 W. This speed can be considered the rated speed, n_{can} = 400 rpm. At this power level, the rotational speed is reduced by a factor of approximately four compared to the synchronous speed (n_0 = 1500 rpm).

The blocked-rotor curve of the can motor, i.e., starting torque as a function of supply voltage U at n = 0, was also measured (Figure 8).

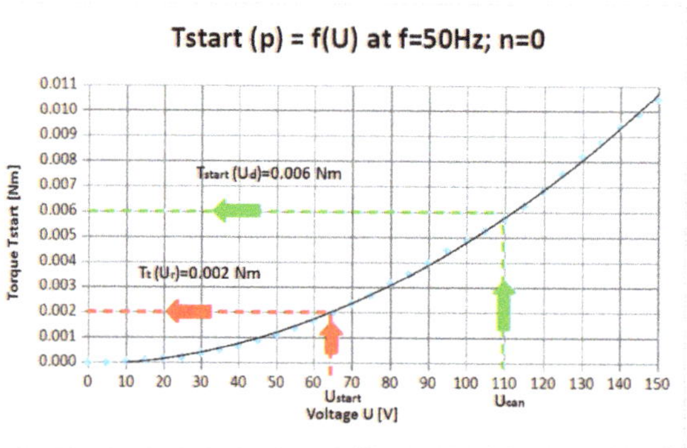

Figure 8. Blocked-rotor curve of the can-type motor.

This curve can be used to determine the starting torque (green line), T_{start} = 0.006 Nm, as well as the mechanical friction torque (red line), T_t = 0.002 Nm. The friction torque was also verified through direct measurement of the unpowered can-type motor.

Based on the useful torque–speed curve and the friction torque value, the electromagnetic torque–speed curve was reconstructed (Figure 9). The friction torque equals approximately 30% of the starting electromagnetic torque.

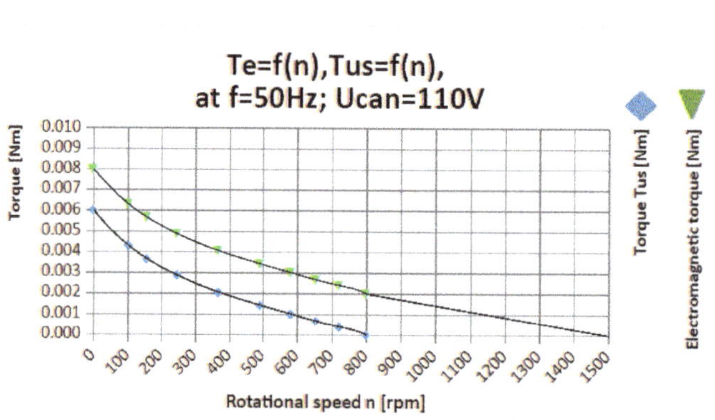

Figure 9. Electromagnetic torque and useful (load) torque as functions of speed.

3.2. Equivalent Circuit of the Can-Type Motor and Physical Interpretation of Measurement and Calculation Results

In the next stage, an equivalent circuit with shunt and series branch parameters for the three-phase can-type motor was determined. At the beginning, the no-load and rotor-blocked characteristics of the base three-phase squirrel-cage motor were measured. Then, after removing the squirrel-cage rotor, the "cavity characteristic" was determined. Another measurement was taken after inserting the internal ferromagnetic core into the stator cavity of the base motor. Finally, after mounting the can-type rotor on the internal core, the prototype of the can-type motor was tested by recording its no-load and rotor-blocked characteristics. These four current–voltage curves are shown in Figure 10. As can be seen, introducing a stationary ferromagnetic core into the rotor cavity significantly reduced the magnetization current. For example, at a supply voltage of U = 50 V, the magnetization current dropped from 0.42 A to 0.27 A, a nearly twofold decrease.

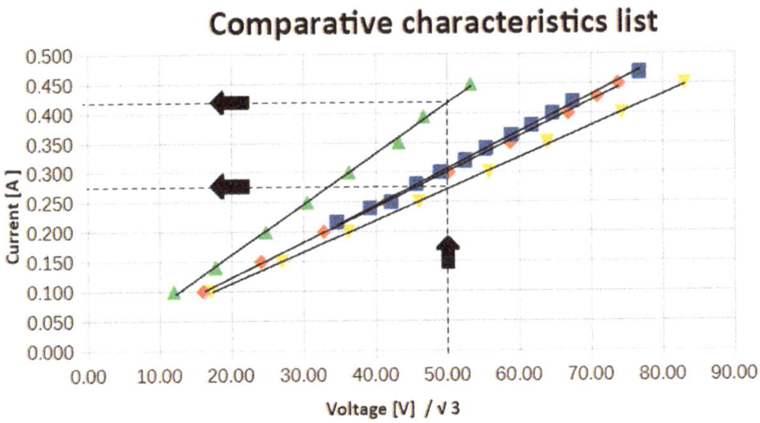

Figure 10. Comparison of current–voltage curves of the can-type motor: the magnetization curve with both the inner stator and the can removed (▲); the magnetization curve with the inner stator installed and the can removed (▼); the blocked-rotor curve (◆); and no-load curve of the complete can-type motor (■).

It is worth noting that stator current waveforms are practically sinusoidal. This demonstrates the correct choice of the motor supply voltage (no effects related to magnetic circuit saturation). The frequency spectrum of the stator currents also does not show any harmonics related to higher MMF (magnetomotive force) and permeance spatial harmonics. This is because the considered motor with can-type rotor has a thick (compared to the depth of the stator slots) air gap which is slotted only on the stator side (the so-called one-sided slotting) and is smooth on the rotor side. The can-shaped rotor also does not generate higher MMF spatial harmonics [13,14].

Based on the characteristics of the base three-phase squirrel-cage motor and the prototype of the three-phase can-type motor, the parameters of the equivalent circuit (Figure 11) were determined using a standard methodology [1–3,11,12].

These include stator resistance R_1 = 105 Ω, stator leakage reactance $X_{\sigma 1}$ = 40 Ω, magnetization reactance X_m = 255 Ω, equivalent core-loss resistance R_{Fe} = 1800 Ω, rotor resistance (referred to the stator) R'_2 = 6200 Ω, and rotor leakage reactance $X'_{\sigma 2}$ = 0 Ω. It was assumed that the calculated and measured results would match at zero speed (n = 0).

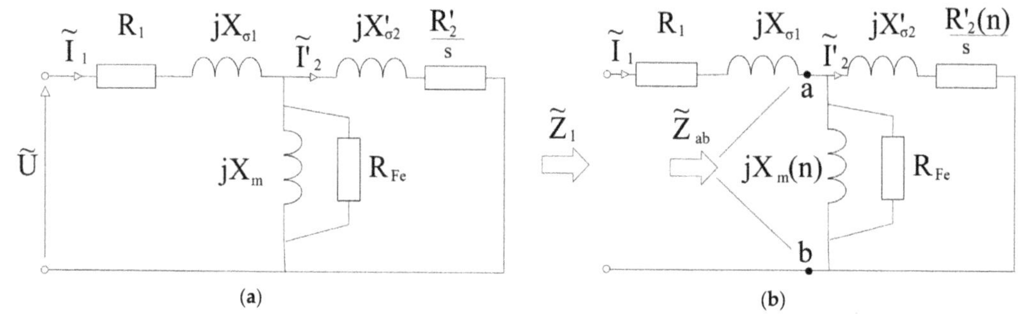

Figure 11. Equivalent circuit of the three-phase can-type induction motor (**a**) with speed-independent parameters; (**b**) with speed-dependent parameters (where s—slip, n—rotor speed).

Assuming constant values of equivalent circuit parameters (independent of rotational speed-Figure 11a), the electromagnetic torque and stator current as functions of speed were calculated according to the following expressions:

$$\tilde{Z}_{ab} = \frac{\frac{R_{Fe}\cdot jX_m}{R_{Fe}+jX_m}\cdot\frac{R_2'}{s}}{\frac{R_{Fe}\cdot jX_m}{R_{Fe}+jX_m}+\frac{R_2'}{s}} \tag{1}$$

$$\tilde{Z}_1 = (R_1 + jX_{\sigma 1}) + \tilde{Z}_{ab} \tag{2}$$

$$\tilde{I}_1 = \frac{\tilde{U}}{\tilde{Z}_1} \tag{3}$$

$$\tilde{I}_2' = \frac{\tilde{I}_1\cdot\tilde{Z}_{ab}\cdot s}{R_2'} \tag{4}$$

$$T_e = \frac{3p}{2\pi f}P\psi = \frac{3p}{2\pi f}\cdot\frac{R_2'}{s}\cdot I_2'^2 \tag{5}$$

where $P\psi$ is air gap power.

A comparison of the calculated and measured results (presented earlier in Figures 7 and 9) is shown in Figure 12.

As expected, there are some discrepancies between the measured and calculated results. These discrepancies arise because as the rotor's speed increases, the frequency of the eddy currents induced in the can decreases. Consequently, the depth of magnetic field penetration into the aluminium can also changes. This phenomenon can be interpreted as a gradual increase in the effective air gap length, which causes a decrease in the magnetization reactance and an increase in the magnetization current. Simultaneously, rotor resistance increases [6,8–10,15]. To align the calculated and measured results, it is necessary to introduce variable parameters into the equivalent circuit by accounting for the dependence of magnetization reactance and rotor resistance on speed, $X_m = f(n)$ and $R_2' = f(n)$, as shown in Figure 11b. Corresponding expressions for the electromagnetic torque and stator current as functions of speed are, as follows:

$$\tilde{Z}_{ab}(n) = \frac{\frac{R_{Fe}\cdot jX_m(n)}{R_{Fe}+jX_m(n)}\cdot\frac{R_2'(n)}{s}}{\frac{R_{Fe}\cdot jX_m(n)}{R_{Fe}+jX_m(n)}+\frac{R_2'(n)}{s}} \tag{6}$$

$$\tilde{Z}_1(n) = (R_1 + jX_{\sigma 1}) + \tilde{Z}_{ab}(n) \quad (7)$$

$$\tilde{I}_1 = \frac{\tilde{U}}{\tilde{Z}_1(n)} \quad (8)$$

$$\tilde{I}'_2 = \frac{\tilde{I}_1 \cdot \tilde{Z}_{ab}(n) \cdot s}{R'_2(n)} \quad (9)$$

$$T_e = \frac{3p}{2\pi f} P\psi = \frac{3p}{2\pi f} \cdot \frac{R'_2(n)}{s} \cdot I'^2_2 \quad (10)$$

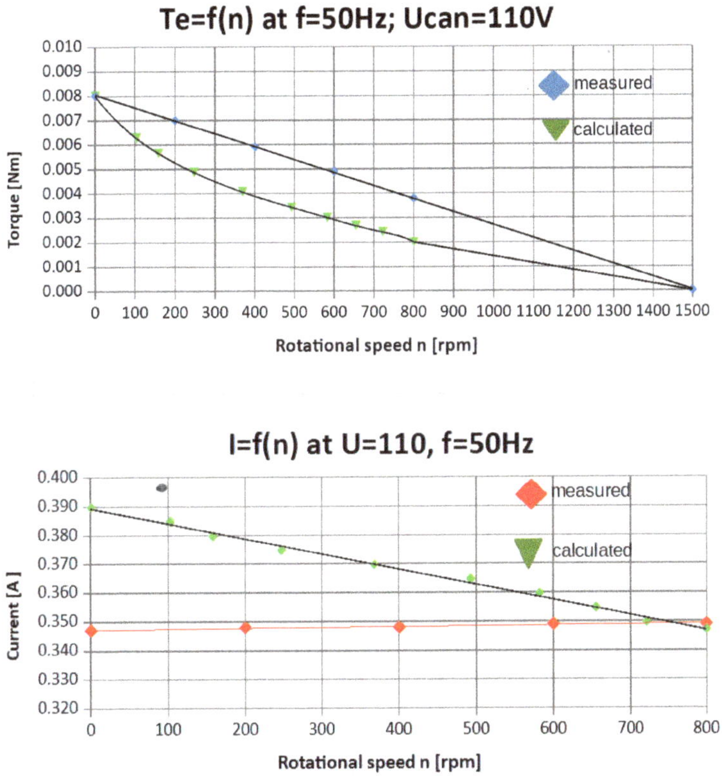

Figure 12. Comparison of measured and calculated curves of electromagnetic torque and stator current calculations based on the equivalent circuit with constant parameters—Figure 11a.

The determined dependencies: $X_m = f(n)$ and $R'_2 = f(n)$ are presented in Figure 13. They can be analytically described by the following linear functions:

$$R'_2(n) = 6.25\, n + 5880 \quad (11)$$

$$X_m(n) = -0.0415 + 253.8 \quad (12)$$

The shape of the magnetization reactance curve as a function of speed, shown in Figure 13, explains the intriguing phenomenon of decreasing stator current when moving from no-load to rotor standstill. As the magnetization reactance decreases, the magnetization current increases, and this component has the most significant effect on the value of the stator current (the rotor current is negligible in comparison). This phenomenon is theoretically interesting, and it could serve as an example in electrical engineering education by illustrating how the depth of magnetic field penetration into a metal layer increases as the supply frequency decreases.

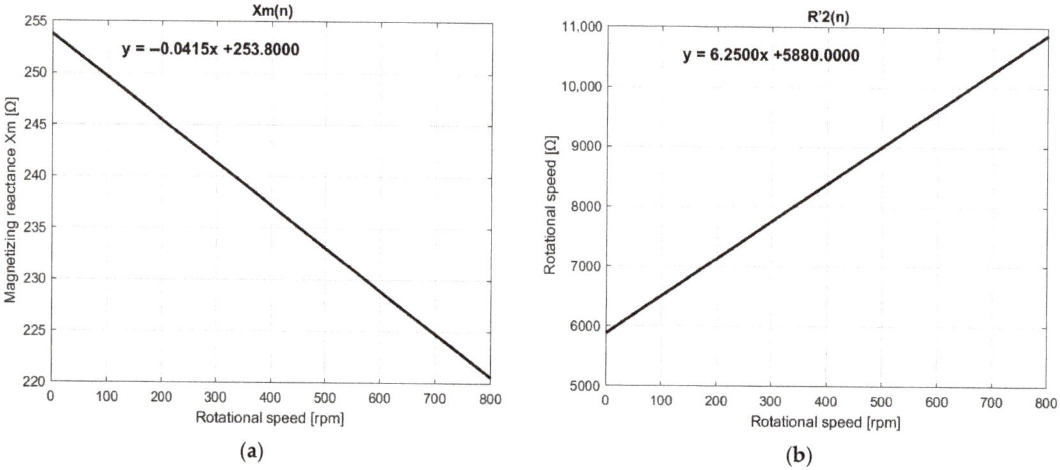

Figure 13. Variation in equivalent circuit parameters with speed: (**a**) magnetization reactance; (**b**) rotor resistance.

It is also worth noting that the can-type motor has a lower moment of inertia than the base squirrel-cage motor. This reduced inertia allows the can-type motor to accelerate and reach its operating speed faster. This difference in dynamic behaviour is clearly illustrated by the start-up speed curves over time for the squirrel-cage motor (red line) and the can-type motor (blue line), as shown in Figure 14 [11,13].

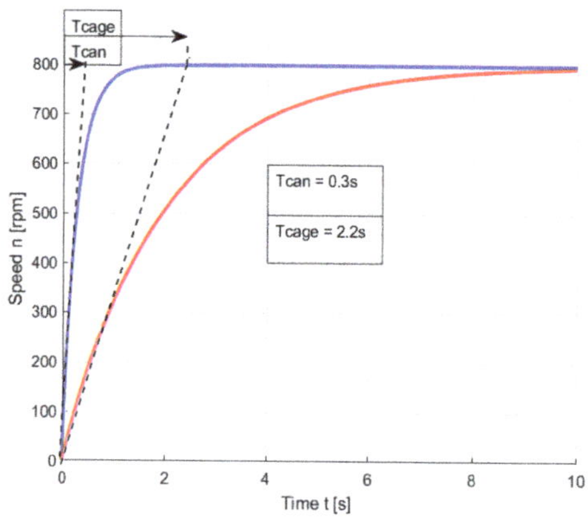

Figure 14. Start-up speed curves for the squirrel-cage motor and the can-type motor.

The electromechanical time constant of the can-type motor is over seven times smaller than that of the squirrel-cage motor, an additional advantage of this design.

Results from thermal imaging camera are presented in Figure 15. The steady-state rotor temperatures at standstill and no-load speed are remarkably less than in the base three-phase motor. Also, temperatures in the rotor of the prototype motor are favourable in relation to the stator temperatures.

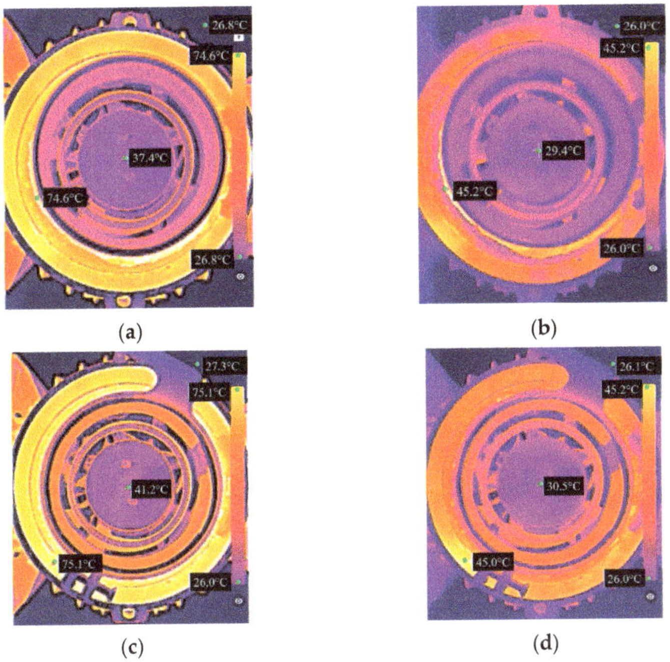

Figure 15. Thermographs from thermal imaging camera (**a**) for rotor speed: n = 800 rpm at thermal steady-state (30 min after switching on); (**b**) for rotor speed: n = 800 rpm, 3 s after switching on; (**c**) for rotor speed: n = 0 rpm (blocked rotor) at thermal steady-state (30 min after switching on); (**d**) for rotor speed: n = 0 rpm (blocked rotor), 3 s after switching on.

4. Discussion and Conclusions

Based on a three-phase squirrel-cage motor, a prototype of a three-phase induction motor with a can-type rotor and a stationary inner ferromagnetic core was built. To the best of the authors' knowledge, this motor has not been previously described in the technical literature. The motor was thoroughly tested, allowing for the determination of its operational characteristics and the most relevant parameters.

The measurements confirmed theoretical predictions and enabled the formation of an equivalent circuit of the motor with shunt and series branch parameters, among which the magnetization reactance and rotor resistance were dependent on rotational speed. The proposed motor has extremely simple rotor construction in the form of an aluminium can. The rotor contains no easily damageable components, such as distributed windings (with insulated wires, slot insulation, and front connections) or a squirrel-cage winding, where common failures results from poor slot filling during the aluminium casting process or mechanical cracking at the connections between the bars and the end rings of the cage [14,16].

The electromagnetic torque in the can-type motor results from the interaction between the stator current and the eddy currents induced in the rotor, which have unrestricted spatial distribution and intensity (current density). The motor exhibits interesting operational characteristics. Notably, the developed electromagnetic torque increases as rotor speed decreases and reaches its maximum at standstill. This increase in electromagnetic torque is accompanied by a slight reduction in stator current, an intriguing effect related to the changing penetration depth of the magnetic field into the can-type rotor. As rotor speed decreases, both the frequency and amplitude of the induced eddy currents increase, strengthening the so-called shielding effect. This behaviour of the stator current as a function of speed indicates that the motor can operate safely and continuously across the entire torque range, from no-load speed down to zero speed.

Another advantage of the motor is its ability to operate at low speeds when powered directly from the grid without the need for mechanical gearboxes or power electronic control systems. The operating range (i.e., the value of the no-load current) can be easily modified during the design stage by adjusting the can wall thickness, changing the can material (its electrical conductivity), or altering the length of the air gaps. We tested a prototype motor with an extremely thin rotor can, achieving a no-load speed of 180 rpm. In this case, the rated rotor speed, corresponding to the maximum useful power of the motor, was more than 16 times lower than the synchronous speed. This significant reduction in rated speed was accompanied by a reduction in the motor power.

An important benefit is the rotor's very low mass and low inertia, which allows the motor to respond quickly to changes in load torque or supply voltage. The electromechanical time constant of the can-type rotor motor is seven times lower than that of the mass-produced base motor. This low time constant enables, for example, the generation of oscillatory motion of the rotor via small supply voltage oscillations.

A disadvantage of the can-type motor is its low efficiency and low power-to-mass ratio, which limits its suitability for general use. Instead, it may find application in special-purpose drives where reliable performance and specific operational properties are required. However, the power-to-mass ratio of the rotor itself is favourable. Continued research and theoretical development are recommended.

Author Contributions: Conceptualization, K.K.; methodology, K.K.; software, K.S.; validation, K.S.; formal analysis, K.K. and K.S.; investigation, K.S.; data curation, K.S.; writing—original draft preparation, K.S.; writing—review and editing, K.K.; visualisation, K.S.; supervision, K.K. All authors have read and agreed to the published version of the manuscript.

Funding: This research was conducted at the Faculty of Electrical and Computer Engineering at the Department of Infotronics and Cybersecurity.

Data Availability Statement: The original contributions presented in this study are included in the article material. Further inquiries can be directed to the corresponding authors.

Conflicts of Interest: The authors declare no conflicts of interest. The funders had no role in the design of the study; in the collection, analyses, or interpretation of data; in the writing of the manuscript; or in the decision to publish the results.

References

1. Nasar, S.A.; Unnewehr, L.E. *Electromechanics and Electric Machines*; John Wiley & Sons, Inc.: New York, NY, USA, 1983.
2. Guru, B.S.; Hiziroglu, H.R. *Electric Machinery and Transformers*; Oxford University Press: New York, NY, USA, 2001.
3. Jordan, H.; Klima, V.; Kovacs, K.P. *Asynchronmaschinen Funktion, Theorie, Technisches*; Akademiai Kiado: Budapest, Hungary, 1975.
4. Basta, J. *Badanie Maszyn Elektrycznych (Testing of Electrical Machines)*; Wydawnictwa Naukowo-Techniczne (WNT): Warszawa, Poland, 1959.

5. Godlewski, Z. *Uszkodzenia Maszyn Elektrycznych (Damage to Electrical Machines)*; Wydawnictwo Naukowe PWN: Warszawa, Poland, 1967.
6. Dąbrowski, M.; Gieras, J. *Maszyny Indukcyjne o Wirniku Masywnym (Induction Machines with a Massive Rotor)*; Państwowe Wydawnictwo Naukowe (PWN): Warszawa, Poland, 1977.
7. Gieras, J.F.; Saari, J. Performance calculation for a high speed solid-rotor induction motor. *IEEE Trans. Ind. Electron.* **2011**, *59*, 2689–2700. [CrossRef]
8. Jamieson, R.A. Eddy-current effects in solid, unslotted iron rotors. *Proc. Inst. Electr. Eng.* **1968**, *115*, 793–800. [CrossRef]
9. McConnell, H.; Sverdrup, E. The induction machine with solid iron rotor. *Trans. Am. Inst. Electr. Eng.* **1955**, *74*, 641–650.
10. Szczygieł, M. Modelowanie Silnika Indukcyjnego o Dwóch Stopniach Swobody Ruchu oraz Badanie jego Właściwości Eksploatacyjnych Przy Wykorzystaniu Magnetoreologicznego Hamulca Obrotowego-Liniowego (Modeling of a Two-Degree-of-Freedom Induction Motor and Testing of Its Operating Properties Using a Magnetorheological Rotary-Linear Brake). Ph.D. Thesis, Silesian Uniwersity of Technology, Gliwice, Poland, 2011.
11. Sochocki, R. *Mikromaszyny Elektryczne (Electric Micromachines)*; Wydawnictwa Naukowo-Techniczne WNT: Warszawa, Poland, 1996.
12. Turowski, J. *Obliczenia Elektromagnetyczne Elementów Maszyn i Urządzeń Elektrycznych (Electromagnetic Calculations of Elements of Electrical Machines and Devices)*; Wydawnictwa Naukowo-Techniczne (WNT): Warszawa, Poland, 1982.
13. Kluszczyński, K.; Spałek, D. *Step-by-Step Analysis of Induction Machines Allowing for Slotting*; Polisch Society for Theoretical and Applied Electrical Engineering (Under the Auspices of Electrical Engineering Committee of Polish Academy of Sciences): Warsaw, Poland, 2002.
14. Kluszczyński, K.; Miksiewicz, R. *Modelowanie 3-Fazowych Maszyn Indukcyjnych Przy Uwzględnieniu Wyższych Harmonicznych Przestrzennych Przepływu (Modeling of 3-Phase Induction Machines Taking into Account Higher MMF Spatial Harmonics)*; Zeszyty Naukowe Politechniki Śląskiej "Elektryka" z.142; Silesian University of Technology: Gliwice, Poland, 1995.
15. Turowski, J. *Elektrodynamika Techniczna (Technical Electrodynamics)*; Wydawnictwo Naukowe PWN: Warszawa, Poland, 2014.
16. Wach, P. *Dynamics and Control of Electrical Drives*; Springer: Berlin/Heidelberg, Germany, 2011.

energies

MDPI

Article

Design of a Boomerang-Type Rotor for Achieving IE4 Efficiency in a 37 kW LS-SynRM

Choung-Seo Kim [1], Chan-Bae Park [1], Jae-Bum Lee [2], Seong-Hwi Kim [3] and Hyung-Woo Lee [1,*]

1 Department of Railroad Convergence System, Korea National University of Transportation, Uiwang-si 16106, Republic of Korea; gmailgoogle@naver.com (C.-S.K.); cbpark@ut.ac.kr (C.-B.P.)
2 Department of Railroad Electric Engineering, Korea National University of Transportation, Uiwang-si 16106, Republic of Korea; leejb@ut.ac.kr
3 Department of Electric Engineering, Hanyang University, Seoul 04763, Republic of Korea; nyen0578@naver.com
* Correspondence: krhwlee@ut.ac.kr; Tel.: +82-31-460-0554

Abstract

Motors are among the most energy-consuming devices worldwide. With growing interest in eco-friendly solutions, minimum efficiency regulations for industrial motors are being enforced. In response to continuously rising minimum efficiency requirements, research on improving the efficiency of motors is actively underway. In the case of induction motors, which are the most widely used industrial electric motors, rotor ohmic loss occurs due to their operating characteristics. In contrast, line-start synchronous reluctance motors (LS-SynRMs) have a significant advantage in efficiency because once they reach synchronous speed, no eddy currents are generated by the fundamental current waveform. This leads to a sharp decrease in rotor ohmic losses, greatly enhancing efficiency. In this paper, a rotor design is carried out to improve the efficiency of LS-SynRMs. To support the rotor design, the torque characteristics of LS-SynRMs were analyzed under both asynchronous and synchronous state operations, and improvement directions for enhancing efficiency were identified. For rotor type selection, two bar-type rotors with linear flux barriers and two boomerang-type rotors with curved flux barriers were designed. The electromagnetic characteristics of these designs were compared using finite element analysis. Among them, the boomerang-type rotor that exhibited the best electromagnetic performance was selected as the final rotor type. Its final geometry was derived through detailed design, considering the mechanical safety of the rotor. Finally, experimental validation was conducted to verify the effectiveness of the proposed rotor design.

Keywords: line-start synchronous reluctance motor; minimum energy performance standard; efficiency improvement

1. Introduction

As interest in environmental sustainability continues to grow, the implementation of minimum energy performance standards (MEPS) for electric motors is underway, which represent a prominent energy-consuming device, constituting approximately 53% of total global electricity consumption. The efficiency of motors is classified into five levels, ranging from the lowest efficiency level, IE1, to the highest, IE5 [1]. Figure 1 shows the IE rating of four-pole motors. Currently, the production of motors with efficiency levels of IE3 and above is mandated, with active research aimed at raising the minimum efficiency rating to IE4.

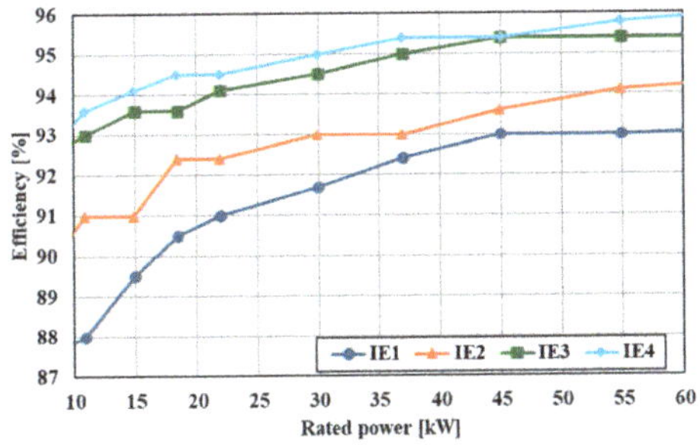

Figure 1. IE1 to IE4 efficiency classes for 4-pole 10 kW to 60 kW motors.

Induction motors (IMs), the most commonly used type in industrial applications, offer advantages such as easy production and low cost [2,3]. However, due to their operational characteristics, they incur rotor ohmic losses, accounting for approximately 25% of the total losses in IMs [4–8]. Line-start synchronous reluctance motors (LS-SynRMs) offer advantages such as ease of manufacturing, lower production costs, and the ability to utilize the benefits of induction motors without requiring inverter-based control. Additionally, once they reach a steady state, LS-SynRMs do not generate rotor ohmic losses from the fundamental components of the load current, making them advantageous for reducing losses [9]. However, unlike induction motors, where appropriate slip is determined based on the load, LS-SynRMs can only operate as motors when they reach a synchronous speed [10,11].

Therefore, active research is underway to address the limitations of LS-SynRMs and comply with increasingly stringent minimum efficiency regulations. H. C. Liu studied the optimal design of the rotor, additional losses that may occur during the rotor manufacturing process, and the resulting changes in synchronous performance [12,13]. H. Kim researched optimal rotor design techniques for LS-SynRMs, considering both efficiency and power factor, as well as ensuring start-up performance by accounting for the maximum allowable moment of inertia for the rotor [9,14]. Additionally, M. Farhadian explored methods to minimize the use of rotor conductors [15]. However, these studies mainly focused on small motors with a power rating of under 4 kW. In the United States, small-capacity motors rated at 15 kW or below account for approximately 77% of all installed industrial motors, yet their share of total electricity consumption is less than 18%. Medium-capacity motors, rated from 15 kW to 75 kW, represent the next largest group, comprising 18.7% of installations but accounting for approximately 30% of total power consumption. Therefore, considering both motor demand capacity and electricity consumption, improving the efficiency of medium-capacity motors is more effective than enhancing small-capacity motors [16].

When designing LS-SynRMs, it is important to consider both the characteristics of IMs and SynRMs due to the nature of their operational features. The rotor bars, which significantly influence asynchronous state performance, should be designed with IM principles in mind to ensure start-up performance. Research has been conducted on the optimal combinations of the number of stator slots and rotor bars to generate uniform magnetic torque during IM operation, with specific recommendations depending on the number of poles. These combinations typically assume uniform spacing of the conductor bars [17–21]. However, in the synchronous state, when the motor operates as a SynRM, uniformly spaced conductor bars can obstruct the flow of the d-axis flux, potentially leading to a significant decline in performance. Therefore, to maximize the performance of LS-SynRMs, it is beneficial to prioritize synchronous state performance during the initial design and then adjust the conductor bar design within a range that does not significantly affect the synchronous state performance.

When conducting electromagnetic analysis for motor design, both the steady-state and transient modes can be used. The analysis using steady-state mode employs a current source approach, allowing for the motor's characteristics to be quickly evaluated. However, industrial motors do not directly input the desired current. Instead, the current characteristics are influenced by the load applied to the motor. Predicting the load current characteristics based on load and rotor shape during motor design is challenging. Current characteristics, such as magnitude, phase angle, and harmonic components, have a significant impact on the motor's output power and losses. Therefore, analysis using steady-state mode, which uses a current source, cannot accurately assess the motor's characteristics. Instead, to accurately analyze the motor's characteristics, it is essential to use transient mode, which allows for voltage source analysis and can accommodate varying current characteristics under changing conditions. Therefore, all finite element analyses (FEA) conducted in this study were performed using transient mode.

This paper aims to design a four-pole 37 kW LS-SynRM rotor that meets IE4 efficiency for industrial applications. The IE4 efficiency for a 37 kW motor is 95.4%. To achieve this aim, we first analyzed the operating principle of the LS-SynRM. Subsequently, we designed and compared the characteristics of two types of rotors: bar-type rotors with straight flux barriers and boomerang-type rotors with curved flux barriers. Among the four rotor designs, the one with the best efficiency was selected for further refinement, considering manufacturability. Additional design considerations were made to ensure the start-up performance necessary to reach synchronous speed, as well as manufacturability and mechanical rigidity. The mechanical safety of the rotor during operation was verified through FEA. Finally, the validity of the LS-SynRM design was analyzed through performance testing.

Through this paper, we extended the scope of research beyond previous works that focused only on small-capacity LS-SynRMs with a power rating of under 15 kW and conducted a high-efficiency design for a medium-capacity 37 kW motor. Using 2D FEA, the advantages of the boomerang-type rotor over the bar-type rotor were demonstrated, and a design methodology that simultaneously considers efficiency improvement and mechanical safety of the boomerang-type rotor is proposed. This approach not only fills a gap in the field of medium-capacity LS-SynRM design but also provides important insights into the future development of high-efficiency LS-SynRMs.

2. Operating Principle

The LS-SynRM operates similarly to an IM in an asynchronous state and operates like a SynRM once it reaches synchronous speed. Therefore, the operating principle of LS-SynRM should be analyzed separately for both the synchronous and asynchronous states.

2.1. Synchronous State Operating

When the rotor of an LS-SynRM reaches a synchronous speed, it operates in the same way as a standard SynRM. Therefore, the torque equation can be expressed as shown in Equation (1) [9,19].

$$T_s = \frac{3}{2}p(L_d - L_q)i_d i_q \tag{1}$$

Here, p is the number of pole pairs, L_d and L_q are the inductance of the d-axis and q-axis, respectively, and i_d and i_q are the current of the d-axis and q-axis, respectively. LS-SynRMs operate based on a constant input voltage without directly controlling the input current.

Therefore, to analyze its characteristics, Equation (1) should be converted into an expression that utilizes the input voltage and reactance, as shown in Equation (2) [19].

$$T_s = \frac{3}{2}p\frac{V_a^2}{2}\frac{(L_d - L_q)}{\left(R_s^2 + X_dX_q\right)^2}\left\{\left(X_dX_q - R_s^2\right)\sin 2\delta_v + R_s\left(X_d - X_q\right) - R_s\left(X_d - X_q\right)\cos 2\delta_v\right\} \tag{2}$$

Here, V_a is the input voltage, R_s is the stator resistance, X_d and X_q are the reactance of the d-axis and q-axis, respectively, and δ_v is the voltage angle. In Equation (2), there are two methods to increase the maximum torque. One method is to reduce the stator resistance, and the other is to maximize the gap between the d-axis and q-axis inductances. Due to the operating characteristics of the LS-SynRM, a load current is generated according to Equation (1) when load torque acts on the system. Therefore, when the load torque is the same, the inductance gap between the d-axis and q-axis becomes inversely proportional to the load current. To improve motor efficiency in synchronous state operation, maximizing this inductance difference is important. In Section 3 of this paper, design parameters are selected to determine the appropriate width and length of the rotor flux circuit in order to maximize the difference between the d-axis and q-axis inductances, and these parameters are incorporated into the rotor design.

2.2. Asynchronous State Operating

In the asynchronous state, while the LS-SynRM operates on a principle similar to that of an IM, it differs from the IM by having the inductance gap between the rotor's d-axis and q-axis. Therefore, the torque in the asynchronous state can be expressed as shown Equation (3) [4,9,22].

$$T_{as} = T_{avg} + T_{rel}\sin(2s\omega_e t + \alpha) \tag{3}$$

Here, T_{as} is the asynchronous torque, T_{avg} is the electromagnetic torque generated by the rotor conductor bars, and T_{rel} is the magnitude of the reluctance torque. s is the slip, and ω_e is the angular frequency of the synchronous speed, while α is the phase angle of the reluctance torque. The reluctance torque is generated due to the inductance gap. Although the average value of the reluctance torque is zero, it induces torque ripple in the asynchronous state. Figure 2 shows the torque–speed curve of the LS-SynRM. As the motor operates, a load torque is applied to the rotor, resulting in start-up with a large load current that generates a significant instantaneous torque. Over time, as the load current stabilizes, the torques resulting from T_{avg} and T_{rel} become apparent.

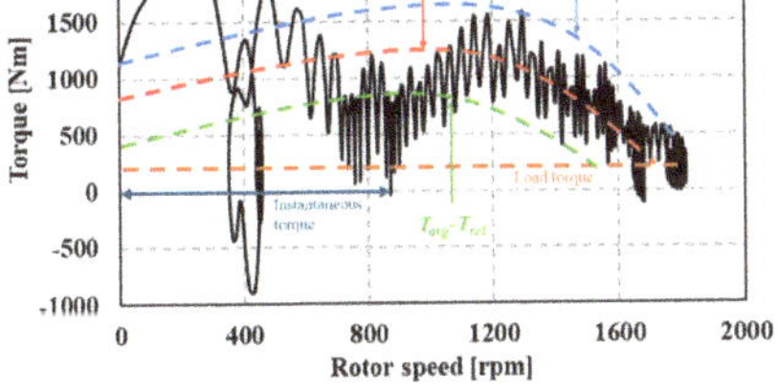

Figure 2. Torque–speed curve of the LS-SynRM.

To ensure that the LS-SynRM reaches synchronous speed safely, T_{avg} must exceed the load torque near the synchronous speed. Considering the characteristics of the rotor, which has the inductance gap, T_{avg} can be expressed by decomposing it into d-axis and q-axis components, as shown in Equation (4) [9,18].

$$T_{avg} = \frac{3p}{4}\left(\frac{V_a}{\omega_e}\right)^2 \left\{ \frac{R_{dr} L_{dm}^2}{L_{ds}^2 R_{dr}^2 + s^2\omega_e^2 (L_{ds} L_{dr} - L_{dm}^2)} + \frac{R_{qr} L_{qm}^2}{L_{qm}^2 R_{qr}^2 + S^2\omega_e^2 \left(L_{qs} L_{qr} - L_{qm}^2\right)} \right\} \quad (4)$$

Here, R_{dr} and R_{qr} are the rotor's d-axis and q-axis resistance, respectively, L_{dm} and L_{qm} are the d-axis and q-axis magnetizing inductance, respectively, L_{dr} and L_{qr} are the rotor's d-axis and q-axis inductance, respectively, and L_{ds} and L_{qs} are the stator's d-axis and q-axis inductance respectively. As the rotor speed approaches a synchronous speed, the slip approaches zero. At this point, the average torque T_{avg} can be expressed as shown in Equation (5) [9,19].

$$T_{avg} = \frac{3p}{4}\left(\frac{V_a}{\omega_e}\right)^2 \left(\frac{L_{qm}^2}{L_{qs}^2 R_{qr}} + \frac{L_{dm}^2}{L_{ds}^2 R_{dr}}\right) \quad (5)$$

Through rotor design, it is important to reduce the values of R_{dr} and R_{qr} in order to increase the magnitude of T_{avg}. The rotor resistance is closely related to the size of the rotor bar area. Therefore, to ensure start-up performance, it is crucial to secure sufficient rotor conductor bars during the design process.

3. Basic Design of LS-SynRM Rotors

The bar-type rotor has the advantage of being relatively easy to manufacture due to its simple shape. On the other hand, the boomerang-type rotor has the drawback of being more difficult to manufacture due to the curvature of the flux barriers. However, the boomerang-type rotor reduces magnetic resistance by shortening the magnetic path along the d-axis, which increases the inductance gap between the d-axis and q-axis. This allows for sufficient inductance gaps with a smaller number of flux barrier layers. If the inductance difference remains the same but the number of flux barriers is reduced, the width of the rotor core forming the flux path becomes wider, leading to a lower magnetic flux density in the rotor core. This reduction in flux density results in decreased iron loss, which can increase efficiency. Therefore, we designed both rotors and compared their characteristics to select the superior rotor type.

Figure 3 shows the rotor design algorithm for achieving IE4 efficiency in an LS-SynRM. To initiate the rotor design process, it is crucial first to determine the stator shape, stator winding specification, and rotor size limit. Additionally, the material properties of the motor core and rotor conductor bars, along with factors such as the operating temperature of the stator and the magnitude of mechanical losses and stray load losses at the rated power, must be selected and incorporated into the design process.

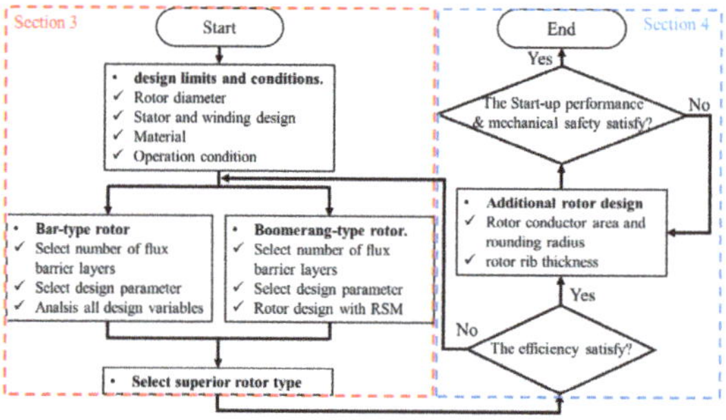

Figure 3. Design algorithm for a 4-pole 37 kW LS-SynRM.

In Section 4, the basic rotor design is conducted, and the superior rotor type is selected. Figure 4a shows the shape of the bar-type rotor, which has straight flux barriers. Two rotors were designed, one with and one without core notches. Figure 4b shows the boomerang-type rotor, which features curved flux barriers. Two cases of the boomerang-type rotor were designed based on selected design variables.

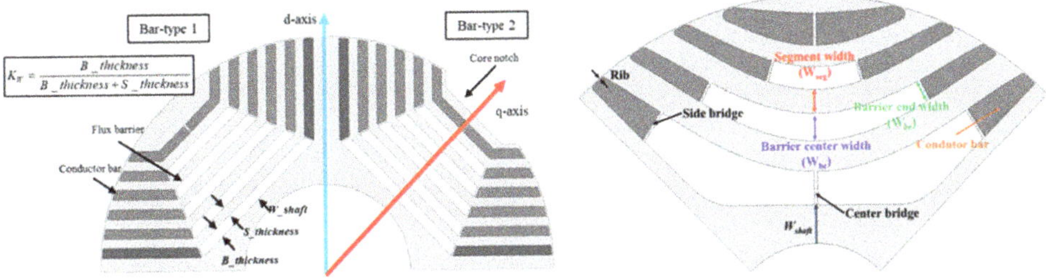

(**a**) (**b**)

Figure 4. Shape of the LS-SynRM rotor and its design variables: (**a**) bar-type rotor; (**b**) boomerang-type rotor.

Due to the fewer design variables, all possible configurations of the bar-type rotor were analyzed to determine the optimal model. In contrast, the boomerang-type rotor design involved three variables. Given the large number of possible configurations for the boomerang-type rotor, the response surface methodology (RSM) was used to reduce the design time required. Due to the characteristics of LS-SynRMs, which start up with the squirrel cage, the motor must be capable of reaching a synchronous state even when accounting for the maximum external inertia moment. The maximum external inertia moment can be calculated using Equation (6) [4,23].

$$J = 0.03P^{0.9}p^{2.5} \quad (6)$$

Here, J is the external inertia moment, and P is the rated power of the motor. The iron losses of the motor can be separated into hysteresis losses and eddy current losses. For calculating iron losses through transient FEA, Bertotti's loss model was employed. Equation (7) represents the expression developed by Lin et al. based on Bertotti's iron loss model [24].

$$P_{iron\ loss} = K_h f B_m^2 + K_c f^2 B_m^2 \quad (7)$$

K_h is the hysteresis loss coefficient, K_c is the eddy current loss coefficient, f is the supply frequency, and B_m is the peak magnetic flux density within the core material. For a conservative motor design, the operating temperature was increased by 10 °C compared to the operating temperature of a four-pole, 37 kW IM that achieves IE3 efficiency. Additionally, a 10% margin was applied to the stray losses and mechanical losses in the design process to calculate rotor damping. Mechanical and stray load losses were experimentally estimated as 1.35% of the rated output. Table 1 shows the design specifications for the LS-SynRM, including the stator core and winding details, and Figure 5 shows the shape of the stator and windings.

Table 1. 37 kW LS-SynRM design specifications and conditions.

Item	Value	Unit
Target efficiency	95.4	%
Maximum external inertia moment	4.376	kg·m^2

Table 1. *Cont.*

Item	Value	Unit
Input voltage	380	V
Frequency	60	Hz
Stator outer diameter	345	mm
Stator inner diameter	235	mm
Air gap	0.9	mm
Number of slots	48	-
Turn for the slot	5	-
Core material	50PN470	-
Iron loss coefficient (Kh/Ke)	152.60/1.28	
Core yield strength	275	MPa
Conductor bar material	Aluminum	-
Phase resistance @60 °C	0.04	Ohm
Rated load torque	196.3	Nm
Mechanical + stray loss	500	W

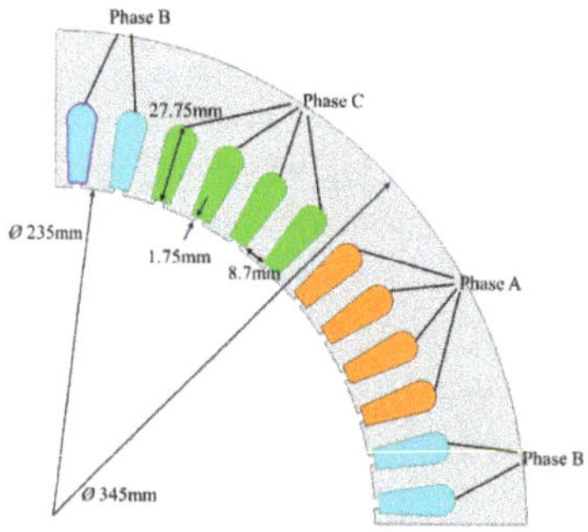

Figure 5. Shape of the stator core and windings for 37 kW LS-SynRM FEA.

3.1. Bar-Type Rotor Design

Prior research exists on the design of bar-type rotors with straight flux barriers and consideration of maximum external inertia. In this paper, the design of the bar-type rotor was carried out by referencing the rotor shape and design techniques from previous studies [9,25,26]. To design the SynRM rotor, the number of flux barriers must be determined. If there are too few flux barriers, it may be difficult to create a sufficient inductance gap, and the lack of rotor conductor bars can make it challenging to achieve start-up performance. On the other hand, if there are too many flux barriers, the thickness of the flux barriers and segments decreases, leading to increased saturation of the rotor core and higher leakage flux through the flux barriers, which ultimately reduces the inductance gap.

Due to its operating characteristics, the LS-SynRM exhibits different load currents at the same output depending on the rotor geometry. In this case, the load current is influenced by the difference between the d-axis and q-axis inductances, as expressed in Equation (1). However, the actual inductance during motor operation is affected by the degree of rotor flux saturation, which varies with the load current. Therefore, predicting performance solely based on rotor geometry cannot accurately reflect the motor's behavior during operation. To address this issue, this study compared characteristics through transient FEA by varying the number of rotor flux barriers, thereby enabling more accurate prediction of motor performance under operating conditions.

Throughout the analysis, the quantity of rotor conductors remained constant, and the ratio between the thickness of the flux barriers and the thickness of the segments between flux barriers was also maintained. Figure 6 shows the rotor shape and flux density of the bar-type model according to the flux barrier layers, and Table 2 presents the variations in efficiency, power factor, and inductance gap according to the number of flux barrier layers. It was confirmed that as the number of flux barriers increases, the inductance gap becomes larger, while the load current decreases inversely with the inductance difference. Consequently, the reduction in load current leads to improvements in both efficiency and power factor. The flux density distribution in Figure 6 shows that the average flux density decreases as the number of rotor flux barriers increases. A saturated flux density above approximately 1.8 T extends the effective length of the d-axis flux path and increases the magnetic reluctance. In the cases of the four-layer and five-layer rotors, local saturation occurred in the inner flux path of the rotor, which increased the d-axis inductance and adversely reduced the inductance gap. Accordingly, for both bar-type designs, a six-layer rotor was selected.

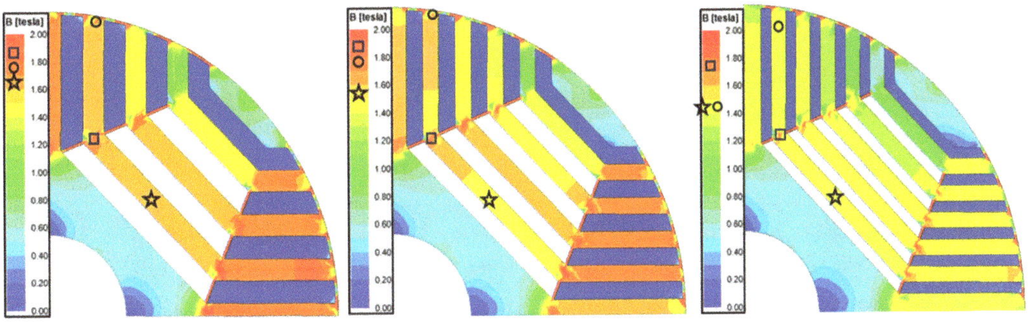

(**a**) (**b**) (**c**)

Figure 6. Rotor shape and flux density of bar-type rotor models according to flux barrier layers: (**a**) 4-layer; (**b**) 5-layer; (**c**) 6-layer.

Table 2. Bar-type rotor characteristics according to the number of flux barrier layers.

Num. of Flux Barrier Layers	Load Current $[A_{rms}]$	Efficiency [%]	Power Factor [-]	$L_d - L_q$ [mH]
4 layers	91.0	94.49	0.65	5.35
5 layers	85.4	94.91	0.69	6.15
6 layers	81.0	95.12	0.72	7.54

Two design parameters were selected for the optimal design of the rotor. The first parameter is the distance from the first flux barrier to the motor shaft (W_{shaft}), and the other is the ratio between the thickness of the flux barrier and the segment (k_w). Since there are only two design variables, analyses were conducted across the entire possible design range. The design range for W_{shaft} is from 11 mm to 18 mm, and the design range for k_w is from 0.4 to 0.6. Figure 7 shows the efficiency and PF according to the design variables. Bar-type models both without and with core notch rotors show similar trends in efficiency and PF. Therefore, the point at which both models achieve the highest efficiency and PF values was selected as the bar-type model's final design point. The final design results of the bar-type rotor show that the outcomes of the two models are identical, with a W_{shaft} of 16 mm and a K_w of 0.6.

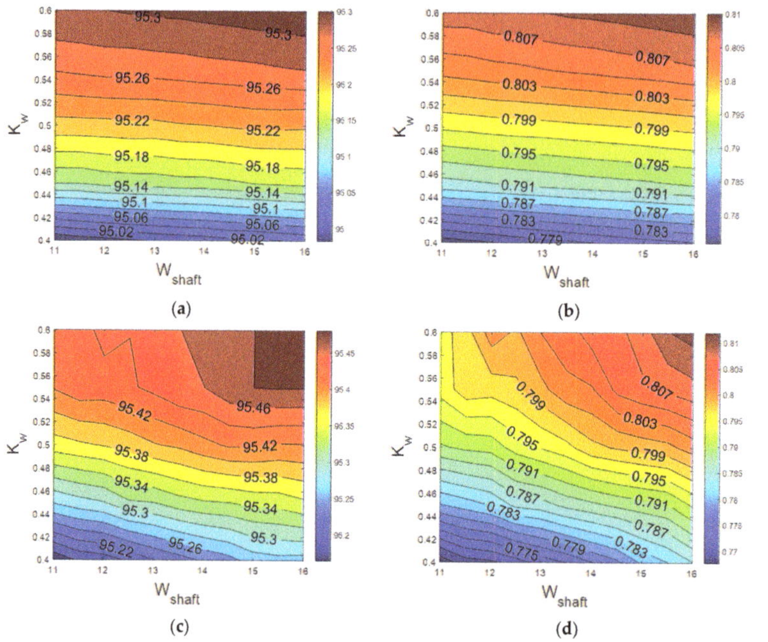

Figure 7. Efficiency and power factor according to the design variables: (**a**) efficiency of the bar-type model without a notch; (**b**) PF of the bar-type model without a notch; (**c**) efficiency of the bar-type model with a notch; (**d**) PF of the bar-type model with a notch.

3.2. Boomerang-Type Rotor Design

To select the boomerang-type rotor's number of flux barrier layers, we analyzed the characteristic changes corresponding to the number of flux barriers. Due to the geometric characteristics of the boomerang-type rotor, the number of flux barriers was reduced by one layer compared to the bar-type model, so only four-layer and five-layer models were analyzed. In order to assess the characteristic changes based on the number of flux barriers, the distance between the first and last flux barriers was kept constant, and the total thickness of the flux barriers was uniformly maintained across all models. Figure 8 shows the flux density and flux line of boomerang-type rotor models according to flux barrier layers, and Table 3 presents the resulting characteristic changes for the boomerang-type motor based on the number of flux barriers. Unlike the bar-type rotor, the boomerang-type rotor shows a tendency for the inductance gap to decrease as the number of flux barrier layers increases. As shown in Figure 8, the thickness at the ends of the rotor conductor bars decreases as the number of flux barrier layers increases, which indicates an increase in leakage current in the ribs. As the leakage current increases, the inductance gap decreases, leading to an increase in load current and saturation of the motor core. Consequently, the increase in core saturation of the rotor and stator teeth is observed in the five-layer rotor compared to the four-layer rotor. As shown in Table 3, the difference in the inductance gap increases the load current under the same load operation, which results in a decrease in efficiency and power factor. Therefore, for the boomerang-type rotor, the four-layer rotor was selected, which demonstrated superior electromagnetic characteristics.

A sensitivity analysis was conducted to select the design parameters for the boomerang-type rotor. The design variables subjected to the sensitivity analysis include four parameters: segment width (W_{seg}), barrier end width (W_{be}), barrier center width

(W_{bc}), and the width between the shaft and the first barrier (W_{shaft}), as shown in Figure 4b. The curvature of the flux barriers in the boomerang-type rotor is determined by the combination of these parameters. Table 4 presents the design variable values of the baseline model used for the sensitivity analysis, while Figure 9 shows the changes in electromagnetic characteristics resulting from variations in the design variables.

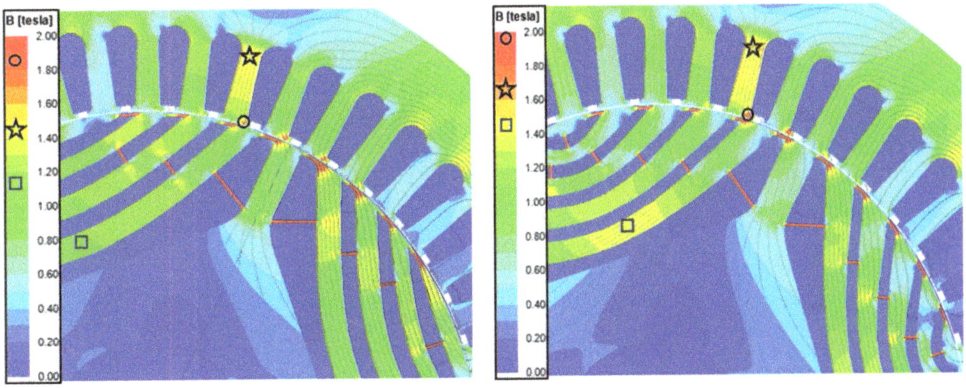

(**a**) (**b**)

Figure 8. Flux density and flux line of boomerang-type rotor models according to flux barrier layers: (**a**) 4-layer model; (**b**) 5-layer model.

Table 3. Boomerang-type rotor characteristics according to the number of flux barrier layers.

Num. of Flux Barrier Layers	Load Current [A_{rms}]	Efficiency [%]	Power Factor [-]	$L_d - L_q$ [mH]
4 layers	75.6	94.90	0.79	14.25
5 layers	78.2	94.77	0.75	12.99

Table 4. Default values of the basic model for the sensitivity analysis of the boomerang-type rotor.

Wshaft [mm]	Wbc [mm]	Wbe [mm]	Wseg [mm]
20	7.5	7	9

Among the design variables, W_{bc} and W_{be} exhibit the highest sensitivity. As W_{bc} increases, the q-axis inductance decreases, leading to an increase in the inductance gap, which in turn decreases the load current and improves efficiency, as shown in Equation (1). In the case of W_{be}, while it did not have a significant impact on the inductance gap compared to Wbc, increasing W_{bc} widened the rotor conductor bar, reducing leakage in the rib and resulting in a slight increase in the inductance gap. Consequently, the load current decreased, which led to improved efficiency. On the other hand, when W_{shaft} increased from 10 mm to 25 mm, the length of the d-axis flux path decreased, thereby increasing the inductance gap. However, beyond 25 mm, saturation occurred at the front of the last flux barrier, increasing leakage flux and ultimately resulting in a decrease in the inductance gap and efficiency. Similarly, when W_{seg} exceeded 6 mm, the d-axis inductance could no longer increase, limiting further increases in the inductance gap, and beyond 9 mm, the saturation at the front of the flux barrier increased, leading to a decline in efficiency.

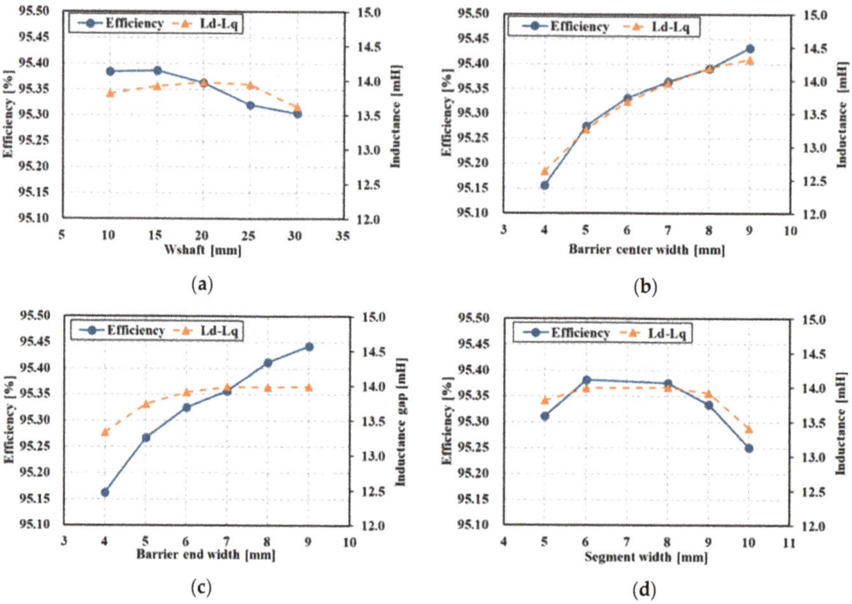

Figure 9. Result of sensitivity analysis: (**a**) width between shaft and first flux barrier; (**b**) barrier center width; (**c**) barrier end width; (**d**) segment width.

The variable with the highest sensitivity is Wbc; however, adjusting Wshaft and Wseg is necessary to secure Wbc. Additionally, as Wbe significantly affects efficiency, a detailed design that includes all design variables is essential. Therefore, the following two cases, which encompass all design variables, were selected:

1. XA1: W_{shaft}, XB1: $W_{bc}/(W_{bc} + W_{seg})$, XC1: W_{be}/W_{bc};
2. XA2: W_{shaft}, XB2: W_{bc}, XC2: W_{seg}, W_{be} is same with W_{bc}.

Since the LS-SynRM requires transient FEA for characteristic analysis, including asynchronous operation, reducing computational cost is essential. Therefore, the boomerang-type rotor was designed using the RSM, which is advantageous for multi-objective optimization with a limited number of data samples [27–32]. Table 5 shows the analysis range of the design variables for RSM. Equation (8) represents the second-order regression function used for RSM, with the constants and coefficients of the regression function presented in Table 6 [33–35].

$$y = \beta_0 + \sum_{i=A}^{C} \beta_i X_i + \sum_{i \leq j}^{C} \beta_{ij} X_i X_j \quad (8)$$

Table 5. Range of design variables for the RSM.

Case	XA	XB	XC
1	15–20 mm	0.35–0.55	0.4–0.8
2	15–20 mm	6–9 mm	6–9 mm

The aim of the detailed design using RSM is to maximize both efficiency and power factor. Figure 10 and Table 7 show the results of the RSM design, comparing the predicted design outcomes from RSM with the electromagnetic analysis results of the designed models obtained through FEA. When comparing the predicted values from RSM with the results from FEA, the error in efficiency was 1%p, and the error in power factor was less than 2.5%, confirming the validity of the design predictions made using RSM.

Table 6. Constants and coefficients of the second-order regression function for RSM.

	Case 1		Case 2	
i/*ij*	**Effciency**	**Power Factor**	**Effciency**	**Power Factor**
0	89.26	0.166	85.64	0.310
A	0.39	0.042	−0.042	0.036
B	4.89	0.516	1.80	0.106
C	4.65	0.394	1.13	0.122
AA	-6×10^{-3}	-7×10^{-4}	-9×10^{-3}	−0.002
AB	−4.26	−0.428	−0.37	−0.022
AC	−1.10	−0.056	−0.25	−0.005
BB	−0.10	−0.012	0.13	0.008
BC	−0.20	−0.022	−0.06	−0.002
CC	0.89	0.131	0.07	0.001

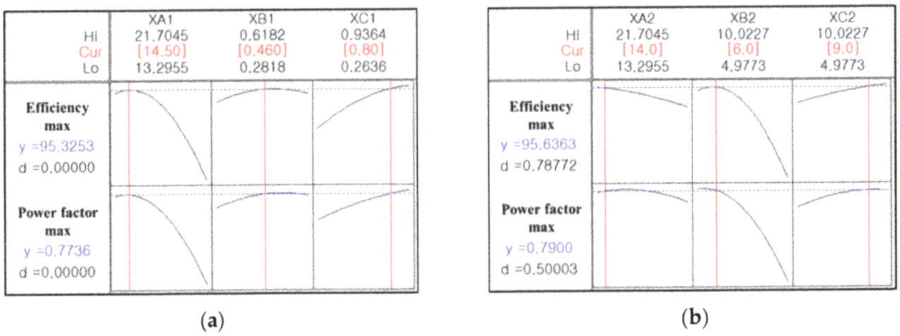

Figure 10. RSM results for the boomerang-type rotor: (**a**) Case 1; (**b**) Case 2.

Table 7. Results of the RSM and its comparison with FEA.

	Parameter			Efficiency [%]		Power Factor [-]	
Case	**XA**	**XB**	**XC**	**RSM**	**FEA**	**RSM**	**FEA**
1	14.5 mm	0.46	0.8	95.3	95.2	0.77	0.79
2	14 mm	6 mm	9.5 mm	95.6	95.7	0.79	0.80

Table 8 presents the comparison between the performance predicted by the RSM and FEA, analyzing the prediction accuracy. Considering the differences in the design variable and characteristics unit, efficiency, and power factor, the Normalized Mean Absolute Error (NMAE) was employed to evaluate the prediction error, while the Coefficient of Determination (R^2) was used to assess the accuracy of trend analysis between the design parameters and performance. The NMAE used in this paper was calculated according to Equation (9) [36,37].

$$NMAE = \frac{\sum_{i=1}^{n} |y_i - \hat{y}_i|}{n(y_{max} - y_{min})} \quad (9)$$

Here, y is the FEA result, while $\hat{y}$ is the predicted value. A larger NMAE indicates a greater error between predicted values and FEA results. Empirically, values below 0.1 are regarded as highly accurate predictions, while values between 0.1 and 0.2 are considered practically acceptable. In Case 1, the prediction of efficiency and power factor was not at a highly accurate level, but it was still at a practically useful level. In contrast, Case 2 showed NMAE values below 0.1 for both efficiency and power factor, indicating highly accurate predictions.

Table 8. Analysis of the prediction of RSM.

Item	Case 1		Case 2	
	Efficiency	PF	Efficiency	PF
NMAE	0.122	0.116	0.052	0.021
R^2	0.817	0.801	0.996	0.995

For R^2, values above 0.9 indicate excellent explanation for trends, while values between 0.7 and 0.9 suggest moderate predictive capability. In Case 1, the R^2 value was around 0.8, indicating a moderate level of prediction. Case 2, however, yielded results very close to the ideal value of 1, confirming that the trend of performance variation with parameter changes was very well predicted.

3.3. Design Results and Selection of the Final Rotor Type

Figure 11 shows the flux density of the four designed four rotors, while Figure 12 shows the flux density and flux lines of the boomerang-type rotor model's stator teeth and rotor rib, and Table 9 presents the characteristics calculated through electromagnetic FEA of these models.

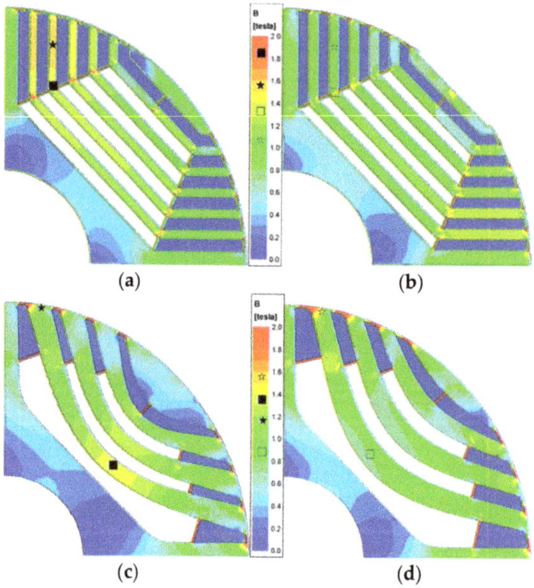

Figure 11. Flux density of four rotor types in LS-SynRMs: (**a**) bar-type without notch rotor; (**b**) bar-type with notch rotor; (**c**) boomerang-type Case 1 rotor; (**d**) boomerang-type Case 2 rotor.

Upon reviewing the design results of the bar-type rotor, it is evident that the saliency ratio and the inductance gap remain minimal, regardless of the presence of core notches. This leads to a negligible variation in load current magnitude, resulting in similar losses across different components, which ultimately fails to produce a significant impact on the efficiency and power factor.

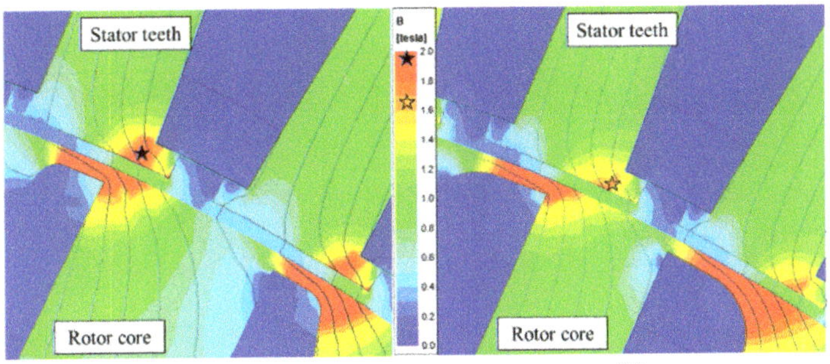

(**a**) (**b**)

Figure 12. Flux density and flux lines of the boomerang-type rotor model's stator teeth and rotor rib: (**a**) boomerang-type Case 1 rotor; (**b**) boomerang-type Case 2 rotor.

Table 9. Electromagnetic characteristics of the rotor types.

Item	Bar-Type		Boomerang-Type	
	Without Notch	With Notch	Case 1	Case 2
Rotor inertia moment (kg·m^2)	0.38	0.37	0.46	0.42
Load current (A_{rms})	72.7	72.5	75.9	73.0
Stator ohmic loss (W)	632	628	689	637
Iron loss (W)	438	427	391	330
Rotor ohmic loss (W)	282	225	310	191
Efficiency (%)	95.3	95.4	95.2	95.7
Power factor (-)	0.81	0.81	0.79	0.80
Saliency ratio (-)	2.9	2.9	2.6	2.8
$L_d - L_q$ (mH)	14.7	14.8	14.0	14.7
Current THD (%)	10.5	7.4	11.5	7.0

Conversely, the boomerang-type rotor demonstrates a marked difference between the Case 1 and Case 2 models. Although the Case 2 model contains fewer flux barriers compared to the bar-type rotor, it achieves a comparable inductance difference while maintaining lower saturation in the d-axis flux path, thereby reducing iron losses. In comparison to the Case 2 model, the Case 1 model exhibits a rapid reduction in the thickness of the rotor conductor bars as they approach the rib. A decrease in the thickness of the rotor conductor bars results in increased leakage flux at the ribs, which subsequently reduces the inductance gap and elevates the load current. The rise in load current intensifies the saturation of the stator teeth. Increased saturation of the stator teeth leads to a higher total harmonic distortion (THD) of the current, which, in turn, increases ohmic losses in the rotor conductor bars, ultimately leading to a reduction in efficiency.

Among the four rotor models, the ones that met the target efficiency were the bar-type with notch model and the boomerang-type Case 2 model. Considering manufacturability, the bar-type with notch model, which utilizes straight flux barriers, appears to be a more favorable choice. However, considering the target efficiency of 95.6% with a margin and the potential efficiency reduction caused by the conductor bar design discussed in Section 4, the boomerang-type Case 2, which exhibited the highest efficiency, was finally selected.

4. Detailed Design of the Boomerang-Type Rotor

Through the rotor designs, the rotor with the highest efficiency was selected. However, for the actual manufacturing of the motor, it is necessary to verify not only the performance at a synchronous state but also the start-up performance, and the mechanical safety regarding stresses during production and operation must also be confirmed. Figure 13 shows the design variables for the detailed design, focusing on start-up performance and mechanical safety.

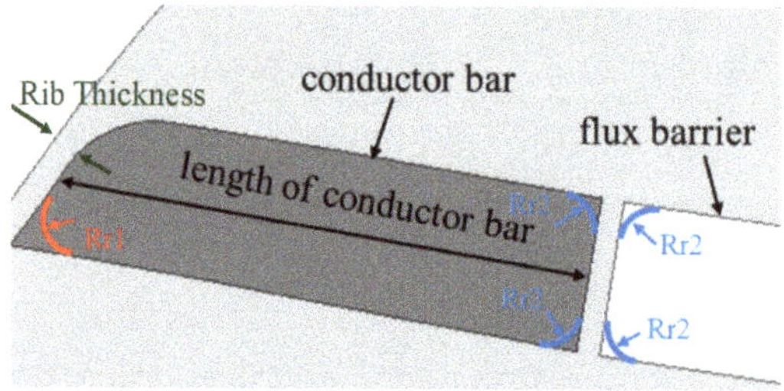

Figure 13. Design variables for start-up performance and mechanical safety.

4.1. Length of the Conductor Bar

To ensure that the LS-SynRM reaches a synchronous speed, it is crucial to reduce the rotor resistance, which inherently means increasing the area of the rotor conductors. This can be achieved by adjusting the thickness and length of the conductor bars. However, increasing the thickness of the rotor bars narrows the d-axis flux path, leading to higher saturation in the rotor core, which in turn reduces d-axis inductance and negatively impacts electromagnetic characteristics. Therefore, only the length of the conductor bars should be adjusted to achieve the desired synchronous performance [9,25].

To analyze the start-up performance relative to the length of the conductor bars, simulations were conducted under a load torque 1.25 times the rated torque and with the maximum external inertia moment. The range of analysis spanned 15–30 mm. Figure 14 presents the start-up analysis results based on the length of the conductor bars. When the length of the rotor conductor bars is less than 25 mm, the motor exhibits pulsations below synchronous speed, whereas it reaches synchronous speed at lengths of 25 mm or greater. Although the time to reach synchronous speed decreases as the length of the rotor conductor bars increases, considering manufacturability and cost, the minimum length of 25 mm was selected.

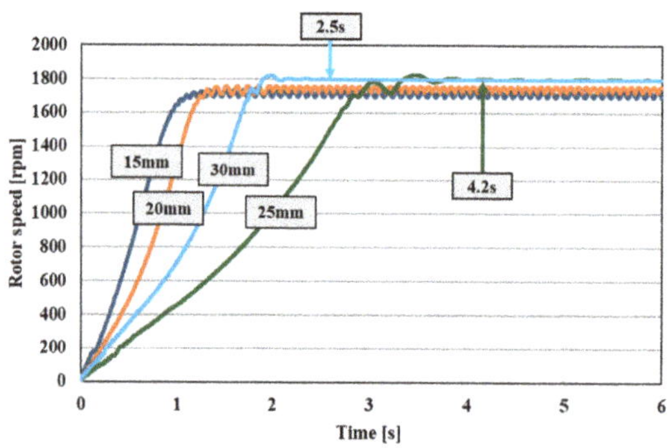

Figure 14. Start-up performance according to rotor conductor bar length.

4.2. Conductor Bar Rounding and Rib Thickness

To ensure the mechanical safety of the LS-SynRM rotor, a detailed design was conducted focusing on the rounding radius of the conductor bars adjacent to the rib (Rr1) and the thickness of the rib (Rt), which experience the highest stress during rotor rotation. An analysis of Rr1 was conducted for values ranging from 0.5 to 2 mm, while Rt was analyzed for values between 0.5 and 1.2 mm. Figure 15 presents the results of the electromagnetic analysis, highlighting the variations in load current and rotor ohmic loss, which are most affected by changes in Rr1 and Rt. In Figure 15a, it can be observed that while the change in load current with respect to Rr1 is minimal, the variation with respect to Rt is significantly larger, with an increase in Rt leading to a rise in load current. Conversely, Figure 15b shows that the rotor's ohmic loss is more sensitive to changes in Rr1 than in Rt, with a larger Rr1 resulting in a reduction in rotor ohmic loss.

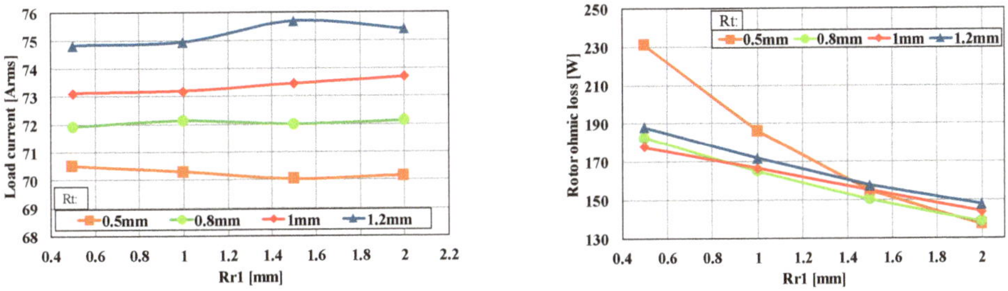

(a) (b)

Figure 15. Changes in electromagnetic characteristics according to Rt and Rr adjacent to the rib: (**a**) load current; (**b**) rotor ohmic loss.

Figure 16 shows the efficiency of the LS-SynRM according to the Rr1 and Rt, while Figure 17 shows the stiffness analysis results for the motor with the final selected values of Rr and Rt. The stiffness analysis was conducted at 2000 rpm, a speed higher than the rated speed, to ensure a conservative safety evaluation. Additionally, the density of the rotor conductor bar was increased by 1.5 times in the analysis to account for the rotor's end-ring structure. The safety factor (*SF*), which is used to assess mechanical safety, was calculated using Equation (10), with a target SF value of greater than 2 [22].

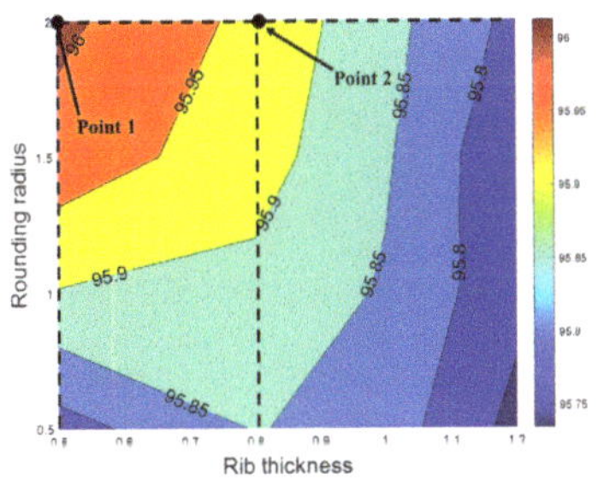

Figure 16. Changes in electromagnetic characteristics according to Rt and Rr adjacent to the rib.

$$SF = \frac{\sigma_{tenssile}}{\sigma_{max}} \tag{10}$$

Here, $\sigma_{tenssile}$ is the tensile yield strength, and σ_{max} is the maximum equivalent stress. In Figure 16, the highest efficiency is observed at the point at which Rr is 2 mm and Rt is

0.5 mm. However, as shown in the stiffness analysis results in Figure 17a, *SF* at this point is 1.5, which falls short of the target *SF*. To address the insufficient safety margin, a new point with a larger Rt value was selected. The second chosen point is at an Rr of 0.8 mm and an Rt of 2 mm. The stiffness analysis results for the second model are shown in Figure 17b, where the SF is 2.6, meeting the target value. Therefore, the final selection for the Rr and Rt values was made based on the second model.

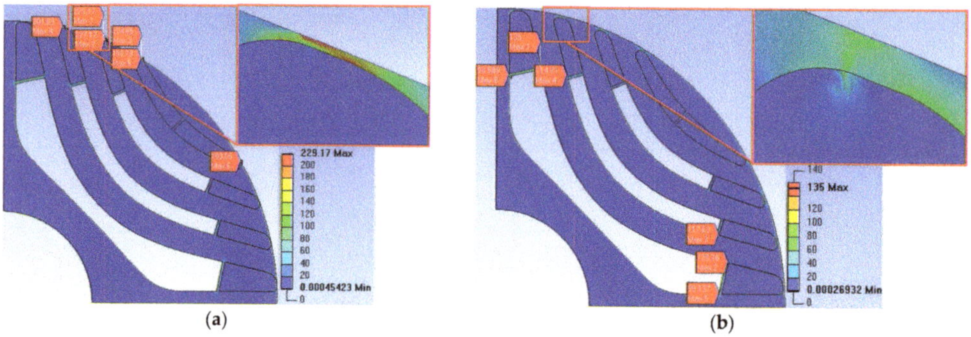

Figure 17. Mechanical stress distribution: (**a**) Rt 0.5 mm, Rr 2 mm; (**b**) Rt 0.8 mm, Rr 2 mm.

To prevent rotor damage during the insertion of the conductor into the rotor, it is also necessary to round the edges of the conductor bars and flux barriers adjacent to the side bridge (Rr2). Figure 18 shows the changes in load current and efficiency according to the Rr2 of the conductor bars and flux barriers adjacent to the side bridge. The design range for Rr2 was set to 0.5–2.5 mm, taking into account the thickness of the conductor bars. Across this design range, the variation in load current was observed to be a maximum of 0.35 Arms, and the efficiency varied by only 0.02%p, indicating minimal impact on the electromagnetic characteristics due to changes in Rr. Therefore, considering manufacturability, the Rr adjacent to the side bridge was finalized at 2.5 mm.

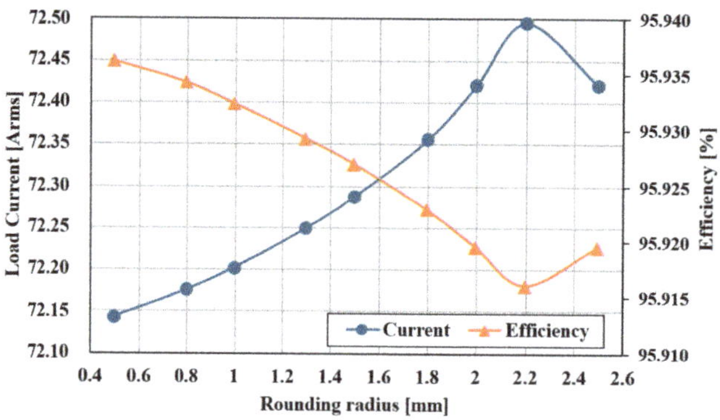

Figure 18. Changes in electromagnetic characteristics according to Rt and Rr2.

4.3. 37 kW LS-SynRM Final Model

Figure 19 shows the rotor shape, magnetic flux density distribution, and structural analysis results obtained using the previously selected conductor bar length, Rr, and Rt values. When analyzed under the same conditions as the previous simulations, the SF reached 3.1, exceeding the target of 2. Additionally, as shown in Figure 19c, it can be confirmed that the maximum deformation of the rotor is less than 0.016 mm. Figure 20 shows the efficiency and power factor of the final model under various load conditions, while Table 10 details the electromagnetic characteristics at a rated load. Both Figure 20 and Table 10 account for the previously identified margins for load current and iron losses. The efficiency at the rated load is 95.6%, achieving the target IE4 efficiency rating with a margin.

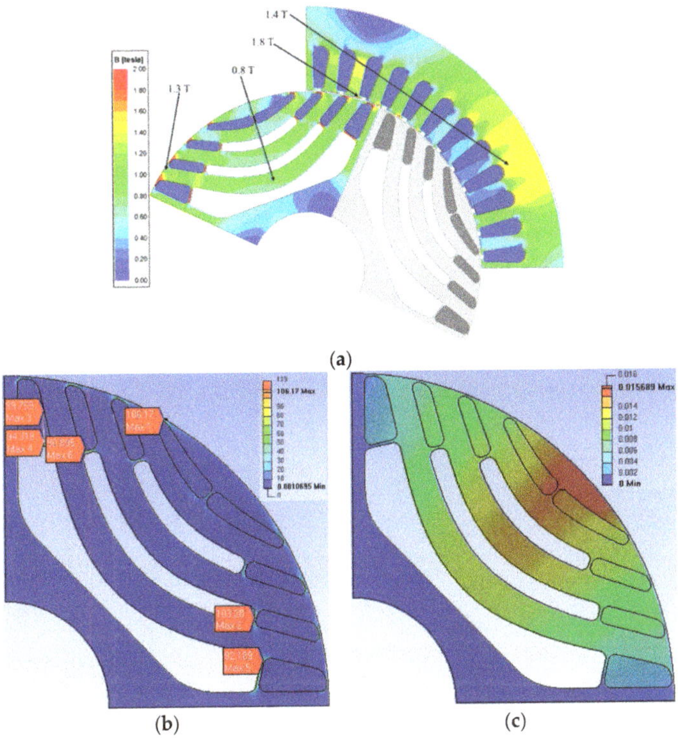

Figure 19. Rotor shape and flux density of the final model: (**a**) shape and flux density; (**b**) result of stiffness analysis; (**c**) result of deformation analysis.

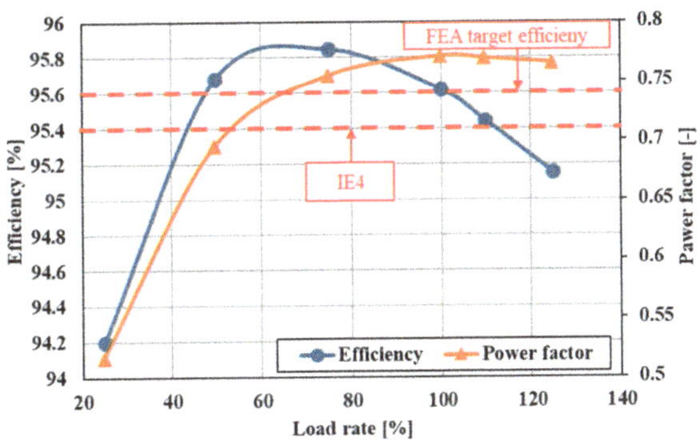

Figure 20. Efficiency and power factor according to the load rate.

Table 10. Electromagnetic characteristics of the final model at the rated power.

Item	Value	Unit
Rotor inertia moment	0.423	$kg{\cdot}m^2$
Load current	76.2	A_{rms}
Stator ohmic loss	694	W
Iron loss	392	W
Rotor ohmic loss	141	W
Efficiency	95.6	%
Power factor	0.77	-
Current THD	6.2	%

5. Experimental Results

An experimental test was conducted to validate the LS-SynRM design. Figure 21 shows the experimental test setup constructed for evaluating the performance of the 37 kW LS-SynRM and the rotor core of the experimental prototype. A servo motor and torque sensor were connected for load application and output power measurement, while a power meter was employed to measure the motor's load current, efficiency, and power factor. To minimize measurement errors in motor characteristics, ten repeated tests were carried out at the rated load, and the averaged results were obtained. A comparison between these experimental results and the FEA predictions is summarized in Table 11.

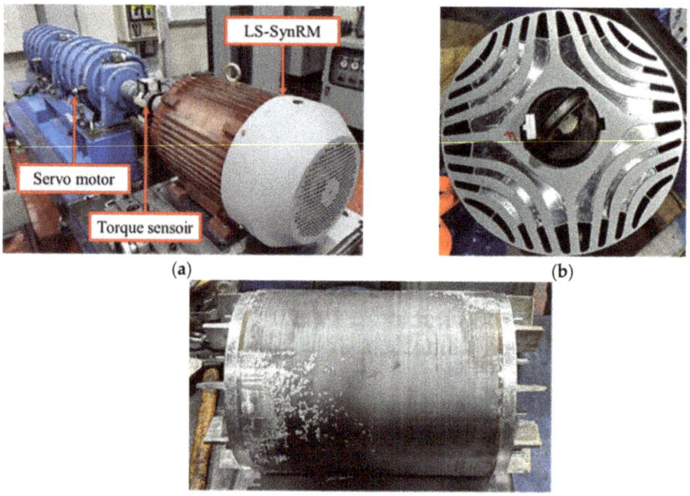

(a) (b) (c)

Figure 21. Experimental test setup: (**a**) test bed; (**b**) rotor core of prototype; (**c**) rotor prototype completed rotor conductor bar die casting.

Compared with the measured experimental values, the FEA results showed a 1.7% lower load current and a 2% higher power factor. However, the experimentally measured efficiency was approximately 0.2%p higher than the FEA prediction. In Figure 22, the current and voltage waveforms measured using an oscilloscope are compared with those obtained from the FEA. Unlike the FEA, where an ideal sinusoidal input voltage without harmonics is applied, the experimental results show that the input voltage contains harmonics. Although the fundamental component of the load current was larger in the experimental measurements, the peak value obtained from the FEA was higher due to the influence of harmonics. As mentioned earlier, the FEA assumes an ideal input voltage, and differences in damping caused by mechanical and stray losses can lead to discrepancies in both the magnitude of the current and the harmonic content. Nonetheless, the motor satisfied the target IE4-class efficiency requirement of at least 95.4%.

Table 11. Experimental results and comparison with FEA results.

Test Number	Speed [rpm]	Torque [Nm]	Output Power [W]	Current [A]	Voltage [V]	Power Factor [-]	Input Power [W]	Efficiency [%]
1	1800	195.9	36,926	77.43	380.35	0.753	38,450	96.04
2	1801	196.6	37,078	77.89	380.01	0.757	38,810	95.54
3	1800	196.3	37,001	77.36	380.33	0.754	38,460	96.21
4	1800	196.2	36,982	77.46	380.38	0.754	38,570	95.89
5	1800	196.2	36,982	77.27	380.51	0.755	38,460	96.16
6	1801	196.1	36,984	77.89	380.23	0.757	38,850	95.20
7	1800	196.1	36,964	77.67	380.52	0.756	38,710	95.49
8	1801	196.3	37,022	77.51	380.34	0.757	38,660	95.76
9	1800	196.4	37,020	77.43	380.21	0.756	38,590	95.93
10	1800	196.4	37,020	77.31	380.44	0.756	38,510	96.13
Average	1800	196.3	36,998	77.52	380.33	0.756	38,607	95.83
FEA	1800	196.3	37,001	76.20	380.00	0.771	38,703	95.60
Error [%]	-	-	-	1.70	-	1.98	0.24	0.02

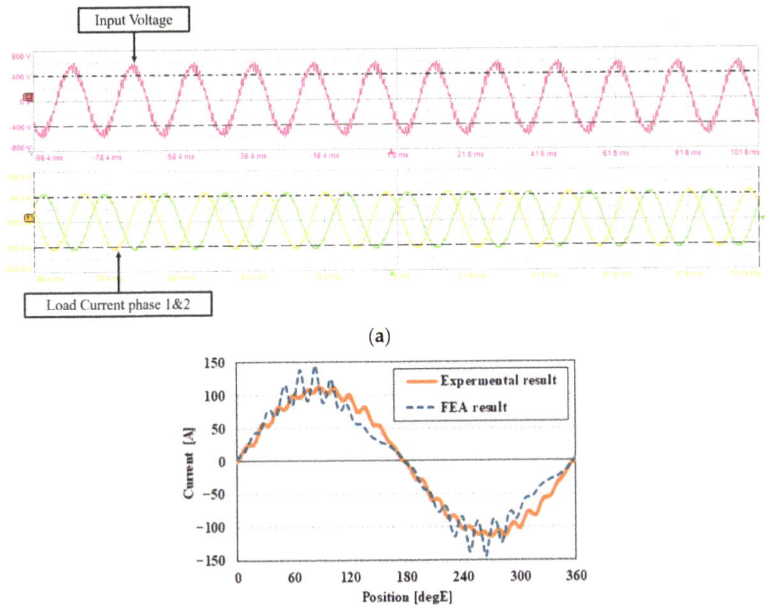

Figure 22. Comparison between experimental measurements and FEA results: (**a**) input voltage and load current measured through an oscilloscope; (**b**) load current comparison between measured waveform and FEA waveform.

6. Conclusions

This study aims to design a four-pole 37 kW LS-SynRM for industrial applications that meets IE4 efficiency standards. To achieve this goal, we first analyzed the characteristics of the motor by categorizing its operating conditions into synchronous and asynchronous states and identified parts requiring improvement to enhance efficiency. Subsequently, we designed two bar-type rotors and two boomerang-type rotors based on different flux barrier geometries and selected the superior rotor type configuration based on efficiency criteria. It was observed that, due to their geometric properties, boomerang-type rotors possess fewer flux barriers than bar-type rotors but facilitate a greater difference in d-axis and q-axis inductances owing to their shorter d-axis magnetic path lengths. To ensure start-up performance and mechanical reliability, detailed designs were conducted concerning the Rr of conductor bars and flux barriers, as well as the Rt. As a result, the 2D FEA of the designed motor achieved an efficiency of 95.6% at the rated power, exceeding the IE4 standard with a margin, while the SF also satisfied the target value of above 2. Furthermore, performance testing confirmed that the motor efficiency exceeded the target efficiency, thereby validating the effectiveness of the motor design.

Through this study, it was confirmed that the boomerang-type rotor is more advantageous than the bar-type rotor in securing a sufficient inductance gap. In addition, rotor design parameters for the boomerang-type rotor were derived with high prediction accuracy using the RSM. In future work, torque ripple—an important factor in the highly nonlinear characteristic—will be added as a design objective to develop a more sophisticated optimal design methodology.

Author Contributions: Methodology, C.-S.K.; Formal analysis, C.-S.K. and S.-H.K.; Investigation, C.-B.P.; Resources, C.-B.P. and J.-B.L.; Writing—original draft, C.-S.K.; Project administration, H.-W.L. All authors have read and agreed to the published version of the manuscript.

Funding: This work was supported by the Korea Institute of Energy Technology Evaluation and Planning (KETEP) and the Ministry of Trade, Industry & Energy (MOTIE) of the Republic of Korea (Project No. RS-2023-00232593/Development of common base technology for medium power industrial motors).

Data Availability Statement: The original contributions presented in this study are included in the article. Further inquiries can be directed to the corresponding author.

Conflicts of Interest: The authors declare no conflict of interest.

References

1. *KS C IEC 60034-30-1*; Rotating Electrical Machines—Part 30-1: Efficiency Classes of Line Operated AC Motors (IE Code). Korean Agency for Technology and Standards: Eumseong, Republic of Korea, 2024.
2. Gregor, R. *Induction Motors: Applications, Control and Fault Diagnostics*; IntechOpen: Rijeka, Croatia, 2015. [CrossRef]
3. Patel, M. *Shipboard Electrical Power Systems*, 2nd ed.; CRC Press: Boca Raton, FL, USA, 2021. [CrossRef]
4. Kersten, A.; Liu, Y.; Pehrman, D.; Thiringer, T. Rotor design of line-start synchronous reluctance machine with round bars. *IEEE Trans. Ind. Appl.* **2019**, *55*, 3685–3696. [CrossRef]
5. Xie, Y.; Pi, C.; Li, Z. Study on Design and Vibration Reduction Optimization of High Starting Torque Induction Motor. *Energies* **2019**, *12*, 1263. [CrossRef]
6. Fuchsloch, J.F.; Finley, W.R.; Walter, R.W. Next Generation NEMA Premium® Motors Substantially Lower Operating Costs. In Proceedings of the 2006 Record of Conference Papers—IEEE Industry Applications Society 53rd Annual Petroleum and Chemical Industry Conference, Philadelphia, PA, USA, 11–15 September 2006; pp. 1–7.
7. Almeida, A.T.; Ferreria, F.J.T.E.; Fong, J.A.C.; Brunner, C.U. Electric Motor Standards, Ecodesign and Global Market Transformation. In Proceedings of the 2008 IEEE/IAS Industrial and Commercial Power Systems Technical Conference, Clearwater Beach, FL, USA, 4–8 May 2008; pp. 1–9.
8. Subramanian, M.; Devarajan, N.; Deivasahayam, S.M.; Ranganathan, G. Review on Efficiency Improvement in Squirrel Cage Induction Motor by using DCR Technology. *J. Electr. Eng.* **2009**, *60*, 227–236.
9. Kim, H.; Kang, J.; Kim, J.; Ahn, J.; Yun, I.; Lee, J.; Noh, Y. Design and Analysis of Line-Start Synchronous Reluctance Motor Considering the Maximum Inertia and Power Factor. *IEEE Trans. Ind. Appl.* **2023**, *59*, 5908–5918. [CrossRef]

10. Hu, Y.; Chen, B.; Xiao, Y.; Shi, J.; Li, X.; Li, L. Rotor Design and Optimization of a Three-Phase Line-Start Synchronous Reluctance Motor. *IEEE Trans. Ind. Appl.* **2021**, *57*, 1365–1374. [CrossRef]
11. Liu, C.-T.; Shih, P.-C.; Cai, Z.-H.; Yen, S.-C.; Lin, H.-N.; Hsu, Y.-W.; Luo, T.-Y.; Lin, S.-Y. Rotor Conductor Arrangement Designs of High-Efficiency Direct-on-Line Synchronous Reluctance Motors for Metal Industry Applications. *IEEE Trans. Ind. Appl.* **2020**, *56*, 4337–4344. [CrossRef]
12. Liu, H.-C.; Lee, J. Optimum Design of an IE4 Line-Start Synchronous Reluctance Motor Considering Manufacturing Process Loss Effect. *IEEE Trans. Ind. Electron.* **2018**, *65*, 3104–3114. [CrossRef]
13. Liu, H.-C.; Hong, H.-S.; Cho, S.; Lee, J.; Jin, C.-S. Bubbles and Blisters Impact on Diecasting Cage to the Designs and Operations of Line-Start Synchronous Reluctance Motors. *IEEE Trans. Magn.* **2017**, *53*, 8202504. [CrossRef]
14. Kim, H.; Park, Y.; Oh, S.-T.; Jang, H.; Won, S.-H.; Chun, Y.-D.; Lee, J. A Study on the Rotor Design of Line Start Synchronous Reluctance Motor for IE4 Efficiency and Improving Power Factor. *Energies* **2020**, *13*, 5774. [CrossRef]
15. Farhadian, M.; Moallem, M.; Fahimi, B.; Dehkordi, B.M.; Sahebzamani, M. Alternate Rotor Design for Line-Start Synchronous Reluctance Motor with Minimum Use of Copper. *IEEE Access* **2024**, *12*, 73–84. [CrossRef]
16. Rao, P.; Chen, Y.; Goldberg, M.L.; Jones, B.; Cropp, J.; Hester, J. *United States Industrial and Commercial Motor System Market Assessment Report, Volume 1: Characteristics of the Installed Base*; Lawrence Berkeley National Laboratory, U.S. Department of Energy: Berkeley, CA, USA, 2021. [CrossRef]
17. Smit, Q.; Sorgdrager, A.; Wang, R.-J. Design and optimisation of a line-start synchronous reluctance motor. In Proceedings of the 24th Southern African Universities Power Engineering Conference, Vereeniging, South Africa, 26–28 January 2016.
18. Aguba, V.; Muteba, M.; Nicolae, D.V. Transient analysis of a start-up synchronous reluctance motor with symmetrical distributed rotor cage bars. In Proceedings of the 2017 IEEE AFRICON, Cape Town, South Africa, 18–20 September 2017; pp. 1290–1295.
19. Kersten, A. Efficiency Investigation of Line Start Synchronous Reluctance Motors. Master's Thesis, Department of Energy and Environment, Chalmers University of Technology, Göteborg, Sweden, 2017.
20. Gamba, M.; Armando, E.; Pellegrino, G.; Vagati, A.; Janjic, B.; Schaab, J. Line-start synchronous reluctance motors: Design guidelines and testing via active inertia emulation. In Proceedings of the 2015 IEEE Energy Conversion Congress and Exposition (ECCE), Montreal, QC, Canada, 20–24 September 2015; pp. 4820–4827.
21. Kim, M. A Study on the Design Method and Operating Characteristics Analysis for EV Traction Induction Motor. Ph.D. Thesis, Department of Energy and Environment, Hanyang University, Seoul, Republic of Korea, 2013.
22. Gamba, M.; Pellegrino, G.; Vagati, A. Design of Non Conventional Synchronous Reluctance Machine. Master's Thesis, Pelitecnico di Torino, Department of Energy, Turin, Italy, 2017.
23. *IEC 60034-12*; Rotating Electrical Machines—Part 12: Starting Performance of Single-Speed Three-Phase Cage Induction Motors. BSI: London, UK, 2024.
24. Lin, D.; Zhou, P.; Fu, W.; Badics, Z.; Cendes, Z. A dynamic core loss model for soft ferromagnetic and power ferrite materials in transient finite element analysis. *IEEE Trans. Magn.* **2004**, *40*, 1318–1321. [CrossRef]
25. Kang, J.; Kim, J.; Park, Y.; Kim, C.; Lee, J.; Kim, H. Optimization of SynRM for High Efficiency Considering Airgap Flux Density Shape. In Proceedings of the 2022 IEEE 20th Biennial Conference on Electromagnetic Field Computation, Denver, CO, USA, 24–26 October 2022; pp. 1–2.
26. Moghaddam, R.R. Synchronous Reluctance Machine Design. Master's Thesis, Royal Institute of Technology (KTH), Stockholm, Sweden, 2007.
27. Ortega-Casanova, J. CFD study on mixing enhancement in a channel at a low Reynolds number by pitching a square cylinder. *Comput. Fluids* **2017**, *145*, 141–152. [CrossRef]
28. Gilmour, S.G. Response surface designs for experiments in bioprocessing. *Biometrics* **2006**, *62*, 323–331. [CrossRef] [PubMed]
29. Mourabet, M.; El Rhilassi, A.; El Boujaady, H.; Bennani-Ziatni, M.; Taitai, A. Use of response surface methodology for optimization of fluoride adsorption in an aqueous solution by Brushite. *Arab. J. Chem.* **2017**, *10*, S3292–S3302. [CrossRef]
30. Khodaei, B.; Sobati, M.A.; Shahhosseini, S. Optimization of ultrasound-assisted oxidative desulfurization of high sulfur kerosene using response surface methodology (RSM). *Clean Technol. Environ. Policy* **2016**, *18*, 2677–2689. [CrossRef]
31. Liu, J.; Wang, J.; Leung, C.; Gao, F. A Multi Parameter Optimization Model for the Evaluation of Shale Gas Recovery Enhancement. *Energies* **2018**, *11*, 654. [CrossRef]
32. Yolmeh, M.; Jafari, S.M. Applications of response surface methodology in the food industry processes. *Food Bioprocess Technol.* **2017**, *10*, 413–433. [CrossRef]
33. Jung, Y.S.; Kwak, Y.S.; Lee, J.; Jin, C.S. Study on the optimal design of PMa-SynRM loading ratio for achievement of ultra-premium efficiency. *IEEE Trans. Magn.* **2017**, *53*, 8001904. [CrossRef]
34. Kim, S.-I.; Hong, J.-P.; Kim, Y.-K.; Nam, H.; Cho, H.-I. Optimal design slotless-type PMLSM considering multiple response by response surface methodology. *IEEE Trans. Magn.* **2006**, *42*, 1219–1222.

35. Saha, S.; Choi, G.-D.; Cho, Y.-H. Optimal rotor shape design of LSPM with efficiency and power factor improvement using response surface methodology. *IEEE Trans. Magn.* **2015**, *51*, 8113104. [CrossRef]
36. Yang, H.; Gong, S.; Liu, Y.; Lin, Z.; Qu, Y. A Multi-task Learning Model for Daily Activity Forecast in Smart Home. *Sensors* **2020**, *20*, 1933. [CrossRef]
37. Piotrowski, P.; Rutyna, I.; Baczyński, D.; Kopyt, M. Evaluation Metrics for Wind Power Forecasts: A Comprehensive Review and Statistical Analysis of Errors. *Energies* **2022**, *15*, 9657. [CrossRef]

energies

MDPI

Article

Speed-Adaptive Droop Control for Doubly Fed Induction Generator-Based Gravity Energy Storage System

Darui He [1], Yan Li [1], Chengjie Cao [1], Yifei Fan [1], Fei Wang [1], Yuanshi Zhang [2] and Chenwen Cheng [2,*]

1 Economic Research Institute, State Grid Jiangsu Electric Power Co., Ltd., Nanjing 210009, China; hdr2908@163.com (D.H.); liyan20211126@163.com (Y.L.); caochengjie@js.sgcc.com.cn (C.C.); fanyf@js.sgcc.com.cn (Y.F.); wangf16@js.sgcc.com.cn (F.W.)

2 School of Electrical Engineering, Southeast University, Nanjing 210018, China; yuanshizhang@seu.edu.cn

* Correspondence: chenwen_cheng@seu.edu.cn

Abstract

In order to ensure the reliable operation of a gravity energy storage system and reduce a converter's cost and power rating, this paper proposes a gravity energy storage system (GESS) based on a doubly fed induction generator (DFIG). To address the issues of low inertia and limited grid-support capability, a speed-adaptive droop control strategy is introduced. The droop curve can be adjusted automatically according to the speed variation. Thereby, the GESS can effectively provide grid support through rotor-speed control. A simulation model of the DFIG-based GESS is developed in MATLAB/Simulink 2024b, and the grid-support capability of the proposed control strategy is verified under various operating conditions.

Keywords: GESS; DFIG; grid-forming control; torque-frequency droop curve; adaptive speed control

1. Introduction

The penetration of renewable energy sources and power electronic devices in modern power systems continues to grow [1,2]. However, renewable energy sources, such as photovoltaic, wind, and ocean energy [3], are highly dependent on natural conditions, resulting in high intermittency and volatility. This significantly increases the complexity of grid operation [4,5], thereby driving the need for energy storage systems to perform peak shaving and valley filling [6,7]. Among various storage technologies, the GESS has attracted growing attention worldwide in recent years [8–10] due to its capability for reversible conversion between gravitational potential energy and electrical energy. Compared with other storage methods, the GESS is less susceptible to environmental fluctuations, demonstrates high stability and reliability, offers a long service life, and entails relatively low maintenance costs.

Existing GESS architectures are predominantly based on synchronous generators [11–13], as illustrated in Figure 1. Nevertheless, these designs face several critical limitations. On the one hand, since the rotor speed of a synchronous generator must remain strictly locked to the grid frequency, the translational velocity of the weight block cannot be controlled independently. On the other hand, the active power delivered to the grid can only be adjusted by engaging or disengaging discrete weights, a method that is inherently inflexible and lacks diversity in control strategies.

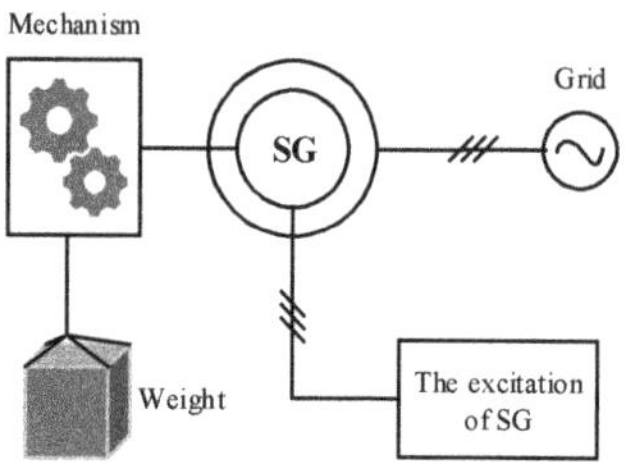

Figure 1. Direct connection of the stator side of the GESS based on a synchronous generator to the grid.

An alternative approach [10,14], shown in Figure 2, connects the stator of the synchronous generator to the load or grid through an AC-DC-AC converter. In this configuration, the stator voltage is rectified to DC by the generator-side converter and then inverted to grid-synchronized AC power by the grid-side converter (GSC). This arrangement eliminates the strict requirement for rotor speed to match the grid frequency, enabling wide-range control of the weight block velocity. However, in GESS architectures that employ an AC-DC-AC converter, the converter must be rated for the full power of the gravity storage system. Consequently, the need for a high-power converter leads to increased manufacturing and maintenance costs.

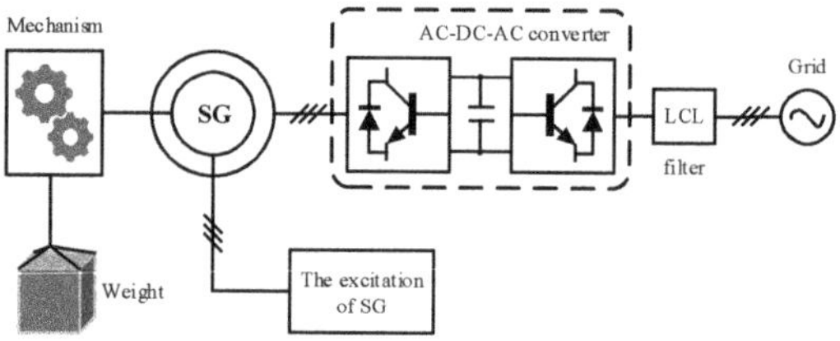

Figure 2. The stator side of the gravity energy storage system based on synchronous generators is connected to the power grid through an AC-DC-AC converter.

Unlike synchronous generators, the rotor speed of a DFIG is not synchronized with the stator frequency. By regulating power through the rotor-side converter (RSC), DFIGs can achieve variable-speed constant-frequency (VSCF) operation. Since the converter is connected to the rotor side, its power rating is only a fraction of the generator's rated power. This partial-scale converter significantly reduces the system cost, representing a clear advantage for GESS applications [15]. Nevertheless, research on DFIG-GESS remains limited. Compared to synchronous generators, DFIGs have lower rotational inertia and thus weaker grid-support capability [16]. To address this issue, Ref. [17] emulates the primary active power-frequency droop characteristic of synchronous generators to provide grid support. More broadly, recent studies on grid support mainly focus on strategies such as droop control [18], virtual synchronous generator (VSG) control [16], virtual oscillator control (VOC) [19], and matching control [20].

In [21] a small-signal model of multiple DFIGs is developed, focusing on the DC-link voltage control dynamics (around 10 Hz). The model reveals mutual interactions among grid-following DFIGs due to terminal voltage coupling under weak grid conditions, as well as the mechanism by which closely matched phase-locked loop (PLL) bandwidths in different units reduce the damping of the dominant mode and can lead to system instability. Ref. [22] develops a small-signal electromechanical model of grid-forming (GFM) controlled DFIG-based wind turbines under weak grid conditions. It identifies their vulnerability to electromechanical oscillations and proposes a hybrid d-q axis voltage reference drivetrain damping control strategy to ensure stable operation. Ref. [23] introduces an adaptive droop control method based on the rate of change of frequency (ROCOF). This method allows the DFIG to dynamically adjust its control coefficients according to system dynamics during frequency support, thereby significantly improving the frequency nadir and mitigating the ROCOF. Ref. [24] investigates the impact of inertial and primary frequency support from DFIG-based wind turbines on power system small-signal stability. The results demonstrate that while primary frequency support enhances the damping of electromechanical oscillations, the effectiveness of the inertial response is highly dependent on control parameters. Ref. [25] proposes a DFIG frequency support control strategy based on kinetic energy release. This strategy rapidly releases the stored kinetic energy of the wind turbine rotor during sudden grid frequency drops, providing both inertia and primary frequency support while ensuring the safe operation of the turbine. However, these DFIG-based GFM strategies are primarily designed for wind power applications, where the control objective is power regulation. In contrast, research on GFM control is specifically tailored for GESS, where speed regulation is the primary objective, remains limited. Comparisons between the proposed method and the existing GFM methods have been listed in Table 1.

Table 1. Comparison table between the proposed method and the existing GFM.

Method	Primary Control Objective	Adaptability Mechanism	Application Context	Notes vs. GESS
[17] (droop for parallel VSIs)	Decentralized real/reactive power sharing P-f, Q-V	Fixed droops; modular outer loops	Stand-alone/parallel inverters, no comms	No mechanical speed objective; not DFIG/GESS-specific.
[23] (ROCOF-adaptive DFIG droop)	Improve nadir and max ROCOF in events	Droop gain increases with df/dt	DFIG wind primary frequency support under disturbance events	Not co-designed with a GESS speed loop; requires additional integration to meet precise speed-tracking needs.
[25] (DFIG ACS + FA)	Seamless islanding/re-sync, feeder-automation compatibility	Mode-adaptive ACS; distant synchronization	Active DN with tie-switches, no DN-DG comms	Robust mode-switching; speed tracking not the central metric.
This work (speed-adaptive droop, DFIG-GESS)	Coupled: precise speed regulation + primary frequency support with partial-scale converter	Droop operating point adapted by the speed loop (torque-referenced)	GESS actuation under grid-forming control	Unifies GESS speed control and grid support; complements frequency-centric/adaptation strands above.

This paper proposes the novel application of a DFIG in the context of a GESS. Although DFIGs have been widely studied in wind turbine applications, their use in gravity energy storage systems is relatively unexplored. The key challenge of applying DFIGs to a GESS lies in overcoming their low rotational inertia and limited grid support capability. Traditional control strategies designed for wind turbines may not be directly applicable in this context due to the unique operational characteristics of gravity-based storage systems.

The speed-adaptive droop control strategy introduced in this study aims to address these challenges by dynamically adjusting the droop curve in response to changes in rotor speed, enabling improved grid support while minimizing the power converter rating. This control method is not only an adaptation of known wind turbine control strategies but also offers distinct advantages for gravity energy storage applications, including more flexible energy conversion and a significant reduction in the power converter size.

The proposed strategy aims to (i) achieve inner-loop regulation of the weight's translational speed using a partial-scale converter and (ii) enhance the system's grid-support capability. The remainder of this paper is organized as follows. Section 2 outlines the DFIG-GESS architecture and presents its mathematical model. Section 3 details the derived RSC control strategy from the model. Section 4 validates the proposed strategy through comprehensive MATLAB/Simulink simulations under various operating conditions. Finally, Section 5 concludes the paper.

2. Structure and Modeling of DFIG-GESS

Figure 3 illustrates the overall architecture of the DFIG-GESS. The system primarily consists of the weight block, mechanical transmission components, a DFIG, RSC, GSC, and an LCL filter. In this configuration, the weight block is mechanically coupled to the DFIG rotor through the transmission system. The gravitational potential energy of the weight drives this system, generating a mechanical torque T_L that acts on the rotor. By regulating the rotor speed ω_r, the translational velocity v of the weight can be precisely controlled during its descent.

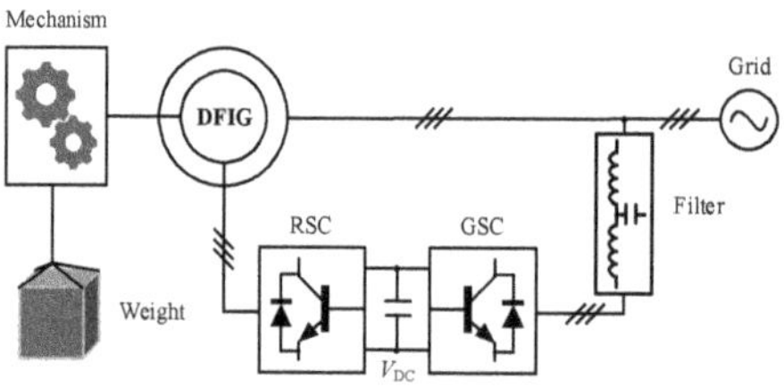

Figure 3. Architecture of gravity energy storage system based on doubly fed asynchronous motor.

2.1. Mechanical Transmission Electronic Control Collaborative System

The mechanical–electrical coordinated transmission system is composed of the mechanical assembly and the DFIG. As illustrated in Figure 4, the mechanical assembly consists of a pulley block, a wheel-and-axle mechanism, and a gearbox.

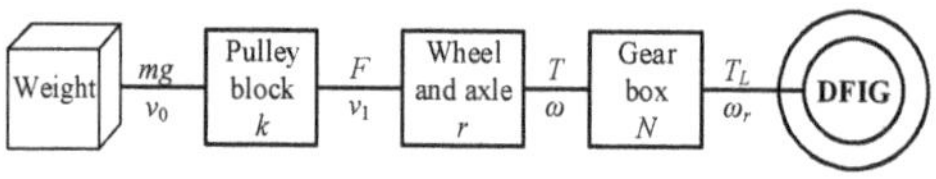

Figure 4. Mechanical transmission electronic control collaborative system mechanical structure.

The pulley block functions as a multi-stage force transmission system. Its combination of fixed and movable pulleys can be flexibly designed according to the application requirements of the GESS, enabling both directional deflection and magnitude scaling of the force vector.

The wheel-and-axle mechanism works as the translational–rotational conversion hub. Through rope winding, it converts the linear displacement of the weight into rotational motion of the shaft, simultaneously transforming the tensile force into torque and mapping translational velocity into angular velocity. The corresponding physical relationship is expressed as

$$\begin{cases} T = r \cdot F \\ v_1 = r \cdot \omega \end{cases} \tag{1}$$

where r is the drum radius, F is the force transmitted by the pulley block, and v is the rope's linear velocity.

The gearbox acts as the speed-matching unit, adapting the system power flow through either speed reduction with torque amplification or speed increase with torque reduction. Under ideal conditions, the mechanical torque T_L delivered to the DFIG rotor through the mechanism can be expressed as

$$T_{\mathrm{L}} = \frac{kr}{N} mg \tag{2}$$

where k is the pulley block ratio, N is the gear box transmission ratio, m is the mass of the weight, and g is the gravitational acceleration.

When the rotor windings rotate relative to the air-gap magnetic field, the interaction between the rotating field and rotor current produces the electromagnetic torque T_e. Thus, the dynamic coupling model between the rotor electromagnetic field and the mechanical load can be established:

$$T_{\mathrm{L}} - T_{\mathrm{e}} = J\frac{d\omega_{\mathrm{r}}}{dt} \tag{3}$$

where J is the rotor inertia of the DFIG, and ω_r is the rotor angular velocity.

The DFIG regulates its rotor speed through the rotor-side converter, which adjusts the rotor excitation current to control the speed of the air-gap magnetic field. Consequently, the rotor speed ω_r—which is equivalent to the output shaft speed of the gearbox—is controlled. Since the angular velocity of the wheel-and-axle mechanism is strictly synchronized with the gearbox input shaft, and the translational velocity satisfies Equation (1), regulating the DFIG rotor speed directly controls the translational velocity of the weight:

$$v_0 = \frac{Nr}{k}\omega_{\mathrm{r}} \tag{4}$$

where v_0 denotes the translational velocity of the weight.

In non-ideal conditions, additional dynamics such as weight inertia and damping D must be considered:

$$F = m\frac{dv_0}{dt} \tag{5}$$

$$F = D \cdot v \tag{6}$$

where v is the relative translational velocity. The friction force $F_{friction}$ can be modeled as

$$\begin{cases} F_{\mathrm{friction}} = \sqrt{2\mathrm{e}}(F_{\mathrm{brk}} - F_{\mathrm{C}}) \cdot \exp\left(-\left(\frac{v}{v_{\mathrm{St}}}\right)^2\right) \cdot \frac{v}{v_{\mathrm{St}}} \\ \qquad + F_{\mathrm{C}} \cdot \tan\mathrm{h}\left(\frac{v}{v_{\mathrm{Coul}}}\right) \\ v_{\mathrm{St}} = v_{\mathrm{brk}}\sqrt{2} \\ v_{\mathrm{Coul}} = v_{\mathrm{brk}}/10 \end{cases} \tag{7}$$

where F_{C} is the Coulomb friction force, F_{brk} is the breakaway friction force, v_{brk} is the breakaway velocity, and v_{St} is the Stribeck threshold velocity.

Correspondingly, the friction torque $T_{friction}$ is given by

$$\begin{cases} T_{\mathrm{friction}} = \sqrt{2\mathrm{e}}(T_{\mathrm{brk}} - T_{\mathrm{C}}) \cdot \exp\left(-\left(\frac{\omega}{\omega_{\mathrm{St}}}\right)^2\right) \cdot \frac{\omega}{\omega_{\mathrm{St}}} \\ \qquad + T_{\mathrm{C}} \cdot \tan\mathrm{h}\left(\frac{\omega}{\omega_{\mathrm{Coul}}}\right) \\ \omega_{\mathrm{St}} = \omega_{\mathrm{brk}}\sqrt{2} \\ \omega_{\mathrm{Coul}} = \omega_{\mathrm{brk}}/10 \end{cases} \tag{8}$$

where T_C is the Coulomb friction torque, T_{brk} is the breakaway torque, ω_{brk} is the breakaway angular velocity, ω_{St} is the Stribeck angular velocity threshold, ω_{Coul} is the Coulomb angular velocity threshold, and ω is the relative angular velocity.

Finally, the torque T_L obtained from the mechanical assembly is regarded as the mechanical input torque T_m to the DFIG.

2.2. System Power Flow Direction and Expression

This section analyzes the energy flow mechanisms of the DFIG-GESS under different operating conditions, thereby laying the foundation for the subsequent development of system control strategies.

The mechanical power generated by the weight and delivered to the mechanical assembly can be expressed as

$$P_{in} = mgv_0 \tag{9}$$

Under steady-state conditions, the magnetic fields of the DFIG stator and rotor are constant; thus, the electromagnetic power associated with magnetic field energy variation, P_{field}, is zero. Neglecting the losses of the AC-DC-AC converters, the total losses P_{loss} of the DFIG-GESS consist of the mechanical losses in the transmission structure (P_{mec}), the stator copper loss ($P_{Cu,s}$), and the rotor copper loss ($P_{Cu,r}$):

$$P_{loss} = P_{mec} + P_{Cu,s} + P_{Cu,r} \tag{10}$$

Accordingly, the active power delivered by the DFIG-GESS to the grid can be written as

$$P_e = P_{in} - P_{loss} \tag{11}$$

Through the coordinated mechanical–electrical transmission system, the DFIG-GESS converts the gravitational potential energy of the weight into electrical energy, which is then fed into the grid. The direction of active power flow depends on the translational velocity of the weight, which is determined by the rotor speed. When operating in generating mode, the electromagnetic active power of the DFIG is expressed as

$$\begin{cases} P_{es} = \omega_1 T_e \\ P_{er} = -\omega_{slip} T_e \\ P_e = \omega_r T_e \end{cases} \tag{12}$$

where ω_1 is the synchronous speed, $\omega_{slip} = \omega_1 - \omega_r$ is the slip speed, P_{es} is the stator-side electromagnetic power, P_{er} is the rotor-side electromagnetic power, and P_e is the total electromagnetic power.

Equation (12) indicates that in the hypersynchronous mode ($\omega_r > \omega_1$), the rotor-side power satisfies $P_{er} > 0$; the rotor delivers active power, and the system power flow is illustrated in Figure 5, where P_L denotes the mechanical power transmitted through the mechanism into the DFIG. In contrast, in the sub-synchronous mode ($\omega_r < \omega_1$), the rotor-side power satisfies $P_{er} < 0$; the rotor absorbs active power, and its electromagnetic power dynamically adjusts with the rotor speed to maintain constant stator electromagnetic power. The corresponding system power flow in this case is shown in Figure 6.

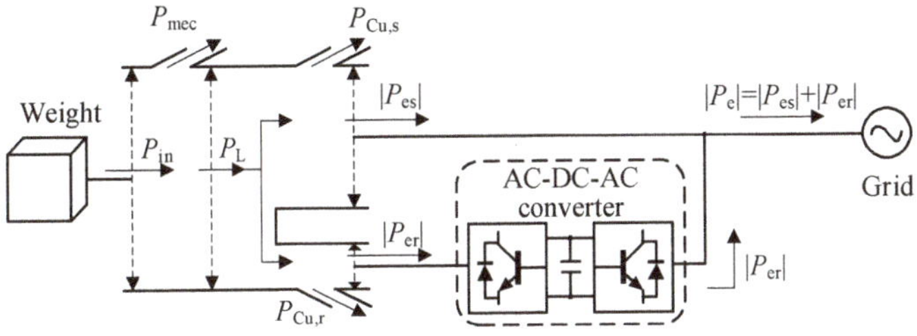

Figure 5. Power flow direction of the system in hypersynchronous mode.

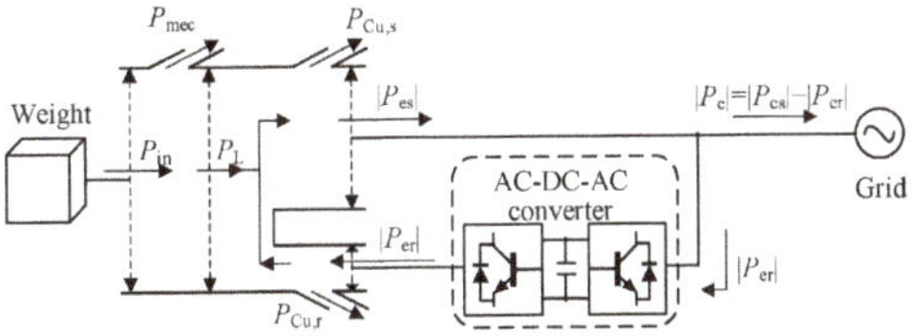

Figure 6. Power flow direction of the system in sub-synchronous mode.

3. Control Strategies

The control architecture of the AC-DC-AC converter in the DFIG-GESS is shown in Figure 7. Let U_{gABC} and I_{gABC} denote the three-phase voltage at the grid side and the three-phase AC-side current of the GSC, respectively. Similarly, let U_{sABC} and I_{sABC} denote the three-phase stator output voltage and current of the DFIG.

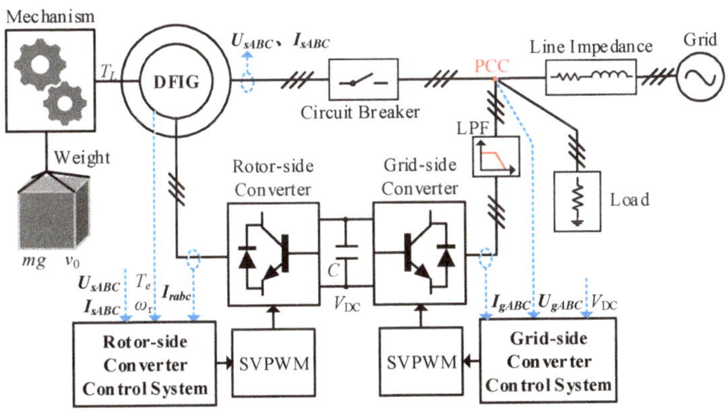

Figure 7. Structure of the doubly fed gravity energy storage system.

The GSC is connected to the point of common coupling (PCC) through an LCL filter. Via the GSC control system and space-vector pulse-width modulation (SVPWM), the three-phase grid supply is actively rectified to a DC voltage, providing a stable DC-link for the RSC. The RSC controls the DFIG using a speed-adaptive droop strategy combined with stator voltage vector-oriented control. This provides the system with active grid-support capability and enables wide-range controllable rotor-speed operation. When the voltages on both sides of the circuit breaker are synchronized—that is, their magnitudes, frequencies, and phases are matched—the breaker closes, allowing the stator-side output power to be delivered to the PCC.

3.1. Speed-Adaptive Droop Control Strategy

According to the rotor motion Equation (3), the rotor speed ω_r dynamics are governed by the difference between the mechanical and the electromagnetic torque. Since the mechanical torque is determined by the weight and transmission mechanism, controlling the rotor speed requires direct regulation of the electromagnetic torque. To achieve this, a closed-loop speed control scheme is designed, as shown in Figure 8.

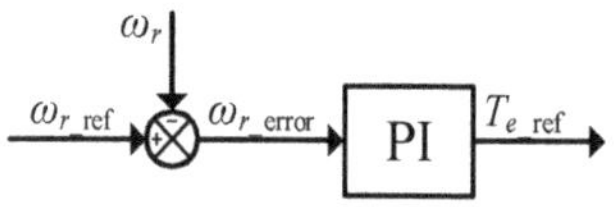

Figure 8. Speed control loop structure.

As shown in Figure 8, the reference speed ω_{r_ref} is compared with the measured rotor speed ω_r. The resulting error ω_{r_error} is processed by a proportional–integral (PI) controller to generate the reference electromagnetic torque T_{e_ref}.

Based on the power flow analysis in Figure 6, the active power injected into the grid corresponds to the stator electromagnetic power, which is regulated by the RSC. Furthermore, from Equation (12) and neglecting the influence of stator voltage frequency variations, the stator electromagnetic power depends only on the electromagnetic torque and is directly proportional to it. Inspired by the active power–frequency (P–T_e) droop characteristic of synchronous generators, a droop relationship between electromagnetic torque and frequency is established, as illustrated in Figure 9.

$$f = f_0 + K_{T_e}(T_{e\_ref} - T_e) \tag{13}$$

where f_0 is the rated grid frequency, T_{e_ref} is the reference electromagnetic torque (from the speed control loop), T_e is the actual electromagnetic torque, K_{Te} is the electromagnetic torque droop coefficient, and f is the frequency reference obtained from droop control.

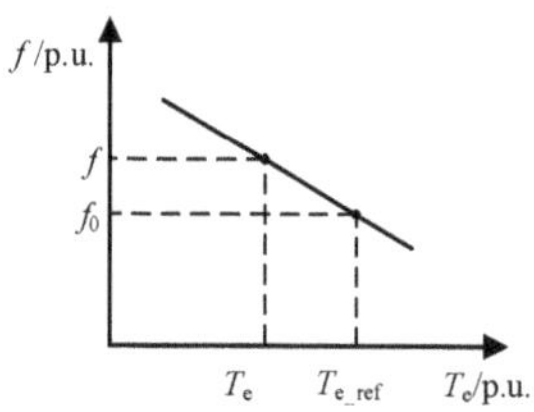

Figure 9. P-f droop characteristic curve.

In parallel, the conventional reactive power–voltage (Q-V) droop characteristic is adopted to generate the reference stator voltage amplitude E, as shown in Figure 10.

$$E = E_0 + K_Q(Q_{ref} - Q) \tag{14}$$

where E_0 is the rated grid voltage amplitude, Q_{ref} is the reference stator reactive power, Q is the actual stator reactive power, K_Q is the reactive power droop coefficient, and E is the reference stator voltage amplitude.

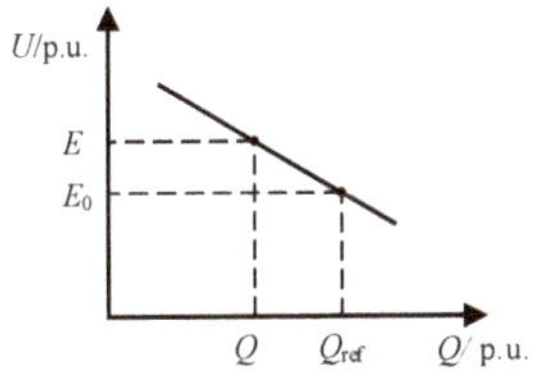

Figure 10. *Q-V* droop characteristic curve.

The frequency and voltage amplitude references obtained from the above droop laws are synthesized into a three-phase voltage reference signal E_{ABC} for the stator side. This enables the DFIG-GESS to exhibit the characteristics of a voltage source, similar to conventional droop-based grid-forming control. Consequently, when a fault at the PCC causes a sudden voltage disturbance, the stator reference signal E_{ABC} remains stable, providing inherent grid-support capability and allowing standalone operation for local loads. In contrast, grid-following control relies on a PLL to track the grid voltage and thus behaves as a current source. In such cases, voltage disturbances at the PCC propagate directly to the system output, resulting in limited grid-support capability.

Furthermore, from Equation (13), the intercept b of the torque-frequency droop curve can be expressed as

$$b = f_0 + K_{T_e} T_{e\_ref} \tag{15}$$

Unlike conventional droop control, where P_{ref} is predefined, the proposed speed-adaptive droop control dynamically sets T_{e_ref} based on the output of the speed loop. Consequently, the droop curve shifts vertically in real time according to the rotor speed, achieving adaptive adjustment.

The overall structure of the proposed speed-adaptive droop control strategy is illustrated in Figure 11.

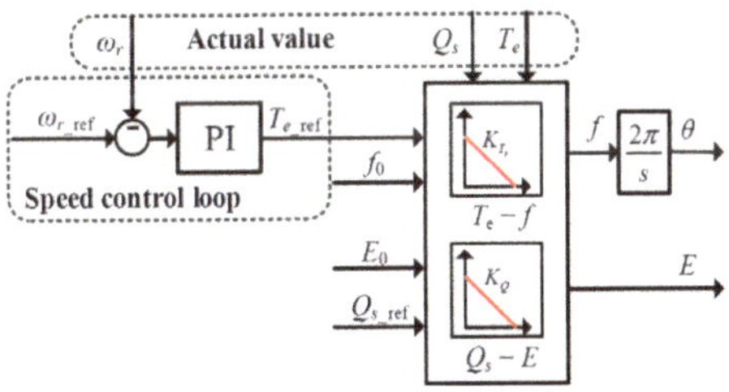

Figure 11. Adaptive control structure for full speed range.

3.2. RSC Control Strategy

The control scheme of the RSC in the DFIG is illustrated in Figure 12. The stator is directly connected to the grid or a local load, while the rotor side is connected to the converter. The converter is supplied by a regulated DC bus, which is maintained by the GSC. The gravitational potential energy of the suspended mass is converted into a mechanical torque T_L through the mechanical system and applied to the DFIG. The rotor speed ω_r and electromagnetic torque T_e of the machine are measured for feedback control.

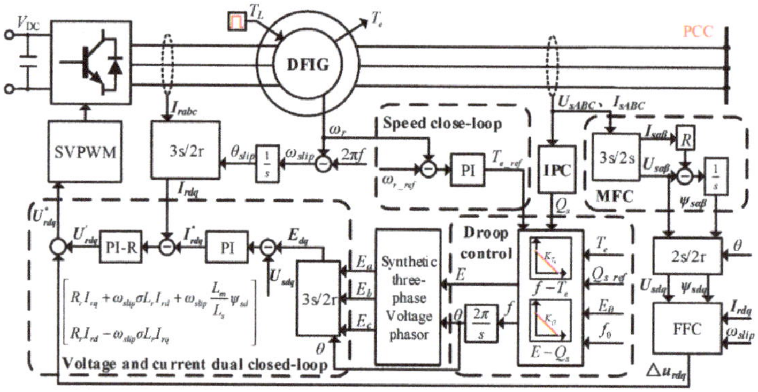

Figure 12. Block diagram of rotor-side converter control.

The RSC control system integrates multiple modules, including the speed closed-loop control, magnetic flux calculation (MFC), coordinate transformation, feed-forward calculation (FFC), instantaneous power calculation (IPC), droop control, three-phase voltage phasor synthesis, and voltage-current dual-loop control. The speed loop combined with the droop control constitutes the core proposed speed-adaptive droop control strategy. The control procedure is detailed below.

The measured rotor speed ω_r is compared with its reference speed ω_{r_ref}. The resulting speed error is processed by the speed controller to generate the reference electromagnetic torque T_{e_ref}, which is then fed to the droop control module.

The droop control module generates the frequency reference f. Combined with the voltage amplitude reference E from the Q-V droop, it synthesizes the three-phase stator reference voltage E_{ABC}.

This reference voltage signal undergoes Clark and Park transformations. The transformed signal is then used as the input to the voltage–current dual closed loops, which compute the reference rotor-side voltage in the dq reference frame.

Finally, the SVPWM algorithm generates the switching signals for the RSC power devices, thereby regulating the DFIG rotor speed and providing grid-supporting capability.

4. Simulation Verification and Case Analysis

A simulation model of DFIG-GESS was established in MATLAB/Simulink. The system performance was tested under both islanded operation and grid-connected operating conditions. For islanded operation, the objectives were to verify the droop characteristics and the speed-tracking performance of the proposed control strategy. In grid-connected mode, the evaluation focused on the active power injection through both the stator and rotor channels, as well as the system's capability to regulate the grid frequency through adjustments in mechanical torque and rotor speed.

In the proposed control strategy, the torque-frequency droop coefficient K_{Te} is selected to ensure precise speed tracking, provide primary frequency support without exceeding converter or current limits, and maintain small-signal damping of the grid-forming loop. The selection process involves mapping performance to bounds (defining maximum frequency deviation and torque error), respecting converter constraints (adjusting K_{Te} to avoid current limits), and verifying small-signal damping. A smaller K_{Te} is used for islanded operation to tighten frequency regulation, while a larger K_{Te} is chosen for grid-connected operation to avoid unnecessary frequency bias.

PI parameters are tuned sequentially, starting with the inner loops (current and voltage control) and then the outer speed loop. The current loop is first adjusted by increasing the proportional gain for a fast, stable response. Once stabilized, the integral gain is adjusted to minimize steady-state error. After stabilizing the current loop, the voltage loop is tuned for stable output, and then the speed loop is adjusted to ensure precise speed tracking. This method ensures stable and precise control across the system.

Based on the above parameter-tuning methodology, the final settings of the simulation parameters are shown in Table 2.

Table 2. Simulation parameters.

Parameter	Value	Parameter	Value
P_o	2 MW	R_s	0.0024 Ω
U_{sABC}	690 V	R_r	0.002 Ω
V_{DC}	1150 V	$L_{\sigma s}$	0.0604 mH
L_m	4.4 mH	$L_{\sigma r}$	0.0827 mH
f_0	50 Hz	g	9.8 m/s^2
v_0	20.408 m/s	n	1500 rad/s
m	10,000 kg	T_e and T_L	−212.206 N·m
$k\_\omega p$	2	$k\_\omega i$	8
K_{Te}	1	K_Q	4
k_up	1	k_ui	5
k_ip	9	k_ii	20

4.1. Islanded Operation

Under islanded conditions, the speed reference was varied to verify the consistency between the frequency of the stator voltage f and the electromagnetic torque T_e, as defined by the constructed T_e-f droop curve. The mechanical torque from the suspended mass was fixed at −0.8 p.u., and the stator supplied a constant local load.

Figure 13 shows the time-domain waveforms of the reference and actual rotor speeds. The rotor speed ω_r (blue curve) accurately tracks the reference ω_{r_ref} (red curve) with a negligible steady-state error across a wide speed range, demonstrating the effectiveness of the speed closed-loop control. At 20 s, the reference speed changes from a constant value to a ramp signal, and the settling time corresponding to a rotor-speed deviation of less than 0.1% is 0.5 s. At 30 s, the reference speed changes from the ramp signal to a constant value, with an overshoot of 0.2% and a settling time of 0.4 s.

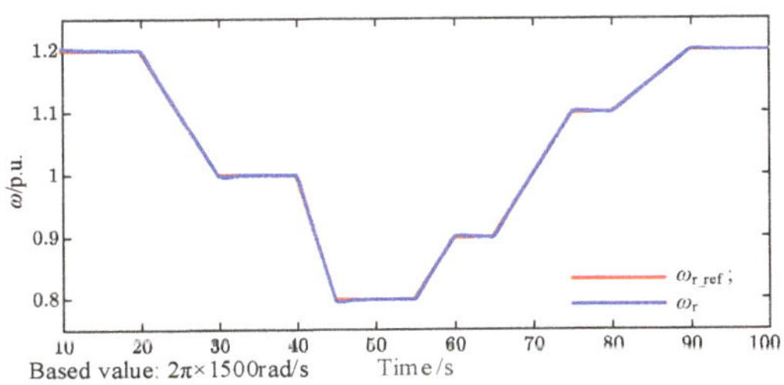

Figure 13. Waveform plot of the reference and actual value of rotor speed at the unit value.

Figure 14 presents the measured electromagnetic torque and rotor-speed waveforms. The filtered electromagnetic torque T_{e_LPF} (red) closely tracks the applied mechanical torque T_L when the rotor speed ω_r is stable, which validates the torque-speed dynamics predicted by Equation (3).

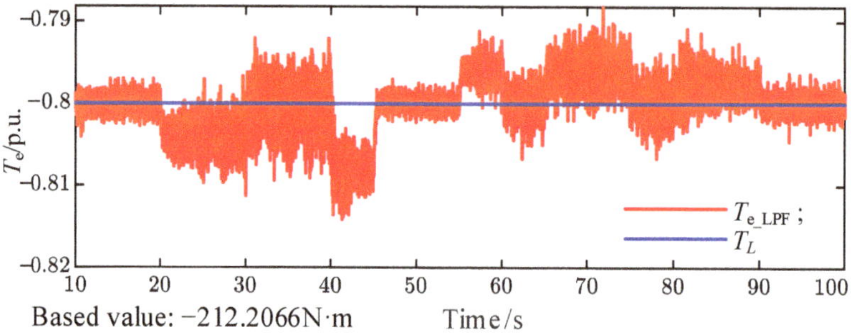

Figure 14. Waveform diagram of electromagnetic torque and mechanical torque.

Furthermore, Figure 15 compares the electromagnetic torque reference T_{e_ref} with the actual torque T_e, together with the stator frequency measured by a PLL. Clearly, the torque tracking error is directly correlated with the frequency deviation, which is consistent with the implemented droop control law. A linear fit of the measured data produces the T_{eo}-f droop curve shown in Figure 16, confirming that the adaptive mechanism shifts the droop characteristic based on the dynamic value of T_{e_ref}.

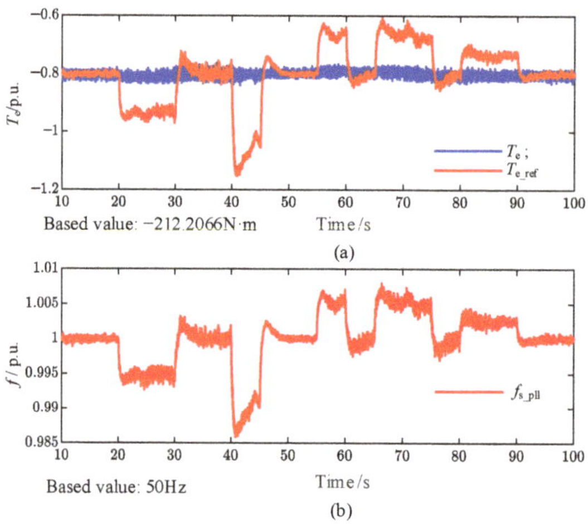

Figure 15. (**a**) Waveform diagram of the reference and actual value of electromagnetic torque; (**b**) the actual frequency of the stator-side voltage is measured by the PLL.

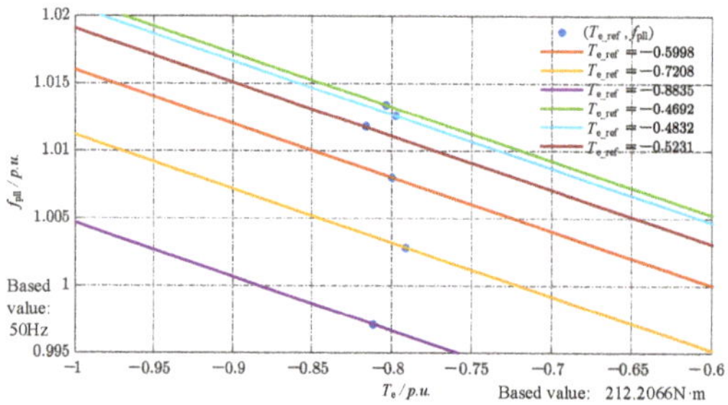

Figure 16. A primary curve fitting the difference between the measured frequency and the set and actual values of the electromagnetic torque.

4.2. Grid-Connected Operation

In the grid-connected scenario, a step change in mechanical torque was applied at $t = 37$ s by engaging an additional mass, causing T_L to change from −0.8 p.u. to −1.0 p.u.

The speed waveform is shown in Figure 17. At 37.028 s, the speed reaches its maximum value of 0.814 p.u., and at 37.154 s, it drops to its minimum value of 0.7598 p.u., with an overshoot of 4.02%. It illustrates that the rotor speed exhibits only minor transients during the torque step and subsequently maintains accurate tracking of its reference t, demonstrating the robustness of the proposed control strategy.

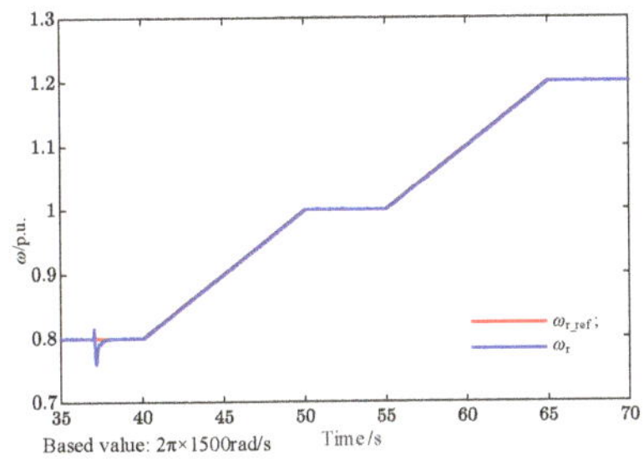

Figure 17. Waveform diagram of rotational speed.

The system's active power responses are shown in Figure 18. The total electromagnetic power P_e (blue) closely matches the mechanical input power P_{in} (purple), with small discrepancies attributable to damping, friction, and machine inertia. The stator power P_s remains independent of rotor speed, while the rotor power P_r varies linearly with ω_r. Specifically, when ω_r exceeds the synchronous speed, the rotor absorbs active power, whereas below the synchronous speed, it delivers active power.

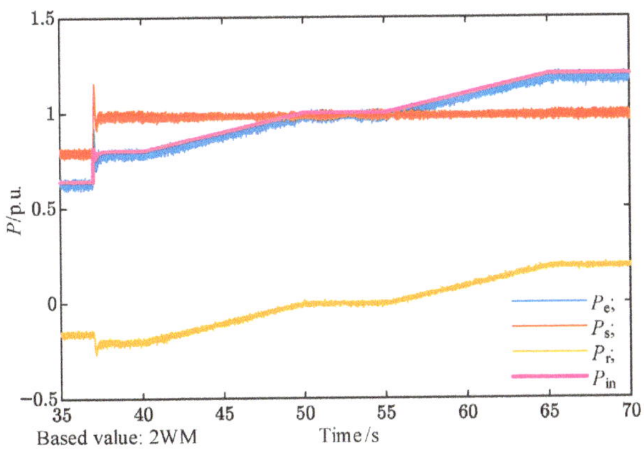

Figure 18. GESS active-power waveform diagram.

Moreover, as shown in Figure 19, the system provides effective grid-frequency support. The PCC frequency, measured by a PLL, increases from 49.68 Hz to 49.95 Hz following the adjustments in mechanical torque and rotor speed. By adding or removing the heavy load, the frequency at the PCC stabilizes at 49.75 Hz after 37 s. At 37.077 s, the PCC frequency reaches its maximum value of 49.8 Hz, and at 37.227 s, it drops to its minimum value of 49.72 Hz, with an overshoot of 1%. By adjusting the rotor speed, the frequency at the PCC stabilizes at 49.85 Hz after 50 s, and at 49.95 Hz after 65 s.

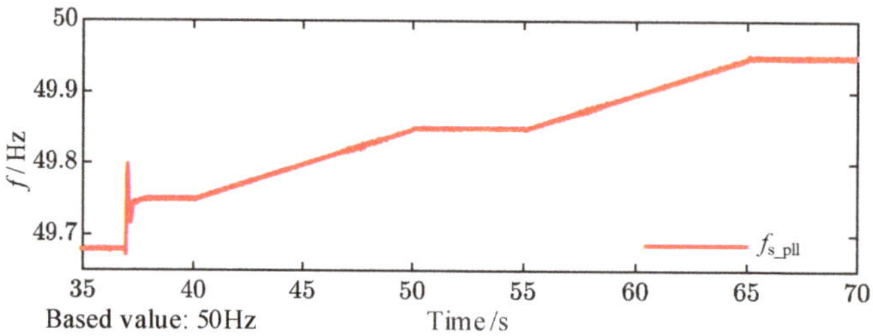

Figure 19. Waveform diagram of stator-side voltage frequency measured by PLL.

The proposed control strategy still ensures precise control of the rotor speed under grid fault conditions. At the 21 s mark, a grid fault is applied (phase A grid voltage drops from 1 p.u. to 0.8 p.u.), and at the 37 s mark, an additional mass is abruptly added, causing a step change from 0.8 p.u. to 1 p.u. The rotor-speed waveform is shown in Figure 20, where ω_{r_ref} (the red line) represents the rotor-speed reference value, and ω_r (the blue line) represents the actual rotor speed. As can be seen, after the grid fault, the rotor speed spikes, but the system continues to follow the reference speed command. Additionally, under grid fault conditions, when the mass is suddenly added, the actual rotor speed stabilizes immediately after a brief fluctuation.

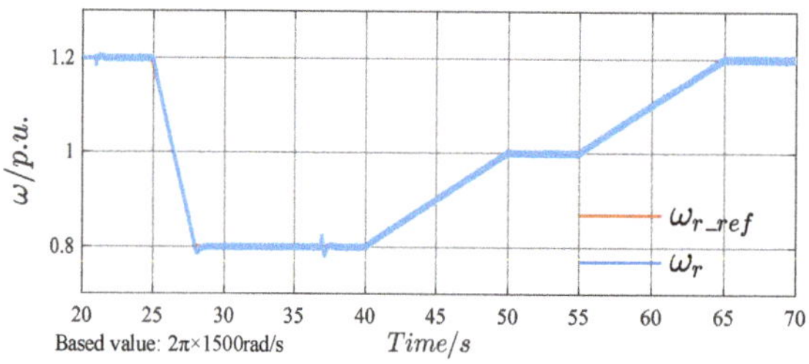

Figure 20. Waveform diagram of rotational speed.

4.3. Analysis of DFIG-GESS Operation Under Three-Phase Rapid Voltage Sag

The following simulation analyzes the operation of the DFIG-GESS system under a grid voltage sag from 1 p.u. to 0.3 p.u. The rotor-speed waveform is shown in Figure 21, where ω_{r_ref} (the red line) represents the rotor-speed reference value, and ω_r (the blue line) represents the actual rotor speed. The time instances t_1, t_2, t_3, and t_4 correspond to the settings under different operating conditions.

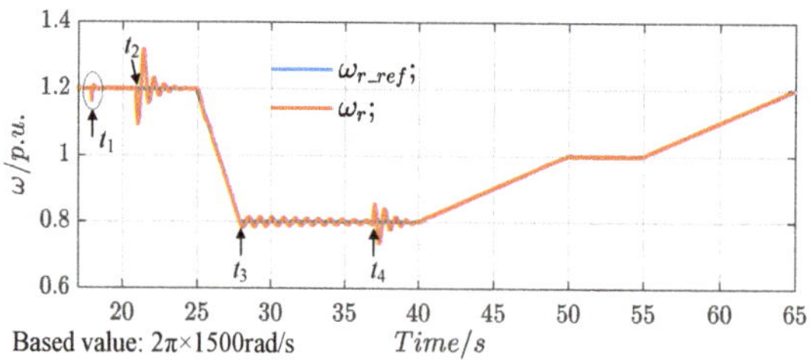

Figure 21. Waveform diagram of rotational speed.

At t_1 (18 s), the load at the PCC is disconnected, causing the active power generated by the DFIG-GESS to be entirely transmitted to the grid through the line impedance. The stator voltage and current, rotor current, stator power, and electromagnetic torque are shown in Figure 22. The sudden load disconnection at t_1 (18 s) causes fluctuations in the stator voltage and current, which stabilize after half a cycle. Additionally, due to the presence of line impedance, the increase in active power fed into the grid raises the PCC voltage magnitude from the original 1 p.u. to 1.253 p.u., demonstrating that, under weak grid conditions, the system can support the grid voltage by increasing the active power output, thereby providing voltage support at the PCC.

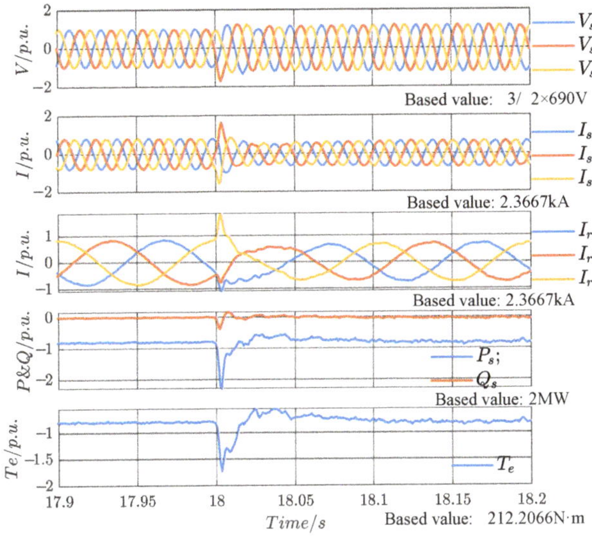

Figure 22. Stator voltage and current, rotor current, stator power, and electromagnetic torque waveforms.

The electromagnetic torque drops to its minimum value of −0.522 p.u. at 18.026 s, with an overshoot of 27% and a settling time of 0.05 s. The speed waveform is shown in Figure 23 below. At 18.02 s, the speed drops to its minimum value of 1.165 p.u., and at 18.138 s, it reaches its maximum value of 1.209 p.u., with an overshoot of 3.5%.

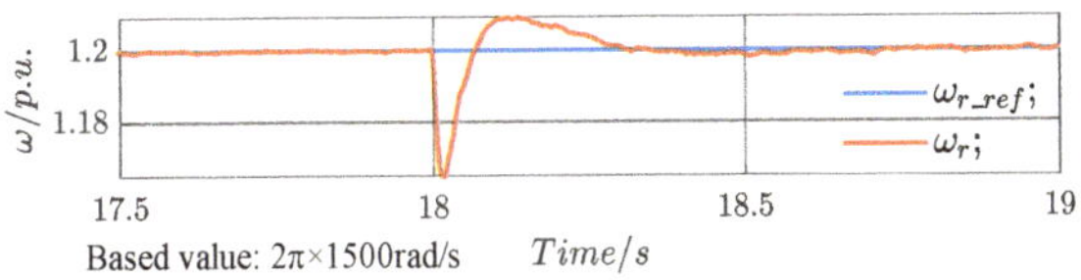

Figure 23. Waveform diagram of rotational speed.

At t_2 (21 s), a sudden three-phase fault occurs in the grid, causing the grid voltage to drop from the original 1 p.u. to 0.3 p.u., with no recovery. The stator voltage and current, rotor current, and stator power are shown in Figure 24. Due to the grid fault, the stator voltage drops from 1.253 p.u. to 0.7624 p.u. Compared to the 70% drop in the grid voltage, the PCC voltage only drops by 49%. After the fault, the grid voltage is 0.3 p.u., and the PCC voltage is 0.7624 p.u., demonstrating that the system can still provide voltage support under grid fault conditions.

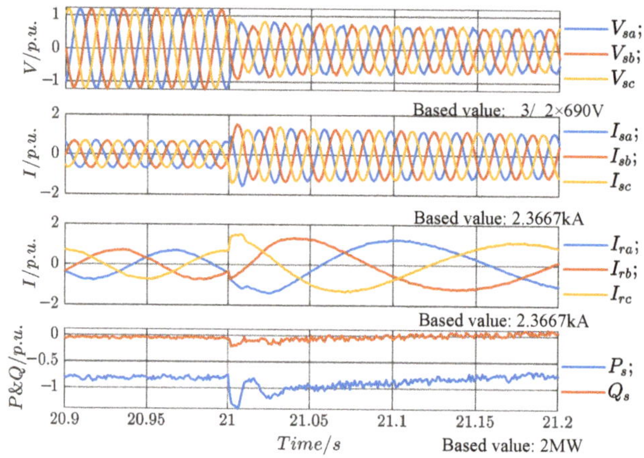

Figure 24. Stator voltage and current, rotor current and stator power waveforms.

The electromagnetic torque and speed waveforms are shown in Figure 25. The electromagnetic torque drops to its minimum value of −1.612 p.u. at 21.002 s, with an overshoot of 81.2%, and enters the ±5% error band at 22.388 s, with a settling time of 1.388 s. At 21.14 s, the speed drops to its minimum value of 1.094 p.u., and at 21.52 s, the speed reaches its maximum value of 1.318 p.u., with an overshoot of 11%. The speed enters the ±5% error band at 21.653 s, with a settling time of 0.653 s.

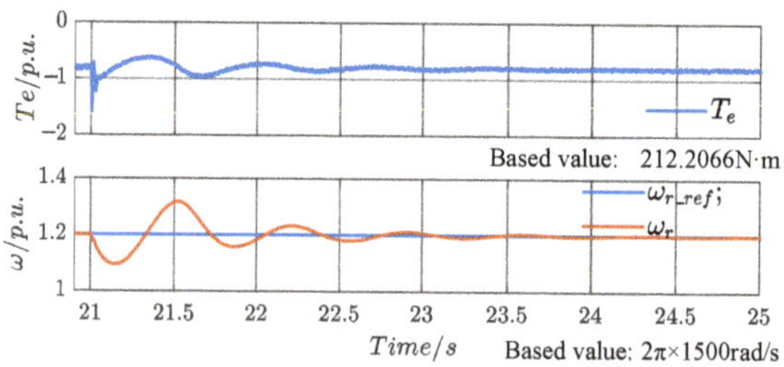

Figure 25. Waveform diagram of electromagnetic torque and rotational speed.

At t_3 (28 s), the speed reference changes from a ramp signal to a constant. Figure 26 shows the stator voltage and current, rotor current, and stator power waveforms, all of which are in steady-state operation and unaffected by the change in the speed reference. However, during the interval from t_2 to t_3, the ramp change in speed causes variations in the active power fed to the grid by the rotor, resulting in a reduction in the total active power delivered to the grid. As a result, the PCC voltage magnitude gradually decreases in this interval, as shown in the figure. After t_3 (28 s), the measured PCC voltage magnitude during the subsequent cycle is 0.6628 p.u., which represents a 9.96% decrease compared to the 0.7624 p.u. at t_2. This demonstrates that the system can regulate the rotor speed to modify the total active power delivered to the grid, thereby adjusting the PCC voltage magnitude.

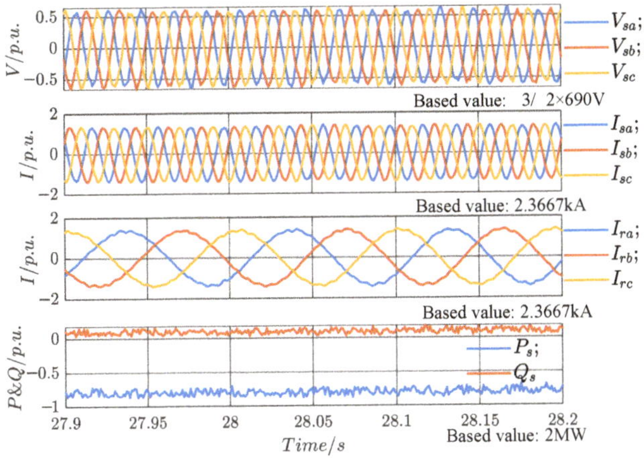

Figure 26. Stator voltage and current, rotor current, and stator power waveforms.

The speed waveform is shown in Figure 27. At 28.22 s, the speed drops to its minimum value of 0.7829 p.u., and at 28.624 s, it reaches its maximum value of 0.8148 p.u., with an overshoot of 1.71%.

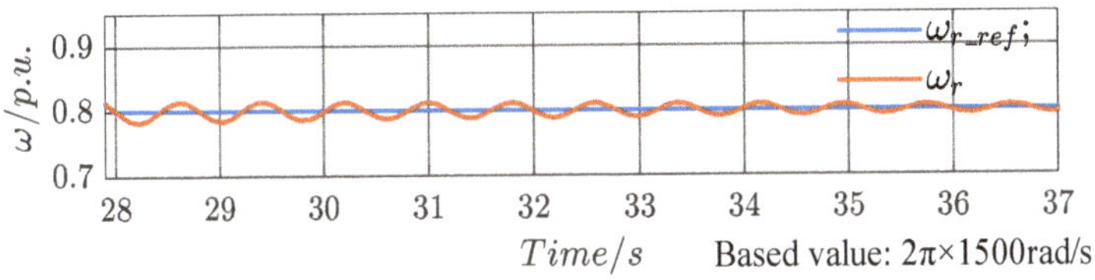

Figure 27. Waveform diagram of rotational speed.

At t_4 (37 s), a sudden addition of a heavy load causes the load torque to drop abruptly from 0.8 p.u. to 0.2 p.u. After stabilization, the PCC voltage magnitude is measured at 0.704 p.u., which represents a 4.12% increase compared to the 0.6628 p.u. at t_3. This demonstrates that the system can regulate the PCC voltage magnitude by adding or removing heavy loads, thus adjusting the total active power delivered to the grid.

The stator power, electromagnetic torque, and speed waveforms are shown in Figure 28. The electromagnetic torque reaches its minimum value of −1.18 p.u. at 37.243 s and further drops to −0.8771 p.u. at 37.582 s, with an overshoot of 18%. It enters the ±5% error band at 38.402 s, with a settling time of 1.402 s. The speed reaches its maximum value of 0.854 p.u. at 37.11 s, and drops to its minimum value of 0.7356 p.u. at 37.408 s, with an overshoot of 6.44%. The speed enters the ±5% error band at 37.498 s, with a settling time of 0.498 s.

Based on Figure 21 and the above analysis, it can be concluded that under a three-phase symmetrical fault in the grid, the DFIG-GESS is still capable of precise rotor-speed control and voltage support.

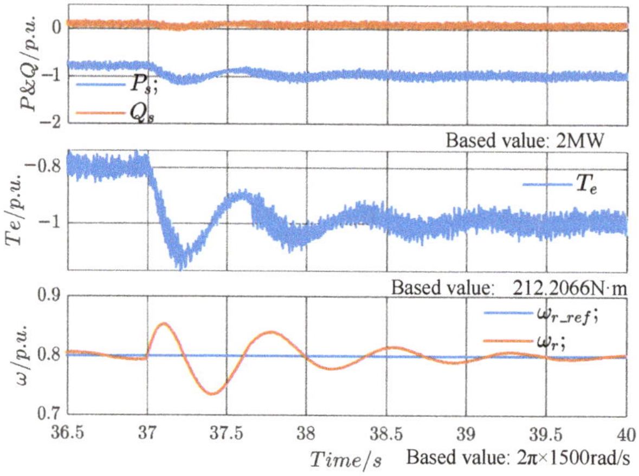

Figure 28. Waveform diagram of stator power, electromagnetic torque and rotational speed.

5. Conclusions

This paper proposed a GESS architecture based on a doubly fed induction machine and developed a speed-adaptive droop control strategy. The proposed architecture facilitates efficient energy conversion between gravitational potential energy and electrical energy, while enabling wide-range, high-precision speed control of the suspended mass. A key advantage of the proposed control strategy is its enhancement of the system's grid-support capability, achieved under low converter cost by combining speed adaptability with the benefits of conventional droop control. Simulation results demonstrate the effectiveness of the proposed control strategy in both islanded and grid-connected operating modes.

While the simulation results indicate that the proposed strategy effectively improves grid support, it is acknowledged that further improvements are needed in robustness testing under real-world disturbances, such as faults and voltage sags. Future work will focus on addressing these practical challenges by incorporating fault ride-through capabilities and conducting experimental validation, either through hardware-in-the-loop (HIL) testing or real-world implementation. Additionally, more detailed quantitative performance metrics, such as overshoot, settling time, and RMS error, will be provided to offer a more comprehensive comparison with traditional droop control.

Overall, this research contributes to the development of advanced control strategies for DFIG-based GESS, offering a promising solution for enhancing grid stability and supporting the integration of renewable energy sources into modern power grids.

Author Contributions: Software, Y.F. and F.W.; Formal analysis, C.C. (Chengjie Cao); Investigation, Y.L.; Data curation, Y.Z.; Writing—original draft, D.H.; Writing—review & editing, C.C. (Chenwen Cheng). All authors have read and agreed to the published version of the manuscript.

Funding: This research was funded by Science and Technology Project of State Grid Jiangsu Electric Power Co., Ltd. grant number J2024225.

Data Availability Statement: The original contributions presented in this study are included in the article. Further inquiries can be directed to the corresponding author.

Conflicts of Interest: Authors Darui He, Yan Li, Chengjie Cao, Yifei Fan and Fei Wang were employed by the company State Grid Jiangsu Electric Power Co., Ltd. The authors declare that this study received funding from State Grid Jiangsu Electric Power Co., Ltd. The funder was not involved in the study design, collection, analysis, interpretation of data, the writing of this article or the decision to submit it for publication. The remaining authors declare that the research was conducted in the absence of any commercial or financial relationships that could be construed as a potential conflict of interest.

References

1. Zhang, Y.; Xu, T.; Chen, T.; Hu, Q.; Chen, H.; Hu, X.; Jiang, Z. A high-resolution electric vehicle charging transaction dataset with multidimensional features in China. *Sci. Data* **2025**, *12*, 643. [CrossRef]
2. Zhang, Y.; Qian, W.; Ye, Y.; Li, Y.; Tang, Y.; Long, Y.; Duan, M. A novel non-intrusive load monitoring method based on ResNet-seq2seq networks for energy disaggregation of distributed energy resources integrated with residential houses. *Appl. Energy* **2023**, *349*, 121703. [CrossRef]
3. Zhang, Y.; Yuan, C.; Du, X.; Chen, T.; Hu, Q.; Wang, Z.; Lu, J. Capacity configuration of hybrid energy storage system for ocean renewables. *J. Energy Storage* **2025**, *116*, 116090. [CrossRef]
4. Pavlík, M. Ever More Frequent Negative Electricity Prices: A New Reality and Challenges for Photovoltaics and Wind Power in a Changing Energy Market—Threat or Opportunity, and Where Are the Limits of Sustainability? *Energies* **2025**, *18*, 2498. [CrossRef]
5. Li, F.; Wang, H.; Wang, D.; Liu, D.; Sun, K. A Review of Wind Power Prediction Methods Based on Multi-Time Scales. *Energies* **2025**, *18*, 1713. [CrossRef]
6. Haghighi, R.; Van-Hai, B.; Wang, M.; Su, W. Survey of Reliability Challenges and Assessment in Power Grids with High Penetration of Inverter-Based Resources. *Energies* **2024**, *17*, 5352. [CrossRef]
7. Wei, L.; Li, Y.; Xie, B.; Xu, K.; Meng, G. Two-Stage Optimal Scheduling Based on the Meteorological Prediction of a Wind–Solar-Energy Storage System with Demand Response. *Energies* **2024**, *17*, 1286. [CrossRef]
8. Li, J.; Wan, J.; Xia, Y.; Zhao, S.; Song, G.; He, Y. A Feasibility Study on Gravity Power Generation Technology by Virtue of Abandoned Oil-Gas Wells in China. *Energies* **2023**, *16*, 1575. [CrossRef]
9. Zhang, Y.; Zou, B.; Jin, X.; Luo, Y.; Song, M.; Ye, Y.; Hu, Q.; Chen, Q.; Zambroni, A.C. Mitigating Power Grid Impact from Proactive Data Center Workload Shifts: A Coordinated Scheduling Strategy Integrating Synergistic Traffic—Data—Power Networks. *Appl. Energy* **2025**, *377*, 124697. [CrossRef]
10. Wang, R.; Zhang, L.; Shi, C.; Zhao, C. A Review of Gravity Energy Storage. *Energies* **2025**, *18*, 1812. [CrossRef]
11. Energy Vault [EB/OL]. Available online: https://energyvault.com/ (accessed on 23 June 2024).
12. Yang, G.; Chai, Y.; Wang, D.; Yan, K.; He, X. Optimal Configuration of Microgrid with Gravity Energy Storage based on Improved Whale Algorithm. In Proceedings of the IEEE International Conference on Electrical Engineering and Mechatronics Technology (ICEEMT), Qingdao, China, 2–4 July 2021; pp. 650–655. [CrossRef]
13. Guo, H.; Cui, J.; Xu, M.; Ge, Y.; Li, D.; Zhao, P. Optimal Capacity Allocation of Combined Output of Tower Gravity Energy Storage Power Station and Wind Farm. In Proceedings of the Power System and Green Energy Conference (PSGEC), Shanghai, China, 25–27 August 2022; pp. 23–27. [CrossRef]
14. Hunt, J.D.; Zakeri, B.; Jurasz, J.; Tong, W.; Dąbek, P.B.; Brandão, R.; Patro, E.R.; Đurin, B.; Filho, W.L.; Wada, Y.; et al. Underground Gravity Energy Storage: A Solution for Long-Term Energy Storage. *Energies* **2023**, *16*, 825. [CrossRef]
15. Liu, J.; Wang, C.; Zhao, J.; Tan, B.; Bi, T. Simplified Transient Model of DFIG Wind Turbine for COI Frequency Dynamics and Frequency Spatial Variation Analysis. *IEEE Trans. Power Syst.* **2024**, *39*, 3752–3768. [CrossRef]
16. Cai, S.; Li, Y.; Wang, W.; Zhan, M. Modeling and Cross-Timescale Mechanism Analysis of Voltage Control Impact on Electromechanical Dynamics of VSG-Controlled DFIG-Based WT. *IEEE Trans. Energy Convers.* **2025**, *40*, 2508–2519. [CrossRef]
17. Chandorkar, M.; Divan, D.; Adapa, R. Control of parallel connected inverters in standalone AC supply systems. *IEEE Trans. Ind. Appl.* **1993**, *29*, 136–143. [CrossRef]
18. Sevostyanov, N.A.; Gorbunov, R.L. Control Strategy to Mitigate Voltage Ripples in Droop-Controlled DC Microgrids. *IEEE Trans. Power Electron.* **2023**, *38*, 15377–15389. [CrossRef]
19. Awal, M.A.; Rachi, M.R.K.; Yu, H.; Husain, I.; Lukic, S. Double Synchronous Unified Virtual Oscillator Control for Asymmetrical Fault Ride-Through in Grid-Forming Voltage Source Converters. *IEEE Trans. Power Electron.* **2023**, *38*, 6759–6763. [CrossRef]
20. Guo, Z.; Wu, W. Matching Synchronous Machine Control for Improving Active Support of Grid-Forming PV Systems with Enhanced DC Voltage Dynamics. *J. Mod. Power Syst. Clean Energy* **2025**, *13*, 179–189. [CrossRef]
21. Wang, D.; Chen, H.; Huang, Y.; Deng, X.; Zhu, G. Modeling and Stability Analysis of Weak-Grid Tied Multi-DFIGs in DC-Link Voltage Control Timescale. *Energies* **2020**, *13*, 3689. [CrossRef]
22. Li, M.; Xie, Z.; Xu, S.; Yang, S.; Zhang, X. Electromechanical Oscillation Analysis and Suppression of Grid Forming DFIG-Based Wind Turbines Under Weak Grid. *IEEE Trans. Energy Convers.* **2025**, *40*, 1365–1377. [CrossRef]

23. Xu, Y.; Chen, P.; Zhang, X.; Yang, D. An Improved Droop Control Scheme of a Doubly-Fed Induction Generator for Various Disturbances. *Energies* **2021**, *14*, 7980. [CrossRef]
24. Pepiciello, A.; Domínguez-García, J.L.; Vaccaro, A. The Impact of Frequency Support by Wind Turbines on the Small-Signal Stability of Power Systems. *Energies* **2022**, *15*, 8470. [CrossRef]
25. Tian, P.; Li, Z.; Hao, Z. A Doubly-Fed Induction Generator Adaptive Control Strategy and Coordination Technology Compatible with Feeder Automation. *Energies* **2019**, *12*, 4463. [CrossRef]

energies

MDPI

Article

Common-Mode Choke Design to Eliminate Electrostatic Discharge Machining Bearing Currents in Wide-Bandgap Inverter-Fed Motor Drives

Mustafa Memon and Mohamed Diab *

School of Mechanical, Electrical and Manufacturing Engineering, Loughborough University, Loughborough LE11 3TU, UK; m.memon@lboro.ac.uk
* Correspondence: m.diab@lboro.ac.uk

Abstract

The electrification of mobility sectors, including automotive, aerospace, and robotics, has accelerated the need for compact and high-efficiency power converters in electric motor drives. Wide-bandgap (WBG) semiconductor–based inverters offer significant advantages over conventional silicon IGBT inverters by enabling higher switching speeds, elevated switching frequencies, and improved power conversion efficiency. However, the adoption of high-frequency switching introduces several challenges, particularly increased motor neutral point voltage stress, originating from inverter common-mode (CM) voltage. The increased neutral point voltage directly elevates motor bearing voltage, the primary driver of motor bearing currents, among which electrostatic discharge machining (EDM) bearing current is the primary cause of bearing degradation in low-power motors. This paper experimentally investigates the root causes of the EDM phenomenon and identifies the key factors influencing its occurrence and severity in WBG-based drive systems. The conventional CM choke designs effectively attenuate motor CM currents and EMI; however, they are ineffective in suppressing EDM bearing currents. In this paper, an alternative CM choke design methodology is proposed to eliminate EDM bearing currents by optimizing the choke inductance to shift the motor CM antiresonance frequency below the inverter switching frequency, thereby ensuring that nearly all source CM voltage is absorbed by the choke. This design approach effectively minimizes the voltage appearing at the motor neutral point and across the bearings, thereby suppressing EDM bearing current spikes without affecting motor DM performance. The choke parameters are mathematically derived for optimal performance and validated through experimental testing on a 2.2 kW three-phase star-connected induction motor powered by a wide-bandgap two-level voltage-source inverter.

Keywords: antiresonance frequency; common-mode choke; common-mode voltage; electrostatic discharge machining bearing current; motor drives; neutral-to-ground voltage; variable speed drives; wide-bandgap devices

1. Introduction

The past decade has witnessed a paradigm shift in the transportation and mobility sectors, primarily driven by stringent greenhouse gas emission regulations and rapid advancements in battery technologies. The resultant decrease in battery manufacturing

 https://doi.org/10.3390/en18215804

costs has enabled widespread adoption of electric propulsion systems across industries such as automotive, aerospace, and advanced robotics. This transformation has created a pressing need for compact, lightweight, and energy-efficient power electronic converters, which are central to the operation of electric motor drives.

In mobile platforms, reducing the size and weight of motor drives is not merely a design preference but a crucial factor in system-level efficiency. Lower power losses directly translate to reduced energy drawn from the battery pack, enabling downsizing of the battery itself. Such cascading benefits—lighter battery packs reducing total vehicle mass and, in turn, lowering energy consumption—underscore the strategic importance of highly efficient motor drive systems.

Recent advances in semiconductor technology have positioned WBG devices, particularly SiC and GaN, as transformative solutions for next-generation power converters [1]. Compared to traditional silicon switches, WBG devices exhibit superior material characteristics, including higher breakdown voltages, elevated thermal conductivity, and the ability to operate at significantly higher switching frequencies. These attributes enable drastic reductions in conduction and switching losses, increased power density, and the use of smaller passive components, which collectively enhance drive performance in weight-sensitive applications [2]. A notable example is the use of SiC MOSFETs in the powertrain of Tesla Model 3 electric vehicle, which achieves an estimated 2–3% improvement in inverter efficiency while significantly reducing system weight to approximately 4.8 kg—less than half that of similar silicon-based inverters [3].

Despite these advantages, the deployment of WBG devices introduces nontrivial challenges that must be addressed to ensure reliable long-term operation. The ultra-fast switching transitions inherent to SiC and GaN transistors produce steep voltage slew rates (high dv/dt), which intensify electromagnetic interference (EMI) and introduce complex insulation stress mechanisms within electric machines. In systems employing inverter-motor power cables, traveling wave reflections can lead to terminal overvoltage that may approach or even exceed twice the applied dc-link voltage [4]. These repetitive overvoltages accelerate the partial discharge phenomenon and degrade stator winding insulation, significantly shortening motor lifetime.

Moreover, high dv/dt CM voltage generated by WBG inverters contributes to the overvoltage appearing at the motor stator neutral point. The latest research in [5,6] indicates that motor neutral point overvoltage can pose greater risks than motor terminal overvoltage to motor windings as well as bearings, which in turn leads to premature motor failure. Motor neutral point overvoltage is caused by the propagation of inverter CM voltage through machine windings. Since the motor neutral point is typically floating, the CM voltage encounters an infinite impedance at the neutral point (due to a very small winding-to-frame parasitic capacitance), where the impedance mismatch between the motor surge impedance and the neutral point parasitic capacitive reactance results in overvoltage oscillations at the motor neutral point, a phenomenon known as the reflected wave effect [5]. It is pertinent to note that the motor neutral point overvoltage oscillations are characterized by the 1st antiresonance frequency of the motor, which is solely the function of machine CM winding inductance and its parasitic capacitances. Hence, the neutral point overvoltage, unlike motor terminal overvoltage, is independent of the cable length where this type of overvoltage can be observed without any cable connected between the inverter and motor [5]. Given that the 1st antiresonance frequency of the motor typically falls within the kHz range [6], the use of high-frequency WBG devices can enable the drive switching frequency or its harmonics to coincide with the motor 1st antiresonance frequency. Unlike inverter CM voltage, the magnitude of motor neutral point voltage is influenced by the relationship between drive switching frequency and machine antiresonance frequency. As the switching frequency approaches the machine antiresonance frequency, the neutral point voltage increases. Furthermore, if both frequencies align, the overvoltage can increase significantly, exceeding CM voltage by several times due to motor resonance excitation [6].

This resonance effect is typically negligible with conventional silicon IGBT inverters, as their switching frequency is practically limited to 20 kHz, while motor antiresonance frequency is usually above 40 kHz [7]. However, with the rise of WBG devices, switching frequencies can now reach up to 200 kHz with SiC MOSFETs and even higher with GaN devices [8], increasing the likelihood of switching frequency or its harmonics matching the motor antiresonance frequency. As a result, an elevated neutral point voltage can impose considerable electrical stress on the winding turns closer to the stator neutral point, increasing the risk of insulation failure. Additionally, it directly affects the magnitude of bearing voltage, resulting in higher bearing current levels and accelerated bearing degradation.

CM chokes are widely utilized as a simple and highly effective approach to improve the motor drive performance by attenuating the CM noise and EMI in conventional variable speed drives. The conventional CM choke inductance is selected to create a high impedance in the CM current path within the desired frequency range, limiting the CM current flow to motor windings and frame. This ensures that CM current and EMI levels remain within the limits of safety standards. However, conventional choke designs have negligible impact on bearing current mitigation, as they do not address the fundamental causes of bearing currents, i.e., neutral point and bearing voltage. Figure 1 shows a typical motor drive system with the connection of a CM choke at the inverter terminals. It also demonstrates the CM current path and overvoltage oscillations between motor neutral point and grounded frame. In Figure 1, the two-level voltage-source inverter (VSI) supplies a three-phase star-connected motor through a four-core power cable. The three output terminals of the VSI (A, B, and C) are connected to the corresponding motor terminals (a, b, and c) through a CM choke and power cable, while the dc-link midpoint (O) is connected to the grounded motor frame (g). Since the CM current I_{cm} flows in the same direction in all three phases, the CM fluxes add up in the choke core and provide a high impedance to I_{cm}, while allowing the line current to pass through with negligible impedance.

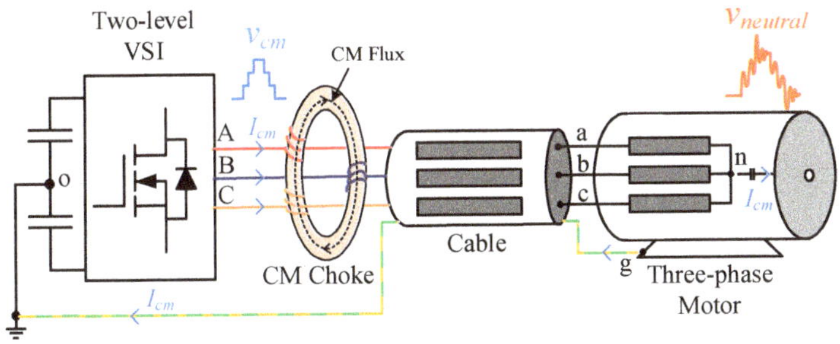

Figure 1. A typical motor drive with choke connected at inverter terminals.

While CM current and EMI mitigation remains important, studies have reported that 67% of motor failures are linked to bearing-related issues [9], resulting in costly repairs, operational downtime, and lost production in industries. This challenge has become even more critical with the increasing adoption of WBG inverter-fed motor drives in electric vehicles (EVs). In recent years, various motor bearing failures attributed to bearing voltages and EDM currents have been reported in EVs [10]. These failures contribute to bearing instability, excessive vibration, and noise, which not only degrade the performance but can also escalate into severe mechanical failures, posing significant reliability concerns for such advanced systems where the highest level of reliability is essential. Therefore, effective strategies to mitigate bearing currents are important to ensure the longevity and reliability of wide-bandgap inverter-fed motor drives, particularly in critical applications such as EVs.

The conventional CM choke design to reduce the CM noise in variable-speed drives has been extensively studied in the literature for many years [11]. Various aspects of CM choke inductor design and performance have also been analyzed [12]. However, conventionally designed CM chokes are unable to mitigate motor neutral point voltage and corresponding bearing voltage and EDM bearing current. Experiments conducted in [13] evaluate the impact of different CM chokes on CM voltage and current, bearing voltage, and EDM bearing current. While the results confirm that CM chokes can reduce CM current by up to 90%, their influence on CM voltage, bearing voltage, and EDM current remains negligible. Additionally, in [14], various CM voltage mitigation techniques are evaluated, including the use of CM chokes. The analysis considers both the physical size and placement of the CM chokes within the motor drive system. The experimental results demonstrate that CM chokes are ineffective in reducing CM voltage, irrespective of their positioning.

EDM bearing current events are produced when the bearing voltage exceeds the dielectric breakdown threshold of the bearing lubricant. As the bearing voltage magnitude is predominantly governed by neutral point voltage, which in turn is largely driven by the inverter-generated CM voltage, it is worth exploring the CM voltage reduction techniques reported in the literature as potential solutions for mitigating EDM bearing currents. These include modified PWM schemes, modified inverter topologies, active noise cancellation circuits, soft switching inverters, and multilevel converters. The modified PWM strategies, such as active zero state PWM (AZSPWM) and near state PWM (NSPWM), reshape switching sequences to eliminate zero voltage vector generation [15–17], whereas modified inverter topologies commonly referred to as H7 and H8 architectures [18,19] or multilevel converters such as T-Type or neutral point clamped [20] inherently synthesize output voltages with reduced CM voltage through additional switching states and voltage levels, respectively. Furthermore, active noise cancellation circuits inject compensating voltages to counteract CM disturbances [21], while soft-switching inverters mitigate high dv/dt transitions, thereby lowering CM voltage excitation [22]. A summary of existing methods and their limitations is presented in Table 1.

It is worth noting that the aforementioned CM voltage reduction methods are limited in their ability to reduce the CM voltage to $\pm V_{dc}/6$ without having additional operational constraints. Although reducing the CM voltage by 66% can suppress the bearing currents, EDM current spikes can still be observed due to the presence of overvoltage at the motor neutral point, especially when the switching frequency approaches the machine antiresonance frequency. Therefore, the CM voltage reduction methods may not be effective in eliminating the EDM current when applied to WBG inverters.

This paper proposes an alternate approach to design a CM choke for eliminating EDM bearing current. Since the added CM choke increases the inductance of the motor CM circuit, it is leveraged to shift the motor antiresonance frequency below the switching frequency to lower the voltage stress at the motor neutral point and bearings. Furthermore, by adjusting the ratio between antiresonance and switching frequency, motor bearing voltage can be reduced to a negligible value, thereby eliminating the EDM bearing current spikes. Additionally, the proposed CM choke design inherently reduces CM current magnitude and EMI by lowering the motor neutral point voltage, which serves as the primary source of CM currents. It is worth noting that the proposed CM choke only modifies the CM circuit of the motor, whereas the motor DM circuit remains unaffected. Design equations for selecting the optimal CM choke inductance and core are presented, while the effectiveness of the proposed strategy is validated through experimental testing on a 2.2 kW three-phase induction motor.

The contributions of this paper are as follows:

- Comprehensive explanation of non-linear behavior of bearing voltage and the corresponding EDM bearing current events in relation to drive switching frequency, offering deeper insight into the underlying mechanisms.
- Analysis of the limitation of conventional choke design approach.
- An alternative choke design approach to prevent the machine's antiresonance frequency from approaching or coinciding with the drive switching frequency, thereby avoiding the excitation of motor resonance that exacerbates bearing current.
- The proposed approach eliminates the EDM currents under all tested scenarios, regardless of switching frequency or dc-link voltage.
- Experimental evaluation of the proposed choke design approach against existing CM voltage reduction methods to highlight its superior performance.

The rest of the paper is structured as follows. Section 2 highlights the limitation of conventional CM choke design. Section 3 defines the bearing voltage and experimentally analyzes its influencing factors. In Section 4, the different types of bearing currents are demonstrated, while EDM bearing current and its influencing factors are experimentally elaborated. Section 5 presents the design equations for selecting CM choke inductance and core to eliminate EDM bearing current. Section 6 outlines the employed experimental approach in this work, detailing the methodologies used to evaluate the proposed choke design. Finally, Section 7 concludes the paper.

Table 1. Existing And Proposed EDM Current Mitigation Methods and Their Limitations.

Mitigation Method	Mechanism	EDM Current Effect	Cost/Complexity	Limitations
PWM schemes	Reduce CM voltage via modulation	Reduced	Low	Cannot eliminate EDM current across all conditions
Modified inverter topologies	Reduce switching stress and CM voltage	Reduced	Medium	Increased circuit complexity
Multilevel inverters	Lower dv/dt and CM voltage	Reduced	High	High cost; complex control
Sine wave/dv/dt filters/CM chokes	Smooth output voltage waveform and attenuate CM EMI	Unaffected	Medium	Ineffective for EDM current; increases losses

2. Limitation of Conventional CM Choke Design Approach

Conventional CM chokes are designed to offer large impedance to high-frequency CM currents. The inductance of the choke is selected based on the peak amplitude of the CM current harmonics and the desired level of attenuation, while its self-resonant frequency is chosen to be higher than the targeted harmonic frequencies for effective attenuation. However, these designs overlook motor neutral point voltage and its dependence on the relationship of motor CM antiresonance frequency and drive switching frequency. Since the added CM choke inductance can lower motor antiresonance frequency, it can bring the antiresonance frequency closer to or even match it with the drive switching frequency. This results in elevated motor neutral point voltage and corresponding bearing voltage and current.

2.1. Motor Neutral Point Voltage

Motor neutral point voltage primarily stems from the CM voltage generated by power inverters in a motor drive system. The CM voltage is the average of the three pole voltages of the inverter relative to the dc-link midpoint. With conventional sinusoidal PWM, two-level VSIs yield four distinct CM voltage levels (i.e., $\pm V_{dc}/6$ and $\pm V_{dc}/2$), where V_{dc} is the dc-link voltage. CM voltage pulses reflect at the stator neutral point due to impedance mismatch between winding surge impedance and motor neutral-to-frame impedance, causing voltage oscillations at the neutral point. This voltage increases with rising switching frequency and reaches its maximum when the switching frequency aligns with the machine antiresonance frequency. Beyond the antiresonance point, a further increase in switching frequency results in a reduction of the neutral point voltage.

This nonlinear behavior of motor neutral point voltage can be analyzed using the high-frequency CM equivalent circuit of the motor drive, as shown in Figure 2, where V_{cm} and L_e are the inverter CM voltage and motor winding inductance, respectively. As observed, motor neutral point voltage V_{ng} is primarily capacitive in nature, as it is influenced by the capacitance between the neutral point and grounded frame, which is half of the motor winding-to-frame capacitance (i.e., $3C_{wf}/2$) (π-model), and the bearing voltage V_b is observed across parallel combinations of rotor-to-frame and bearing capacitances. According to Figure 2, the bearing voltage, which is the primary cause of EDM bearing currents, is directly proportional to the neutral point voltage V_{ng} rather than the inverter CM voltage, where it can be determined by the capacitive voltage divider circuit within the motor CM equivalent circuit. The bearing voltage can therefore be calculated by using the bearing voltage ratio (BVR) as follows:

$$\frac{V_b}{V_{ng}} = \frac{C_{wr}}{C_{wr} + C_{rf} + 2C_b} \quad (1)$$

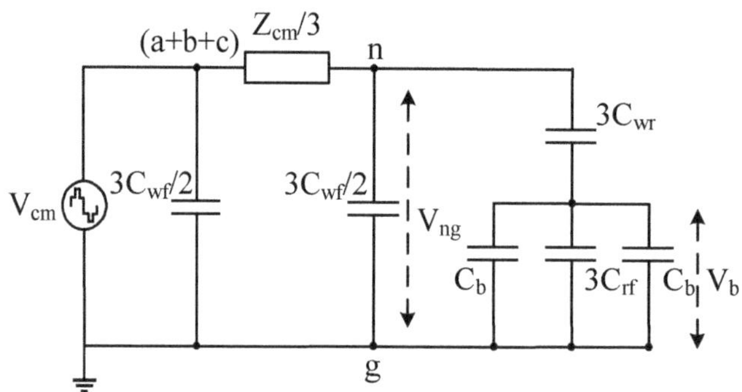

Figure 2. Motor CM equivalent circuit.

In a simple LC series circuit, the capacitive reactance and corresponding voltage exceed the inductive reactance and voltage when the supply frequency is below the circuit's antiresonance frequency and vice versa, as graphically demonstrated in Figure 3a, where the difference in both voltages is equal to the supply voltage, as shown in Figure 3b. Since the motor CM equivalent circuit behaves similarly, the neutral point voltage is higher than the CM inductive voltage when the drive switching frequency is below the motor antiresonance frequency. On the other hand, when the switching frequency exceeds the antiresonance frequency, the motor neutral point voltage becomes lower than the CM inductive voltage, while the difference between the two equals the CM voltage. Since bearing voltage is a capacitive voltage divider of the neutral point voltage, it follows the same non-linear trend with switching frequency as the neutral point voltage.

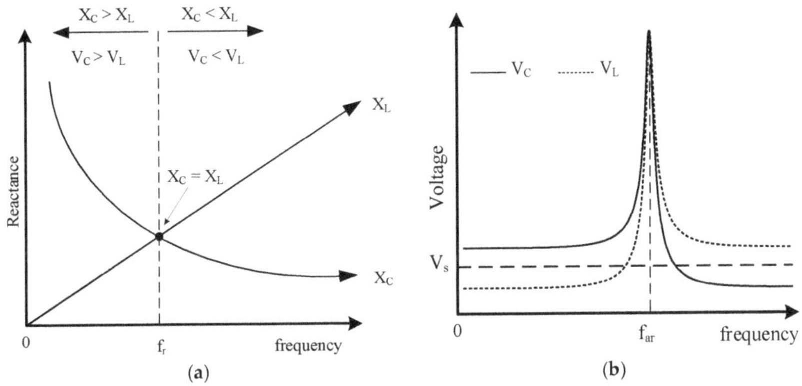

Figure 3. Frequency response of a series LC circuit. (**a**) Variation of inductive and capacitive reactance. (**b**) Variation of inductive and capacitive voltage.

By way of example, a motor drive setup is simulated using the motor parameters provided in [7]. The antiresonance frequency f_{ar} of the motor model is 55 kHz. The motor is supplied with a three-phase VSI using conventional SPWM at a dc-link voltage of 400 V and modulation index of 0.3, while the switching frequency f_{sw} is varied. Figure 4 shows the effect of different switching frequencies on motor neutral point voltage. At 5 kHz (see Figure 4a), the V_{ng} exhibits damped voltage oscillations. This oscillatory response arises due to the wave reflection phenomenon, resulting in 1.4 pu (CM voltage peak as base) peak voltage magnitude at the motor neutral point. In subsequent subplots, as the switching frequency increases, the motor neutral point voltage becomes sinusoidal, and oscillation frequency matches the switching frequency. The peak motor neutral point voltage varies non-linearly with switching frequency, where it increases to 1.9 pu when switching frequency approaches machine antiresonance frequency (see Figure 4b) and becomes maximum (i.e., 2.4 pu) when both frequencies coincide (see Figure 4c). However, when switching frequency becomes higher than antiresonance frequency, the voltage decreases to 1.6 pu (see Figure 4d). This demonstrates that the magnitude of motor neutral point voltage depends upon the relationship between switching frequency and antiresonance frequency of the motor.

2.2. *Effect of Conventional CM Choke Design on Motor Neutral Point Voltage and Bearing Voltage*

Conventional CM chokes are designed to introduce high impedance at targeted frequency ranges (typically in the MHz range) to suppress the EMI. However, incorporating such CM chokes into the drive system increases the overall CM inductance, thereby shifting the CM antiresonance frequency of the motor drive to lower values. These CM chokes are typically used with Si IGBT-based motor drives having practically limited switching frequency of up to 20 kHz, while the motor antiresonance frequency is typically higher than 40 kHz [7]. As a result, although the antiresonance frequency is lowered with CM choke inclusion, it typically remains above the switching frequency. This causes motor neutral point voltage to remain higher than the CM voltage. Nonetheless, the downward shift in motor CM antiresonance frequency may bring it closer to the switching frequency, potentially resulting in an increase in motor neutral point voltage (see Figure 3b).

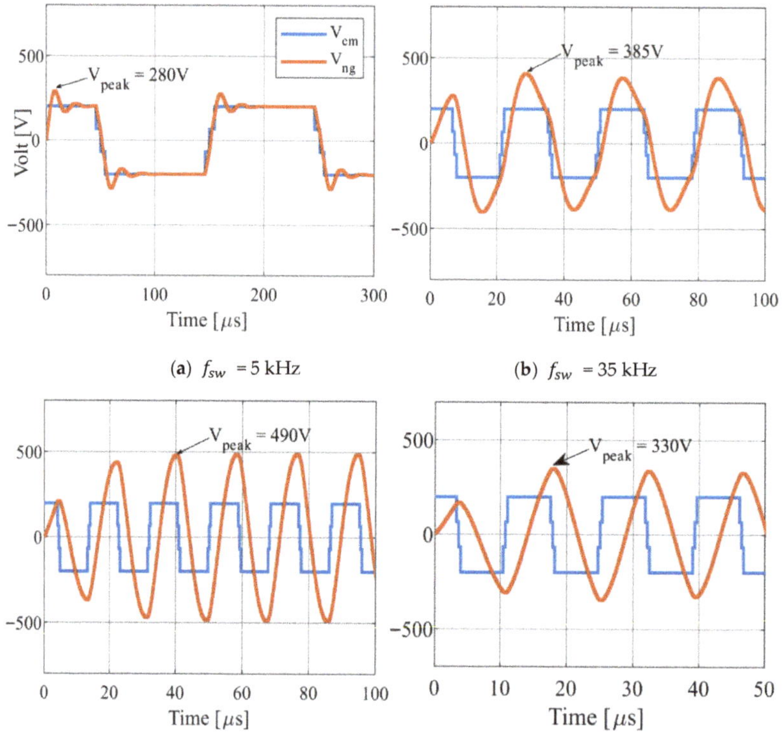

(a) f_{sw} = 5 kHz (b) f_{sw} = 35 kHz (c) f_{sw} = 55 kHz (d) f_{sw} = 75 kHz

Figure 4. Impact of switching frequency on motor neutral point voltage.

As an example, the motor drive simulation model used previously is simulated with and without the conventional CM choke at a switching frequency of 20 kHz. The CM choke inductance is determined based on the CM current profile of the motor drive, as shown in Figure 5a. Without a CM choke, the CM current reaches approximately 75–80 dBµA in the EMI frequency range, which exceeds the permissible limits as described in DO-160F EMI standards (i.e., 72–40 dBµA for 150 kHz–2 MHz and 40 dBµA for 2 MHz–30 MHz) [23]. To meet the standards, a minimum attenuation of 35–40 dBµA is required. Using the insertion loss formula, an attenuation of 35–40 dB corresponds to a CM impedance of approximately 2.8 kΩ at 150 kHz, which converts to a CM choke inductance of 3 mH. Hence, introducing a 3 mH CM choke effectively reduces the CM current harmonics within the EMI range (see Figure 5a). However, the inclusion of the CM choke shifts the motor CM antiresonance frequency from 55 kHz to 40 kHz, as shown in Figure 5b. Although the shifted antiresonance frequency remains higher than the switching frequency, the gap between them is reduced.

Figure 6a,b illustrate the impact of CM choke connection on both the motor neutral point and bearing voltages, respectively. Without a CM choke, the peak neutral point and bearing voltages are 314 V (1.5 pu) and 13 V, respectively. With a 3 mH CM choke, the voltages increase to 357 V (1.7 pu) and 15 V, respectively. This rise in motor neutral point voltage is attributed to the reduced gap between motor antiresonance frequency and switching frequency. Since motor bearing voltage is directly proportional to the neutral point voltage, it increases proportionally. Such a high bearing voltage can cause an electrical breakdown of the motor bearing oil film, producing EDM current spikes. Moreover, a high neutral-to-ground voltage can also be detrimental to the winding insulation, where it can induce partial discharges that accelerate insulation aging and ultimately result in the degradation of organic coatings on the motor winding.

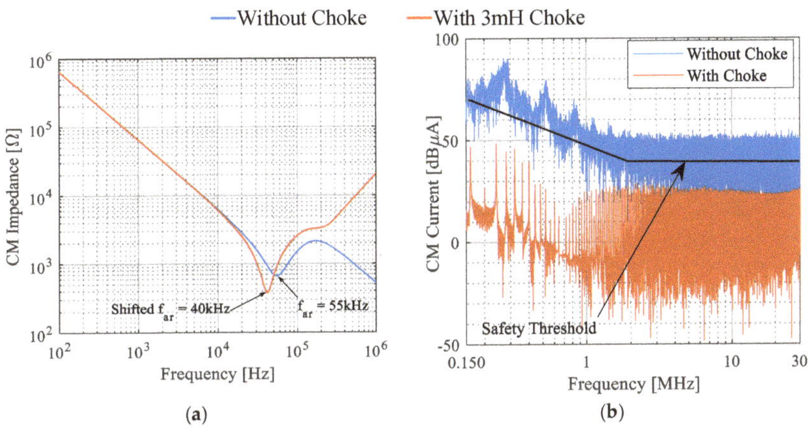

Figure 5. Impact of conventional CM choke on (**a**) EMI (**b**) CM impedance.

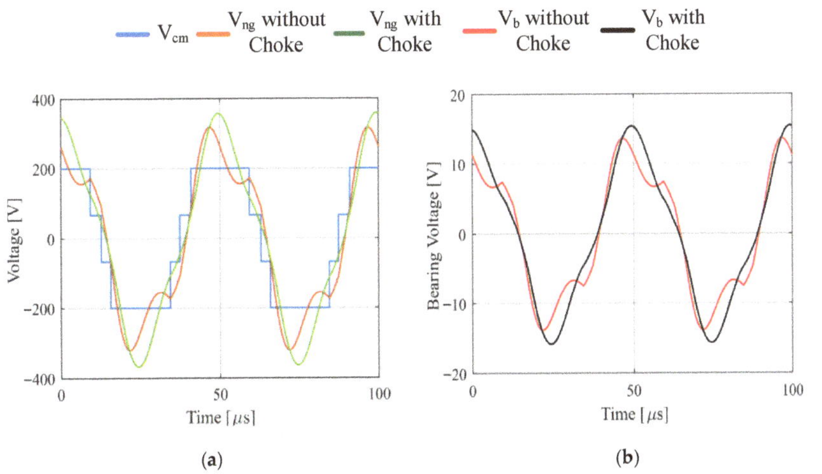

Figure 6. Impact of conventional CM choke connection on (**a**) neutral point voltage; (**b**) bearing voltage.

Overall, the conventional CM chokes are effective in attenuating the high-frequency CM current harmonics. However, they do not reduce motor neutral point voltage or its associated adverse effects. On the contrary, their inclusion may increase both neutral point and bearing voltages, thereby posing a significant risk to the motor's lifetime and overall reliability.

3. Motor Bearing Voltage and Its Influencing Factors

Bearing voltage, defined as the potential difference between the inner and outer races of motor bearings, has long been recognized as a critical reliability concern in electric motor drives [24]. The adoption of WBG semiconductor devices has further exacerbated its severity, as the higher switching frequencies enabled by WBG inverters intensify neutral point overvoltage. This results in an increase in the dielectric stress across the bearing lubricant layer, promoting harmful discharge currents that accelerate surface degradation and significantly increase the risk of premature bearing and motor failure.

Electric motors inherently exhibit various parasitic capacitances distributed across different components, including the stator windings, rotor, and bearings. In high-switching-frequency motor drives, these capacitances can provide a low-impedance path, resulting in leakage currents flowing through both the motor frame and the bearings. Furthermore, the winding-to-frame capacitance couples the CM voltage to the motor frame, creating the motor neutral-point potential relative to ground. Simultaneously, the bearing capacitance establishes a capacitive path between the rotor and stator frame, resulting in a voltage difference across the inner and outer bearing races. Figure 7 shows the cross-sectional view of a three-phase star-connected motor, highlighting the different parasitic capacitances distributed throughout the machine. It also highlights the voltages that arise between different motor components, which contribute to the generation of bearing currents. The parasitic capacitances present within the motor are the capacitance from stator winding to grounded frame C_{wf}, stator winding to rotor C_{wr}, rotor to grounded frame C_{rf}, and bearing capacitance C_b. The motor bearing capacitance C_b consists of C_{b_I} and C_{b_O}, which represent the capacitances from the inner and outer bearing race to the rolling element (ball), respectively. These inner and outer bearing capacitances combine to form the total bearing capacitance, expressed as C_b (i.e., $C_{b\_I} + C_{b\_O} = C_b$).

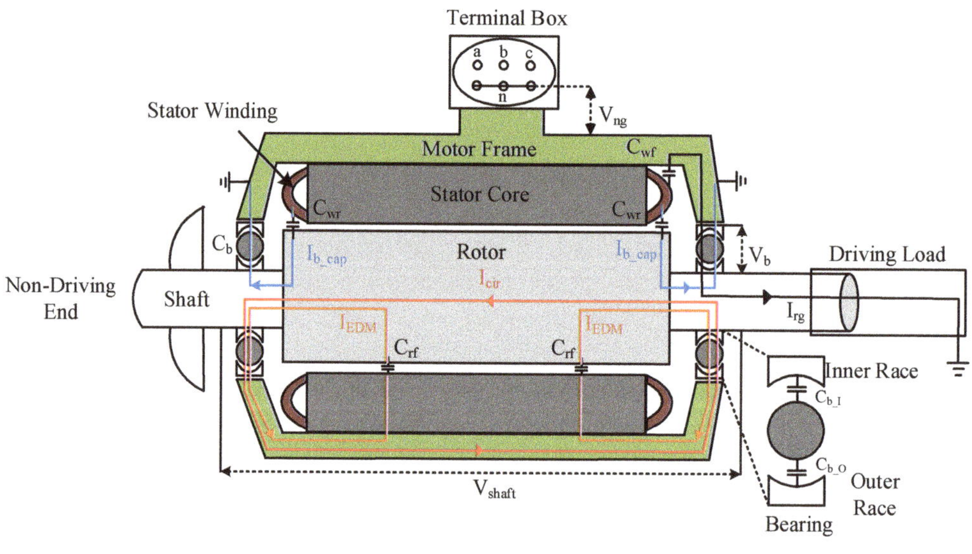

Figure 7. Cross-sectional view of a typical three-phase star-connected motor.

In the existing literature, the terms "bearing voltage" and "shaft voltage" are used interchangeably. However, there is a significant difference between the two voltages. The bearing voltage V_b is defined as the potential difference between the inner and outer races of the bearing (i.e., between shaft and grounded frame), while the shaft voltage V_{shaft} represents the potential difference between the driving end and non-driving end bearings [24]. These voltages are directly affected by neutral point voltage V_{ng}, which is the voltage between the stator winding neutral point and the grounded frame of the motor, as shown in Figure 7. This voltage is typically assumed to be equal to the inverter CM voltage. However, in the previous section, it was demonstrated that it can be several times higher than the CM voltage, depending on the drive switching frequency. This observation underscores the need for a detailed investigation into the influence of switching frequency on bearing voltage formation.

3.1. Impact of Switching Frequency

To observe the effect of switching frequency on bearing voltage, experimental analysis is carried out on a 2.2 kW three-phase four pole star-connected induction motor supplied by a three-phase SiC two-level VSI, as shown in the experimental setup in Figure 8a. Experimental parameters of the inverter, motor, and measurement probes are listed in Table 2. The CM voltage is calculated by averaging the three pole voltages of the inverter, while the neutral point voltage is measured between stator neutral point and grounded motor frame. On the other hand, the bearing voltage is measured between the rotor shaft and ground by using a carbon brush and a V-shaped brush holder, as shown in Figure 8b. The inverter is supplied with a constant dc-link voltage of 600 V at a modulation index of 0.3. The switching frequency f_{sw} is increased from 10 kHz to 60 kHz, and the corresponding effect on V_{cm}, V_{ng}, and V_b is analyzed. The results are presented in Figure 9, showing how the switching frequency variations impact the motor neutral point voltage and bearing voltage, where they have distinct behaviors, exhibiting peaks that increase/decrease under certain changes in the switching frequency, whereas the CM voltage peak remains constant at 300 V. At low switching frequency, i.e., 10 kHz (see Figure 9a), the bearing voltage and neutral point voltage traverse in different voltage levels almost equivalent to the CM voltage. However, as the switching frequency increases, as shown in subsequent subplots of Figure 9, V_{ng} and V_b become almost sinusoidal due to the increasing filtering effect of motor parasitic capacitances. This again proves that the bearing voltage is directly influenced by the neutral point voltage rather than the CM voltage. Additionally, it can be observed from Figure 10, which summarizes the data collected from Figure 9, that the peak values of both V_{ng} and V_b increase as the switching frequency rises from 10 kHz to 40 kHz, after which these values begin to decrease.

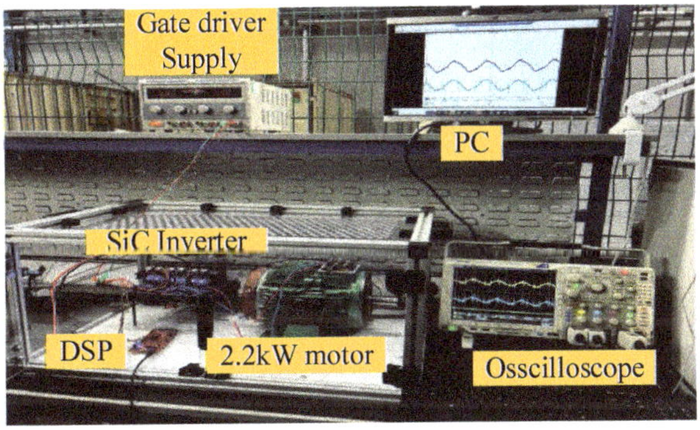

(a)

Figure 8. *Cont.*

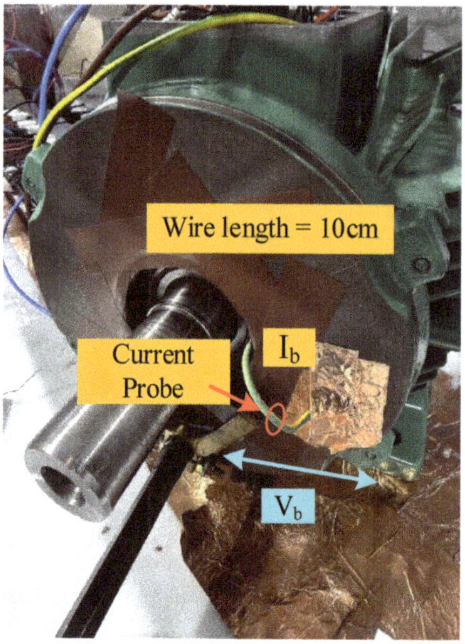

(b)

Figure 8. Experimental setup. (**a**) Hardware view. (**b**) Bearing voltage and current measurement setup.

Table 2. Parameters of Experimental Setup.

Inverter Parameter	**Description**
Inverter type	Three-phase VSI
Voltage levels	Two-level
Switching devices	SiC single N-channel MOSFETs (Wolfspeed C2M0040120D)
Switch rating	1200 V/60 A
Gate resistance	25 Ω
Switching rise/fall time	52 ns/34 ns
DSP	TMS320F28379D
Modulation technique	Sinusoidal PWM
Motor Parameter	**Description**
Rated line voltage	400 V
Rated line current	4.9 A
Rated Power	2.2 kW
Rated rotor speed	1445 RPM
Measurement	**Description**
Voltage measurement	200 MHz Tektronix (THDP0200) voltage probe
Current measurement	120 MHz Tektronix (TCP0030A) current probe
Oscilloscope sampling rate	200 K/s
Oscilloscope edge count threshold	100 mA
Oscilloscope record length	20 million points

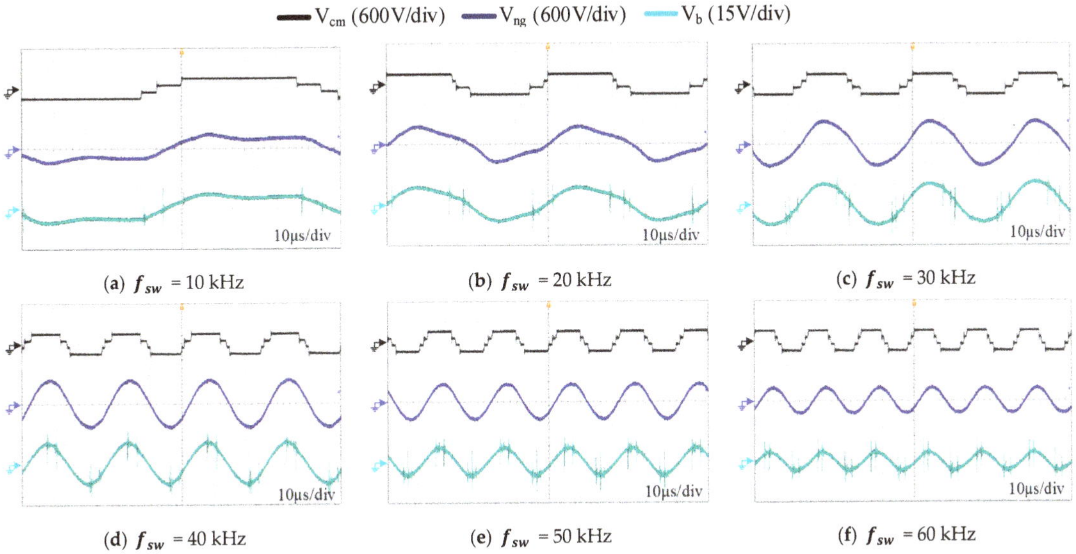

Figure 9. Impact of switching frequency on neutral point voltage and bearing voltage.

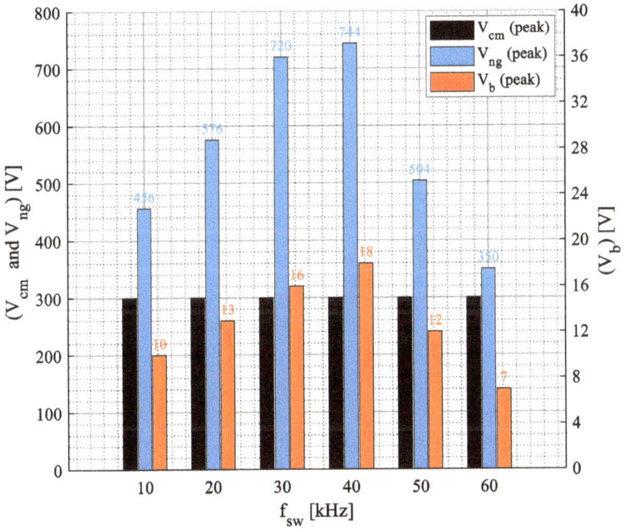

Figure 10. Variation in motor neutral point voltage and bearing voltage at different switching frequencies and constant dc-link voltage.

To explain this phenomenon, it is important to consider the machine antiresonance frequency, which can be determined by measuring machine CM impedance between the parallel-connected three phases and ground wire of the motor. To measure the CM impedance of the motor under test, an impedance analyzer Soletron (1260 + 1287) is utilized, and the corresponding CM impedance graph is shown in Figure 11. The machine antiresonance frequency is observed at 41.6 kHz. The motor neutral point voltage, due to its capacitive nature, increases when the switching frequency is below the machine antiresonance frequency (i.e., f_{sw} < 41.6 kHz) (see Figures 9a–d and 10) and decreases when the switching frequency exceeds the antiresonance point (i.e., f_{sw} > 41.6 kHz) (see Figures 9e,f and 10). A similar trend can be observed in the bearing voltage. The reduction in bearing voltage after the antiresonance point inherently leads to a corresponding reduction in EDM currents, as it is unable to break down the bearing oil film.

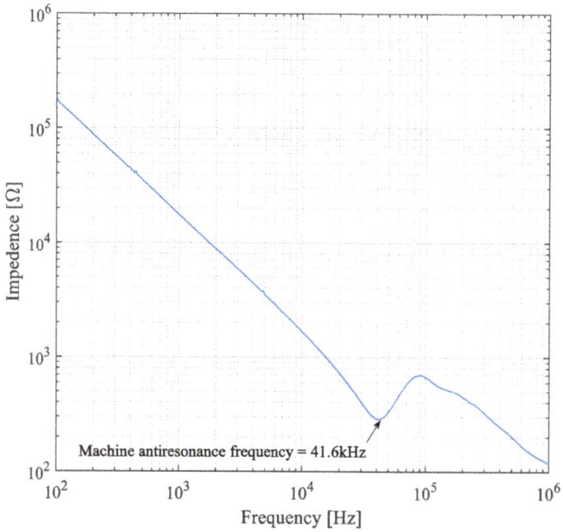

Figure 11. CM impedance of the 2.2 kW motor under test.

3.2. Impact of Dc-Link Voltage

To evaluate the effect of dc-link voltage on bearing voltage, the motor drive is operated at a fixed switching frequency of 20 kHz and a modulation index of 0.3, while the dc-link voltage is varied from 200 V to 700 V. The motor neutral point voltage and bearing voltage are measured and illustrated in Figure 12. Furthermore, the peak neutral point and bearing voltage values at different dc-link voltages are summarized in Figure 13. As observed, both the neutral point voltage and bearing voltage are directly proportional to the dc-link voltage. This phenomenon is attributed to the neutral point voltage being derived from the CM voltage generated by the inverter. As the dc-link voltage increases, the CM voltage correspondingly rises, causing an increase in both the neutral point and bearing voltages. At a low dc-link voltage of 200 V (see Figures 12a and 13), the peak neutral point voltage and peak bearing voltage are approximately 195 V and 5 V, respectively (i.e., $V_{ng(peak)}/V_{cm(peak)}$ = 1.95 pu). As the dc-link voltage is increased in the subsequent subplots in 100 V steps up to 700 V (see Figures 12b–f and 13), the neutral point and bearing voltages are observed to increase almost linearly from 195 V to 620 V and 5 V to 15 V, respectively. Since the switching frequency, i.e., 20 kHz, is kept constant at a value far away from motor antiresonance frequency, i.e., 41.6 kHz, the ratio of peak neutral point voltage to peak CM voltage remains almost constant at approximately 1.8–1.9 pu across all dc-link voltages, whereas, in the case of switching frequency matching the machine antiresonance frequency, this ratio was observed as approximately 2.5 pu (see Figure 10). Notably, even at the lowest dc-link voltage of 200 V (i.e., 100 V CM voltage) and a low switching frequency of 20 kHz (away from motor antiresonance frequency), the peak bearing voltage reaches 5 V, which is sufficient to compromise the bearing lubricant film and initiate EDM currents depending upon the bearing geometry and size. Therefore, any mitigation strategy must be designed to consider the influence of dc-link voltage alongside switching frequency to effectively reduce EDM bearing currents.

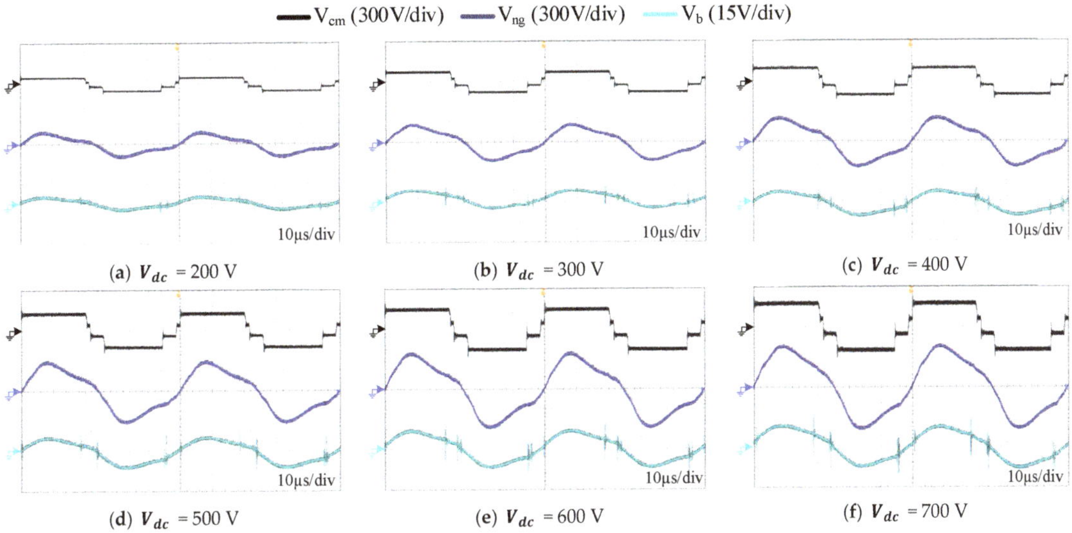

Figure 12. Impact of dc-link voltage on neutral point voltage and bearing voltage.

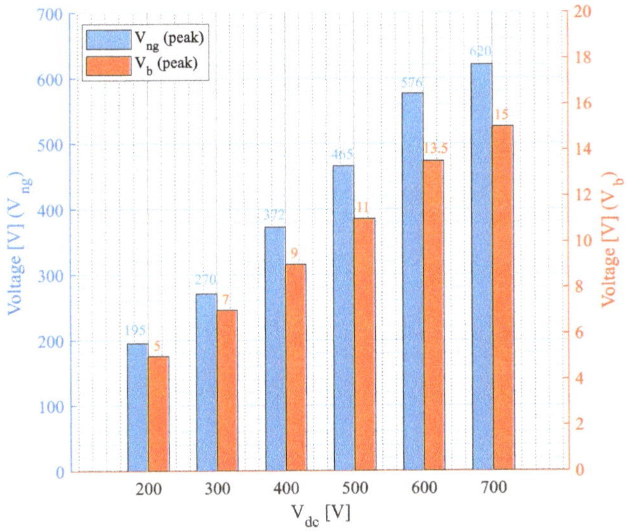

Figure 13. Variation in motor neutral point voltage and bearing voltage at different dc-link voltages and constant switching frequency.

4. Electrostatic Discharge Machining Bearing Currents

Conventional CM choke designs primarily target the mitigation of CM grounding currents flowing from the stator winding to the grounded motor frame through parasitic capacitances. However, additional grounding currents, known as bearing currents, also exist within the motor. These currents flow from the stator winding to the rotor, and subsequently from the rotor shaft to the grounded motor frame via the motor bearings.

In this research, the choke design is specifically focused on reducing bearing currents. An added benefit of the proposed design is its inherent ability to reduce CM grounding currents as well, achieved through the reduction of the motor neutral point voltage and bearing voltage, which acts as the primary driver of CM currents.

Various circulating and non-circulating bearing currents exist within the motor, as demonstrated in Figure 7. High-frequency circulating and EDM bearing currents flow within a closed loop, whereas rotor grounding and dv/dt capacitive bearing currents originate from the motor windings and discharge to the grounded motor frame. The high-frequency circulating bearing current I_{cir} circulates between the stator frame and rotor shaft through the bearings. Similarly, the EDM current I_{EDM} forms a loop beginning at the rotor shaft, traveling through the bearings to the stator frame, continuing to the stator core, and then reaching the rotor core via the capacitance C_{rf} before completing its path back to the rotor shaft. Additionally, a portion of the total CM grounding current may be conducted as rotor ground current I_{rg}, particularly when the rotor is grounded through the driven load. On the other hand, the dv/dt capacitive bearing current I_{b_cap} flows from the winding to the rotor, then to the bearing, and finally to the grounded frame through the parasitic capacitances C_{wr} and C_b. It is worth noting that the EDM bearing current directly penetrates the lubricating film within the bearing instead of flowing through the bearing capacitances. That is why the EDM current is a primary contributor to bearing degradation compared to the other types of bearing currents.

In the case of healthy lubrication, the bearings maintain lubricant film thickness ranging from 0.1 to 1.4 μm, capable of withstanding voltages between 1.5 and 21 V [25]. However, when this voltage threshold is exceeded, EDM occurs within the motor bearings. This can be visualized by measuring the bearing current of the motor. There are different methods to measure bearing currents, such as direct intrusion measurement, the use of an additional bearing, or radio frequency antenna-based methods. In this paper, a direct intrusive measurement method is employed, as it allows accurate measurement of all types of bearing currents as compared to other methods.

In this method, the motor's endcap is modified to accommodate the measurement circuit, as shown in Figure 14. The motor end cap is widened to incorporate a 1 mm thick metal layer and a 1 mm thick insulation layer. The metal layer is inserted around the bearing, and the insulation layer is positioned between the bearing and the motor frame afterwards to prevent the bearing current from flowing into the motor frame. A short wire is connected to the metal layer and taken out to allow the bearing current to bypass the original internal current flow path, thereby enabling direct measurement of the current passing through the wire. Due to the random occurrence of EDM currents, an "Edge count" function is utilized in the oscilloscope to count the number of EDM spikes. The threshold for EDM current detection is set as 100 mA which is significantly higher as compared to dv/dt bearing currents.

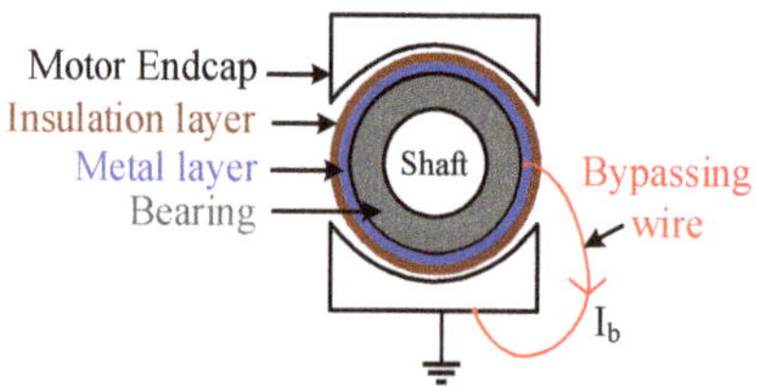

Figure 14. Employed bearing current measurement method.

Figure 15 shows the bearing voltage along with the associated bearing current measured at 600 V dc-link voltage and 20 kHz switching frequency. As seen, during the EDM event, the voltage across the bearing drops to zero as the lubricant film ruptures. This rupture causes the energy stored in the rotor-to-frame capacitance to discharge through the bearings, resulting in an impulse current. The released energy generates an arc that can melt the surface of the bearing raceway. The molten material is subsequently rolled over by the rolling elements, forming a grey trace, and in severe cases, leading to a series of flutings. Some of this molten material may also mix with the bearing lubricant, impairing its insulating properties. Additionally, the rupture of the lubricant creates high temperatures that can cause the lubricant to evaporate. This transient breakdown process lasts only for a few hundred nanoseconds, before which the dielectric lubricant flows back into the gap created by the discharge, reforming a new insulating film. Consequently, the bearing capacitance recharges, and the EDM current ceases. This cycle of breakdown and recovery occurs repeatedly, impacting the bearing voltage waveform after each breakdown. Referring to Figure 15, each time the bearing voltage drops to zero during a breakdown, the voltage waveform shifts upward or downward. A downward shift occurs if the EDM event takes place during the positive half-cycle of the bearing voltage, and vice versa. As a result of this shift, the voltage trajectory deviates from its original path during the subsequent switching event.

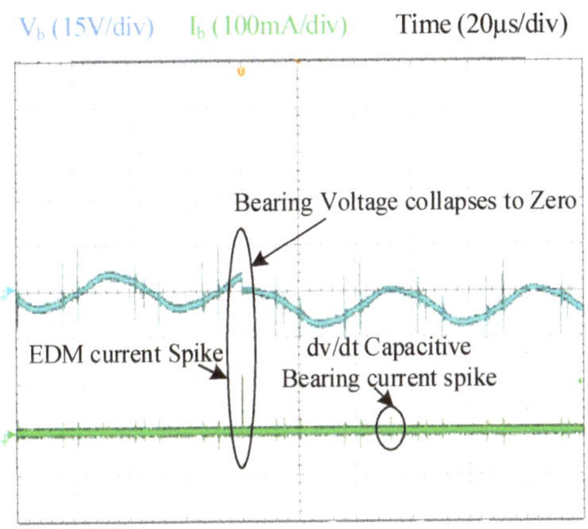

Figure 15. Experimental bearing voltage and bearing current waveforms.

On the other hand, the capacitive dv/dt bearing current is generated with each switching event, as shown in Figure 15. The frequency of occurrence of capacitive bearing current is significantly higher than that of EDM current. However, due to its relatively low amplitude (less than 100 mA), this dv/dt bearing current is generally not considered harmful to the bearing raceway. In contrast, EDM currents can range from 200 mA to 1.4 A [26], posing a serious risk of damage to the bearing.

4.1. Impact of Switching Frequency on EDM Bearing Current

To observe the effect of switching frequency on EDM events, the motor under test is supplied by a fixed dc-link voltage of 600 V and a modulation index of 0.3, while the switching frequency is varied from 10 kHz to 60 kHz (10 kHz increment). This switching frequency range is selected to cover the antiresonance frequency of the motor under test (i.e., 41.6 kHz). The results presented in Figure 16 demonstrate that the number of EDM current spikes, within 1 min, rises from 622 at 10 kHz to 1887 at 40 kHz, and then decreases to 146 at 60 kHz. This non-linearity can be explained by considering the machine antiresonance frequency and switching frequency of the drive.

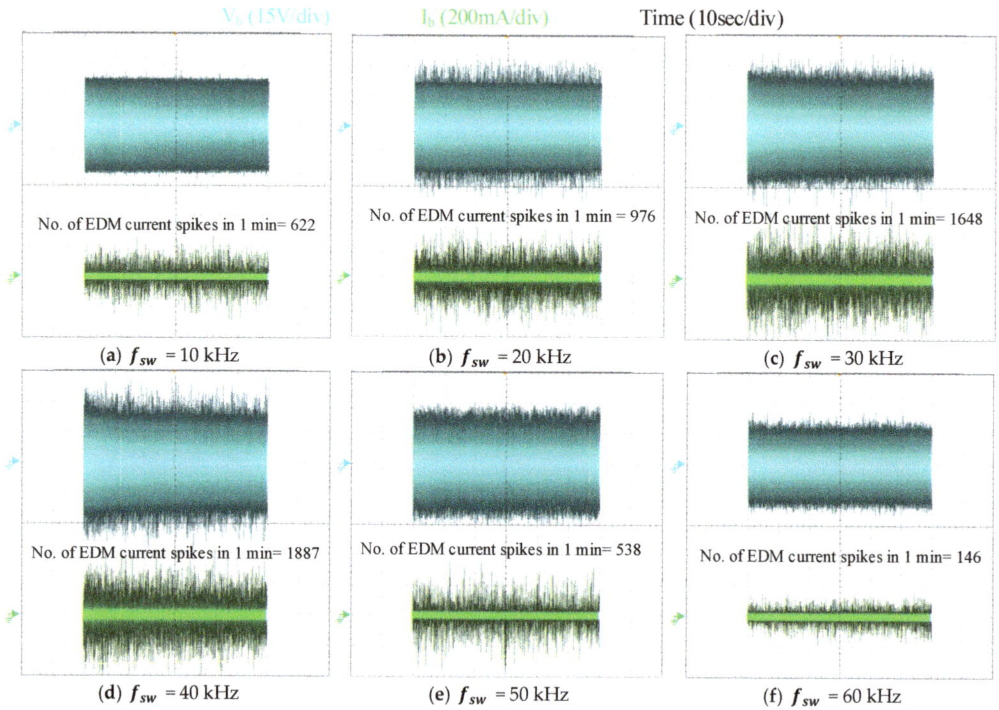

Figure 16. Impact of switching frequency on EDM event count.

As demonstrated in Sections 2 and 3, the bearing voltage, which acts as the primary source of EDM current generation, exhibits a direct proportionality to the motor neutral point voltage, whereas the neutral point voltage itself depends on the relationship between the machine antiresonance frequency and switching frequency. Since the antiresonance frequency of the motor under test is measured as 41.6 kHz, the motor neutral point voltage increases when the switching frequency increases up to this antiresonance point.

However, when the switching frequency exceeds the machine antiresonance frequency, the neutral point voltage subsequently decreases. Since the bearing voltage directly reflects the behavior of the motor neutral point voltage, it follows a similar pattern, as demonstrated in Figure 10. Given that EDM current occurrence is governed by the bearing voltage, the increase in switching frequency initially results in an increased EDM current count. Beyond the point where the switching frequency surpasses the machine antiresonance frequency (see Figure 9e,f), the decrease in bearing voltage leads to a corresponding reduction in EDM current events.

4.2. Impact of Dc-Link Voltage on EDM Bearing Current

To evaluate the effect of the dc-link voltage on the EDM current, the motor is supplied at a fixed switching frequency and modulation index of 20 kHz and 0.3, respectively, while the dc-link voltage is varied from 200 V to 700 V in an increment of 100 V. The results presented in Figure 17 show that the number of EDM events, within 1 min, increases from 35 counts at 200 V to 1034 counts at 700 V. This indicates that the increase in dc-link voltage significantly influences the EDM current events, where they increase linearly with increasing dc-link voltage even at low modulation index.

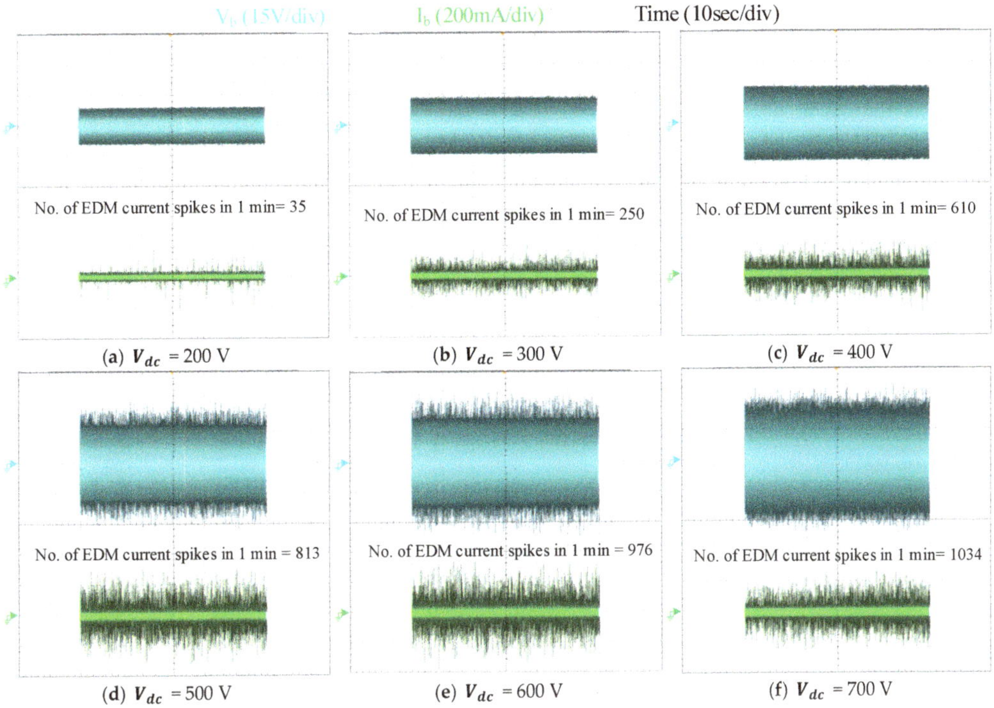

Figure 17. Impact of dc-link voltage on EDM event count.

Figure 18 illustrates the combined effects of switching frequency and dc-link voltage on neutral point voltage, bearing voltage, and EDM event count. As shown, the neutral point voltage exhibits a nonlinear variation with switching frequency, attributed to its capacitive behavior, and a linear increase with dc-link voltage, owing to its dependence on the inverter CM voltage (see Figure 18a). Since the bearing voltage and the corresponding EDM current event count depends on motor neutral point voltage, both follow a similar trend with switching frequency and dc-link voltage (see Figure 18b,c), that is, with lower values of both observed at low switching frequencies and dc-link voltages. The values increase progressively and reach a maximum when the switching frequency equals the motor antiresonance frequency (i.e., f_{sw} = 40 kHz) at the highest applied voltage (i.e., V_{dc} = 700 V). Beyond this antiresonance frequency, a further increase in switching frequency results in a decrease in these values, although they remain comparatively high at elevated voltage levels.

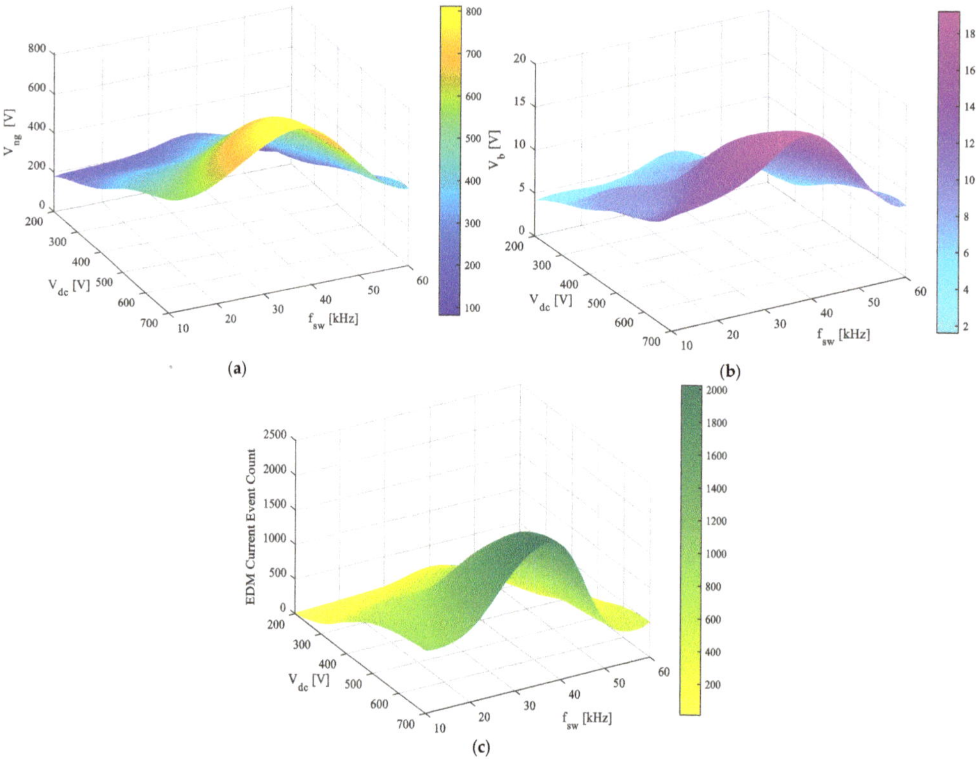

Figure 18. Combined effect of switching frequency and dc-link voltage on (**a**) neutral point voltage, (**b**) bearing voltage, (**c**) EDM current event count.

5. Proposed CM Choke Design Approach

This paper employs a bearing voltage reduction approach to prevent EDM current events. As outlined in Section 4, when the bearing voltage exceeds the breakdown voltage of the bearing lubricant, the lubricant insulation fails and allows the current to pass through the bearing, resulting in EDM current spikes. To eliminate these spikes, it is essential to reduce the bearing voltage below the lubricant breakdown voltage. As detailed in Section 2, the bearing voltage is directly proportional to the motor neutral point voltage.

Therefore, reducing the neutral point voltage will inherently reduce the bearing voltage, which decreases the likelihood and magnitude of EDM current generation.

Since the practical switching frequency of IGBT inverters is limited to 20 kHz and motor antiresonance frequencies are typically higher than 40 kHz, the CM chokes conventionally designed for IGBT inverters typically operate under the condition of $f_{sw} < f_{ar_{shifted}}$, where $f_{ar_{shifted}}$ represents the motor antiresonance frequency shifted to lower values due to the inclusion of the CM choke. In this condition, motor neutral point voltage tends to be higher than inverter CM voltage (see Figure 3a,b). Consequently, the conventional CM choke is ineffective at reducing the neutral point voltage and the corresponding bearing voltage. To effectively mitigate these voltages and, in turn, EDM current events, the shifted antiresonance frequency must be lower than the switching frequency, i.e., $f_{ar_{shifted}} < f_{sw}$. Therefore, this paper proposes a CM choke design that optimally shifts the motor antiresonance frequency below the switching frequency, leading to negligible neutral point and bearing voltages, thereby eliminating the EDM bearing current spikes.

5.1. CM Choke Inductance Selection

In this paper, CM choke inductance is determined based on the principle of shifting motor antiresonance frequency to a lower value relative to switching frequency. Referring to Figure 2, the winding-to-frame capacitance is much higher than other parasitic capacitances; therefore, it acts as the dominant capacitance, and other capacitances can be neglected. As a result, the motor CM equivalent circuit shown in Figure 2 can be simplified into a series LC circuit consisting of three-phase equivalent CM inductance $L_e/3$ and equivalent capacitance between neutral point and grounded frame $3C_g$ (i.e., $3C_g = 3C_{wf}/2$), as shown in Figure 19a. According to Figure 19a, the motor antiresonance frequency f_{ar} is given by

$$f_{ar} = \frac{1}{2\pi\sqrt{(L_e/3) \times (3C_g)}} \quad (2)$$

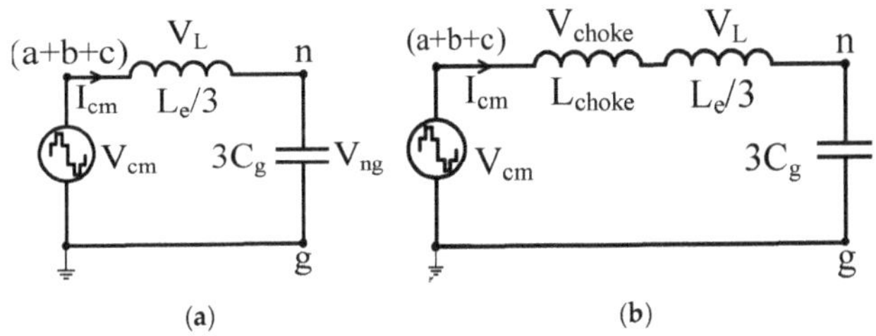

Figure 19. Motor simplified CM equivalent circuit (**a**) without CM choke connection (**b**) with CM choke connection.

By shifting f_{ar} to a lower value than the switching frequency, the capacitive voltage of the motor CM equivalent circuit (i.e., V_{ng}) can be decreased. According to Equation (2), this can be achieved by increasing either L_e, C_g, or both. Due to the series connection of the CM choke with the motor, its inductance is added to the motor CM equivalent inductance, as shown in Figure 19b. Hence, the total CM inductance increases, effectively shifting the motor antiresonance frequency to a lower value. According to Figure 19b, the shifted antiresonance frequency $f_{ar_{shifted}}$ of the motor, influenced by the series addition of the CM choke inductance L_{choke}, is

$$f_{ar_{shifted}} = \frac{1}{2\pi\sqrt{(L_e/3 + L_{choke}) \times (3C_g)}} \quad (3)$$

From Equations (2) and (3), the value of L_{choke} can be calculated as

$$L_{choke} = \frac{f_{ar}^2 L_e}{3f_{ar_{shifted}}^2} - \frac{L_e}{3} \quad (4)$$

where per-phase winding inductance can be determined by rearranging Equation (2) as

$$L_e = \frac{1}{4\pi^2 f_{ar}^2 C_g} \quad (5)$$

Referring to Figure 11, the antiresonance frequency of the utilized motor is observed as 41.6 kHz, whereas C_g can be calculated from the magnitude of CM impedance Z_{cm} in the low-frequency range (i.e., 1 kHz to 10 kHz), as

$$C_g = \frac{C_{wf}}{2} = \frac{1}{6(2\pi f Z)} = 1.35nF \tag{6}$$

where f and Z represent the selected frequency in the low-frequency range and the CM impedance value, respectively. By substituting the determined parameters in Equation (5), the value of L_e for the motor under test is determined as 10.8 mH.

From Equation (4), the value of CM choke inductance required to shift the motor antiresonance frequency to an arbitrary low value can be obtained. However, the optimum shift required in the antiresonance frequency to reduce the bearing voltage to a certain level and eliminate the EDM current spikes requires further calculations.

Motor neutral point can be calculated by applying a voltage divider rule on the CM choke-connected motor CM equivalent circuit shown in Figure 16b. Since the CM choke is designed to reduce the motor antiresonance frequency below the switching frequency, the CM voltage as a function of inductive and capacitive voltage is given as

$$|V_{cm}| = \left|V_{L_{eq}}\right| - |V_{ng}| \tag{7}$$

where ($V_{L_{eq}} = V_L + V_{choke}$). Using the voltage divider rule, V_{ng} can be calculated as

$$|V_{ng}| = \frac{X_{c_g}}{\left(X_{L_{eq}} - X_{c_g}\right)} \times |V_{cm}| \tag{8}$$

Since the RMS value of CM voltage and the corresponding neutral point voltage varies inversely with the modulation index, the proposed approach is designed for the worst-case scenario where the modulation index is low (i.e., high V_{ng}). In the case of lower modulation indices, the three phases switch almost simultaneously. As a result, the inverter CM voltage approximates a square wave, with its RMS value equal to its peak value. Therefore, Equation (8) can be rewritten in terms of peak values rather than RMS values, as

$$\frac{V_{ng(p)}}{\sqrt{2}} = \frac{X_{c_g}}{\left(X_{L_{eq}} - X_{c_g}\right)} \times \frac{V_{dc}}{2} \tag{9}$$

Rearranging Equation (9), the ratio of peak neutral point voltage and peak CM voltage is given by

$$\frac{V_{ng(p)}}{V_{dc}/2} = \frac{\sqrt{2}\,X_{c_g}}{\left(X_{L_{eq}} - X_{c_g}\right)} = k \tag{10}$$

Equation (10) indicates that the motor neutral point voltage can be reduced to any fraction "k" of the inverter CM voltage by adjusting the ratio of X_{c_g} and $X_{L_{eq}}$. Substituting $X_{L_{eq}} = 2\pi f_{sw}\,(L_{choke} + L_e/3)$ and $X_{c_g} = 1/\left[2\pi f_{sw}\,(3C_g)\right]$ into Equation (10) and the values of L_{choke} from Equation (4) and f_{ar} from Equation (2) afterwards, the ratio of the shifted antiresonance frequency and switching frequency can be expressed as

$$\frac{f_{ar_{shifted}}}{f_{sw}} = \frac{1}{\sqrt{1 + 1/\sqrt{2}\,k}} \tag{11}$$

Incorporating Equation (11) into Equation (4), the CM choke inductance required to limit he neutral point voltage to any fraction "k" of the CM voltage is given by

$$L_{choke} = \frac{\left(1 + 1/k\sqrt{2}\right) f_{ar}^{2} L_e}{3f_{sw}^2} - \frac{L_e}{3} \tag{12}$$

Equation (12) provides the required choke inductance to reduce the peak neutral point voltage to any fraction "k" of the inverter CM voltage. To ensure the elimination of EDM current, the bearing voltage must be reduced below a critical threshold. Consequently, the value of "k" can be determined based on the required peak bearing voltage level.

From Equation (10), the peak neutral point voltage can be given as

$$V_{ng_{(p)}} = kV_{cm_{(p)}} \tag{13}$$

The peak neutral point voltage can also be calculated using the BVR of the motor given in Equation (1) as

$$V_{ng_{(p)}} = \frac{V_{b_{(p)}}}{BVR} \tag{14}$$

Equating Equations (13) and (14) yields

$$k = \frac{V_{b_{(p)}}}{V_{cm_{(p)}} \times BVR} \tag{15}$$

From Equation (15), the value of "k" can be calculated for any desired peak value of the bearing voltage. The "k" value directly impacts the choke core design and size, specifically the CM volt-seconds impressed on the choke core. The CM volt-seconds can be given as

$$\lambda = \int_0^{\frac{T_{sw}}{2}} V_{L_{choke}}\, dt \tag{16}$$

where T_{sw} is the switchinng period and $V_{L_{choke}}$ is the CM voltage drop across the CM choke, where the value of $V_{L_{choke}}$ can be determined by converting Equation (7) into peak values and substituting the value of $V_{ng(p)}$ from Equation (13) afterwards, as

$$V_{L_{eq}(p)} = (1+k)V_{cm_{(p)}} \tag{17}$$

Assuming a negligible voltage drop across the motor CM winding to emulate the worst-case scenario, the entire CM voltage is dropped across the choke coil. Therefore, $V_{L_{eq}(p)} = V_{L_{choke}(p)}$, substituting the value $V_{L_{choke}}$ in Equation (16) and solving the integration yields

$$\lambda = \frac{(1+k)V_{cm}}{2f_{sw}} \tag{18}$$

To analyze the impact of the fraction "k" on CM volt-seconds, different values of "k" ranging from 0.1 to 1 are substituted into Equation (18), considering switching frequencies of 40, 50, and 60 kHz and a constant dc-link voltage of 600 V. The corresponding CM choke inductances and volt-seconds experienced by the choke core for varying "k" values are shown in Figure 20. As observed, both the required choke inductance and the core volt-seconds decrease as the switching frequency increases. Additionally, lower "k" values require larger choke inductances compared to higher "k" values; however, the volt-seconds across the core are reduced at lower "k" values. Since the volt-seconds are directly proportional to the maximum flux density B_{max} of the core, as expressed in Equation (19), higher "k" values may cause the choke core to saturate, leading to a loss of inductance capability.

$$B_{max} = \frac{\lambda}{nA_e} \tag{19}$$

where n is the number of turns and A_e is the effective cross-sectional area of the core.

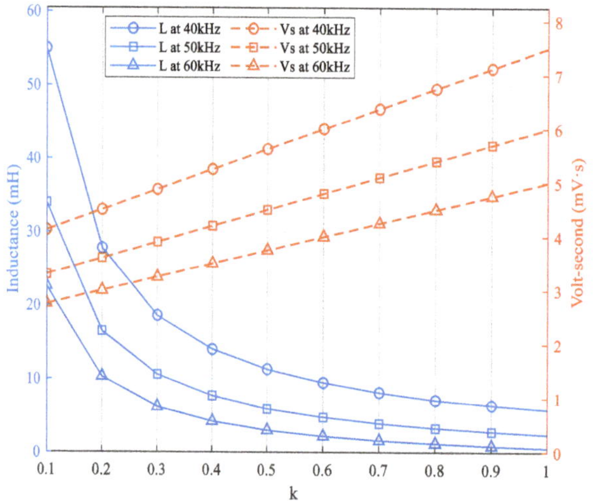

Figure 20. Impact of different "k" values on CM choke core volt-seconds.

Furthermore, at higher "k" values, the reduction in choke inductance corresponds to fewer turns, which again increases the maximum flux density and the risk of core saturation. Therefore, to ensure the practical feasibility of the choke size, the "k" value should be kept as low as possible. Although a higher choke inductance will be required at lower "k" values, the size of the choke will remain practical since the core volt-seconds will be lower.

For the motor under test, selecting a peak bearing voltage of $\leq$1 V to fully eliminate EDM currents results in a calculated "k" value of 0.15 using Equation (15), assuming a peak CM voltage of 300 V and the BVR of the motor as 2.2% (calculated based on the peak values of neutral point voltage and bearing voltage). Based on this "k" value, the required choke inductance at a switching frequency of 40 kHz is calculated as 37 mH using Equation (12).

5.2. CM Choke Core Selection

In this paper, the CM choke is designed to shift the motor antiresonance frequency below the switching frequency, resulting in nearly all the inverter CM voltage being dropped across the CM choke. Consequently, the primary concern is the core's ability to withstand the source CM voltage for a period determined by switching frequency. Therefore, to select the appropriate CM choke core, it is important to first calculate the CM volt-seconds across it.

To ensure that the selected choke core is capable of withstanding maximum voltage stress, a worst-case operating condition is assumed. First, the CM voltage drop across the motor winding inductance is considered negligible (i.e., $V_L \approx 0$), such that the entire inductive CM voltage is imposed across the choke. Additionally, a low modulation index is assumed, representing a scenario in which an RMS CM voltage equal to its peak value (i.e., $V_{dc}/2$) is continuously applied across the choke within a switching cycle. Under these assumptions, given a switching frequency of 40 kHz and dc-link voltage of 600 V, CM volt-seconds across the CM choke core can be calculated for k = 0.15 as $\lambda = 4.31$ m Vs using Equation (18).

Furthermore, a nanocrystalline core material with a saturation flux density B_{sat} of 1.2 T–1.3 T is selected. Therefore, the maximum flux density B_{max} in the core is selected to be less than 1.2 T. According to Equation (19), to prevent core saturation, the product of n and A_e must be greater than or equal to λ / B_{max}. Therefore, by substituting the values for the maximum flux density and maximum volt-seconds into Equation (19), and considering the number of turns as 40, the cross-sectional area of the core should be larger than 0.89 cm^2. Furthermore, determining the inner and outer diameters of the core requires the calculation of the magnetic path length l_m, which can be derived from the CM choke inductance as

$$l_m = \frac{n^2 \mu\, A_e}{L_{choke}} \tag{20}$$

where $\mu = \mu_0 \mu_r$. For nanocrystalline materials, the relative permeability μ_r is approximately 30,000, yielding a total permeability μ = 3.77 $\times$ 10^{-2} H/m. Substituting the calculated values into Equation (20), the magnetic path length can be determined as 14.5 cm. Using the values of A_e and l_m and considering the height of the core as 20 mm (for practicality), the inner and outer diameters of the CM choke core are determined as 41 mm and 50 mm, respectively.

The calculated dimensions indicate that the CM choke can effectively withstand 115% of the inverter CM voltage at V_{dc} of 600 V and switching frequency matching machine antiresonance frequency, even at low modulation index. The resulting CM choke size is practical, and a suitable core can be readily selected from commercially available options.

In this paper, a nanocrystalline (Magnetec NANOPERM M-011, 50 $\times$ 40 $\times$ 20 mm^3) core is selected that achieves a volumetric power density of approximately 8–12 kW/L, which is within the same order of magnitude as advanced traction inverters that integrate planar EMI filters, where system-level power densities of 25–35 kW/L have been reported [27,28]. The high saturation flux density ($\approx$1.3 T) and low-loss properties of the nanocrystalline material enable a 30–50% reduction in volume compared with ferrite equivalents while maintaining thermal stability and magnetic headroom. Although the material cost ($\approx$USD 3–25 per core) is higher than that for ferrites, it is justified by the compactness, reliability, and EDM-current suppression that improve drivetrain durability. Therefore, the proposed CM choke offers a practical balance among performance, cost, and volume for WBG inverter-based applications.

5.3. Evaluation of Saturation Margin of the Selected CM Choke Core

The selected nanocrystalline core has a saturation flux density of 1.3 T. The proposed CM choke core is designed to tolerate up to 115% of the CM voltage, resulting in a maximum flux density of 1.18 T. This corresponds to a nominal saturation margin of approximately 9.23%. Although this margin appears narrow, it represents a highly conservative case established to capture worse-case operating conditions. Specifically, the flux density and volt-second stress are evaluated at an extremely low modulation index ($\approx$0), where the CM voltage exhibits its highest magnitude. However, such low modulation indices are not encountered during normal inverter operation.

In practical operation, as the modulation index (MI) increases, the CM voltage transitions from a bipolar square wave to a quasi-square shape; therefore, its effective RMS value decreases, leading to a lower flux swing and greater magnetic headroom. Since no explicit analytical relationship between modulation index and CM voltage RMS is available in the literature, a rough estimation is obtained using MATLAB/Simulink R2022b for a two-level VSI (V_{dc} = 600 V, f_{sw} = 40 kHz). The simulated CM voltage RMS values for modulation indices from 0.1 to 0.9 are then used to compute the corresponding flux linkage (λ), core flux density (B_{max}), and saturation margins relative to the nanocrystalline core's saturation flux density.

To emulate transient and surge conditions, a +20% increase in dc-link voltage is applied in all calculations, and the resulting margins are presented in Figure 21. The results indicate that the lowest margin occurs near MI = 0.1, where $B_{max} \approx 1.28$ T ($\approx$1.4% margin). At MI = 0.3, which is the operating point used in this paper, the saturation margin is approximately 27.5%, even with the 20% surge and 1.15 $\times$ V_{cm} stress considered. The margin increases progressively with modulation index, exceeding 80% at M = 0.9.

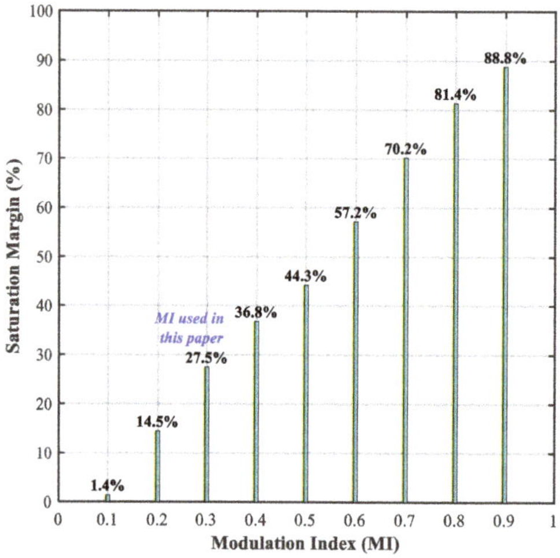

Figure 21. Variation of the saturation margin of the proposed nanocrystalline CM choke core with modulation index, considering a 20% dc-link voltage surge.

These results confirm that the choke remains well below the saturation limit under both nominal and transient high-voltage conditions. The corner-case evaluation thus validates that the design is conservative and possesses sufficient magnetic headroom for all expected inverter operating scenarios.

6. Experimental Validation

Experimental analysis is carried out on a 2.2 kW three-phase four-pole star-connected induction motor supplied by a three-phase SiC two-level VSI, as shown in the experimental setup in Figure 8. The designed choke and the selected choke core parameters used in the experiment are shown in Figure 22 and Table 3, respectively. The inverter is supplied with a constant dc-link voltage of 600 V at a modulation index of 0.3 and switching frequency of 40 kHz.

Table 3. Calculated and Experimental CM Choke Parameters.

Designed Choke Parameters	Calculated	Experimental
Choke inductance	37 mH	38 mH
Outer diameter	50 mm	52.3 mm
Inner diameter	41 mm	37.1 mm
Height	20 mm	22.3 mm
No of turns	40	41

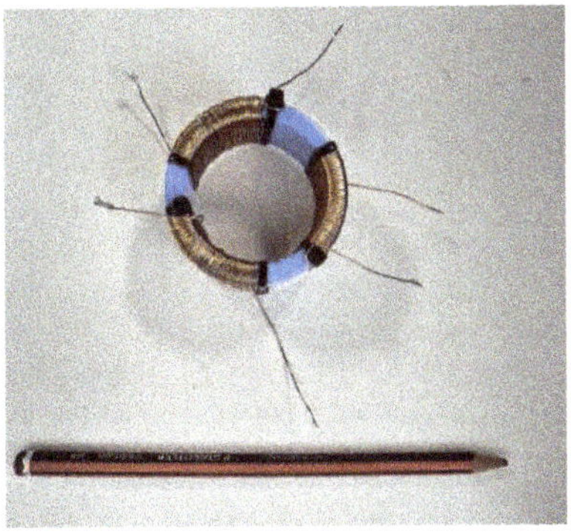

Figure 22. A 38 mH choke designed for the experimental analysis, shown alongside a standard pencil for scale.

6.1. Impact of Proposed CM Choke Connection on Motor Bearing Voltage

To evaluate the effectiveness of the proposed CM choke in mitigating motor neutral point voltage and resulting bearing voltage, measurements are conducted for both voltages with and without the proposed CM choke connection. On the other hand, the CM voltage is calculated by averaging the inverter pole voltages.

Figure 23 shows the results obtained. As shown, without the CM choke connection, the peak amplitude of the neutral point voltage V_{ng} reaches 600 V (i.e., 2 pu) (Figure 23a), while the peak bearing voltage is observed as 17 V (Figure 23b). Such a high motor neutral point voltage and bearing voltage pose a significant risk to motor winding insulation and bearings. However, with the implementation of the proposed CM choke, the peak neutral point voltage drops to 35 V (i.e., 0.12pu), and the peak bearing voltage drops to 1 V. This substantial reduction in motor neutral point and bearing voltage minimizes electrical stress on motor windings and effectively suppresses the risk of EDM-induced damage to bearings.

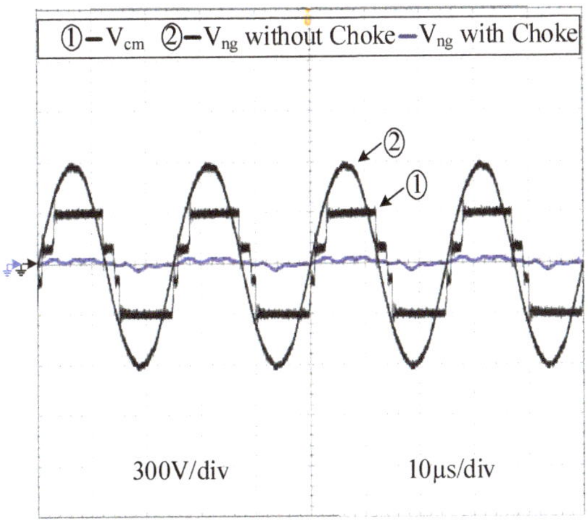

(a)

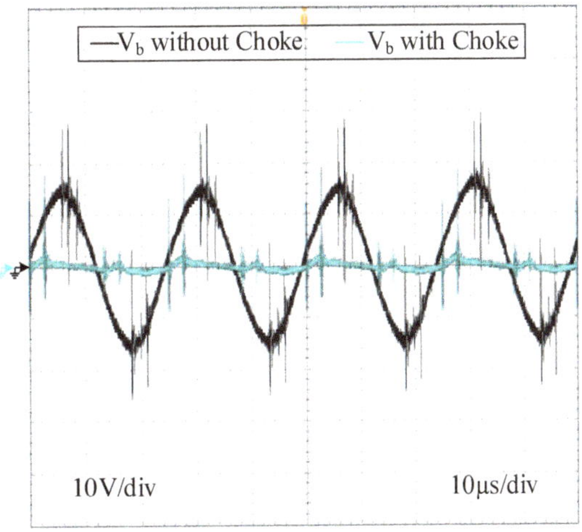

(b)

Figure 23. Impact of proposed CM choke connection on (**a**) neutral point voltage and (**b**) bearing voltage.

6.2. Impact of Proposed CM Choke Connection on EDM Bearing Current

In the case of motor drive operation without any CM choke connected, the highest EDM current count occurred when the switching frequency matched the machine antiresonance frequency, particularly under the high dc-link voltage condition (i.e., at 40 kHz and 600 V) (see Figure 16d). To validate the effectiveness of the proposed approach under these conditions, the bearing current is measured at a switching frequency of 40 kHz, dc-link voltage of 600 V, and modulation index of 0.3 with connection of the proposed CM choke. The experimental results, presented in Figure 24a, show a complete elimination of EDM current spikes over the specified time frame, whereas 1887 EDM current spikes were recorded in the same time frame when no mitigation was considered.

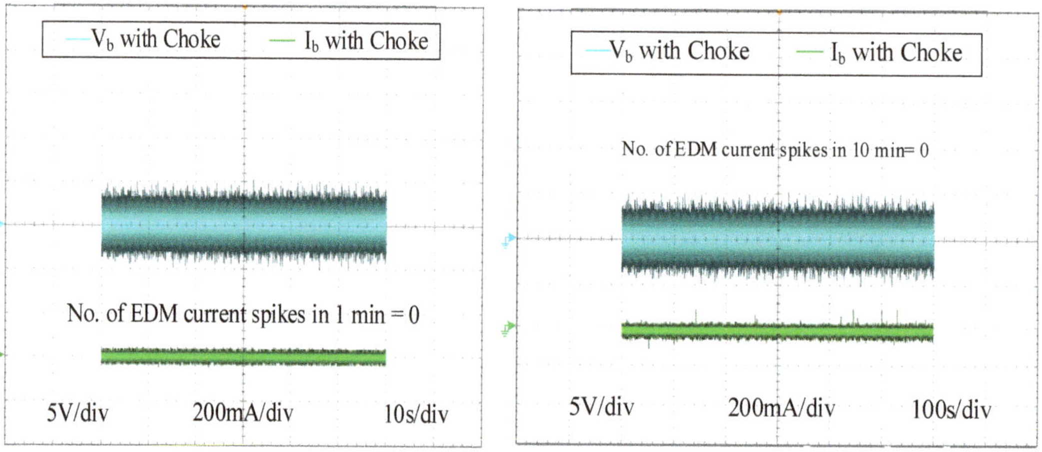

(**a**) In 1 min (**b**) In 10 min

Figure 24. Impact of proposed CM choke connection on EDM bearing current event count.

To further validate the performance of the proposed CM choke, the bearing current is monitored over an extended duration of 10 min, as shown in Figure 24b. It is observed that, with the proposed CM choke connected, no EDM current spikes > 100 mA occur even over the prolonged observation period.

6.3. Impact of Proposed CM Choke Connection on Motor Common-Mode and Line Current

To analyze and compare the CM current I_{cm} of the motor drive system with and without the proposed CM choke connection, measurements are taken using a current probe placed across the three lines of the motor (AC side) and around both positive and negative DC buses (DC side). The proposed CM choke lowers the motor's antiresonance frequency relative to the switching frequency, thereby reducing the neutral point voltage, bearing voltage, and the associated EDM bearing current. As the neutral point voltage is the primary driver of CM current, this reduction decreases CM current flow from the motor winding-to-frame/ground (AC side). The CM current, however, circulates in a closed loop: Inverter → Motor windings/parasitics → Motor frame → Ground → DC source/AC mains → Inverter. Hence, any reduction on the AC side inherently leads to a corresponding reduction on the DC side. Thus, the implementation of the CM choke provides simultaneous suppression of CM current on both sides of the inverter.

The results, presented in Figure 25a, show the CM current measured simultaneously on the AC and DC side of the inverter without the CM choke. As shown, the CM current is similar in magnitude as well as waveshape on both sides of the inverter. Furthermore, without the CM choke, CM current exhibits a sinusoidal waveform with high-frequency harmonics and a peak amplitude of 0.9 A, primarily due to the high neutral point voltage. However, with the optimally designed CM choke in place, the CM current is reduced by 83%, lowering the peak amplitude to 0.15 A while also minimizing harmonic content, as shown in Figure 25b. Figure 25c presents the frequency spectrum of the CM current. Notably, with the proposed CM choke, the overall CM current harmonics are attenuated as compared to the case of without the CM choke. Furthermore, in the higher frequency range (i.e., 10 MHz onwards), it remains around 50 dBμA. Although the proposed choke is designed to eliminate the EDM bearing current, it also attenuates the AC and DC side CM current peak and CM-EMI by reducing the neutral point voltage.

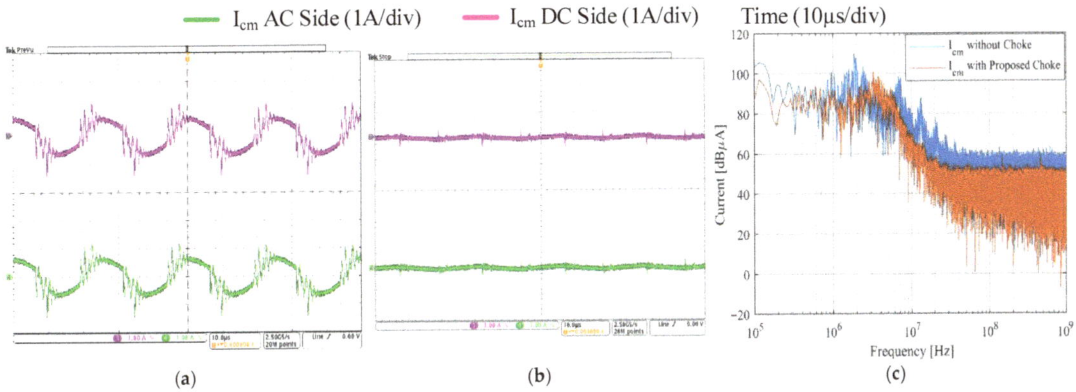

Figure 25. Experimentally measured CM current (**a**) without proposed CM choke connection; (**b**) with proposed CM choke connection; (**c**) DC and AC CM EMI.

The line currents also benefit from the proposed CM choke connection, as shown in Figure 26. Without the CM choke, line currents show a peak of approximately 4 A with a noticeable ripple. With the proposed CM choke installed, the peak line current remains almost the same; however, the ripple amplitude is effectively reduced, resulting in a smoother sinusoidal waveform with total harmonic distortion reduced from 13.4% to 5.3%. This demonstrates that the proposed CM choke not only suppresses CM disturbances on both the AC and DC sides but also improves line current quality, enhancing overall system performance.

Table 4 summarizes the measured impact of the proposed approach on motor parameters in the worst-case scenario. Significant reductions are achieved across all metrics, including a 94% decrease in peak bearing voltage, complete elimination of EDM current spikes, and an 83% reduction in peak CM current. The average EMI level is also lowered from approximately 85 dBμA to 75 dBμA, indicating improved compliance with EMC guidelines and reduced stress on motor bearings.

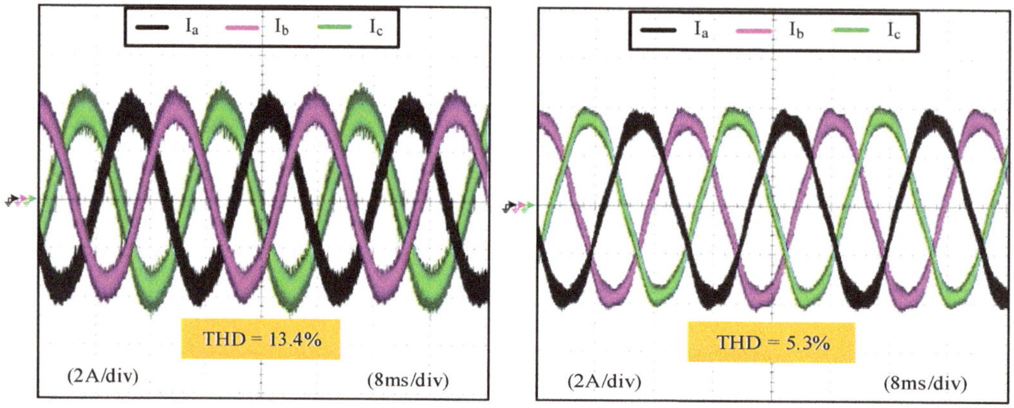

(**a**) (**b**)

Figure 26. Experimentally measured motor line currents (**a**) without proposed CM choke connection; (**b**) with proposed CM choke connection.

Table 4. Performance of Proposed CM Choke Design Approach in Reducing Bearing Voltage, Neutral Point Voltage, EDM Bearing Current, CM Current, and Line Current **at** f_{sw} = 40 kHz and V_{dc} = 600 V.

Parameter	Without Proposed Approach	with Proposed Approach	Percentage Reduction
EDM Current spikes in 1 min	1887	0	100%
Bearing voltage (peak)	17 V	1 V	94%
Neutral point voltage (peak)	600 V	35 V	94%
CM current (peak)	0.9	0.15	83%
EMI (150 kHz–30 MHz)	110–60 dBμA	100–50 dBμA	9–16%
Line current	THD = 13.4%	THD = 5.3%	60%

6.4. Robustness of the Proposed CM Choke Design Under Parasitic Element Variations

The accuracy of parasitic parameter estimation can be influenced by insulation aging, temperature, and variations among motor designs. Published studies report measured trends in stator insulation capacitance and winding inductance under these conditions, although explicit absolute percentage changes are limited. For example, the Electric Power Research Institute (EPRI) report on thermal aging of stator windings shows that winding-to-frame capacitance changes are typically below 2–3% under moderate deterioration [29]. Thermal-aging experiments on enameled coils indicate similar gradual increases with aging [30], while dielectric measurements on traction-motor winding insulation under thermal aging show a ±1–5% change, with typical operating temperature variations [31]. For the magnetic components, manufacturer data show that permeability variations with temperature can lead to inductance changes of up to ±30%, with typical production tolerances of ±10–20% [32].

Based on these reported trends, a winding-to-frame capacitance variation of up to ±30% and a common-mode winding inductance variation of ±50% is considered for robustness analysis. Figure 27 shows the variation in bearing voltage with the variation in motor parasitic element values. These bearing voltage values are obtained when a fixed proposed CM choke of 37 mH is connected for the worst case of 600 V dc-link voltage and 40 kHz switching frequency. Using the design Equation (12), the "k" values are calculated for each parameter variation and then substituted in Equation (15) to estimate the bearing voltages. As observed in the chart, even with the widest ranges of parameter variations, the peak bearing voltage consistently remains below 3 V, and EDM spikes are unlikely at this small bearing voltage.

To further validate this claim, the motor operating points are adjusted such that, with the proposed 37 mH CM choke, the resulting bearing voltage reaches approximately 3 V (to emulate parameter variation). Under operating conditions of 25 kHz switching frequency, 400 V dc-link voltage, and a modulation index of 0.3, the measured bearing voltage is approximately 3 V, as shown in Figure 28a. Figure 28b shows the corresponding EDM bearing current. As seen, there are no EDM spikes observed even when the bearing voltage increases to 3 V. This confirms that the proposed CM choke design remains resilient to both manufacturing tolerances and operational variability.

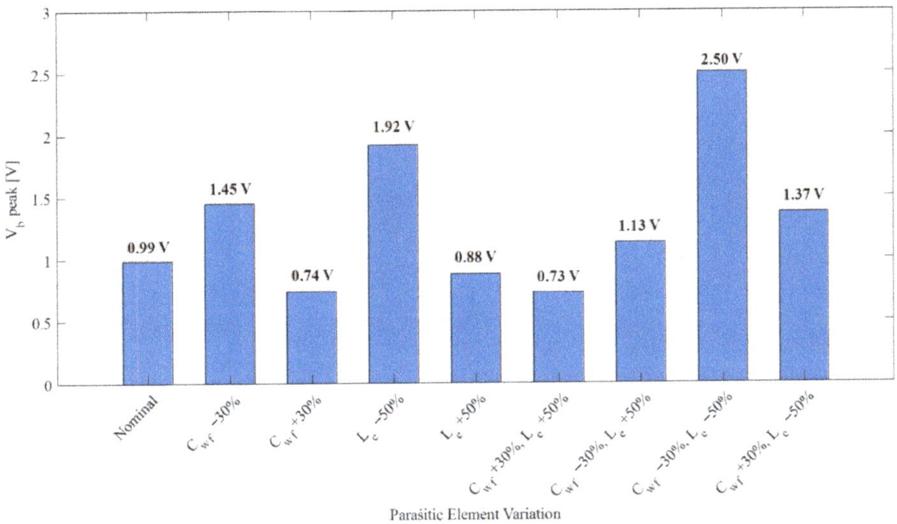

Figure 27. Calculated motor bearing voltages for motor parasitic parameter variations.

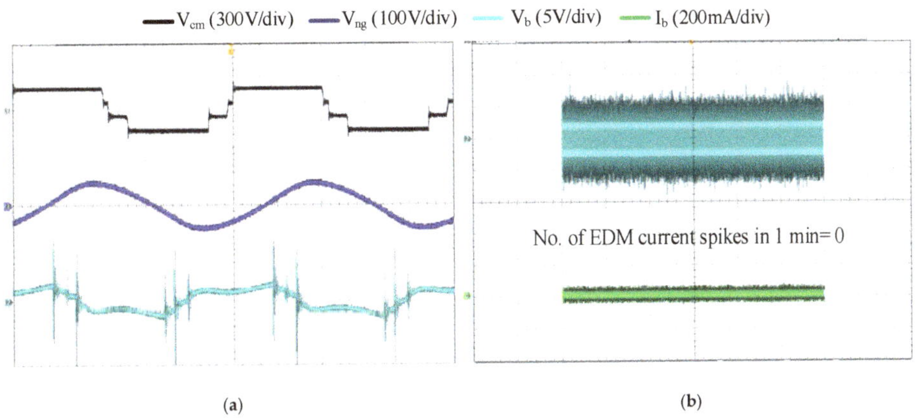

Figure 28. Experimental (**a**) motor bearing voltage and (**b**) EDM bearing current measured using fixed CM choke of 38 mH at switching frequency of 25 kHz and dc-link voltage of 400 V.

6.5. Robustness of the Proposed CM Choke Design Across Different Operating Conditions

To confirm the robustness of the proposed 37 mH CM choke (designed for switching frequency of 40 kHz) across different converter operating points, additional evaluations are conducted using switching frequencies of 30 kHz and 50 kHz, while all other parameters remain constant. Since the CM voltage, neutral point, and the resulting bearing voltage scale with the dc-link voltage, a lower dc-link voltage inherently reduces these voltages. Therefore, the robustness assessment is carried out at the highest test dc-link voltage of 600 V to represent the most severe operating condition.

The proposed CM choke is analyzed using the design Equation (12) to compute the "k" value for each switching frequency while keeping L_{choke} fixed at 37 mH. The calculated "k" values of 0.29 and 0.09 for 30 kHz and 50 kHz, respectively, are used to estimate the bearing voltage from Equation (15) and the λ using Equation (18). With a worst-case CM voltage of 300 V and a BVR of 0.022, the resulting bearing voltages are 1.9 V and 0.6 V for 30 kHz and 50 kHz, respectively, while λ varies inversely, being 6.4 mVs at 30 kHz and 3.1 mVs at 50 kHz. Although λ increases slightly at the lower switching frequency, the experimental modulation index of 0.3 reduces the effective CM voltage below 300 V, which compensates for the rise in λ and ensures that both flux and B_{max} remain within the non-saturating limits of the core.

Figure 29 presents the waveforms of the bearing voltage and EDM current measured at different switching frequencies. As shown, across all operating points, no EDM-related discharge spikes occur because the calculated bearing voltage remains below 2 V. These results demonstrate that the proposed single 37 mH CM choke, optimally designed for a switching frequency 40 kHz, exhibits strong robustness across different switching frequencies (higher and lower than 40 kHz). Even under high dc-link voltage, the magnetic stress and bearing voltage remain within safe limits, and no EDM discharge events are observed. Consequently, a single fixed-value CM choke can be confidently applied across a broad operating range, simplifying practical implementation in motor-drive systems.

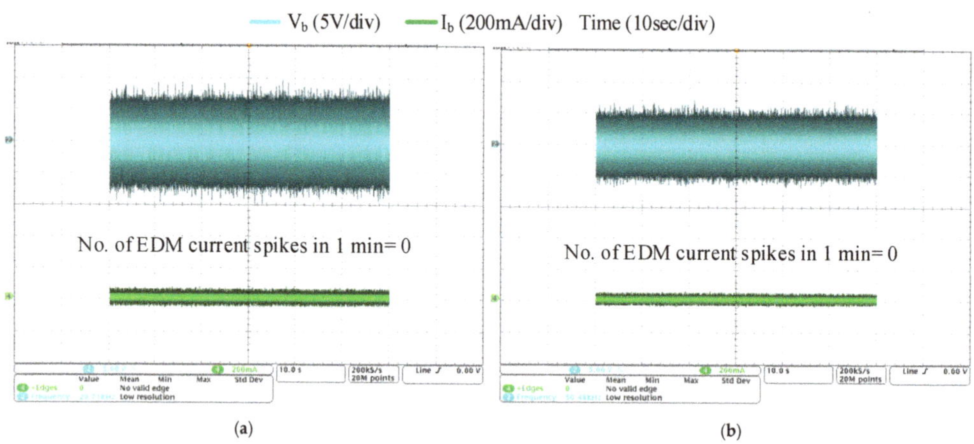

Figure 29. (**a**,**b**) Bearing voltage and EDM bearing current at different switching frequencies using fixed CM choke of 38 mH.

6.6. Core-Loss Estimation and Thermal Validation of the Proposed CM Choke

This section presents the analytical estimation and experimental validation of core losses and thermal behavior for the nanocrystalline core employed in the proposed CM choke. The evaluation of the loss characteristics of the Magnetec NANOPERM M-011 material under high-frequency excitation and confirmation the thermal stability of the design under worst-case operating conditions are performed.

The nanocrystalline toroidal core used for the proposed CM choke (Magnetec NANOPERM M-011) is analyzed for core loss and thermal performance under high-frequency excitation. Core losses are estimated using the improved Generalized Steinmetz Equation (iGSE), which accounts for non-sinusoidal flux excitation. Based on the manufacturer's material constants (k = 2.0×10^{-3}, $\alpha = 1.45$, and $\beta = 1.95$) at 100 °C, the equivalent coefficient k_i is derived and applied to evaluate the volumetric power loss.

For the applied ±300 V CM voltage (square wave) at 40 kHz and 44 turns per phase, the iGSE model predicts a total core loss of approximately 2.5 W, increasing to 4.9 W at 60 kHz for a flux swing of 2.4 T_{pp}. Thermal analysis based on natural convection (h = 10–20 W/m^2K) and the toroid surface area yielded a steady-state temperature rise of 24–48 K, corresponding to a core temperature of 44–68 °C at 40 kHz in a 20 °C ambient environment. Figure 30 shows the experimental infrared thermography under V_{dc} = 600 V and f_{sw} = 40 kHz with a stabilized surface temperature of about 43.3 °C, indicating a loss of 1.2–2.3 W, which is less than the predicted value due to the reduced effective CM voltage (i.e., modulation index of 0.3) used during testing. Overall, the analytical and experimental results confirm low loss levels and thermally stable operation well below the 130 °C limit of the nanocrystalline material. All detailed calculation steps are provided in Appendix A.

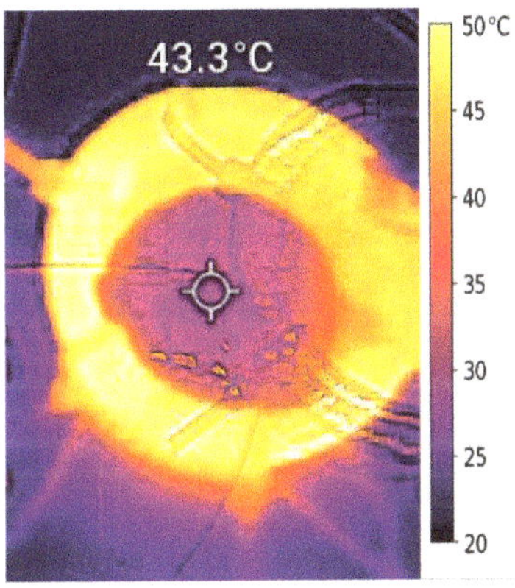

Figure 30. Infrared thermography of the proposed CM choke at 600 V dc-link voltage and 40 kHz switching frequency.

6.7. Experimental Evaluation of Proposed CM Choke Design Against Existing CM Voltage Reduction Methods for EDM Current Elimination

Several active CM voltage reduction methods have been proposed in the literature, as discussed in Section 1. While each method employs a different technique, they all share a common limitation in that the maximum achievable CM voltage reduction is constrained to $\pm V_{dc}/6$. This reduced CM voltage is typically sufficient to eliminate EDM spikes at low switching frequencies. However, WBG inverters enable motor drives to operate at higher switching frequencies, increasing the possibility of switching frequency approaching the machine antiresonance frequency. This can excite machine resonance, resulting in a significant overvoltage at the machine neutral point. As the bearing voltage is directly proportional to the neutral point voltage, it can also increase and exceed the bearing lubrication breakdown threshold, even with the CM voltage being reduced to $\pm V_{dc}/6$. As a result, EDM current spikes can still be observed with reduced CM voltage approaches.

Among the CM voltage reduction methods, the modified PWM schemes, such as the AZSPWM, stand out as the most cost-effective solution. Therefore, the proposed approach is compared with the AZSPWM method to evaluate its effectiveness in eliminating EDM current spikes. For comparison, the motor drive under test is operated under the same proposed approach, effectively eliminating EDM current spikes at all switching frequencies. In contrast, under the AZSPWM1 operating conditions using the AZSPWM1 modulation technique instead of SPWM, measurements are recorded for the inverter CM voltage, motor neutral point voltage, bearing voltage, and bearing current.

Figure 31 shows the experimental results under the AZPWM1 modulation method conducted at a dc-link voltage of 600 V and switching frequency of 40 kHz, while the modulation index is fixed at 0.3. From Figure 31a, it can be observed that the peak CM voltage is limited to ± 100 V (i.e., $\pm V_{dc}/6$). However, despite this reduction, a peak neutral point voltage of 290 V and a corresponding peak bearing voltage of 7 V can be observed. This bearing voltage is sufficiently high, exceeding the lubrication breakdown threshold, causing EDM current spikes (see Figure 31b). As seen, there are 626 EDM current spikes occurring within a 1 min interval, which indicates persistent electrical stress on the motor bearing system. In contrast, the proposed approach demonstrates a complete elimination of EDM current spikes under identical operating conditions, as illustrated in Figure 24.

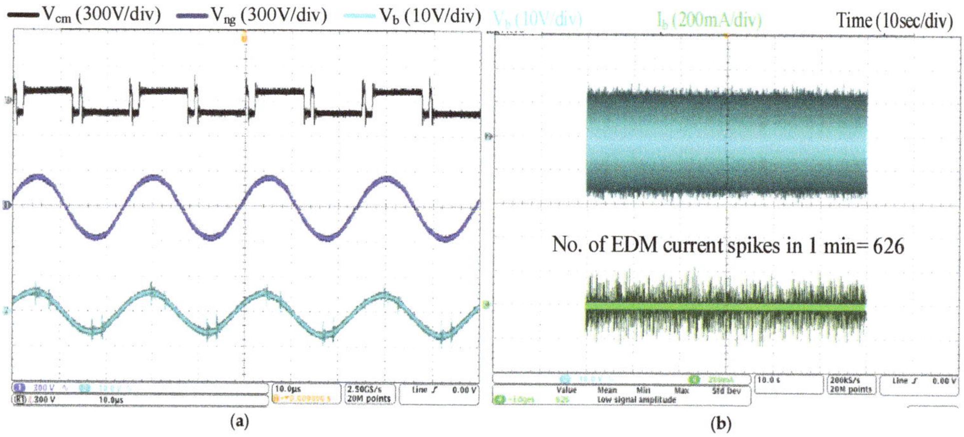

Figure 31. Effect of AZSPWM1 modulation technique on (**a**) CM voltage, neutral point voltage, and bearing voltage. (**b**) EDM event count.

7. Conclusions

In this paper, a new design approach for a CM choke is proposed. The proposed design overcomes the limitation of conventional CM chokes by effectively eliminating EDM bearing current spikes. Additionally, the proposed CM choke can inherently minimize CM current and EMI by reducing the neutral point voltage. The CM choke inductance is optimally selected to shift the motor antiresonance frequency to an optimized lower value relative to the switching frequency. By maintaining an optimized ratio between these frequencies, the motor bearing voltage can be mitigated, resulting in a complete elimination of the EDM bearing current spikes. Since the proposed CM choke connection shifts the motor antiresonance frequency below the switching frequency, most of the source CM voltage is absorbed by the choke core, thereby minimizing its propagation to the load and bearings. As a result, the CM choke core is specifically designed to withstand the source CM voltage without saturation. The calculations indicate that the CM choke's overall size remains practical, enabling the use of commercially available off-the-shelf cores. To validate the effectiveness of this design, extensive experimental testing is conducted under various operating conditions. The experimental results confirm that the bearing voltage is not only significantly reduced but also maintained below the threshold required for EDM current generation. This ensures that EDM current spikes are eliminated across various operating conditions. The robustness of the proposed approach is experimentally validated, demonstrating the elimination of EDM bearing current spikes under various operating conditions using a single CM choke, without requiring reselection. Furthermore, a comparison of the proposed CM choke with an existing CM voltage reduction technique, which reduces the CM voltage to $\pm V_{dc}/6$, is performed to assess the relative effectiveness of both methods in addressing EDM bearing currents. This comparison confirms that the proposed CM choke outperforms conventional CM voltage reduction techniques in eliminating EDM bearing currents.

Author Contributions: Conceptualization, M.M. and M.D.; Methodology, M.M.; Validation, M.M.; Formal analysis, M.M.; Resources, M.D.; Writing—original draft, M.M.; Writing—review and editing, M.D.; Supervision, M.D. All authors have read and agreed to the published version of the manuscript.

Funding: This research received no external funding.

Data Availability Statement: The original contributions presented in this study are included in the article. Further inquiries can be directed to the corresponding author.

Conflicts of Interest: The authors declare that there are no conflicts of interest.

Abbreviations

The following terms and symbols are used in this manuscript.

A_e	Core cross-sectional area
A_s	Core surface area
α	Steinmetz exponent (frequency dependent)
β	Steinmetz exponent (flux dependent)
AZPWM	Active Zero State Pulse Width Modulation
B_{max}	Maximum flux density
BVR	Motor Bearing Voltage Ratio
C_b	Bearing capacitance
C_{b_I}	Bearing capacitance from ball to inner race
C_{b_O}	Bearing capacitance from ball to outer race
C_g	Motor parasitic capacitance between neutral point and frame
GaN	Gallium Nitride
h	Convection heat transfer coefficient
I_b	Capacitive bearing current
I_{CIR}	High frequency circulating current
I_{cm}	Common-mode current
I_{EDM}	Electrostatic discharge machining current
I_{rg}	Rotor grounding current
iGSE	Improved Generalized Steinmetz Equation
k	Steinmetz coefficient
C_{wf}	Motor parasitic capacitance between winding and frame
C_{wr}	Motor parasitic capacitance between winding and rotor
ΔB_{pp}	Peak-to-peak flux density
ΔT	Temperature rise
EDM	Electrostatic Discharge Machining
EMI	Electromagnetic Interference
EV	Electric Vehicle
f	Frequency of excitation
f_{ar}	Motor antiresonance frequency
$f_{ar_{shifted}}$	Shifted antiresonance frequency
f_{sw}	Switching frequency
V_L	Inductive voltage drop
V_{ng}	Neutral point voltage
V_s	Supply voltage
X_L	Inductive reactance
X_{C_g}	Motor neutral-point parasitic capacitive reactance
$X_{L_{eq}}$	Motor winding and choke equivalent inductive reactance
Z_{CM}	Motor common-mode impedance

k_i	Equivalent iGSE constant
l_m	Magnetic path length
L_{choke}	Choke inductance
L_e	Motor winding inductance
λ	Volt-seconds
m	Core mass
N	Number of turns per phase
n	Number of turns (generic)
NSPWM	Near State Pulse Width Modulation
P_{core}	Total core loss (W)
P_{meas}	Measured power loss
P_v	Core loss density (W/kg)
R_{th}	Thermal resistance
SiC	Silicon Carbide
T_a	Ambient temperature
T_{core}	Core temperature
T_{sw}	Switching period
$V_{L_{eq}}$	Motor winding and choke equivalent voltage drop
V_b	Bearing voltage
V_c	Capacitive voltage drop
V_{choke}	Voltage drop across CM choke
V_{cm}	Common-mode voltage
V_{dc}	dc-link voltage
VSI	Voltage Source Inverter
WBG	Wide-Bandgap
X_C	Capacitive reactance

Appendix A

This appendix provides the detailed calculations used to estimate the core losses and thermal performance of the Magnetec NANOPERM M-011 nanocrystalline core.

Core Geometry and Excitation Parameters:

$$A_e = 7.3 \times 10^{-5}\text{m}^2,\ l_e = 0.1403\ \text{m},\ V_c = 1.02 \times 10^{-5}\text{m}^3$$

Material Density = 7250 kg/m^3 → Core Mass = 0.074 kg.

For the applied excitation conditions (±300 V CM voltage, f_{sw} = 40 kHz, n = 44 turns per phase):

$$\Delta B_{pp} = 2.4\ \text{T} \Rightarrow B_{max} = 1.2\ \text{T}$$

The rate of flux change is given by

$$\frac{dB}{dt} = 2\,f\,\Delta B_{pp}$$

iGSE Parameters (from datasheet at 100 °C):

$$k = 2.0 \times 10^{-3},\ \alpha = 1.45,\ \beta = 1.95$$

The equivalent iGSE constant is computed as

$$k_i = \frac{k}{\pi^{\alpha-1}} = \left(2.0 \times 10^{-3}\right)/\pi^{0.45} = 8.6 \times 10^{-4}$$

Loss Density Calculation:

$$P_v = k_i \left(\frac{dB}{dt} \right)^{\alpha} (\Delta B_{pp})^{\beta - \alpha}$$

At 40 kHz:

$$\frac{dB}{dt} = 2\, f\, \Delta B_{pp} = 2 \times 40{,}000 \times 2.4 = 1.92 \times 10^5 \, \frac{\mathrm{T}}{\mathrm{s}}$$

$$P_v = 8.6 \times 10^{-4} \times \left(1.92 \times 10^5\right)^{1.45} \times (2.4)^{0.5} = 33.8 \frac{\mathrm{W}}{\mathrm{kg}}$$

Total Core Loss:

$$P_{core} = P_v \times m = 33.8 \times 0.074 = 2.50 \, \mathrm{W}$$

Similarly, predicted losses are 3.7 W and 4.9 W at 50 kHz and 60 kHz, respectively.
Thermal Model:
The surface area of the toroidal core is approximated as

$$A_s = 0.00524 \, \mathrm{m}^2$$

For natural convection (h = 10–20 W/m^{2}K):

$$R_{th} = \frac{1}{h\, A_s} = 9.5 - 19 \frac{\mathrm{K}}{\mathrm{W}}$$

The steady-state temperature rise is given by

$$\Delta T = P_{core} \times R_{th}$$

At 40 kHz:

$$\Delta T = 2.5 \times (9.5 - 19) = 24 - 48 \mathrm{K}$$

Hence, for ambient temperature T_a = 20 °C:

$$T_{core} = T_a + \Delta T = 44 - 68 \, ^{\circ}\mathrm{C}$$

Validation:
Measured core temperature from infrared thermography = 43 °C → ΔT = 23 K.
Back-calculated loss:

$$P_{meas} = \frac{\Delta T}{R_{th}} = 1.2 - 2.3 \, \mathrm{W}$$

References

1. Husain, I.; Ahmed, S.; Qahouq, J.; Rajashekara, K. Electric Drive Technology Trends, Challenges, and Opportunities for Future Electric Vehicles. *Proc. IEEE* **2021**, *109*, 1039–1059. [CrossRef]
2. Zhou, W.; Yuan, X. Experimental Evaluation of SiC MOSFETs in Comparison to Si IGBTs in a Soft-Switching Converter. *IEEE Trans. Ind. Appl.* **2020**, *56*, 5108–5118. [CrossRef]
3. Tesla's Innovative Power Electronics: The Silicon Carbide Inverter. Available online: https://autotech.news/teslas-innovative-power-electronics-the-silicon-carbide-inverter (accessed on 5 August 2025).
4. Diab, M.S.; Zhou, W.; Yuan, X. Evaluation of Topologies and Active Control Methods for Overvoltage Mitigation in SiC-Based Motor Drives. In Proceedings of the 2021 IEEE Energy Conversion Congress and Exposition (ECCE), Vancouver, BC, Canada, 10–14 October 2021; pp. 4867–4873. [CrossRef]

5. Sundeep, S.; Wang, J.; Griffo, A.; Alvarez-Gonzalez, F. Antiresonance Phenomenon and Peak Voltage Stress Within PWM Inverter Fed Stator Winding. *IEEE Trans. Ind. Electron.* **2021**, *68*, 11826–11836. [CrossRef]
6. Memon, M.; Diab, M.; Yuan, X. Mitigation of Machine Neutral Point Voltage in SiC Motor Drives by Controlling Antiresonance Frequency and Switching Frequency. *IEEE J. Emerg. Sel. Top. Power Electron.* **2025**, *13*, 1514–1527. [CrossRef]
7. Schinkel, M.; Weber, S.; Guttowski, S.; John, W.; Reichl, H. Efficient HF Modeling and Model Parameterization of Induction Machines for Time and Frequency Domain Simulations. In Proceedings of the Twenty-First Annual IEEE Applied Power Electronics Conference and Exposition (APEC 2006), Dallas, TX, USA, 19–23 March 2006; p. 6.
8. Han, D.; Morris, C.T.; Lee, W.; Sarlioglu, B. Comparison Between Output Common-Mode Chokes for SiC Drive Operating at 20- and 200-kHz Switching Frequencies. *IEEE Trans. Ind. Appl.* **2017**, *53*, 2178–2188. [CrossRef]
9. Bonnett, A.H.; Yung, C. A Construction, Performance and Reliability Comparison for Pre-EPAct, EPAct and Premium-Efficient Motors. In Proceedings of the IEEE Industry Applications Society 53rd Annual Petroleum and Chemical Industry Conference, Philadelphia, PA, USA, 11–13 September 2006; pp. 1–7.
10. He, F.; Xie, G.; Luo, J. Electrical Bearing Failures in Electric Vehicles. *Friction* **2020**, *8*, 4–28. [CrossRef]
11. Akagi, H.; Shimizu, T. Attenuation of Conducted EMI Emissions from an Inverter-Driven Motor. *IEEE Trans. Power Electron.* **2008**, *23*, 282–290. [CrossRef]
12. Mei, C.; Balda, J.C.; Waite, W.P.; Carr, K. Analyzing Common-Mode Chokes for Induction Motor Drives. In Proceedings of the Power Electronics Specialists Conference (PESC 2002), Cairns, QLD, Australia, 23–27 June 2002; Volume 3, pp. 1557–1562.
13. Muetze, A.; Binder, A. Experimental Evaluation of Mitigation Techniques for Bearing Currents in Inverter-Supplied Drive-Systems—Investigations on Induction Motors up to 500 kW. In Proceedings of the IEEE International Electric Machines and Drives Conference, Madison, WI, USA, 1–4 June 2003; pp. 1452–1457.
14. Turzyński, M.; Musznicki, P. A Review of Reduction Methods of Impact of Common-Mode Voltage on Electric Drives. *Energies* **2021**, *14*, 4003. [CrossRef]
15. Un, E.; Hava, A.M. Performance Analysis and Comparison of Reduced Common Mode Voltage PWM and Standard PWM Techniques for Three-Phase Voltage Source Inverters. In Proceedings of the Twenty-First Annual IEEE Applied Power Electronics Conference and Exposition (APEC 2006), Dallas, TX, USA, 19–23 March 2006; p. 7.
16. Lai, Y.-S.; Chen, P.-S.; Lee, H.-K.; Chou, J. Optimal Common-Mode Voltage Reduction PWM Technique for Inverter Control with Consideration of the Dead-Time Effects-Part II: Applications to IM Drives with Diode Front End. *IEEE Trans. Ind. Appl.* **2004**, *40*, 1613–1620. [CrossRef]
17. Un, E.; Hava, A.M. A Near-State PWM Method with Reduced Switching Losses and Reduced Common-Mode Voltage for Three-Phase Voltage Source Inverters. *IEEE Trans. Ind. Appl.* **2009**, *45*, 782–793. [CrossRef]
18. Guo, X.; Xu, D.; Wu, B. Three-Phase Seven-Switch Inverter with Common Mode Voltage Reduction for Transformerless Photovoltaic System. In Proceedings of the IECON 2014—40th Annual Conference of the IEEE Industrial Electronics Society, Dallas, TX, USA, 29 October–1 November 2014; pp. 2279–2284.
19. Hota, A.; Agarwal, V. A New H8 Inverter Topology with Low Common Mode Voltage and Phase Current THD for Three-Phase Induction Motor Drive Applications. *IEEE Trans. Ind. Appl.* **2022**, *58*, 6245–6252. [CrossRef]
20. Zhou, W.; Diab, M.; Yuan, X. Comparison of Motor Neutral Point Overvoltage Oscillations in SiC-Based Adjustable Speed Drives Using Two-Level and Three-Level Inverters. In Proceedings of the 2022 IEEE Energy Conversion Congress and Exposition (ECCE), Detroit, MI, USA, 9–13 October 2022; pp. 1–8.
21. Ogasawara, S.; Ayano, H.; Akagi, H. An Active Circuit for Cancellation of Common-Mode Voltage Generated by a PWM Inverter. *IEEE Trans. Power Electron.* **1998**, *13*, 835–841. [CrossRef]
22. Turzynski, M.; Chrzan, P.J. Reducing Common-Mode Voltage and Bearing Currents in Quasi-Resonant DC-Link Inverter. *IEEE Trans. Power Electron.* **2020**, *35*, 9553–9562. [CrossRef]
23. *DO-160F*; Environmental Conditions and Test Procedures for Airborne Equipment. Radio Technical Commission for Aeronautics (RTCA): Washington, DC, USA, 2007.
24. Plazenet, T.; Boileau, T.; Caironi, C.; Nahid-Mobarakeh, B. A Comprehensive Study on Shaft Voltages and Bearing Currents in Rotating Machines. *IEEE Trans. Ind. Appl.* **2018**, *54*, 3749–3759. [CrossRef]
25. Abu-Rub, H.; Malinowski, M.; Al-Haddad, K. Common-Mode Voltage and Bearing Currents in PWM Inverters: Causes, Effects and Prevention. In *Power Electronics for Renewable Energy Systems, Transportation and Industrial Applications*; Wiley-IEEE Press: Hoboken, NJ, USA, 2014; pp. 664–694.
26. Tawfiq, K.B.; Güleç, M.; Sergeant, P. Bearing Current and Shaft Voltage in Electrical Machines: A Comprehensive Research Review. *Machines* **2023**, *11*, 550. [CrossRef]

27. Zhao, X.; Hu, J.; Ravi, L.; Dong, D.; Burgos, R.; Chandrasekaran, S.; Eddins, R. Planar Common-Mode EMI Filter Design and Optimization for High-Altitude 100 kW SiC Inverter/Rectifier System. In Proceedings of the IEEE Energy Conversion Congress and Exposition (ECCE), Detroit, MI, USA, 11–15 October 2020.
28. Stella, F.; Vico, E.; Cittanti, D.; Liu, C.; Shen, J.; Bojoi, R. Design and Testing of an Automotive-Compliant 800 V 550 kVA SiC Traction Inverter with Full-Ceramic DC-Link and EMI Filter. In Proceedings of the IEEE Energy Conversion Congress and Exposition (ECCE), Detroit, MI, USA, 9–13 October 2022.
29. Electric Power Research Institute (EPRI). *Testing of Stator Windings for Thermal Aging*; EPRI Report No. 000000000001009252; Electric Power Research Institute: Palo Alto, CA, USA, 2018. Available online: https://restservice.epri.com/publicdownload/000000000001009252/0/Product (accessed on 13 October 2025).
30. Madonna, V.; Giangrande, P.; Galea, M. Evaluation of Strand-to-Strand Capacitance and Dissipation Factor in Thermally Aged Enameled Coils for Low-Voltage Electrical Machines. *IET Sci. Meas. Technol.* **2019**, *13*, 1341–1349. [CrossRef]
31. Gyftakis, K.N.; Sumislawska, M.; Kavanagh, D.F.; Howey, D.A.; McCulloch, M. Dielectric Characteristics of Electric Vehicle Traction Motor Winding Insulation under Thermal Ageing. In Proceedings of the 2015 IEEE 15th International Conference on Environment and Electrical Engineering (EEEIC), Rome, Italy, 10–13 June 2015; pp. 313–318. [CrossRef]
32. Magnetics Inc. *Ferrite Core Catalog 2021*; Magnetics Inc.: Pittsburgh, PA, USA, 2021. Available online: https://www.mag-inc.com/Media/Magnetics/File-Library/Product%20Literature/Ferrite%20Literature/Magnetics-2021-Ferrite-Catalog.pdf (accessed on 13 October 2025).

energies

MDPI

Article

Development and Design Optimization of a Single-Phase Doubly-Fed Flux-Switching Permanent Magnet Machine

Lijian Wu [1], Usman Tahir [1], Wenting Wang [2,*], Haoyu Zhou [1], Jianglong Chen [1] and Tao Wang [1]

1 College of Electrical Engineering, Zhejiang University, 38 Zheda Road, Hangzhou 310027, China; ljw@zju.edu.cn (L.W.); usmantahir@zju.edu.cn (U.T.); haoyuzhou@zju.edu.cn (H.Z.); 12410048@zju.edu.cn (J.C.); wangtaoee2@zju.edu.cn (T.W.)
2 Power Electronics and Machines Control Research Institute, The University of Nottingham, PEMC Building, Nottingham NG7 2GT, UK
* Correspondence: wenting.wang@nottingham.ac.uk

Abstract

Demand for brushless alternatives to the series universal motors and induction motors in domestic applications and automotive applications is increasing. Among the available candidates, single-phase flux-switching permanent magnet (SP-FSPM) machines have gained attention due to a simpler magnetic structure and control system. However, their torque density remains limited. Therefore, a SP doubly-fed FSPM (SP-DF-FSPM) machine is developed in this paper which features an additional set of armature windings on the rotor. By effectively utilizing the rotor slot area, the proposed SP-DF-FSPM machine enhances electrical loading and torque density while providing inherent fault-tolerant capability, a critical addition compared with conventional SP-FSPM machines. A comprehensive parameter-sensitivity analysis is conducted for a 10-stator-pole/10-rotor-tooth configuration to optimize key geometric parameters for the maximum torque and reliable self-starting operation. The electromagnetic performance of an optimized design is evaluated and compared against a conventional SP-FSPM machine. The results show that the SP-DF-FSPM machine can achieve a 24.75% higher torque output, improved efficiency, and enhanced power factors under the healthy condition. Moreover, the machine can deliver 63.5% and 36.0% torque when operating with only stator and rotor windings, respectively, demonstrating the fault-tolerant capability. Experimental validation via an SP-DF-FSPM prototype shows close agreement with simulation results.

Keywords: dual-armature; doubly-fed; fault-tolerant; flux switching permanent magnet machine; single-phase

1. Introduction

Series universal motors are a common, cost-effective choice for home appliances and power tools; they, along with other existing options like induction or switched-reluctance machines, are limited by torque density, lower efficiency, and durability issues [1–3]. This has created a pressing need for high torque density, high efficiency, and reliable machine alternatives for these applications. Therefore, research has since progressed into permanent magnet (PM) machines, among which the stator-PM machines, such as doubly salient PM machines, flux-reversal PM machines, and flux-switching PM (FSPM) machines, have emerged as a compelling alternative [4–6].

In recent years, FSPM machines have been extensively investigated as one of the most promising stator-PM topologies, owing to its high torque capability and compact structure [7]. Their key advantage is that PMs are embedded in the stator, which houses the windings as well, allowing for a simple, robust salient-pole rotor. Thus, this configuration merges the mechanical durability of SRMs with the high efficiency and power factor of PM machines. Nevertheless, a fundamental design constraint remains: the placement of magnets within the stator core significantly reduces the available slot area for windings, which limits torque density [8,9]. To increase the slot area, partitioned stator topologies are proposed in [10–12] which separates the PMs. These partitioned stator machines have a double air gap, and the rotor is sandwiched between the stator components. This structure better utilizes the inner space of the machine, hence improving the torque density. However, this approach results in complicated machine structure and increase in manufacturing costs. Furthermore, enhancing the fault tolerance of FSPM machines for high-reliability applications often necessitates complex strategies such as increasing the number of phases, single-layer windings, or adding redundant windings [13–17]. These solutions increase costs, control complexity, and inverter requirements. A double-armature configuration was later proposed for poly-phase FSPM machines to simultaneously enhance torque density and fault-tolerant capability [18]. By augmenting the conventional design by integrating an additional armature winding within the rotor slots, this configuration directly leverages the previously unused rotor slot area, thereby increasing the electrical loading, and hence the torque density. Furthermore, the dual-armature winding structure inherently provides fault tolerance, as the stator and rotor windings can operate independently, introducing a layer of operational redundancy. The dual-armature technique has also been successfully applied to other poly-phase stator-PM machines like flux-reversal machines [19] and doubly salient machines [20].

Given that poly-phase dual-armature FSPM machines generally require different numbers of stator poles and rotor teeth and hence two different phase numbers, their control systems become inherently complex and costly. In contrast, SP FSPM machines feature a simpler structure and control scheme, making them more suitable for cost- and reliability-sensitive consumer applications [21,22]. Despite these, SP FSPM machines suffer from relatively low torque density, poor starting capability, and phase redundancy. The self-starting capability can be improved by introducing the chamfered rotor tooth and segmented rotor tooth [23], but the absence of an SP FSPM configuration capable of simultaneously achieving high torque density and fault tolerance remains evident.

To address this gap, this paper develops an SP doubly-fed (DF) FSPM (SP-DF-FSPM) configuration by employing dual-armature configurations. The effects and potential benefits of dual-armature configurations on the torque density and fault tolerance in SP FSPM machines are systematically investigated or quantified for the first time. In addition, analysis is conducted to investigate the influence of design parameters, including the unique rotor tooth chamfer size, on torque density. It should be noted that the requirement to maintain self-starting capability introduces additional design trade-offs during the conducted optimization, which distinguishes SP machines from their poly-phase counterparts. Torque outputs under healthy and faulty conditions, along with other electromagnetic performance, of the optimized SP-DF-FSPM machine are compared with the optimized conventional SP-FSPM machine to demonstrate the superiority.

In this article, initial topology and working principle of the SP-DF-FSPM machine are presented, followed by the sensitivity analysis of key parameters to maximize the torque while maintaining the starting torque capabilities. Afterwards, a detailed electromagnetic analysis is performed for the optimized machines, and their performance is compared.

Finally, a 10-stator-pole/10-rotor-teeth SP-DF-FSPM machine was fabricated and tested for validation.

2. SP-DF-FSPM Machine Topology and Operating Principle

Common topologies of the conventional SP-FSPM and the proposed SP-DF-FSPM machines are presented in Figure 1. The proposed SP-DF-FSPM machine is derived from the conventional SP-FSPM configuration by introducing an additional set of armature windings. The two sets of armature windings in the SP-DF-FSPM machine are powered by separate inverters, allowing them to function both collaboratively and independently. Hereby, this feature adds fault-tolerant capability to the machine, as the remaining set of armature windings can continue to function even if one set fails. It is worth noting that the SP-FSPM machines in Figure 1a,b must comprise the same number of stator poles and rotor teeth, as opposed to poly-phase DF-FSPM machines in Figure 1c, which require different numbers of stator poles and rotor teeth [18]. Moreover, the rotor tooth chamfer is introduced to enhance self-starting capability, while poly-phase DF-FSPM machines can have symmetry tooth tips.

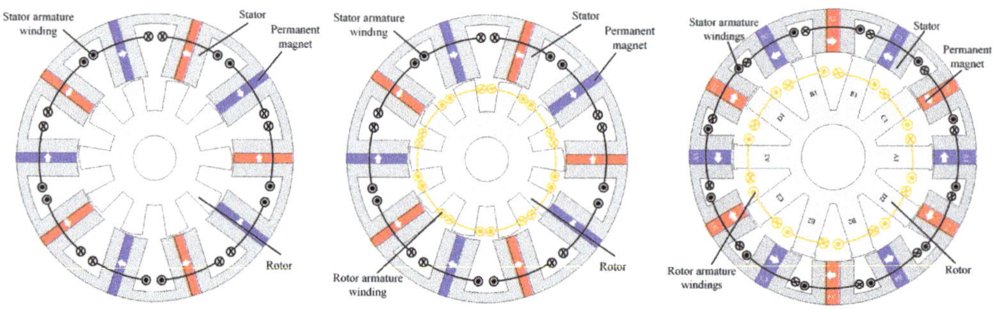

(a) (b) (c)

Figure 1. FSPM machine topologies. (**a**) SP FSPM. (**b**) SP-DF-FSPM. (**c**) Poly-phase DF-FSPM.

The total torque of the SP-DF-FSPM machine mainly originates from two components: the interaction between the PM and stator armature (PM-stator armature) fields, and the interaction between the PM and rotor armature (PM-rotor armature) fields.

The PM-stator armature torque component is equivalent to that of the conventional SP-FSPM machine based on the flux-switching principle. The working principle and the corresponding flux linkage of the stator armature component are presented in Figures 2 and 3. Flux linkage in the stator coil is zero when the rotor tooth is aligned with the PM (Figure 2a) or the stator slot (Figure 2c). When the rotor tooth aligns with either of the stator teeth (Figure 2b,d), positive or negative flux linkage is achieved. The stator coil flux varies with the rotation of the rotor, inducing back-EMFs.

The PM-stator armature interaction exhibits a similar manner to that of the interior PM machine. Rotor flux linkage is zero when the rotor tooth is aligned with the stator PM (Figure 4a,c). Positive and negative maximum flux linkage (Figure 4b,d) is obtained when the rotor tooth is aligned with the stator slot. Rotor rotation varies flux linkage (Figure 5), inducing back-EMFs in rotor armature windings.

According to the explained working principle, the stator flux linkage waveform is periodic over one rotor tooth pitch, whereas the rotor flux linkage waveform repeats over two stator pole pitches. Therefore, the fundamental electrical frequencies of stator f_s and rotor f_r can be determined as:

$$f_s = \frac{P_r n_r}{60} \quad (1)$$

and

$$f_r = \frac{\left(\frac{P_s}{2}\right) n_r}{60} \quad (2)$$

where P_s and P_r are the number of stator poles and rotor teeth, respectively, and n_r is the rotational speed of the rotor. For the SP-DF-FSPM machine having the same number of stator poles and rotor teeth, the rotor fundamental frequency is half of that of the stator fundamental frequency:

$$f_r = \frac{f_s}{2} \quad (3)$$

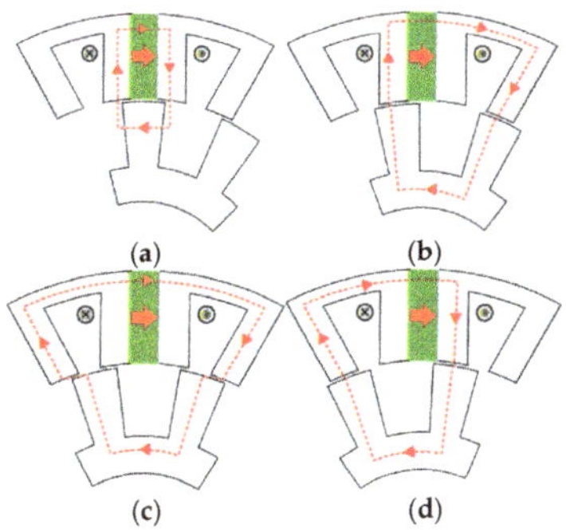

Figure 2. Flux linkage states of the stator armature windings. (**a**) Zero. (**b**) Positive maximum. (**c**) Zero. (**d**) Negative maximum.

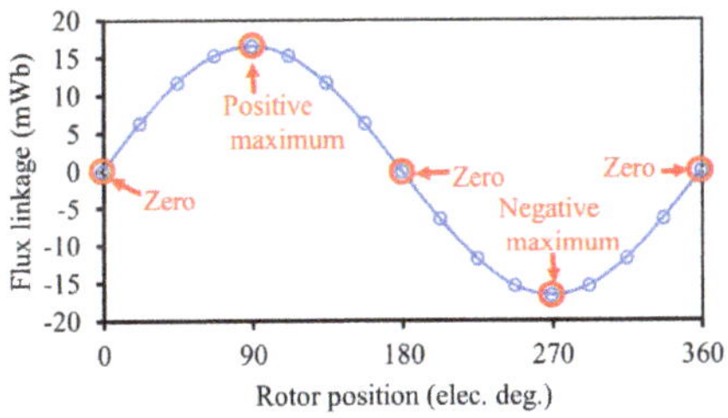

Figure 3. Flux linkage of the stator armature winding.

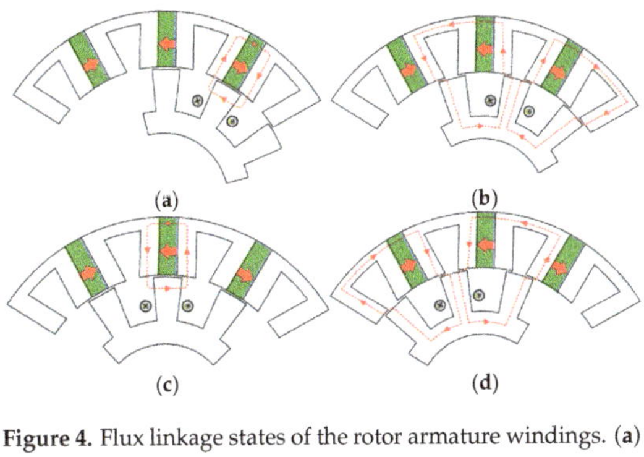

Figure 4. Flux linkage states of the rotor armature windings. (**a**) Zero. (**b**) Positive maximum. (**c**) Zero. (**d**) Negative maximum.

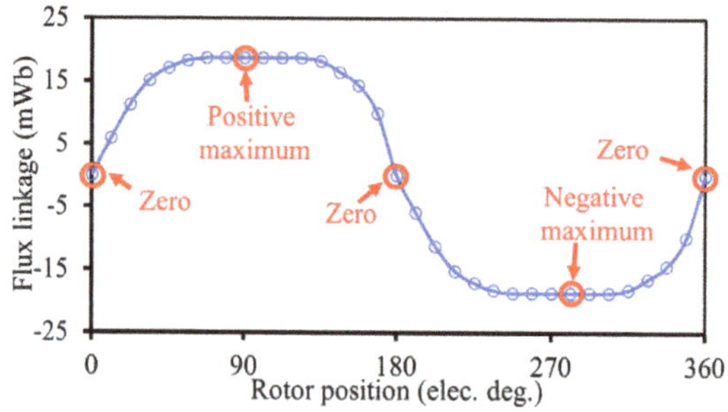

Figure 5. Flux linkage of the rotor armature winding.

3. Parameter Sensitivity Analysis

In this section, parameter sensitivity analysis is carried out to maximize average torque, as well as the minimum starting torque for the proposed SP-DF-FSPM machine. For comparison, a conventional SP-FSPM machine is analyzed as well.

It should be noted that, different from three-phase FSPM machines, conventional SP machines also face a well-documented limitation in their self-starting capability. This is due to the presence of a "dead zone" at certain rotor positions where the starting torque is zero. Consequently, significant research has been devoted to mitigating this problem via geometric optimization of the machine's rotor and stator to introduce magnetic asymmetry in the air gap, including the use of segmented or auxiliary rotors, the application of stator tooth chamfering, and the design of sub-rotors with non-uniform tooth widths.

To enhance the self-starting capability for the investigated design, the rotor tooth chamfer is employed. Figure 6 depicts the major geometric parameters of the SP-FSPM machine, and the initial parameters of the selected 10-stator-slot/10-rotor-slot SP-FSPM and SP-DF-FSPM machines are given in Table 1, where the total armature copper loss is set at 40 W. End effect is not considered during finite element analysis.

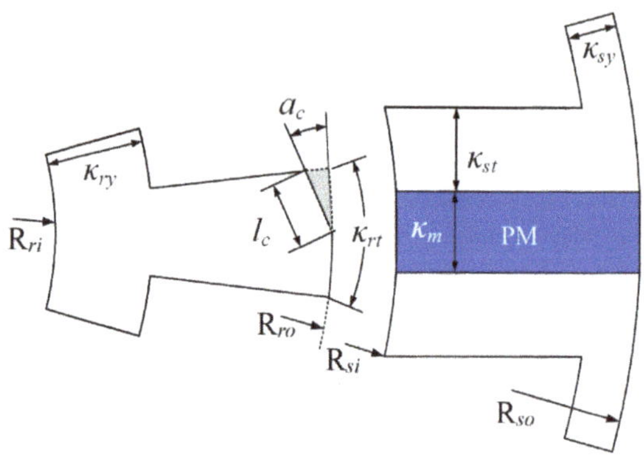

Figure 6. Geometric parameter illustration.

3.1. Copper Loss Ratio

The electromagnetic performance of the SP-DF-FSPM machine is influenced by the distribution of copper losses between its stator and rotor windings. Analysis reveals that the average electromagnetic torque is maximized when the stator-to-rotor copper loss ratio is approximately 1.667, as Figure 7 shows.

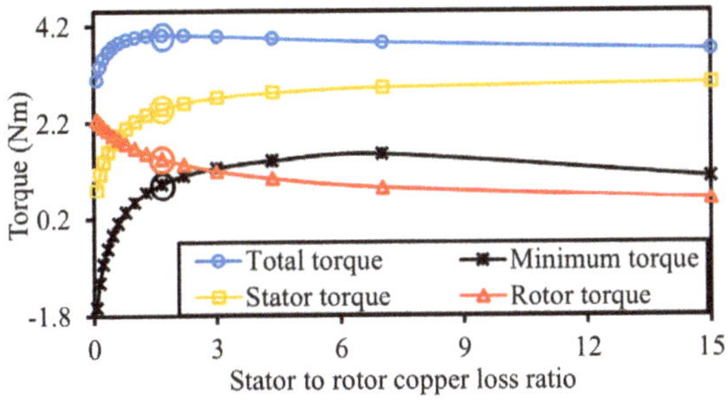

Figure 7. Influence of stator to rotor copper loss ratio on average torque and minimum torque.

Table 1. Initial design specifications of SP-DF-FSPM and SP-FSPM machines.

Parameters	SP-DF-FSPM	SP-FSPM
Number of stator poles	10	
Number of rotor teeth	10	
Number of stator phase	1	
Number of rotor phase	1	-
Active axial length (mm)	50	
Outer radius of stator, R_{so} (mm)	45	
Inner radius of stator, R_{si} (mm)	25.5	
Stator tooth arc to pole pitch ratio, κ_{st}/κ_{spp}	0.27	
Magnet arc to stator pole pitch ratio, κ_m/κ_{pp}	0.19	
Stator yoke thickness to tooth arc ratio, κ_{sy}/κ_{st}	0.7	
Air-gap length, l_g (mm)	0.5	
Rotor inner radius, R_{ri} (mm)	7	
Rotor tooth arc to tooth pitch ratio, κ_{rt}/κ_{rtp}	0.49	
Rotor tooth chamfer length, l_c (mm)	4	
Rotor yoke thickness to tooth arc ratio, κ_{ry}/κ_{rt}	1.08	
Rotor tooth chamfer angle, a_c (degree)	8.5	
Number of stator winding turns	100	
Number of rotor winding turns	90	-
Stator slot opening to pole pitch ratio, $\kappa_{sso}/\kappa_{spp}$	0.33	
Rotor slot opening to tooth pitch ratio, $\kappa_{rso}/\kappa_{rtp}$	0.51	
Magnet remanence (T)	1.2	
Magnet relative permeability	1.05	
Rated copper loss (W)	40	
Rated speed (rpm)	1500	

Since the SP-DF-FSPM machine can retain torque-producing capability even if one set of armature windings fails, the two aforementioned torque components are analyzed individually for fault-tolerant consideration. In an ideal scenario where the stator and rotor contribute equally to torque, approximately half of the rated torque could be maintained. However, a key constraint is the presence of dead zones in SP machines, rotor positions that yield negative torque. For instance, a copper loss ratio of 0.33 results in equal torque contribution from both windings and a total torque output of 90% of the maximum, but this configuration exhibits a negative minimum torque, rendering it susceptible to a dead zone. For self-starting capability and fault-tolerant operation, the stator and rotor windings contribute 62% and 36% of the torque, respectively, when keeping current in the healthy winding unchanged. It should be noted that the remaining small torque arises from the interaction of stator and rotor armature reaction fields, accounting for the slight deviation from 100% total torque contribution.

3.2. Split Ratio

The average torque in both SP-DF-FSPM and SP-FSPM machines exhibits a strong dependence on the split ratio, as Figure 8 demonstrates. A higher split ratio increases the flux linkage but simultaneously reduces the stator slot area in both machine types. This reduction in slot area results in a decrease in the stator armature current under a fixed copper loss. In the SP-DF-FSPM machine, however, the rotor armature current demonstrates a contrasting trend, increasing with the split ratio due to the expansion of the rotor slot area. This increase in rotor current enhances the electromagnetic torque produced by the PM-rotor-armature interaction. Consequently, the SP-DF-FSPM machine has a higher optimal split ratio compared to the SP-FSPM machine.

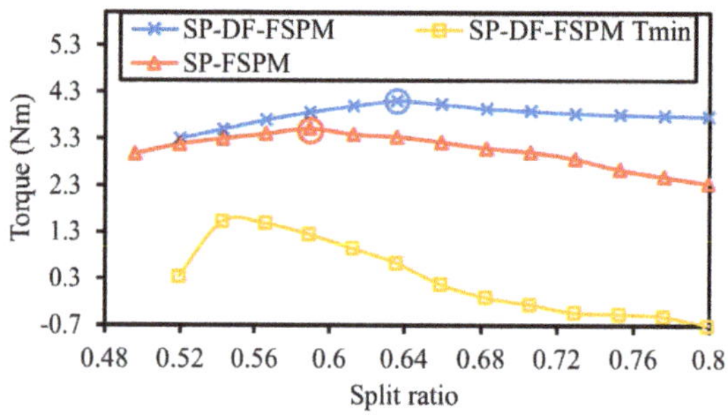

Figure 8. Influence of split ratio on torque.

Furthermore, while wider PMs introduced by larger split ratio can boost the average torque, it also amplifies cogging torque and torque ripple, which can produce a negative minimum torque. Hence, although torque is similar when the split ratio is larger than 0.64 for the SP-DF-FSPM machine, the minimum starting torque drops dramatically and thereby impair the self-starting capability.

3.3. Magnet Arc

Figure 9 depicts the relationship between the magnet arc and the average torque. Increasing the magnet arc enhances the PM flux but simultaneously reduces the stator slot area, thus limiting the stator armature current. It is noteworthy that, in the SP-DF-FSPM machine, the rotor slot area and consequent rotor armature current remain unaffected by this geometric variation. The wider PMs increase flux linkages in the rotor coils and hence increases their electromagnetic torque contribution. Consequently, the SP-DF-FSPM machine is designed with a larger optimal magnet arc than the SP-FSPM machine. This design choice, however, can increase cogging torque and torque ripple. Similarly to the influence of the split ratio, the increased magnet arc negatively affects the minimum torque.

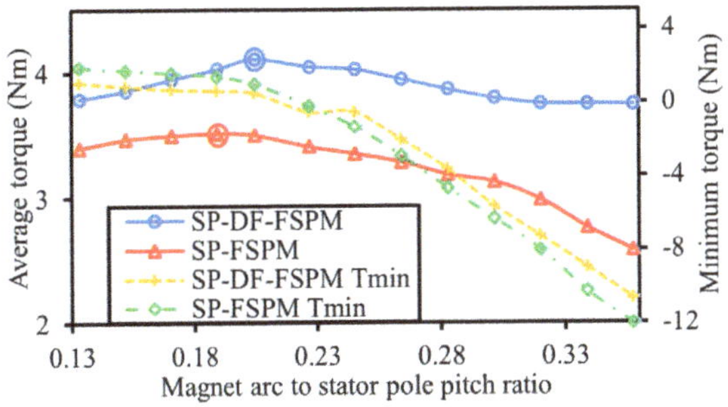

Figure 9. Influence of magnet arc on torque.

3.4. Stator Tooth Arc

Figure 10 illustrates the influence of stator tooth arc on the average torque for both SP-DF-FSPM and SP-FSPM machine topologies. Enlarging the stator tooth arc reduces magnetic reluctance and alleviates saturation in the stator tooth, thereby enhancing the PM flux linkage. However, this geometric variation simultaneously reduces the stator slot area, which limits the armature current. In the SP-DF-FSPM machine, the resultant increase in the PM flux linking the rotor coils increases the torque produced by the PM-rotor-armature part. This additional torque contribution enables the SP-DF-FSPM machine to operate effectively with a slightly larger optimal stator tooth arc than its SP-FSPM counterpart. In terms of the minimum torque, it maintains a positive value when the stator tooth arc is larger than the 0.29, the optimal value for the maximum average torque.

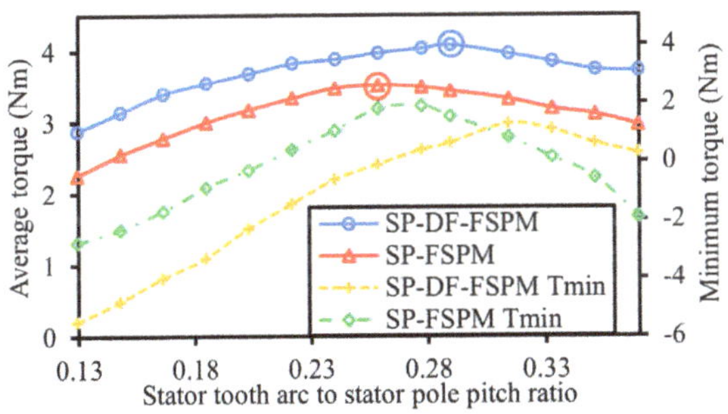

Figure 10. Influence of stator tooth arc on torque.

3.5. Rotor Tooth Arc

As shown in Figure 11, the average torque exhibits a non-monotonic relationship with the rotor tooth arc, increasing to a maximum before declining. The optimal rotor tooth arcs are 0.47 and 0.49 of the rotor tooth pitch for the SP-DF-FSPM and SP-FSPM machines, respectively. Initially, a larger rotor tooth arc increases the overlapping area with the stator teeth, thereby reducing the air-gap reluctance. This reduction enhances the PM flux and consequently the average torque. Beyond a critical point, however, a further increase in the rotor tooth arc causes significant flux leakage at the inner magnet surfaces and slot opening, which reduces the effective PM flux and results in a decline in torque. Additionally, in the SP-DF-FSPM machine, the rotor armature current is also affected by the rotor tooth arc due to the resultant changes in rotor slot area.

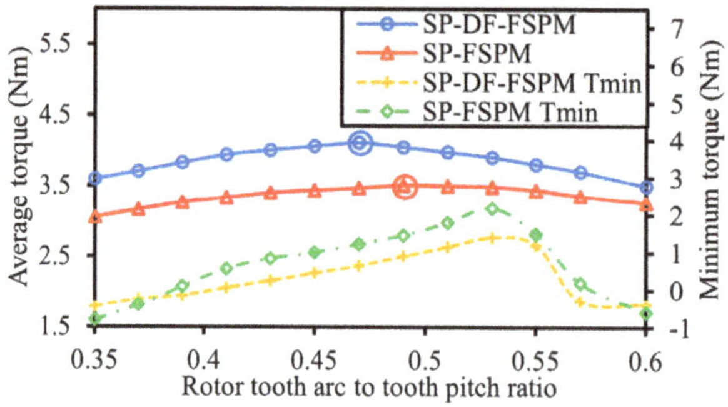

Figure 11. Influence of rotor tooth arc on torque.

3.6. Stator Yoke Thickness

A reduced magnetic flux passes through the yoke in comparison to the stator tooth, which is reflected in the lower flux density. Therefore, the design necessitates a stator yoke thickness that is less than the stator tooth width. As the stator yoke thickness increases, the reduction in reluctance results in a slight increase in the PM flux. Nevertheless, it also leads to smaller stator slots and hence reduced stator currents. As illustrated in Figure 12, the torques of both the SP-FSPM and DF-SP-FSPM machines only increase a bit but then decline markedly. Therefore, the respective optimal yoke thicknesses for the SP-DF-FSPM and SP-FSPM machines are determined to be approximately 0.6 and 0.7 times the stator tooth width. This design optimizes flux densities and maximizes the stator slot area.

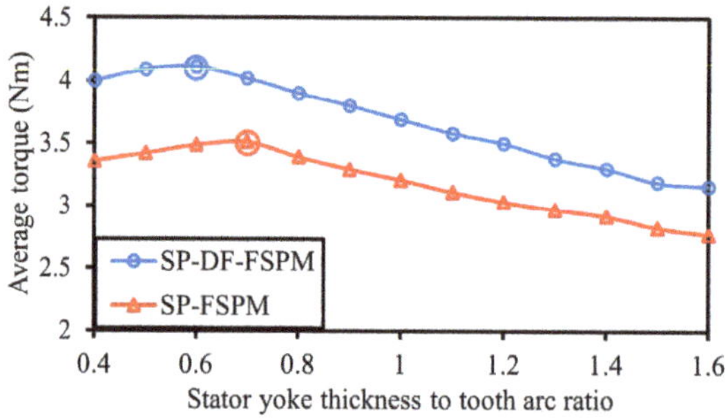

Figure 12. Influence of stator yoke thickness on average torque.

3.7. Rotor Yoke Thickness

As illustrated in Figure 13, the average torque of the SP-FSPM machine remains largely unaffected by rotor yoke thickness, except under extreme geometric conditions. An excessively narrow yoke increases the saturation and reluctance, while an overly thick yoke reduces stator flux linkage and increases flux leakage. In contrast, the average torque of the SP-DF-FSPM machine exhibits a pronounced sensitivity to rotor yoke thickness due to its direct influence on the rotor armature current. Consequently, the optimal rotor yoke thickness for the SP-DF-FSPM machine is approximately 0.6 times the rotor tooth dimension, a value significantly less than that required for the SP-FSPM machine.

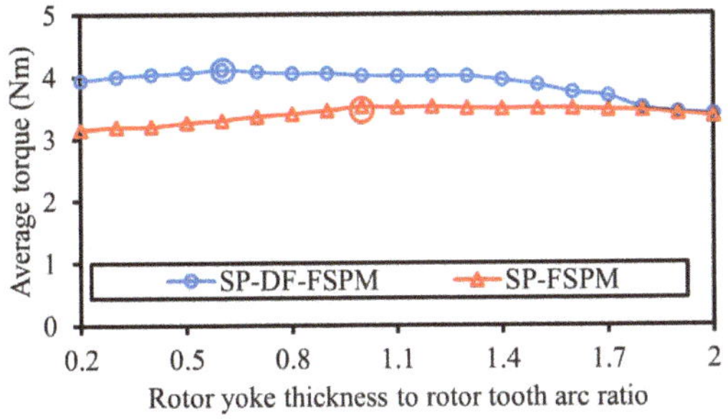

Figure 13. Influence of rotor yoke thickness on average torque.

3.8. Rotor Tooth Chamfer

The rotor tooth chamfer is essential for the self-starting capability of SP FSPM machines, which creates an asymmetrical air gap, resulting in varying air-gap reluctance from one side of the rotor teeth to the other. This variation ensures the unidirectional rotation of the machine. The variations in average torque and minimum torque with the chamfer length l_c and chamfer angle a_c for both the SP-DF-FSPM and SP-FSPM machines is demonstrated in Figure 14. T_{ave_a} and T_{min_a} represent the average and minimum torque at chamfer angle a_c, while T_{ave_l} and T_{min_l} represent the average and minimum torque at chamfer length l_c. From Figure 14, it can be seen that the average torques decrease as the chamfer area gets larger, but the variation is not significant. However, for the minimum torque, almost all combinations of the chamfer dimensions achieve a positive minimum torque for the SP-FSPM machine, while some of them result in a negative minimum torque in the DF-SP-FSPM machine. Therefore, the chamfer dimensions are more critical for DF-SP-FSPM machines and should be carefully chosen for the consideration of self-starting capability.

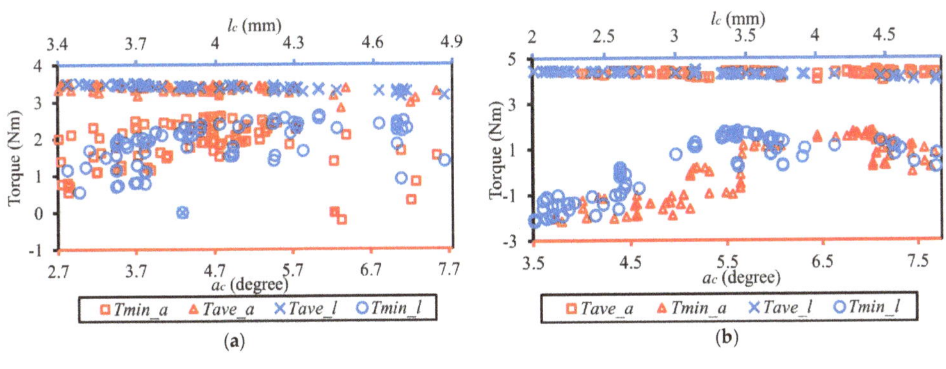

Figure 14. Rotor teeth chamfer optimization for optimal torque performance. (**a**) SP-FSPM. (**b**) SP-DF-FSPM.

4. Electromagnetic Performance

This section presents a comparative analysis of the electromagnetic performance between the proposed SP-DF-FSPM machine and the conventional SP-FSPM machine. Both machines are optimized by a sequential optimization methodology to maximize the output torque. The process involved the iterative refinement of key geometric parameters, namely, the split ratio, magnet arc, stator tooth arc, stator yoke thickness, rotor tooth arc, rotor yoke thickness, and rotor tooth chamfer, where each subsequent parameter was optimized while holding previously adjusted parameters at their optimal values. The resulting optimal parameters are detailed in Table 2 and Figure 15. Discrepancies between the outcomes of this sequential approach and isolated optimizations arise from the strong interdependencies among these parameters. Notably, the optimal split ratio, stator tooth arc to stator pole pitch ratio, and rotor yoke thickness to rotor tooth arc ratio differ substantially between the SP-DF-FSPM and SP-FSPM topologies. However, the optimal ratios for stator yoke thickness to stator pole pitch, magnet arc to stator pole pitch, and rotor tooth arc to rotor tooth pitch are rather consistent across both machine types.

Table 2. Optimized parameters of the SP-FSPM and SP-DF-FSPM machines.

Parameters	SP-FSPM	SP-DF-FSPM
Split ratio	0.59	0.63
κ_{st}/κ_{spp}	0.26	0.29
κ_{m}/κ_{spp}	0.19	0.2
κ_{sy}/κ_{st}	0.7	0.6
κ_{rt}/κ_{rtp}	0.49	0.47
κ_{ry}/κ_{rt}	1	0.6
a_c (°)	4.96	5.95
l_c (mm)	3.79	3.61
Copper loss ratio	-	1.667
Stator peak phase current (A)	14	13
Rotor peak phase current (A)	-	11.5

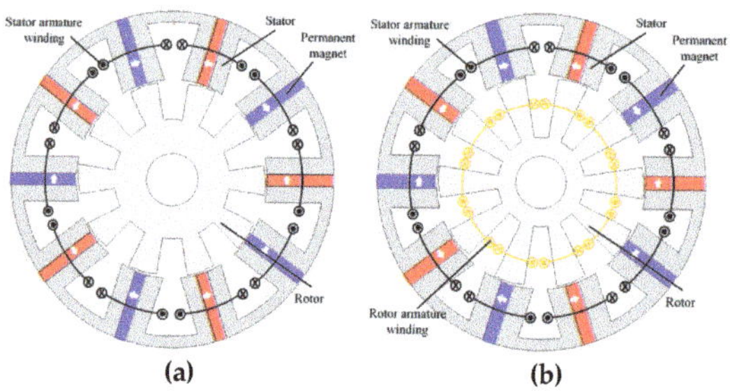

Figure 15. Optimized machine geometries. (**a**) SP-FSPM. (**b**) SP-DF-FSPM.

4.1. Field Distribution and Back-EMF

Figure 16 depicts the no-load magnetic field distribution and flux densities for both configurations. A key geometrical distinction is evident, i.e., the SP-DF-FSPM machine employs a larger split ratio alongside a reduced rotor yoke thickness compared to the SP-FSPM machine. This configuration arises from the spatial requirement to accommodate both the stator winding and the additional rotor winding within the SP-DF-FSPM topology.

Figure 17 illustrates the back EMF of the SP-DF-FSPM and SP-FSPM machines. The back EMF of the SP-FSPM and SP-DF-FSPM stators exhibits a comparable pattern. The stator back EMF resembles sinusoidal waves and exhibits high harmonic content, consisting of both even and odd harmonic components. Notably, there is a significant fifth harmonic that contributes to the torque ripple. Additionally, the rotor back EMF of the SP-DF-FSPM features prominent third, fifth, and ninth harmonic components.

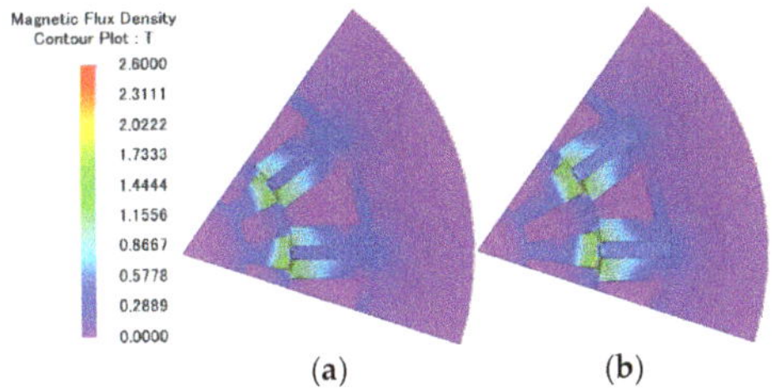

Figure 16. No-load field distribution and flux density. (**a**) SP-FSPM. (**b**) SP-DF-FSPM.

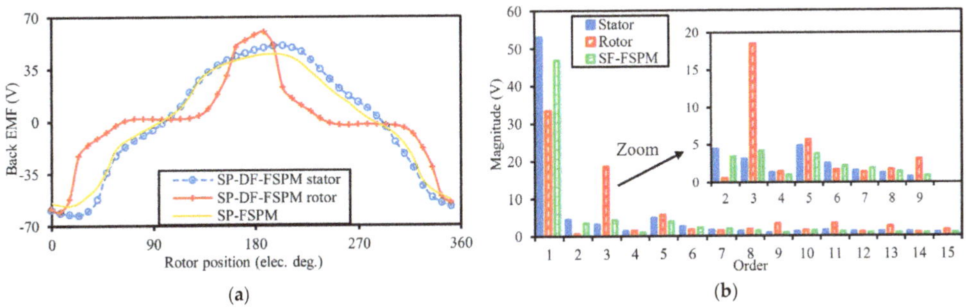

Figure 17. Back EMFs of the SP-FSPM and SP-DF-FSPM machines. (**a**) Waveforms. (**b**) Spectra.

4.2. Inductance

A comparative analysis of self-inductances, presented in Figure 18, reveals that the stator self-inductance (L_s) is nearly identical for both the SP-DF-FSPM and SP-FSPM machines. Notably, within the SP-DF-FSPM topology, the stator self-inductance is observed to be greater than the rotor self-inductance (L_r). Furthermore, a distinct characteristic of the SP-DF-FSPM machine is the presence of magnetic coupling between its stator and rotor windings. However, the coupling is weak, so the average mutual inductance (L_{sr}) is nearly zero in Figure 18.

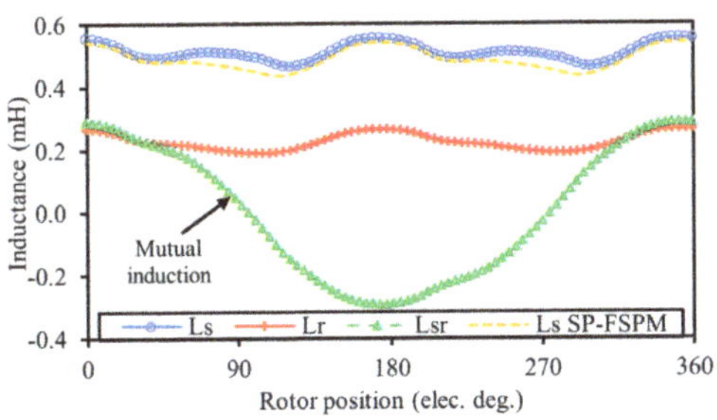

Figure 18. Inductances of the SP-FSPM and SP-DF-FSPM machines.

4.3. Cogging Torque

Figure 19 presents a comparison of the cogging torque for the SP-DF-FSPM and SP-FSPM machines. When identical geometrical parameters are used, both machines exhibit equivalent cogging torque profiles, as their magnetic structures are inherently the same under open-circuit conditions. However, the optimized SP-DF-FSPM machine, with its final parameters detailed in Table 2, incorporates a larger rotor tooth arc and magnet arc than the SP-FSPM machine, resulting in a higher value of the cogging torque peak.

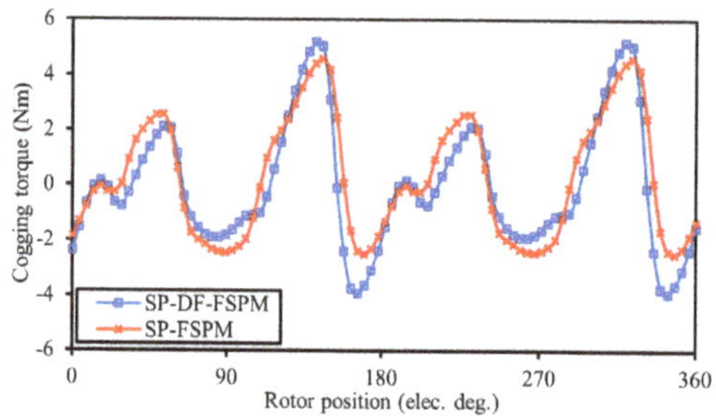

Figure 19. Cogging torque of the SP-FSPM and SP-DF-FSPM machines.

4.4. On-Load and Fault-Tolerant Performance

Figure 20 illustrates the on-load magnetic field distribution and flux densities for both machine topologies operating under a fixed copper loss constraint of 40 W. The flux densities observed throughout the core structures remain below the material's saturation threshold, approximately 1.85 T. Notably, the SP-DF-FSPM machine exhibits no discernible saturation despite its comparatively slimmer rotor yoke.

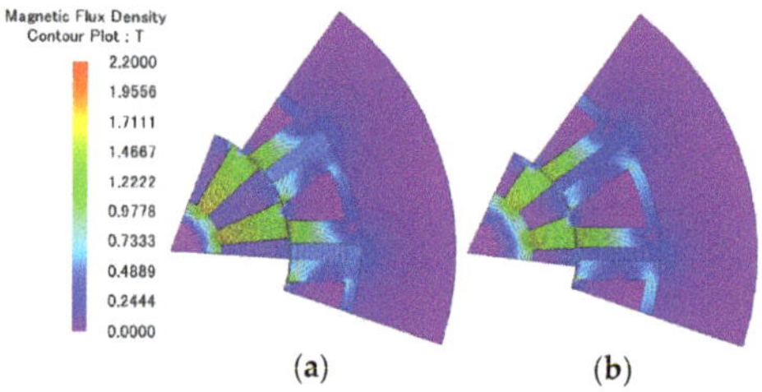

Figure 20. On-load field distribution and flux density. (**a**) SP-FSPM. (**b**) SP-DF-FSPM.

Both stator and rotor armature components employ step load current. The SP-DF-FSPM machine achieves a 53% larger effective slot area, a result of the incorporation of rotor slots. This enhanced area allows higher electrical loading, thereby increasing the electromagnetic torque output for a given copper loss. Consequently, under load conditions, the SP-DF-FSPM machine produces an average torque approximately 24.75% greater than that of the conventional SP-FSPM machine. Moreover, the torque production per unit volume of PMs has increased by approximately 24.57%, showing cost-effectiveness. However, this increased torque is accompanied by a higher torque ripple, as detailed in Figure 21a and Table 3. Torque ripple is defined by

$$T_{rp} = \frac{T_{max} - T_{min}}{T_{ave}} \tag{4}$$

where T_{max}, T_{min}, and T_{ave} are the maximum torque, minimum torque, and average torque, respectively.

Table 3. Full-load torque comparison.

Parameters	SP-DF-FSPM	SP-FSPM
Average torque (Nm)	4.36	3.51
Average torque per magnet volume (kNm/m^3)	144.37	116.22
Torque ripple	2.5	2.23

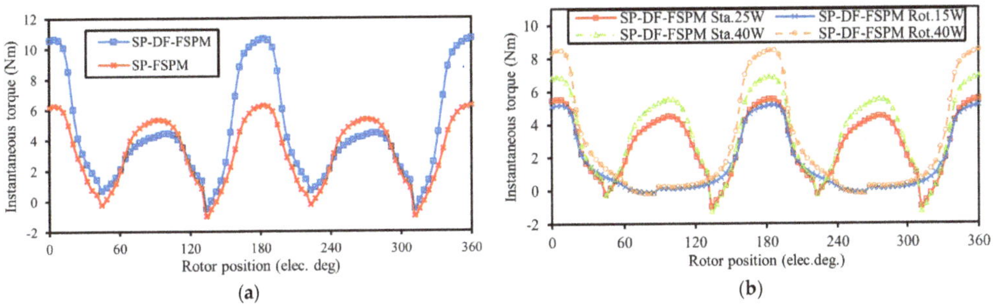

Figure 21. Instantaneous torque waveforms of SP-FSPM and SP-DF-FSPM. (**a**) Normal condition. (**b**) Fault-tolerant.

Under a single-armature winding fault condition, the SP-DF-FSPM machine maintains torque production through its healthy winding. When the healthy winding current is held constant, corresponding to 25 W stator and 15 W rotor copper loss, the resulting torque constitutes 63.5% (stator) and 36% (rotor) of the machine's pre-fault output, equivalent to 78.91% and 45%, respectively, of the healthy SP-FSPM machine's torque. Alternatively, when total copper loss is maintained at 40 W, and the current in the healthy winding is increased, the SP-DF-FSPM achieves 79.12% and 59.4% of the healthy output, which equates to 98.4% (stator-supplied) and 73.73% (rotor-supplied) of the healthy SP-FSPM torque, as summarized in Figure 21b and Table 4.

Table 4. Average torques under different operating conditions.

Condition with Different Copper Loss		Average Output Torque (Nm)	
		SP-DF-FSPM	**SP-FSPM**
Healthy operation (Stator 25 W; Rotor 15 W)		4.36	3.51
Fault-tolerant operation	Stator (25 W)	2.77	-
	Rotor (15 W)	1.57	-
	Stator (40 W)	3.45	-
	Rotor (40 W)	2.59	-

Figure 22 shows the starting torques under healthy and faulty conditions. Under healthy conditions and stator excitation, the machine exhibits self-starting capability. The starting torque can be adversely affected if the rotor stops within this dead zone for rotor excitation only. Nevertheless, the starting toque generated under fault-tolerant conditions is subject to significant torque ripples and may remain in the dead zone at specific rotor positions. Consequently, an alternative control strategy should be implemented to initiate the rotor's movement.

The variation in the average torque with the ratio of copper loss to the rated copper loss is illustrated in Figure 23, where the stator-to-rotor copper loss ratio is maintained at 1.667 in the SP-DF-FSPM machine, and Table 2 contains the corresponding rated currents for the SP-DF-FSPM and SP-FSPM machines. The torque density of the SP-DF-FSPM machine is constantly much higher than that of the SP-FSPM machine throughout the analyzed range. Additionally, the SP-DF-FSPM has a higher overload capability and is less likely to become saturated, as evidenced by the fact that the ratio of average torque generated by the SP-DF-FSPM and SP-FSPM machines increases with phase current.

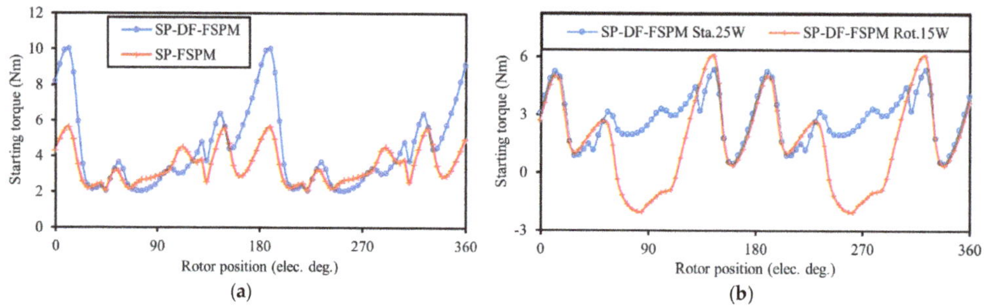

Figure 22. Starting torque. (**a**) Healthy condition. (**b**) Faulty condition.

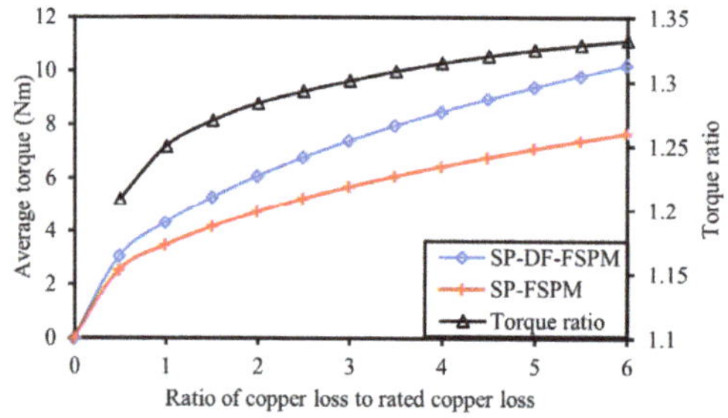

Figure 23. Variation in average torque with copper loss of the SP-FSPM and SP-DF-FSPM machines.

4.5. Efficiency and Power Factor

As illustrated in Table 5, the loss distribution of the SP-DF-FSPM machine shifts compared to the SP-FSPM configuration, characterized by reduced stator core losses but increased rotor core losses. This redistribution stems from the SP-DF-FSPM's rotor armature reaction, which increases magnetic saturation in the rotor core, while its reduced stator armature reaction slightly reduces stator iron saturation. Consequently, the SP-DF-FSPM machine exhibits higher total core losses; however, these remain secondary to copper losses, with stator losses constituting the dominant share at 59% of the total.

Table 5. Efficiency and power factor comparison.

Parameters	SP-DF-FSPM			SP-FSPM		
	Stator	Rotor	Total	Stator	Rotor	Total
Core loss (W)	11.1	10.7	21.8	12.65	8.5	21.1
Copper loss (W)	25	15	40	40	0	40
Efficiency (%)	-	-	90.59	-	-	88.16
Power factor	0.82	0.75	-	-	-	0.78

Despite this, the SP-DF-FSPM machine demonstrates superior overall efficiency, a benefit derived from its enhanced torque density. Furthermore, the stator component achieves higher power factors than that of the SP-FSPM machine, as calculated from the phase difference between the finite-element-predicted voltage and the phase current. The JMAG 23.0 software is employed to conduct 2D time-stepping transient finite element analysis coupled with an external circuit to evaluate the power factor. The machine is excited with rated current, and steady-state phase voltage and current waveforms are extracted. The power factor is then calculated based on the phase angle between the fundamental components of phase current and phase voltage, which are obtained through FFT processing.

5. Experimental Validation

Experimental validation of the preceding FEA results was conducted on an SP-DF-FSPM prototype, constructed in accordance with the optimized parameter set in Table 2. As illustrated in Figure 24, the machine features armature windings on both the stator and rotor elements. To enable external access to the rotor circuit, a slip ring is mounted directly onto the rotating shaft.

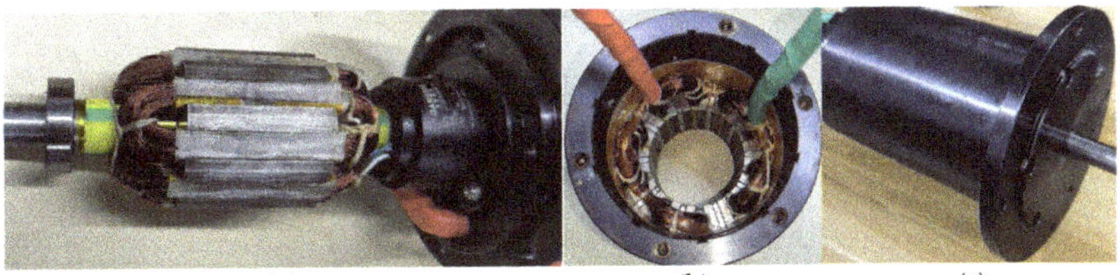

(a) (b) (c)

Figure 24. Prototype of SP-DF-FSPM. (**a**) Rotor with slip rings. (**b**) Stator. (**c**) Assembled machine.

Static torque characteristics are evaluated by adopting the cogging torque measurement technique elaborated in [24]. Figure 25 presents a cogging torque comparison between the measured waveform and the corresponding FEA predictions. The experimental results confirm the simulated results. However, the measured peak torque values are marginally lower than those predicted. This divergence is primarily a consequence of mechanical and assembly tolerances inherent in the prototype.

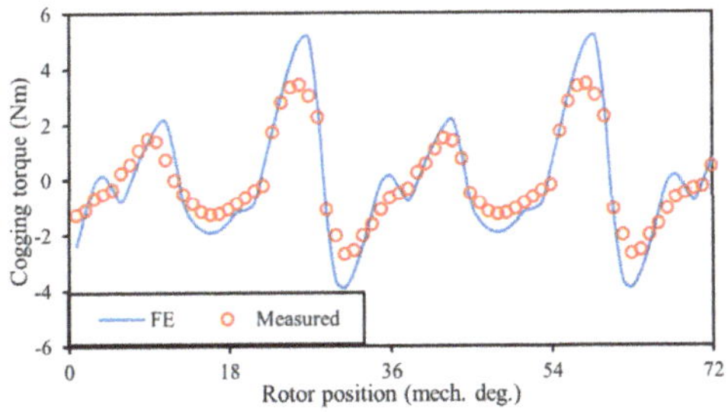

Figure 25. Comparison of cogging torque.

The static torque waveforms under DC excitation of either the stator or rotor armature windings are presented in Figure 26a,b, respectively. The phase current values are set as rated current at time zero. While the measured torque waveforms exhibit some non-sinusoidal characteristics, primarily due to cogging torque, a favorable correlation between the predicted and experimental results is evident. Figure 27 displays the static torque characteristics when both stator and rotor windings are DC-excited. The total torque includes a component produced by the interaction between the stator and rotor magnetic fields. To account for the periodic nature of this interaction, additional experimental data were collected with the rotor position shifted by 72 mechanical degrees. The measured results show general consistency with the finite-element predictions, thereby validating the modeling approach and the design.

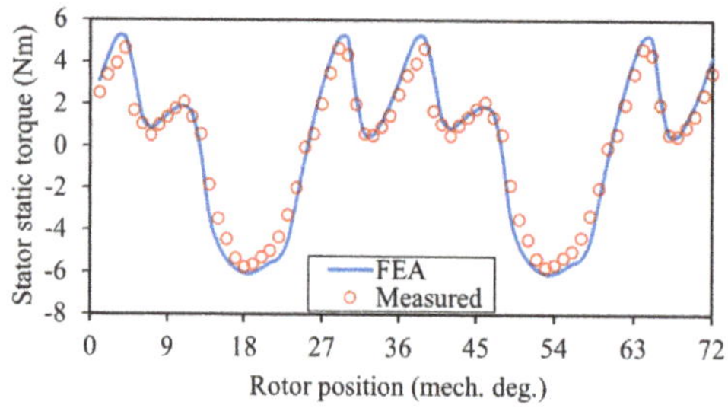

(a)

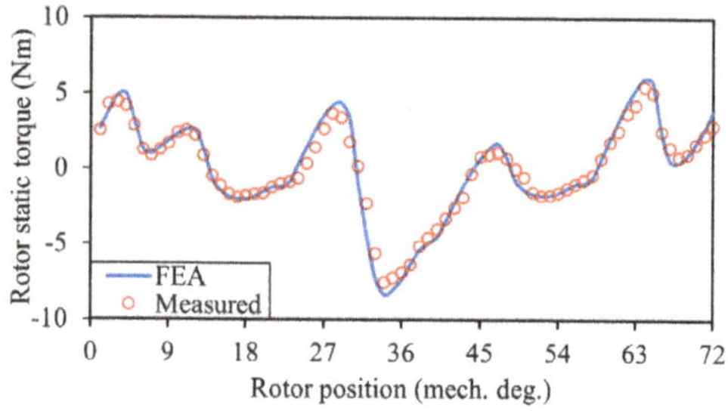

(b)

Figure 26. Comparison of measured and FEA-predicted static torque. (**a**) Only stator current excitation. (**b**) Only rotor current excitation.

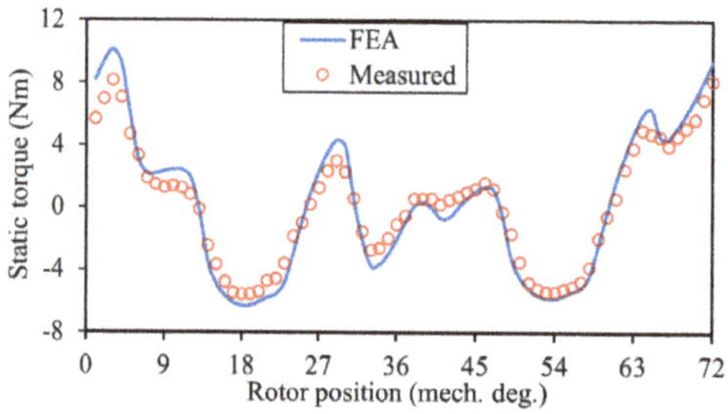

Figure 27. Dual-excitation static torque.

6. Conclusions

This paper presents the development and investigation of an SP-DF-FSPM featuring armature windings on both the stator and rotor teeth to boost torque and enable fault-tolerant operation. A parameter sensitivity analysis is carried out on the SP-DF-FSPM and SP-FSPM machines, and they are optimized to achieve the maximum torque while maintaining self-start capability. While the optimal split ratio, stator tooth arc ratio, and rotor yoke thickness ratio vary significantly between the two topologies, parameters such as stator yoke thickness, magnet arc ratio, and rotor tooth arc ratio remain largely consistent.

A detailed comparative analysis of the electromagnetic performance indicates that the optimized SP-DF-FSPM achieves a 24.75% higher torque than its conventional counterpart when maintaining self-starting capability, while the dual-armature configuration enables 46.5% higher torque in the poly-phase FSPM machine. In addition, the SP-DF-FSPM machine achieves 24.57% higher torque per PM volume than the SP-FSPM, compared to a 39.6% increase for the poly-phase dual-armature FSPM over its counterpart.

Furthermore, under 40 W copper loss condition in healthy winding, the SP-DF-FSPM achieves 79.13% and 59.4% of the healthy torque when only stator and rotor winding is supplied, respectively, demonstrating superior fault-tolerance capability. These values correspond to 98.3% and 73.8% of the healthy torque of the conventional SP-FSPM machines, compared to 93.0% and 120.6% in poly-phase DF-FSPM machines. Under rotor-only fault-tolerant conditions, the self-starting capability is impaired and requires additional control strategies.

A prototype of the SP-DF-FSPM machine was manufactured and tested, effectively validating the analysis.

Author Contributions: Conceptualization, L.W.; Methodology, W.W. and H.Z.; Validation, U.T. and J.C.; Formal Analysis, L.W., U.T., W.W. and H.Z.; Investigation, L.W., U.T. and W.W.; Resources, T.W.; Data Curation, U.T. and J.C.; Writing—Original Draft, U.T.; Writing—Review and Editing, L.W., W.W. and H.Z.; Visualization, U.T.; Supervision, L.W. and T.W.; Project Administration, L.W. and T.W.; Funding Acquisition, L.W. All authors have read and agreed to the published version of the manuscript.

Funding: This research was funded by the National Science Fund for Distinguished Young Scholars under Grant 52225703.

Data Availability Statement: The raw data supporting the conclusions of this article will be made available by the authors on request.

Acknowledgments: The authors have reviewed and edited the output and take full responsibility for the content of this publication.

Conflicts of Interest: The authors declare no conflicts of interest.

References

1. Gerlando, A.D.; Perini, R. Model of the Commutation Phenomena in a Universal Motor. *IEEE Trans. Energy Convers.* **2006**, *21*, 27–33. [CrossRef]
2. Cros, J.; Viarouge, P.; Chalifour, Y.; Figueroa, J. A New Structure of Universal Motor Using Soft Magnetic Composites. *IEEE Trans. Ind. Appl.* **2004**, *40*, 550–557. [CrossRef]
3. Lu, K.; Jakobsen, U.; Rasmussen, P.O. Single-Phase Hybrid Switched Reluctance Motor for Low-Power Low-Cost Applications. *IEEE Trans. Magn.* **2011**, *47*, 3288–3291. [CrossRef]
4. Chen, H.; EL-Refaie, A.M.; Demerdash, N.A.O. Flux-Switching Permanent Magnet Machines: A Review of Opportunities and Challenges—Part I: Fundamentals and Topologies. *IEEE Trans. Energy Convers.* **2020**, *35*, 684–698. [CrossRef]
5. Wu, L.; Ming, G.; Zhang, L.; Fang, Y.; Li, T.; Zheng, W. Comparative Study Between Doubly Salient PM Machine with New Stator/Rotor-Pole Number Combination and Biased Flux PM Machine. *IEEE Trans. Ind. Appl.* **2021**, *57*, 2354–2365. [CrossRef]
6. Bai, J.; Liu, B.; Qiao, G.; Liu, G.; Liu, Y.; Zheng, P. Design and Analysis of a Novel Tubular High-PM-Utilization Transverse-Flux Linear Machine. *IEEE Trans. Magn.* **2022**, *58*, 8202505. [CrossRef]
7. Bangura, J.F. Design of High-Power Density and Relatively High-Efficiency Flux-Switching Motor. *IEEE Trans. Energy Convers.* **2006**, *21*, 416–425. [CrossRef]
8. Thomas, A.S.; Zhu, Z.Q.; Jewell, G.W. Comparison of Flux Switching and Surface Mounted Permanent Magnet Generators for High-Speed Applications. *IET Electr. Syst. Transp.* **2011**, *1*, 111–116. [CrossRef]
9. Fasolo, A.; Alberti, L.; Bianchi, N. Performance Comparison Between Switching-Flux and IPM Machines with Rare-Earth and Ferrite PMs. *IEEE Trans. Ind. Appl.* **2014**, *50*, 3708–3716. [CrossRef]
10. Evans, D.J.; Zhu, Z.Q. Novel Partitioned Stator Switched Flux Permanent Magnet Machines. *IEEE Trans. Magn.* **2015**, *51*, 8100114. [CrossRef]
11. Wu, Z.Z.; Zhu, Z.Q. Analysis of Magnetic Gearing Effect in Partitioned Stator Switched Flux PM Machines. *IEEE Trans. Energy Convers.* **2016**, *31*, 1239–1249. [CrossRef]
12. Awah, C.C.; Zhu, Z.Q.; Wu, Z.Z.; Zhan, H.L.; Shi, J.T.; Wu, D.; Ge, X. Comparison of Partitioned Stator Switched Flux Permanent Magnet Machines Having Single- or Double-Layer Windings. *IEEE Trans. Magn.* **2016**, *52*, 9500310. [CrossRef]
13. Thomas, A.S.; Zhu, Z.Q.; Owen, R.L.; Jewell, G.W.; Howe, D. Multiphase Flux-Switching Permanent-Magnet Brushless Machine for Aerospace Application. *IEEE Trans. Ind. Appl.* **2009**, *45*, 1971–1981. [CrossRef]
14. Raminosoa, T.; Gerada, C.; Galea, M. Design Considerations for a Fault-Tolerant Flux-Switching Permanent-Magnet Machine. *IEEE Trans. Ind. Electron.* **2011**, *58*, 2818–2825. [CrossRef]
15. Li, G.J.; Ojeda, J.; Hoang, E.; Gabsi, M. Thermal-Electromagnetic Analysis of a Fault-Tolerant Dual-Star Flux-Switching Permanent Magnet Motor for Critical Applications. *IET Electr. Power Appl.* **2011**, *5*, 503–513. [CrossRef]
16. Aboelhassan, M.O.E.; Raminosoa, T.; Goodman, A.; De Lillo, L.; Gerada, C. Performance Evaluation of a Vector-Control Fault-Tolerant Flux-Switching Motor Drive. *IEEE Trans. Ind. Electron.* **2013**, *60*, 2997–3006. [CrossRef]
17. Taras, P.; Li, G.-J.; Zhu, Z.Q. Comparative Study of Fault-Tolerant Switched-Flux Permanent-Magnet Machines. *IEEE Trans. Ind. Electron.* **2017**, *64*, 1939–1948. [CrossRef]
18. Wu, L.; Zhu, J.; Fang, Y. A Novel Doubly-Fed Flux-Switching Permanent Magnet Machine with Armature Windings Wound on Both Stator Poles and Rotor Teeth. *IEEE Trans. Ind. Electron.* **2020**, *67*, 10223–10232. [CrossRef]

19. Wu, L.; Zheng, Y.; Fang, Y.; Huang, X. Novel Fault-Tolerant Doubly Fed Flux Reversal Machine with Armature Windings Wound on Both Stator and Rotor Teeth. *IEEE Trans. Ind. Electron.* **2021**, *68*, 4780–4789. [CrossRef]
20. Ming, G.; Wu, L.; Yuan, J.; Ma, S.; Yang, J. Design and Electromagnetic Performance Analysis of Novel Dual-Armature Π-Core Doubly Salient Permanent Magnet Machines. *IEEE Trans. Ind. Appl.* **2024**, *60*, 498–506. [CrossRef]
21. Chen, Y.; Chen, S.; Zhu, Z.Q.; Howe, D.; Ye, Y.Y. Starting Torque of Single-Phase Flux-Switching Permanent Magnet Motors. *IEEE Trans. Magn.* **2006**, *42*, 3416–3418. [CrossRef]
22. Wang, D.; Feng, W.; Wang, B.; Xu, G.; Wang, X. Design, Prototype and Experimental Verification of Single Phase Flux Switching Motor Using Low Cost Magnets. *IEEE Trans. Energy Convers.* **2023**, *38*, 284–295. [CrossRef]
23. Wang, D.; Xu, G.; Wang, B.; Wang, X.; Li, Z.; Wang, X. A New Hybrid Excitation Flux Switching Motor Using Low Cost Ferrites with Flux Regulation Capability. *IEEE Trans. Energy Convers.* **2025**, *40*, 382–393. [CrossRef]
24. Zhu, Z.Q. A Simple Method for Measuring Cogging Torque in Permanent Magnet Machines. In Proceedings of the 2009 IEEE Power & Energy Society General Meeting, Calgary, AB, Canada, 26–30 July 2009; pp. 1–4.

MDPI AG
Grosspeteranlage 5
4052 Basel
Switzerland
Tel.: +41 61 683 77 34

Energies Editorial Office
E-mail: energies@mdpi.com
www.mdpi.com/journal/energies

www.ingramcontent.com/pod-product-compliance
Lightning Source LLC
Chambersburg PA
CBHW061758210926
51850CB00006B/585
9783725862580